THE PICTURE OF THE TAOIST GENII PRINTED ON THE COVER
of this book is part of a painted temple scroll, recent but traditional, given to
Mr Brian Harland in Szechuan province (1946). Concerning these four divinities,
of respectable rank in the Taoist bureaucracy, the following particulars have been
handed down. The title of the first of the four signifies 'Heavenly Prince', that
of the other three 'Mysterious Commander'.

At the top, on the left, is Liu *Thien Chün*, Comptroller-General of Crops and
Weather. Before his deification (so it was said) he was a rain-making magician
and weather forecaster named Liu Chün, born in the Chin dynasty about +340.
Among his attributes may be seen the sun and moon, and a measuring-rod or
carpenter's square. The two great luminaries imply the making of the calendar, so
important for a primarily agricultural society, the efforts, ever renewed, to reconcile
celestial periodicities. The carpenter's square is no ordinary tool, but the gnomon
for measuring the lengths of the sun's solstitial shadows. The Comptroller-General
also carries a bell because in ancient and medieval times there was thought to be
a close connection between calendrical calculations and the arithmetical acoustics
of bells and pitch-pipes.

At the top, on the right, is Wên *Yuan Shuai*, Intendant of the Spiritual Officials
of the Sacred Mountain, Thai Shan. He was taken to be an incarnation of one of
the Hour-Presidents (*Chia Shen*), i.e. tutelary deities of the twelve cyclical characters
(see p. 297). During his earthly pilgrimage his name was Huan Tzu-Yü and he was
a scholar and astronomer in the Later Han (b. +142). He is seen holding an
armillary ring.

Below, on the left, is Kou *Yuan Shuai*, Assistant Secretary of State in the Ministry
of Thunder. He is therefore a late emanation of a very ancient god, Lei Kung.
Before he became deified he was Hsin Hsing, a poor woodcutter, but no doubt an
incarnation of the spirit of the constellation Kou-Chhen (the Angular Arranger),
part of the group of stars which we know as Ursa Minor. He is equipped with
hammer and chisel.

Below, on the right, is Pi *Yuan Shuai*, Commander of the Lightning, with his
flashing sword, a deity with distinct alchemical and cosmological interests. According
to tradition, in his early life he was a countryman whose name was Thien Hua.
Together with the colleague on his right, he controlled the Spirits of the Five
Directions.

Such is the legendary folklore of common men canonised by popular acclamation.
An interesting scroll, of no great artistic merit, destined to decorate a temple wall,
to be looked upon by humble people, it symbolises something which this book has
to say. Chinese art and literature have been so profuse, Chinese mythological
imagery so fertile, that the West has often missed other aspects, perhaps more
important, of Chinese civilisation. Here the graduated scale of Liu Chün, at first
sight unexpected in this setting, reminds us of the ever-present theme of quantitative measurement in Chinese culture; there were rain-gauges already in the Sung
(+12th century) and sliding calipers in the Han (+1st). The armillary ring of
Huan Tzu-Yü bears witness that Naburiannu and Hipparchus, al-Naqqāsh and
Tycho, had worthy counterparts in China. The tools of Hsin Hsing symbolise that
great empirical tradition which informed the work of Chinese artisans and technicians
all through the ages.

SCIENCE AND CIVILISATION IN CHINA

CHANGING to the utmost is the means by which to respond to things; only he who holds fast to the conception of the unity of the universe can do this. Then he can command the myriad things. Then he shines in an equality with the sun and moon, and partakes of the same pattern-principle as heaven and earth....The sage follows after things; therefore he can control them....The sage commands things, and is not commanded by things.

Kuan Tzu
4TH CENTURY B.C.

THE balance and the steelyard, the square and the compasses, are fixed in a uniform and unvarying manner. Neither the people of Chhin nor Chhu can change their properties, neither the northern Hu nor the southern Yüeh can modify their form. These things are for ever the same....A single day formed them; ten thousand generations propagate them.

Huai Nan Tzu
2ND CENTURY B.C.

中國科學技術史

李約瑟 著

冀朝鼎

SCIENCE AND CIVILISATION IN CHINA

BY

JOSEPH NEEDHAM, F.R.S.

FELLOW AND PRESIDENT OF CAIUS COLLEGE
SIR WILLIAM DUNN READER IN BIOCHEMISTRY IN THE
UNIVERSITY OF CAMBRIDGE
FOREIGN MEMBER OF ACADEMIA SINICA

With the collaboration of

WANG LING, PH.D.

TRINITY COLLEGE, CAMBRIDGE
ASSOCIATE RESEARCH FELLOW OF ACADEMIA SINICA

and the special co-operation of

KENNETH GIRDWOOD ROBINSON, B.LITT.

EDUCATION OFFICER, SARAWAK

VOLUME 4

PHYSICS AND PHYSICAL TECHNOLOGY

PART I: PHYSICS

CAMBRIDGE
AT THE UNIVERSITY PRESS
1962

PUBLISHED BY

THE SYNDICS OF THE CAMBRIDGE UNIVERSITY PRESS

Bentley House, 200 Euston Road, London, N.W. 1
American Branch: 32 East 57th Street, New York 22, N.Y.
West African Office: P.O. Box 33, Ibadan, Nigeria

©

CAMBRIDGE UNIVERSITY PRESS

1962

Printed in Great Britain at the University Press, Cambridge
(Brooke Crutchley, University Printer)

*The Syndics of the Cambridge University Press
desire to acknowledge with gratitude certain financial aid
towards the production of this book, afforded by the
Bollingen Foundation*

CONTENTS

LIST OF ILLUSTRATIONS

LIST OF TABLES

LIST OF ABBREVIATIONS

The following abbreviations are used in the text. For abbreviations used for journals and similar publications in the bibliographies, see pp. 336 ff.

B   Bretschneider, E., *Botanicon Sinicum*.

BCFA   Britain–China Friendship Association.

B & M   Brunet, P. & Mieli, A., *Histoire des Sciences* (*Antiquité*).

CLPT   Thang Shen-Wei *et al.* (ed.), *Chêng Lei Pên Tshao* (Reorganised Pharmacopoeia), ed. of +1249.

CPCRA   Chinese People's Association for Cultural Relations with Foreign Countries.

CSHK   Yen Kho-Chün (ed.), *Chhüan Shang-Ku San-Tai Chhin Han San-Kuo Liu Chhao Wên* (complete collection of prose literature (including fragments) from remote antiquity through the Chhin and Han Dynasties, the Three Kingdoms, and the Six Dynasties), 1836.

G   Giles, H. A., *Chinese Biographical Dictionary*.

HY   Harvard–Yenching (Institute and Publications).

K   Karlgren, B., *Grammata Serica* (dictionary giving the ancient forms and phonetic values of Chinese characters).

KCCY   Chhen Yuan-Lung, *Ko Chih Ching Yuan* (Mirror of Scientific and Technological Origins), an encyclopaedia of +1735.

KCKW   Wang Jen-Chün, *Ko Chih Ku Wei* (Scientific Traces in Olden Times), 1896.

MCPT   Shen Kua, *Mêng Chhi Pi Than* (Dream Pool Essays), +1086.

N   Nanjio, B., *A Catalogue of the Chinese Translations of the Buddhist Tripiṭaka*, with index by Ross (3).

P   Pelliot numbers of the Chhien-fo-tung cave temples.

PTKM   Li Shih-Chen, *Pên Tshao Kang Mu* (The Great Pharmacopoeia), +1596.

PWYF   Chang Yü-Shu (ed.), *Phei Wên Yün Fu* (encyclopaedia), +1711.

R   Read, Bernard E., Indexes, translations and précis of certain chapters of the *Pên Tshao Kang Mu* of Li Shih-Chen. If the reference is to a plant, see Read (1); if to a mammal see Read (2); if to a bird see Read (3); if to a reptile see Read (4); if to a mollusc see Read (5); if to a fish see Read (6); if to an insect see Read (7).

RP   Read & Pak, Index, translation and précis of the mineralogical chapters in the *Pên Tshao Kang Mu*.

S   Schlegel, G., *Uranographie Chinoise*; number-references are to the list of asterisms.

SCTS   *Chhin-Ting Shu Ching Thu Shuo* (imperial illustrated edition of the *Historical Classic*), 1905.

T   Tunhuang Archaeological Research Institute numbers of the Chhien-fo-tung cave temples. In the present work we follow as far as possible the numbering of Hsieh Chih-Liu in his *Tunhuang I Shu Hsü Lu* (Shanghai, 1955) but give the other numbers also.

TCKM   Chu Hsi *et al.* (ed.), *Thung Chien Kang Mu* (Short View of the *Comprehensive Mirror* (*of History*) *for Aid in Government*), general history of China, +1189; with later continuations.

TH   Wieger, L. (1), *Textes Historiques*.

TKKW   Sung Ying-Hsing, *Thien Kung Khai Wu* (The Exploitation of the Works of Nature), +1637.

TPYL   Li Fang (ed.), *Thai-Phing Yü Lan* (the Thai-Phing reign-period (Sung) Imperial Encyclopaedia), +983.

TSCC   *Thu Shu Chi Chhêng* (the Imperial Encyclopaedia of +1726). Index by Giles, L. (2).

TT   Wieger, L. (6), *Tao Tsang* (catalogue of the works contained in the Taoist Patrology).

TW   Takakusu, J. & Watanabe, K., *Tables du Taishō Issaikyō* (*nouvelle édition* (*Japonaise*) *du Canon bouddhique chinoise*), Index-catalogue of the Tripiṭaka.

WCTY/CC   Tsêng Kung-Liang (ed.), *Wu Ching Tsung Yao* (*Chhien Chi*) (military encyclopaedia, first section), +1044.

YCLH   Chang Ying (ed.), *Yuan Chien Lei Han* (encyclopaedia), +1710.

YHSF   Ma Kuo-Han (ed.), *Yü Han Shan Fang Chi I Shu* (Jade-Box Mountain Studio Collection of (reconstituted and sometimes fragmentary) Lost Books), 1853.

ACKNOWLEDGEMENTS

LIST OF THOSE WHO HAVE KINDLY READ THROUGH SECTIONS IN DRAFT

The following list, which applies only to this volume, brings up to date those printed in Vol. 1, pp. 15 ff., Vol. 2, p. xxiii, and Vol. 3, pp. xxxix ff.

Mr S. Adler (Cambridge)	Magnetism (Chess).
Dr J. C. Belshé (Cambridge)	Magnetism.
Prof. Derk Bodde (Philadelphia)	Acoustics (Cosmic Tide).
Dr J. A. Clegg (London)	Magnetism.
Mr J. P. B. Dobbs (London)	Acoustics
Prof. W. A. C. H. Dobson (Toronto)	Acoustics.
Prof. V. Elisséeff (Paris)	All subsections.
Mr John Ellison (Amersham)	All subsections.
Dr R. Fraser (The Hague)	Magnetism.
Sir Harry Garner (Beckenham)	Optics (Glass Technology).
Prof. E. Newton Harvey (Princeton, N.J.)	Luminescence.
Dr Lu Gwei-Djen (Cambridge)	All subsections.
Mr J. V. Mills (La Tour de Peilz, Vaud)	Magnetism.
Dr Dorothy M. Needham, F.R.S. (Cambridge)	All subsections.
Prof. G. Owen (Antioch, Ohio)	Waves and Particles.
Prof. Luciano Petech (Rome)	All subsections.
Dr Laurence Picken (Cambridge)	Acoustics.
Prof. Percival Price (Ann Arbor, Mich.)	Acoustics.
Dr Victor Purcell (Cambridge)	Magnetism.
Mrs Juliette Robson (London)	Acoustics.
Prof. Keith Runcorn (Newcastle)	Magnetism.
Prof. Edward H. Schafer (Berkeley, Calif.)	All subsections.
Mr E. S. Shire (Cambridge)	Magnetism.
Dr Dorothea Singer (Par)	All subsections.
Prof. G. C. Steward (Hull)	Optics.
Prof. E. G. R. Taylor (Bracknell)	Magnetism.
Prof. S. Tolansky, F.R.S. (Richmond)	Optics (Magic Mirrors).
Mr P. Tranchell (Cambridge)	Acoustics.
Dr H. J. J. Winter (Exeter)	Optics.
The late Prof. W. P. Yetts (Chesham)	Magnetism.

AUTHOR'S NOTE

PURSUING our exploration of the almost limitless caverns of Chinese scientific history, so much of which has never yet come to the knowledge and recognition of the rest of the world, we now approach the glittering veins of physics and physical technology; a subject which forms a single whole, constituting Volume Four, though delivered to the reader in three separate volumes. First come the physical sciences themselves (Vol. 4, pt. 1), and then their diverse applications in all the many branches of mechanical engineering (Vol. 4, pt. 2), civil and hydraulic engineering, and nautical technology (Vol. 4, pt. 3).

With the opening chapter we find ourselves at a focal point in the present study, for mechanics and dynamics were the first of all the conquests of modern science. Mechanics was the starting-point because the direct physical experience of man in his immediate environment is predominantly mechanical, and the application of mathematics to mechanical magnitudes was relatively simple. But ancient and medieval China belonged to a world in which the mathematisation of hypotheses had not yet brought modern science to birth, and what the scientific minds of pre-Renaissance China neglected might prove almost as revealing as that which aroused their interest and investigation. Three branches of physics were well developed among them, optics (Section 26g), acoustics (26h), and magnetism (26i); mechanics was weakly studied and formulated, dynamics almost absent. We have attempted to offer some explanation for this pattern but without any great conviction, and better understanding of the imbalance must await further research. The contrast with Europe, at least, where there was a different sort of one-sidedness, is striking enough, for in Byzantine and late medieval times mechanics and dynamics were relatively advanced while magnetic phenomena were almost unknown.

In optics the Chinese of the Middle Ages kept empirically more or less abreast of the Arabs, though greatly hampered in theory by the lack of that Greek deductive geometry of which the latter were the inheritors. On the other hand they never entertained that peculiar Hellenistic aberration according to which vision involved rays radiating from, not into, the eye. In acoustics the Chinese proceeded along their own lines because of the particular and characteristic features of their ancient music, and here they produced a body of doctrine deeply interesting but not readily comparable with those of other civilisations. Inventors of the bell, and of a great variety of percussion instruments not known in the West, they were especially concerned with timbre both in theory and practice, developing their unique theories of melodic composition within the framework of a twelve-note gamut rather than an eight-note scale. At the end of the +16th century Chinese mathematical acoustics succeeded in solving the problem of equal temperament just a few decades before its solution was reached in the West (Section 26h, 10). Lastly, Chinese investigation of magnetic phenomena and their practical application constituted a veritable epic. Men were arguing in China about the cause of the declination of the magnetic needle, and using it at sea, before Westerners even knew of its directive property.

Readers pressed for time will doubtless welcome once more a few suggestions. In the chapters which we now present it is possible to perceive certain outstanding traditions of Chinese physical thought and practice. Just as Chinese mathematics was indelibly algebraic rather than geometrical, so Chinese physics was wedded to a prototypic wave-theory and perennially averse to atoms, always envisaging an almost Stoic continuum; this may be seen in Section 26 *b* and followed through in relation to tension and fracture (*c*, 3) and to sound vibrations (*h*, 9). Another constant Chinese tendency was to think in pneumatic terms, faithfully developing the implications of the ancient concept of *chhi* (=*pneuma*, *prāṇa*). Naturally this shows itself most prominently in the field of acoustics (Section 26*h*, 3, 7, etc.), but it was also connected with some brilliant successes in the field of technology such as the inventions of the double-acting piston-bellows and the rotary winnowing-fan (Section 27*b*, 8), together with the water-powered metallurgical blowing-machine (27*h*, 3, 4, direct ancestor of the steam-engine itself). It was also responsible for some extraordinary insights and predictions in aeronautical pre-history (27*m*, 4). Traditions equally strong and diametrically opposite to those of Europe also make their appearance in the purely technical field. Thus the Chinese had a deep predilection for mounting wheels and machinery of all kinds horizontally instead of vertically whenever possible, as may be followed in Section 27 (*h*, *k*, *l*, *m*).

Beyond this point, guidance to the reader is not very practicable since so many different preoccupations are involved. If he is interested in the history of land transport he will turn to the discussion of vehicles and harness (Section 27*e*, *f*); if he delights, like Leviathan, in the deep waters, a whole chapter (29) will speak to him of Chinese ships and their builders. The navigator will turn from the compass itself (26*i*, 5) to its fuller context in the haven-finding art (29*f*); the civil engineer, attracted by a survey of those grand water-works which outdid the 'pyramides of Aegypt', will find it in Section 28*f*. The folk-lorist and the ethnographer will appreciate that 'dark side' of history where we surmise that the compass-needle, most ancient of all those pointer-readings that make up modern science, began as a 'chess-man' thrown on to a diviner's board (26, *i*, 8). The sociologist too will already find much of interest, for besides discussing the place of artisans and engineers in feudal-bureaucratic society (27*a*, 1, 2, 3), we have ventured to raise certain problems of labour-saving invention, manpower, slave status and the like, especially with regard to animal harness (*f*, 2), massive stone buildings (28*d*, 1), oared propulsion (29*g*, 2), and water-powered milling and textile machinery (27*h*).

Many are the ways in which these volumes link up with those which have gone before. We shall leave the reader's perspicacity to trace how the *philosophia perennis* of China manifested itself in the discoveries and inventions here reported. We may point out, however, that mathematics, metrology and astronomy find numerous echoes: in the origins of the metric system (Section 26*c*, 6), the development of lenses (*g*, 5), and the estimation of pitch-pipe volumes (*h*, 8)—or the rise of astronomical clocks

(Section 27*j*), the varying conceptions of perspective (28*d*, 5) and the planning of hydraulic works (*f*, 8). Similarly, much in the present volume points forward to chapters still to come. All uses of metal in medieval Chinese engineering imply what we have yet to say on metallurgical achievements; in the meantime reference may be made to the separate monograph *The Development of Iron and Steel Technology in China*, published as a Newcomen Lecture[a] in 1958. In all mentions of mining and the salt industry it is understood that these subjects will be fully dealt with at a later stage. All water-raising techniques remind us of their basic agricultural purpose, the raising of crops.

As for the discoveries and inventions which have left permanent mark on human affairs, it would be impossible even to summarise here the Chinese contributions. Perhaps the newest and most surprising revelation (so unexpected even to ourselves that we have to withdraw a relevant statement in Vol. 1) is that of the six hidden centuries of mechanical clockwork which preceded the clocks of + 14th-century Europe. Section 27*j* is a fresh though condensed treatment of this subject, incorporating still further new and strange material not available when the separate monograph *Heavenly Clockwork* was written in 1957 with our friend Professor Derek J. de Solla Price, now of Yale University.[b] It still seems startling that the key invention of the escapement should have been made in a pre-industrial agrarian civilisation among a people proverbially supposed by bustling nineteenth-century Westerners to take no account of time. But there are many other equally important Chinese gifts to the world, the development of the magnetic compass (Section 26*i*, 4, 6), the invention of the first cybernetic machine (27*e*, 5), both forms of efficient equine harness (27*f*, 1), the canal lock-gate (28*f*, 8, iv) and the iron-chain suspension-bridge (28*e*, 4). The first true crank (Section 27*b*, 4), the stern-post rudder (Section 29*h*), the man-lifting kite (Section 27*m*)—we cannot enumerate them all.

In these circumstances it seems hardly believable that writers on technology have run up and down to find reasons why China contributed nothing to the sciences, pure or applied. At the beginning of a recent popular *florilegium* of passages on the history of technology one comes across a citation from the + 8th-century Taoist book *Kuan Yin Tzu*,[1] given as an example of 'oriental rejection of this world and of worldly activity'. It had been culled from an interesting essay on religion and the idea of progress, well-known in the thirties and still stimulating, the author of which, led astray by the old rendering of Fr Wieger, had written: 'It is obvious that such beliefs can afford no basis for social activity and no incentive to material progress.' He was, of course, concerned to contrast the Christian acceptance of the material world with 'oriental' otherworldliness, in which the Taoists were supposed to participate. Yet in almost every one of the inventions and discoveries we here describe the Taoists and Mohists were intimately involved (cf. e.g. Sections 26*c*, *g*, *h*, *i*, 28*e*, etc.). As it happened, we had ourselves studied the same *Kuan Yin Tzu* passage and given parts

---

[a] Needham (32), cf. (31).   [b] Needham, Wang & Price (1), cf. Needham (38).

[1] 關尹子

of it in translation at an earlier stage;[a] from this it can be seen that Wieger's version[b] was no more than a grievously distorted paraphrase. Far from being an obscurantist document, denying the existence of laws of Nature (a concept totally unheard-of by the original writer)[c] and confusing reality with dream, the text is a poem in praise of the immanent Tao, the Order of Nature from which space and time proceed, the eternal pattern according to which matter disperses and reassembles in forms ever new; full of Taoist relativism, mystical but in no way anti-scientific or anti-techno-logical, on the contrary prophesying of the quasi-magical quasi-rational command over Nature which he who truly knows and understands the Tao will achieve. Thus upon close examination, an argument purporting to demonstrate the philosophical impotence of 'oriental thought', turns out to be nothing but a figment of occidental imagination.

Another method is to admit that China did something but to find a satisfying reason for saying nothing about it. Thus a recent compendious history of science published in Paris maintains that the sciences of ancient and medieval China and India were so closely bound to their peculiar cultures that they cannot be understood without them. The sciences of the ancient Greek world, however, were truly sciences as such, free of all such subordination to their cultural matrix and fit subjects with which to begin a story of human endeavour in all its abstract purity. It would be much more honest to say that while the social background of Hellenistic science and technology can be taken for granted because it is quite familiar to us from our schooldays onwards, we do not yet know much about the social background of Chinese and Indian science, and that we ought to make efforts to get acquainted with it. In fact, of course, no ancient or medieval science and technology can be separated from its ethnic stamp,[d] and though that of the post-Renaissance period is truly universal, it is no better understandable historically without a knowledge of the milieu in which it came to birth.

Finally, many will be desirous of looking into questions of intercultural contacts, transmission, and influences. Here we may only mention examples still puzzling of inventions which occur almost simultaneously at both ends of the Old World, e.g. rotary grinding (Section 27 d, 2) and the water-mill (h, 2). Parallels between China and ancient Alexandria often arise (for instance in Section 27 b) and the powerful influence of Chinese technology on pre-Renaissance Europe appears again and again (26 c, h, i; 27 b, d, e, f, g, j, m; 28 e, f; 29 j). In the realm of scientific thought, as usual, influences were less marked, but one may well wonder whether the implicit wave-conceptions of China did not exert some effect on Renaissance Europe.

In a brilliant *ponencia* at the 9th International Congress of the History of Science at Barcelona in 1959 Professor Willy Hartner raised the difficult question of how far anyone can ever anticipate anyone else. What does it mean to be a precursor or a predecessor? For those who are interested in intercultural transmissions this is a vital point. In European history the problem has assumed acute form since the

---

[a] Vol. 2, pp. 449 and 444.
[c] Cf. Section 18 in Vol. 2 above.

[b] Originally (4), p. 548.
[d] Cf. Vol. 3, p. 448.

school of Duhem acclaimed Nicolas d'Oresme and other medieval scholars as the precursors of Copernicus, Bruno, Francis Bacon, Galileo, Fermat and Hegel. Here the difficulty is that every mind is necessarily the denizen of the organic intellectual medium of its own time, and propositions which may look very much alike cannot have had quite the same meaning when considered by minds at very different periods. Discoveries and inventions are no doubt organically connected with the milieu in which they arose. Similarities may be purely fortuitous. Yet to affirm the true originality of Galileo and his contemporaries is not necessarily to deny the existence of precursors, so long as that term is not taken to mean absolute priority or anticipation; and in the same way there were many Chinese precursors or predecessors who adumbrated scientific principles later acknowledged—one thinks immediately of Huttonian geology (Vol. 3, p. 604), the comet tail law (p. 432) or the declination of the magnetic needle (Section 26i). So much for science more or less pure; in applied science we need hesitate less. For example, the gaining of power from the flow and descent of water by a wheel can only have been first successfully executed once. Within a limited lapse of time thereafter the invention may have occurred once or twice independently elsewhere, but such a thing is not invented over and over again. All subsequent successes must therefore derive from one or other of these events. In all these cases, whether of science pure or science applied, it remains the task of the historian to elucidate if possible how much genetic connection there was between the precursor and the great figures which followed him. Did they know certain actual written texts? Did they work by hearsay? Did they first conceive their ideas alone and find them unexpectedly confirmed? As Hartner says, the variations range from the certain to the impossible.[a] Often hearsay seems to have been followed by a new and different solution (cf. Section 27j, 1). In our work here presented to the reader he will find that we are very often quite unable to establish a genetic connection (for example, between the suspension of Ting Huan and that of Jerome Cardan, in Section 27d, 4; or between the rotary ballista of Ma Chün and that of Leonardo, in Sections 27a, 2 and 30h, 4), but in general we tend to assume that when the spread of intervening centuries is large and the solution closely similar, the burden of proof must lie on those who desire to maintain independence of thought or invention. On the other hand the genetic connection can sometimes be established with a high degree of probability (for example, in the matters of equal temperament, Section 26h, 10; sailing-carriages, Section 27e, 3; and the kite, the parachute and the helicopter, Section 27m). Elsewhere one is left with strong suspicions, as with regard to the water-wheel escapement clock (Section 27j, 6).

Although every attempt has been made to take into account the most recent research in the fields here covered, we regret that it has generally not been possible to mention work appearing after March 1960.

[a] Many a surprise is still in store for us. After the discovery by Al-Ṭaṭāwī in 1924 that Ibn al-Nafīs (+1210 to +1288) had clearly described the pulmonary circulation (cf. Meyerhof, 1, 2; Haddad & Khairallah, 1), it was long considered extremely unlikely that any hint of this could have reached the Renaissance discoverer of the same phenomenon, Miguel Servetus (cf. Temkin, 2). But now O'Malley (1) has found a Latin translation of some of the writings of Ibn al-Nafīs published in +1547.

We have not printed a contents-table of the entire project since the beginning of Vol. 1, and it has now been felt desirable to revise it in prospectus form.[a] So much work has now been done in preparation for the later volumes that it is possible to give their outline subheadings with much greater precision than could be done seven years ago. More important, perhaps, is the division into volumes. Here we have sought to retain unaltered the original numbering of the successive Sections, as for the needs of cross-referencing we must. Vol. 4, as originally planned, included physics, all branches of engineering, military and textile technology and the arts of paper and printing. As will be seen, we now entitle Vol. 4 *Physics and Physical Technology*, Vol. 5 *Chemistry and Chemical Technology* and Vol. 6 *Biology and Biological Technology*. This is a logical division, and Vol. 4 concludes very reasonably with Nautics (29), for in ancient and medieval times the techniques of shipping were almost entirely physical. Similarly Vol. 5 starts with Martial Technology (30), for in this field and in those times the opposite was the case; the chemical factor was essential. We found not only that we must embody iron and steel metallurgy therein (hence the slight but significant change of title), but also that without the epic of gunpowder, the fundamental discovery of the first known explosive and its development through five pre-occidental centuries, the history of Chinese military technique could not be written. With Textiles (31) and the other arts (32) the same argument was found to apply, for so many of the processes (retting, fulling, dyeing, ink-making) allied them to chemistry rather than to physics. Of course we could not always consistently adhere to this principle; for instance, no discussion of lenses was possible without some knowledge of glass technology, and this had therefore to be introduced at an early stage in the present volume (26*g*, 5, ii). For the rest, it is altogether natural that Mining (36), Salt-winning (37) and Ceramic Technology (35) should find their place in Vol. 5. The only asymmetry is that while in Vols. 4 and 6 the fundamental sciences are dealt with at the beginning of the first part, in Vol. 5 the basic science, chemistry, with its precursor, alchemy, is discussed in the second part. This probably matters the less because in ready response to the critics who found Vol. 3 too heavy and bulky for comfortable meditative evening reading, the University Press has decided to produce the present volume in three physically separate parts, each being as usual independent and complete in itself. One more point remains. In Vol. 1, pp. 18 ff. we gave details of the plan of the work (conventions, bibliographies, indexes, etc.) to which we have since closely adhered, and we promised that in the last volume a list would be given of the editions of the Chinese books used. It now seems undesirable to wait so long, and thus for the convenience of readers with knowledge of the Chinese language we propose to append to the last part of this Volume an interim list of these editions down to the point then reached.

China to Europeans has been like the moon, always showing the same face— a myriad peasant-farmers, a scattering of artists and recluses, an urban minority of scholars, mandarins and shopkeepers. Thus do civilisations acquire 'stereotypes' of one another. Now, raised upon the wings of the space-ship of linguistic resource and

[a] An extract of this contents-table relevant to the present volume will be found on pp. 432–4 below.

riding the rocket of technical understanding (to use an Arabic trope), we intend to see
what is on the other side of the disc, and to meet the physicists and engineers, the ship-
wrights and the metallurgists of China's three-thousand year old culture.

In our note at the beginning of Vol. 3 we took occasion to say something of the
principles of translation of old scientific texts and of the technical terms contained
in them.[a] Since this is the first volume largely devoted to the applied sciences we are
moved to insert a few reflections here on the present position of the history of
technology, a discipline which has suffered even more perhaps than the history of
science itself from that dreadful dichotomy between those who know and those who
write, the doers and the recorders. If men of scientific training, with all their handi-
caps, have contributed far more than professional historians to the history of science
and medicine (as is demonstrably true), technologists as a whole have been even less
well equipped with the tools and skills of historical scholarship, the languages, the
criticism of sources, and the use of documentary evidence. Yet nothing can be more
futile than the work of a historian who does not really understand the crafts and
techniques with which he is dealing, and for any literary scholar it is hard to acquire
that familiarity with things and materials, that sense of possibilities and probabilities,
that understanding of Nature's ways, in fact, which comes (in greater or lesser measure)
to everyone who has worked with his hands whether at the laboratory bench or in
the factory workshop. I always remember once studying some medieval Chinese texts
on 'light-penetration mirrors' (*thou kuang chien*[1]), that is to say, bronze mirrors which
have the property of reflecting from their polished surfaces the designs executed in
relief on their backs. A non-scientific friend was really persuaded that the Sung
artisans had found out some way of rendering metal transparent to light-rays, but
I knew that there must be some other explanation and it was duly found (cf. Section
26g, 3). The great humanists of the past were very well aware of their limitations in
these matters, and sought always, so far as possible, to gain acquaintance with what
my friend and teacher, Gustav Haloun, used half-wistfully half-ironically to call the
realia. In a passage we have already quoted (Vol. 1, p. 7), another outstanding
sinologist, Friedrich Hirth, urged that the Western translator of Chinese texts must
not only translate, he must identify, he must not only know the language but he must
also be a collector of the objects talked about in that language. The conviction was
sound, but if porcelain or cloisonné could (at any rate in those days) be collected and
contemplated with relative ease, how much more difficult is it to acquire an understanding of machinery, of tanning or of pyrotechnics, if one has never handled a lathe,
fitted a gear-wheel or set up a distillation.

What is true of living humanists in the West is also true of some of the Chinese
scholars of long ago whose writings are often our only means of access to the techniques
of past ages. The artisans and technicians knew very well what they were doing, but
they were liable to be illiterate, or at least inarticulate (cf. the long and illuminating
text which we have translated in Section 27a, 2). The bureaucratic scholars, on the
other hand, were highly articulate but too often despised the rude mechanicals whose

---

[a] Cf. Needham (34).   [1] 透光鑑

activities, for one reason or another, they wrote about from time to time. Thus even the authors whose words are now so precious were often more concerned with their literary style than with the details of the machines and processes which they mentioned. This superior attitude was also not unknown among the artists, back-room experts (like the mathematicians) of the officials' yamens, so that often they were more interested in making a charming picture than in showing the precise details of machinery when they were asked to limn it, and now sometimes it is only by comparing one drawing with another that we can reach certainty about the technical content. At the same time there were many great scholar-officials throughout Chinese history from Chang Hêng in the Han to Shen Kua in the Sung and Tai Chen in the Chhing who combined a perfect expertise in classical literature with complete mastery of the sciences of their day and the applications of these in artisanal practice.

For all these reasons our knowledge of the development of technology is still in a lamentably backward state, vital though it is for economic history, that broad meadow of flourishing speculation. In a recent letter, Professor Lynn White, who has done as much as anyone else in the field, wrote memorable words with which we fully agree: 'The whole history of technology is so rudimentary that all one can do is to work very hard, and be happy when one's errors are corrected.' On every hand pitfalls abound. On a single page of a recent most authoritative and admirable collective treatise, one of our best historians of technology can first suppose Heron's toy windmill to be an Arabic interpolation, though the *Pneumatica* never passed to us through that language, and a moment later assert that Chinese travellers in +400 saw wind-driven prayer-wheels in Central Asia, a story based on a mistranslation now just 125 years old. The same authoritative treatise says that Celtic wagons of the −1st century had hubs equipped with roller-bearings, and we ourselves at first accepted this opinion. We learnt in time, however, that examination of the actual remains preserved at Copenhagen makes this highly improbable, and that reference to the original paper in Danish clinches the matter—the pieces of wood which came out from the hub-spaces when disinterred were flat strips and not rollers at all. We have often been saved from other such mistakes only by the skin of our teeth, so to say, and it is not in a spirit of criticism, but rather to demonstrate the difficulties of the work, that we draw attention to them.

Certain safeguards one can always try to obtain. There is no substitute for actually seeing for oneself in the great museums of the world, and the great archaeological sites; there is no substitute for personal intercourse with the practising technicians themselves. To be sure the scholarly standard of any particular work must necessarily depend upon the ground which is covered. Only the specialist using intensive methods—a Rosen elucidating the tangled roots of ophthalmic lenses or a Drachmann exploring Roman oil-presses—can afford the time to go into a matter *au fond* and bring truth wholly out of the well. We have tried to do this only in very few fields, such as that of medieval Chinese clockwork, because our aim is essentially extensive and pioneering. There is no escape, much must be taken on trust. If we are deficient in our knowledge of the objects of occidental archaeology, it is because we have laboured to study *in situ* those of the Chinese culture-area, our primary responsibility. If we had been able to visit the museum in Copenhagen where the Dejbjerg wagons are kept we might have been more wary of accepting current statements about them, but— ὁ βίος βραχύς, ἡ δὲ τέχνη μακρή, the craft is long but life is short. On the other side of the scale a deep debt of gratitude is owing to the President and Council of Academia Sinica for generous facilities which enabled me in 1958, together with Dr Lu Gwei-Djen, to visit or revisit many of the great museums and archaeological sites in China.

But not with archaeologists only must one converse. One must follow the example of little Dr Harvey (of Caius College). In the seventeenth century John Aubrey tells us of a conversation he had with a sow-gelder, a countryman of little learning but much practical experience and wisdom. He told him that he had met Dr William Harvey, who had conversed with him above two or three hours, and 'if he had been', the man remarked, 'as stiff as some of our starched and formall doctors, he had known no more than they'. A Kansu carter threw light upon the harness not only of our own time but indirectly of the Han and Thang, Szechuanese iron-workers were well able to help our understanding of how Chhiwu Huai-Wên in +545 made co-fusion steel, and a Peking kite-maker could reveal with his simple materials those secrets of the cambered wing and the airscrew which lie at the heart of modern aeronautical science. Nor may the technicians of one's own civilisation be neglected, for a traditional Surrey wheelwright can explain how wheels were 'dished' by the artisans of the State of Chhi two thousand years and more ago. A friend in the zinc industry disclosed to us that the familiar hotel cutlery found today all over the world is made essentially of the medieval Chinese alloy paktong,[1] a nautical scholar from Greenwich demonstrated the significance of the Chinese lead in fore-and-aft sailing, and it took a professional hydraulic engineer to appreciate at their true value the Han measurements of the silt-content of river-waters. As Confucius put it, *San jen hsing, pi yu wo shih*,[2] 'Where there are three men walking together, one or other of them will certainly be able to teach me something'.[a]

The demonstrable continuity and universality of science and technology prompts a final observation. Some time ago a not wholly unfriendly critic of our previous volumes wrote, in effect: this book is fundamentally unsound for the following reasons. The authors believe (1) that human social evolution has brought about a gradual increase in man's knowledge of Nature and control of the external world, (2) that this science is an ultimate value and with its applications forms today a unity into which the comparable contributions of different civilisations (not isolated from each other as incompatible and mutually incomprehensible organisms) all have flowed and flow as rivers to the sea, (3) that along with this progressive process human society is moving towards forms of ever greater unity, complexity and organisation. We recognised these invalidating theses as indeed our own, and if we had a door like that of Wittenberg long ago we would not hesitate to nail them to it. No critic has subjected our beliefs to a more acute analysis, yet it reminded us of nothing so much

[a] *Lun Yü*, VII, xxi.

[1] 白銅   [2] 三人行必有我師

as that letter which Matteo Ricci wrote home in +1595 to describe the various absurd ideas which the Chinese entertained about cosmological questions:[a] (One), he said, they do not believe in solid crystalline celestial spheres, (Item) they say that the heavens are empty, (Item) they have five elements instead of the four so universally recognised as consonant with truth and reason, etc. But we have made our point.

A decade of fruitful collaboration came to an end when early in 1957 Dr Wang Ling[1] (Wang Ching-Ning[2]) departed from Cambridge to the Australian National University, where he is now Associate Professor in Chinese Language and Literature. Neither of us will ever forget the early years of the project, when our organisation was finding its feet, and a thousand problems had to be solved (with equipment much less adequate than now) as we went along. In the present volume, Dr Wang's activity was mainly exercised in subsections c, g, and i. The essential continuity of day-to-day collaboration with Chinese scholars was, however, happily preserved at his departure by the arrival of a still older friend, Dr Lu Gwei-Djen,[3] late in 1956. Among other posts, Dr Lu had been Research Associate at the Henry Lester Medical Institute, Shanghai, Professor of Nutritional Science at Ginling College, Nanking, and later in charge of the Field Cooperation Offices Service in the Department of Natural Sciences at UNESCO headquarters in Paris. With a basis of wide experience in nutritional biochemistry and clinical research, she is now engaged in pioneer work for the biological and medical part of our plan (Vol. 6). Probably no single subject in our programme presents more difficulties than that of the history of the Chinese medical sciences. The volume of the literature, the systematisation of the concepts (so different from those of the West), the use of ordinary and philosophical words in special senses so as to constitute a subtle and precise technical terminology, and the strangeness of certain important branches of therapy—all demand great efforts if the result is to give, as has not yet been given, a true picture of Chinese medicine. It is very fortunate that time permits our excavations to commence from the bedrock upwards. At the same time Dr Lu has participated in the revision of the present volume for press.

A year later (early in 1958) we were joined by Dr Ho Ping-Yü,[4] Reader in Physics at the University of Malaya, Singapore. Primarily an astro-physicist by training, and the translator of the astronomical chapters of the *Chin Shu*, he was happily willing to broaden his experience in the history of science by devoting himself to the study of alchemy and early chemistry, helping thus to lay the foundations for the relevant volume (Vol. 5). Such work had been initiated some years earlier by yet another friend, Dr Tshao Thien-Chhin,[5] when a Research Fellow of Caius College, before his return to the Biochemical Institute of Academia Sinica at Shanghai. Dr Tshao had been one of my wartime companions, and while in Cambridge made a most valuable study of the alchemical books in the *Tao Tsang*.[b] Dr Ho Ping-Yü was able to extend this work with great success in many directions. Although Dr Ho is now back at his substantive post in Singapore, it is my earnest hope that he will be able to rejoin us

---

[a] Cf. Vol. 3, p. 438. [b] Cf. Vol. 1, p. 12.

[1] 王鈴 [2] 王靜寧 [3] 魯桂珍 [4] 何丙郁 [5] 曹天欽

in Cambridge for the final preparation of the volume on chemistry and chemical technology.

It is good to record that already a number of important subsections of both these volumes (5 and 6) have been written. The publication of some of these in draft form facilitates criticism and aid by specialists in the different fields.

Lastly, an occidental collaborator appears with us on the title-page of the first part of this volume, Mr Kenneth Robinson, one who combines most unusually sinological and musical knowledge. Professionally he is an educationalist, and with a Malayan background in teachers' training, now as Education Officer in Sarawak frequents the villages and long-houses of the Dayaks and other peoples, whose remarkable orchestras seem to him to evoke the music of the Chou and Han. We were fortunate indeed that he was willing to undertake the writing of the Section on the recondite but fascinating subject of physical acoustics, indispensable because it was one of the major interests of the scientific minds of the Chinese middle ages. He is thus the only participator so far who has contributed direct authorship as well as research activity.

Once again it is a pleasure to offer public gratitude to those who have helped us in many different ways. First, our advisers in linguistic and cultural fields unfamiliar to us, notably Mr D. M. Dunlop for Arabic, Dr Shackleton Bailey for Sanskrit, Dr Charles Sheldon for Japanese, and Mr G. Ledyard for Korean. Secondly, those who have given us special assistance and counsel, Dr H. J. J. Winter in medieval optics, Dr Laurence Picken in acoustics, Mr E. G. Sterland in mechanical engineering, Dr Herbert Chatley in hydraulic engineering and Mr George Naish in nautics. Thirdly, all those whose names will be found in the adjoining list of readers and kind critics of Sections in draft or proof form. But only Dr Dorothy Needham, F.R.S., has weighed every word in these volumes and our debt to her is incalculable.

Once again we renew our warmest thanks to Mr Derek Bryan, O.B.E., and Mrs Margaret Anderson for their indispensable and meticulous help with press work, and to Mr Charles Curwen for acting as our agent-general with regard to the ever-increasing flood of current Chinese literature on the history and archaeology of science and technics. Miss Muriel Moyle has continued to provide her very detailed indexes, the excellence of which has been saluted by many reviewers. As the enterprise continues, the volume of typing and secretarial work seems to grow beyond expectation, and we have had many occasions to recognise that a good copyist is like the spouse in Holy Writ, precious beyond rubies. Thus we most gratefully acknowledge the help of Mrs Betty May, Miss Margaret Webb, Miss Jennie Plant, Miss June Lewis, Mr Frank Brand, Mrs W. M. Mitchell, Miss Frances Boughton, Mrs Gillian Rickaysen and Mrs Anne Scott McKenzie.

The part played by publisher and printer in a work such as this, considered in terms either of finance or technical skill, is no less vital than the research, the organisation and the writing itself. Few authors could have more appreciation of their colleagues executive and executant than we for the Syndics and the Staff of the Cambridge University Press. Among the latter formerly was our friend Frank Kendon, for many years Assistant Secretary, whose death has occurred since the appearance of our previous volume. Known in many circles as a poet and literary scholar of high achievement, he was capable of divining the poetry implicit in some of the books which passed through the Press, and the form which his understanding took was the bestowal of infinite pains to achieve the external dress best adapted to the content. I shall always remember how when *Science and Civilisation in China* was crystallising in this way, he 'lived with' trial volumes made up in different styles and colours for some weeks before arriving at a decision most agreeable to the author and his collaborators—and what was perhaps more important, equally so to thousands of readers all over the world.

To the Master and Fellows of the Hall of the Annunciation, commonly called Gonville and Caius College, a family of immediate colleagues, I can offer only inadequate words. I do not know where else conditions so perfect for carrying out an enterprise such as this could be found, a peaceful workshop in the topographical centre of the University and all its libraries, between the President's apple-tree and the Porta Honoris. The daily appreciation and encouragement of everyone in the Society helps us to surmount all the difficulties of the task. Nor can I omit meed of thanks to the Head of the Department of Biochemistry and its Staff for the indulgent understanding which they show to a colleague seconded, as it were, to another universe.

The financing of the research work for our project has always been difficult and still presents serious problems. We are nevertheless deeply indebted to the Wellcome Trust, whose exceptionally generous support has relieved us of all anxiety concerning the biological and medical volume. We cannot forbear from offering our deepest gratitude for this to the Chairman, Sir Henry Dale, O.M., F.R.S. An ample benefaction by the Bollingen Foundation, elsewhere acknowledged, has assured the adequate illustration of the successive volumes. To Dato Lee Kong-Chian of Singapore we are beholden for a splendid contribution towards the expenses of research for the chemical volume, and Dr Ho's work towards this was made possible by sabbatical leave from the University of Malaya. Here we wish to pay a tribute to the memory of a great physician and servant of his country, Wu Lien-Tê, of Emmanuel College, already Major in the Chinese Army Medical Corps before the fall of the Chhing dynasty, founder long ago of the Manchurian Plague Prevention Service and pioneer organiser of public health work in China. During the last year of his life Dr Wu exerted himself to help in securing funds for our work, and his kindness in this will always be warmly remembered. Some kind well-wishers of our enterprise have now grouped themselves together in a committee of 'Friends of the Project' with a view to securing further necessary financial support, and to our old friend Dr Victor Purcell, C.M.G., who has most kindly accepted the honorary secretaryship of this committee, our best appreciation is offered. At various periods during the studies which see the light in these volumes we have also received financial help from the Universities' China Committee and from the Managers of the Ocean Steamship Company acting as Trustees of funds bequeathed by members of the Holt family; for this we record most grateful thanks.

26. PHYSICS

(a) INTRODUCTION

THOUGH physics has often been regarded as the fundamental science, it was a branch of natural knowledge in which Chinese traditional culture was never strong. This is in itself a striking fact. We may find it of special significance when placed in the context of a general discussion of the factors in East Asian society which inhibited the auto-chthonous rise of modern science.[a] Nevertheless, there is no lack of material for the present Section. Already we have seen certain aspects of classical Chinese physical thought in the Section on fundamental ideas,[b] and this volume will in the first place continue the exposition of the work of the Mohist school in the −4th and −3rd centuries. It will then appear that at no time did the concept of atoms become important in Chinese thinking. Just as Chinese mathematics was algebraic rather than geometrical, and Chinese philosophy organic rather than mechanical, so we shall find that Chinese physical thought (one can hardly speak of a developed science of physics) was dominated by the notion of waves rather than particles.

Most important in this Section will be, of course, the development of knowledge about magnetism, in particular the discovery and exploitation of the directive property of the lodestone. The Chinese were so much in advance of the Western world in this matter that we might almost venture the speculation that if the social conditions had been favourable for the development of modern science, the Chinese might have pushed ahead first in the study of magnetism and electricity, passing to field physics without going through the stage of 'billiard-ball' physics. Had the Renaissance been Chinese and not European, the whole sequence of discoveries would probably have been entirely different. Next to magnetism, there was not a little interest in China in optics,[c] again beginning with the Mohists; there was some statics and hydrostatics, though, as in Europe, very little study of heat. Another striking difference between the two civilisations seems to be that China had no parallels to the medieval Western students of motion. Chinese literature seems to contain no discussion of the trajectory of missiles or the free fall of bodies;[d] at least we have not come upon any traces of it. There is no one to correspond to the so-called 'precursors of Galileo', men such as Philoponus and Buridan, Bradwardine and Nicolas d'Oresme; and hence no dynamics or cinematics. We may find it possible to hazard a suggestion as to why this was so. It is, however, important to note that this theoretical vacuum did not in the least

---

[a] A preliminary discussion of the significance of the rise of modern physics, with its mathematised hypotheses, in Renaissance Europe, especially in the early 17th century, has already been given at the end of the Section on mathematics above, Vol. 3, pp. 154 ff.

[b] Vol. 2, pp. 171 ff., 185 ff., 232 ff., 273 ff., 371 ff.

[c] It has been said that one reason perhaps for the Arabic pre-eminence in optics was the prevalence of eye diseases in subtropical climates, and this may hold good for China also. Cf. below, Section 44, on ophthalmology.

[d] Strangely, in view of the interest of the Mohist school in military technology.

inhibit the development of engineering in China,[a] which before +1500 was frequently much superior to anything which Europe could show.

The literature gives us very little help. The sinologists have done nothing.[b] No Chinese scholar has written at any length on the subject.[c] Occidental historians of physics have been mainly occupied with post-Renaissance physics,[d] less so with that of the Middle Ages,[e] and perhaps still less so with antiquity.[f] None of them has taken into account any Chinese contribution.[g]

Some general propositions of a physical nature are contained in the Mohist Canon (the *Mo Ching*),[h] and others will be given in connection with atomic ideas, mass, and mechanics. In the text, composed not much before −300, we find:[i]

Cs 39/252/81.37. *Duration*

C   Duration (*chiu*[1]) includes all the different times.

CS Former times, the present time, the morning and the evening, are combined together to form duration. (auct.)

Cs 65/—/32.58. *Volume*

C   The enclosure (*ying*[2]) of a certain thing means that every part of that thing is enclosed.

CS If nothing can be enclosed, there can be no volume. (auct.)

Cf. Proposition Cs 55, in the Section on geometry, Vol. 3, p. 94 above.

Cs 67/—/36.60. *Contact and coincidence*

C   Contact (*ying*[3]) means two bodies mutually touching.

CS Lines placed in contact with one another will not (necessarily) coincide (since one may be longer or shorter than the other). Points placed in contact with one another will coincide (because they have no dimensions). If a line is placed in contact with a point, they may or may not coincide (*chin*[4]); (they will do so if the point is placed at the end of the line, for both have no thickness; they will not do so if the point is placed at the middle of the line, for the line has length while the point has no length). If a hard white thing is placed in contact with another hard white thing, the hardness and whiteness will coincide mutually (*hsiang chin*[5]); (since the hardness and the whiteness are qualities diffused throughout the two objects, they may be considered to permeate

---

[a] As Hermann (1) has pointed out.

[b] Such articles as there are, e.g. Edkins (11), are concerned purely with commonplaces about the Five Elements and the Yin and Yang.

[c] With the exception of Wu Nan-Hsün (1), whose interesting little book, privately circulated in Wuhan University, did not become available to us until after the present Section was completed. I am grateful to my old friend Dr Kao Shang-Yin for presenting us with a copy.

[d] Hoppe (1); Buckley (1); Gerland (1); Gerland & Traumüller; Cajori (5).

[e] Maier (1–7); Dugas (1).   [f] A. Heller (1); Seeger (1).

[g] As a companion for the student of physics in China, the book of L. W. Taylor (1) may be mentioned.

[h] See Vol. 2, pp. 171 ff. above. Modern Chinese scholars continue to take a great interest in the Mohist school. Since Wu Yü-Chiang (1) edited the collected commentaries during the last war, many books have appeared, some concentrating chiefly on the philosophy and logic, such as that of Chan Chien-Fêng (1), but others dealing also with the scientific propositions, such as that of Luan Tiao-Fu (1). Wu Nan-Hsün (1) discusses most of those which have to do with physics. In the present connection, see esp. pp. 16 ff.

[i] For explanation of the conventions identifying the passages of text and ancient commentary, see Vol. 2, p. 172 above.

¹ 久   ² 盈   ³ 攖   ⁴ 盡   ⁵ 相盡

the new larger object formed by the contact of the two smaller ones). But two (material) bodies (*thi*[1]), placed in contact (*ying*[2]), cannot mutually coincide (because of the mutual impenetrability of material solids). (auct.)

Cs 64/—/30.57. *Cohesion*

C   A discontinuous line (*lu*[3]) includes empty spaces.

CS The meaning of 'empty' is like the spaces between two apposed pieces of wood. In those spaces there is no wood. (I.e. surfaces cannot be absolutely smooth and cannot therefore fully cohere.) (auct.)

Some of these statements verge on the borderline of physics and geometry; they illustrate well the tendencies of Mohist thought.

(b) WAVES AND PARTICLES

It must now be made clear, as a further preliminary, how Chinese physical thinking was dominated throughout by the concept of waves rather than of atoms. This is only one aspect of that great debate which has been going on at all stages in the history of human thought between the proponents of continuity and those of discontinuity. Atomism is one of the most familiar features of European and Indian theorising, yet although at various times some Chinese thinkers watered its seeds, the idea never took root among them, presumably because it was out of harmony with those organic presuppositions on which Chinese thought was based. First let us look at some of the passing appearances of atoms on the Chinese stage. As has already been suggested,[a] the fundamental idea of atoms could be expected to arise in all civilisations independently, since everywhere men were engaged in cutting up lengths of wood, and the question would inevitably arise as to what would happen if successive cuttings were to go on until the uncuttable was reached.

Logically strictest in this sense were the Mohists, with their atomic definition of the geometrical point.[b] The word used for this in the *Mo Ching* is *tuan*.[4] But they also seem to have considered instants of time in an atomic sense. For example:

Cs 43/—/88.41. *Instants of time*

C   The 'beginning' (*shih*[5]) means an (instant of) time.

CS Time sometimes has duration (*chiu*[6]) and sometimes not, for the 'beginning' point of time has no duration. (auct.)

This makes it clear that the notion of atomic instants of time can hardly have come into China with Buddhism; and to believe that the Mohists were influenced by Indian thought is much more difficult. There is a cosmogonic background to this passage, as in the many passages where *shih*[5] is discussed in *Chuang Tzu*. Here the moment of beginning is thought of like the point at the end of a line.

[a] Vol. 1, p. 155; Vol. 3, p. 92.   [b] Already given, Vol. 3, p. 91 above.

[1] 體   [2] 攖   [3] 纑   [4] 端   [5] 始   [6] 久

When Indian atoms of time did become current in China they were termed *chha-na*[1] (Skr. *kṣaṇa*), a usage which we may find in the *Shu Shu Chi I*[2] (Memoir on some Traditions of Mathematical Art) by Hsü Yo,[3] probably written in the Later Han. The idea seems much more Indian than Greek, but it passed westwards, and curiously the only occurrence of the word 'atom' in the New Testament is with the meaning of a moment of time.[a] According to Gandz (5), Mar Samuel, the great Hebrew polymath of Nehardea in Babylonia (+165 to +257), Hsü Yo's contemporary, reckoned 56,848 atoms to the hour, one atom (*rega'*) being equivalent to two twinklings of an eye (*heref 'ayyin*). The late Roman agrimensores also spoke of *athomi* as very small durations of time,[b] and hence the conception found its way down to Honorius of Autun (+11th century) and Bartholomaeus Anglicus (+13th).

The word *shih*[4] used by the Mohists derives from a graph (K976*e'*, *f'*) which represents the birth of a child,[c] while *tuan*,[5] used for 'point', was originally (K168*b*, *d*)

K976*f'*   K168*b*   K584*b*   K547*b*

a graph showing the first visible sprouts of a plant.[d] Three further words, however, call also for consideration, first *wei*[6] (something very minute), then *khuai*[7] (a very small lump of something), and then *chi*[8] (a germ of something). The first of these derives from a graph showing two hands holding something small (K584*b*), the second is of purely phonetic origin, and the third (K547*b*) once represented two little embryos or other small living things held in the hands.

The word *wei*[6] has invited the translation of 'atom' by sinologists, but if acceptable in literary versions, this can hardly be admitted as correct for scientific purposes. Thus Waley,[e] translating one of the odes of Sung Yü[9],[f] gives 'an atom secretly nurtured in the heart of the indivisible' for this word—the poets were competing for a fief which the king had promised to give to whoever could describe the smallest thing in the world. So also le Gall,[g] translating a passage from the *Chu Tzu Chhüan Shu*,[h] uses the expression 'tout passe de la puissance à l'acte, de l'état atomique (imperceptible) à l'apparence distincte', when the text says:

But there is only one Great Source, and from its undifferentiated energy all specific actions come forth, following through from their minute (beginnings) and so reaching manifestation (*tshung wei erh chih chu*[10]).[i]

[a] I Cor. xv, 52.

[b] There were 47 *athomi* to 1 *untia*, 12 *untiae* to 1 *momentum*, 10 *momenta* to 1 *punctum*, and 5 *puncta* to the hour. Hence the *athomus* was of the order of 0·1 second in our reckoning.

[c] One can see a woman, an aiding hand, the foetus, and the new mouth. The character is close to *thai*,[11] which still means uterus or (loosely) foetus.

[d] The ground is represented in cross-section so that one can see the roots.

[e] (11), p. 27.

[f] Said to have been Chhü Yuan's nephew; lived *c.* −320 to −260; G1841.

[g] (1), p. 102.

[h] Ch. 49, p. 10*b*.

[i] Tr. auct.

[1] 剎那   [2] 數術記遺   [3] 徐岳   [4] 始   [5] 端   [6] 微
[7] 塊   [8] 幾   [9] 宋玉   [10] 從微而至著   [11] 胎

Chu Hsi must of course have been familiar with the Buddhist use of the term, which will shortly be mentioned, but this never had scientific definition or application.

Another word in which some have seen an atomic conception is *khuai*;[1] we have met with it already in the passage of the *Lieh Tzu* book in which Chhang Lu Tzu laughs at the man who is afraid that the heavens may fall down upon him.[a] It will be remembered that there the heavens were said to be nothing but 'piled-up chhi' (*chi chhi*[2]) and the earth 'piled-up khuai' (*chi khuai*[3]).[b] Just how much emphasis can be laid upon this hint of a particulate conception of vaporous and solid matter is not too obvious. In after times the word occurs infrequently, if ever, in this sense.

A related expression has already been noted, however, in the Neo-Confucian account of the formation of the world by the centrifugal grinding of matter to a sedimentary state;[c] the theory which we have called the 'Centrifugal Cosmogony'. The nearest that the text comes to speaking of particles is simply sediment (*cha tzu*[4]). Here again one feels disinclined to suppose that Chu Hsi and his followers had anything seriously like atoms in mind.

More biological is the term *chi*.[5] As already mentioned,[d] it occurs in the Great Appendix of the *I Ching* (Book of Changes) with the sense of the minute imperceptible beginnings of things out of which come good and evil. But earlier than the probable date of this text (perhaps −2nd century), the word had been used by Chuang Chou about −290 in his famous passage concerning biological change, even evolution.[e] 'All species contain "germs".' 'All things come from the "germs" and return to the "germs".' But it would be stretching the meaning much too far to insist on a strictly atomic significance for this passage.

Chinese translators of Buddhist texts in the Thang used the words *wei*[6] or *wei chü*[7] for Skr. *sūkṣma* or *aṇu*, a 'molecule' seven times larger than an atom, *chi wei*,[8] *paramāṇu*.[f] Earlier translators had used for this the phrase *lin hsü*,[9] 'nearest to nothingness', equivalent to Skr. *upākāśa*.[g] The *Shu Shu Chi I* has *chi wei*[10] for particles. But there seems no reason to suppose that these primarily philosophical speculations had much influence on Chinese scientific thinking. Modern Chinese has adopted entirely new words, such as *fên tzu*[11] for molecule, *yuan tzu*[12] for atom, and *tien tzu*[13] for electron.

Sometimes the minutest of weights, such as *hao*[14] and *li*,[15] are referred to in senses almost atomic, particularly in the Han medical literature. Thus in the *Huang Ti*

[a] Wieger (7), p. 79; L. Giles (4), p. 29; see Vol. 2, p. 41.
[b] *Lieh Tzu*, ch. 1, p. 16*b*.
[c] Vol. 2, pp. 372ff.   [d] Vol. 2, p. 80.
[e] *Chuang Tzu*, ch. 18. Cf. *Lieh Tzu*, ch. 4, p. 12*b* (L. Giles (4), p. 79). See Vol. 2, p. 78.
[f] These terms are those of the Vaiśeṣika school, which goes back at least to the +1st century and may be somewhat older (cf. Renou & Filliozat (1), vol. 2, p. 73, and Vol. 1, p. 154 above). An *aṇu* was equal to the sixth part of a mote of dust visible in a sunbeam.
[g] I owe thanks to Dr A. Waley for checking these terms. Cf. Fêng Yu-Lan (1), vol. 2, p. 386.

[1] 塊	[2] 積氣	[3] 積塊	[4] 渣滓	[5] 幾	[6] 微
[7] 微聚	[8] 極微	[9] 隣虛	[10] 積微	[11] 分子	[12] 原子
[13] 電子		[14] 毫	[15] 釐		

Nei Ching Su Wêng[1] (The Yellow Emperor's Manual of Internal Medicine; the Plain Questions) there is the following:[a]

To know the supreme Tao is embarrassingly (difficult). To communicate it is frighteningly (difficult). Who can know its chief characteristics? Worrying about the exactness of the descriptions, who can decide what may best be said? Numbering (the things) which are too small to be distinctly seen, starts from (the smallest units) *hao* and *li*—these are (conceptions which) arise from measurement. But agglomerating together by thousands and ten thousands, they become larger and larger, until shapes[b] emerge.

Veith's use of the word 'atoms' in translating this passage[c] would almost seem justified, if it did not introduce something more definite than the original language will bear.

In contrast to these rare examples of atomistic thinking, we find that the texts speak as one voice with regard to the wave-like progression of the Yin and Yang forces,[d] reciprocally rising and falling. Throughout Nature there is a tidal flow of the two elementary influences. To insist on the wave-like character of Yin and Yang is unnecessary for anyone at all acquainted with Chinese writings,[e] yet a few examples must be given. Our oldest may be taken from the *Kuei Ku Tzu* book,[f] some of which may go back to the −4th century, though this passage will presumably not much antedate Tsou Yen.

The Yang returns cyclically to its beginning; the Yin attains its maximum and gives place to the Yang. (*Yang huan chung shih, Yin chi fan Yang.*[2])[g]

In the −2nd century, the *Huai Nan Tzu* says[h] that

Yang is born at the cyclical sign Tzu (i.e. due north, where Yin, dark and cold, is at its maximum). Yin is born at the cyclical sign Wu (i.e. due south, where Yang, bright and hot, is at its maximum).[i]

Liu An's contemporary, Tung Chung-Shu, has much to say on the subject in his *Chhun Chhiu Fan Lu*. Thus ch. 51 ('The Unitary Nature of the Tao of Heaven'):

The unchanging Tao of Heaven is that things of opposing nature are not permitted to start at the same time. The result is that this Tao is unitary, that is, it is one and not two, this being the process of Heaven in action. Yin and Yang are of opposing nature; hence when one comes out the other goes in, if one is on the right the other is on the left....If the Yin flourishes the Yang declines; if the Yang flourishes the Yin declines.[j]

[a] Ch. 8, p. 33*b*, tr. auct.

[b] The context shows that, since diseases are being discussed, this word carries also the sense of 'syndrome'.

[c] (1), p. 134, where, however, the general sense of the passage seems to be missed.

[d] Cf. above, in Sect. 13 on fundamental ideas, Vol. 2, pp. 273ff.

[e] Martin (5) drew attention to this nearly a century ago (in 1867). Of the Yang he wrote: 'It is curious to see light connected with motion. Did the Chinese anticipate the undulatory theory, and the modern doctrine of thermodynamics?'

[f] Ch. 1 (Pai Ho), p. 6*b*.   [g] Tr. Forke (13), p. 486.

[h] Ch. 3, p. 8*a*.

[i] Tr. auct. Note the dialectical quality of the thought here; everything carries within it the seeds of its own decay.

[j] Tr. Fêng Yu-Lan (2), p. 120, et auct.

[1] 黃帝內經素問   [2] 陽還終始陰極反陽

In other words when the Yang wave is at the height of its crest, the Yin wave is at the depth of its trough, and vice versa. They 'advance and retreat in classifiable (predictable) fashion'.[a] Wang Chhung (about +80) says:[b]

The Yang having reached its climax retreats in favour of Yin; the Yin having reached its climax retreats in favour of Yang. (*Yang chi fan Yin, Yin chi fan Yang.*[1])[c]

A very explicit statement is found in the *Liu Tzu*[2] of Liu Chou[3] about +550, who says:[d]

When the Yang has reached its highest point the Yin begins to rise, and when the Yin has reached its highest point the Yang begins to rise. Just as when the sun has reached its greatest altitude it begins to decline, and when the moon has waxed to its full it begins to wane. This is the changeless Tao of Heaven. When forces have reached their climax, they begin to weaken (*shih chi tsê sun*[4]), and when natural things have become fully agglomerated they begin to disperse (*tshai chü pi san*[5]).[e] After the year's fullness follows decay, and the keenest joy is followed by sadness. This (too) is the changeless condition of Man.

So dominant in Chinese thought was the conception of wave-motion that it seems sometimes to have acted in an inhibitory way upon the advance of scientific knowledge. Traditional natural philosophy in China conceived of the whole universe as undergoing slow pulsations of its fundamentally opposed but mutually necessary basic forces. As the radiating mutual influences of individual things were pulsatile also, it was entirely in accord with the grain of Chinese philosophical thinking to envisage intrinsic rhythms in natural objects. For, as has already been said, 'the harmonious cooperation of all beings arose, not from the orders of a superior authority external to themselves, but from the fact that they were all parts in a hierarchy of wholes forming a cosmic pattern, and what they obeyed were the internal dictates of their own natures'.[f] But this very organic world-view was not altogether propitious for scientific investigation, at any rate in physics, since chains of causes always led back to individual objects the intrinsic rhythms of which were liable to remain inscrutable.

Thus one is not at all surprised to find as late as the Sung, about +1140, a writer (Chhen Chhang-Fang[6]) denying the reflection of the sun's light by the moon, on the ground that the periodical rise and fall of the Yin force was a much better explanation for the moon's phases.[g] In this he was only following a tradition which had been well established in the Han. Wang Chhung in the +1st century, as we have seen,[h] argued vigorously against the correct theory of eclipses, preferring the view that the sun and moon had intrinsic rhythms of brightness of their own. About +274 the astronomer

---

[a] Ch. 57; cf., Vol. 2, p. 282 above.   [b] *Lun Hêng*, ch. 46.

[c] Tr. Forke (4), vol. 2, p. 344.

[d] Ch. 2, p. 10a, tr. Forke (12), p. 258; eng. mod. auct.

[e] The words used can of course also be translated 'human influence' and 'wealth' or 'riches'. There is thus at least a moralising undertone.

[f] Vol. 2, p. 582; cf. p. 287.

[g] *Pu Li Kho Than*[7] (Discussions with Guests at Pu-li), ch. 2, p. 5a.

[h] The passage has been given in full in Vol. 3, pp. 411 ff.

[1] 暘極反陰陰極反暘   [2] 劉子   [3] 劉晝   [4] 勢積則損
[5] 財聚必散   [6] 陳長方   [7] 步里客談

Liu Chih[1] was dissatisfied with it too. In his *Lun Thien*[2] (Discourse on the Heavens)
he maintained that eclipses of the sun could not be due to the obstruction of the sun's
rays by the moon, for the moon must necessarily follow the Tao of ministers and could
not dare to cover the face of the princely sun. So also eclipses of the moon could not
be due to the obstruction of the sun's rays by the earth for the shadow of the earth
would be insufficiently large. But these reactionary arguments do not forfeit Liu Chih
our sympathy for he goes on to expound his own world-conception in words which
demonstrate a whole series of other valid scientific ideas in their 'prehistoric' form—
wave-motion, action at a distance, intrinsic rhythm and the like.

Someone then said: 'According to your own arguments there must be a great shadow,
but since the moon is on the opposite side from the sun, how could it have any light?'
I answered that the Yin contains (always) some Yang, and thus can be bright.[a] It does not
have to wait for the Yang to shine upon it. The Yin and the Yang respond (*ying*[3])[b] to one
another; what is pure receives light, what is cold receives warmth—such communication
needs no intermediary (*wu mên erh thung*[4]). They can respond to each other in spite of the
vast space which separates them (*sui yuan hsiang ying*[5]). When a stone is thrown into the
water, the (ripples) spread forth one after another—this is the propagation of the *chhi* of the
water.[c] Mutual echoing means mutual receptivity. There is no bound which (the mutual
influences of things) cannot attain; there is no barrier which can stand in their way.[d]

(It is in this manner that) the purest substance (i.e. the moon) receives the light of the
Yang (i.e. the sun), occupying different positions because of the roundness of the heavens,
sometimes facing, sometimes turning the back, sometimes slanting. Such is the pattern-
principle (*li*[6]) of the light of the moon. (When) Yin and Yang receive from each other, if the
one flourishes the other must decay.[e] Therefore the sun and moon contend together for
brightness. When the sun is weak the moon appears by day. If there were only a reflection
of light (between the sun and moon), and no question of the mutual radiation and reception
of *chhi*, then the brightness of the Yin ought to flourish most when the Yang is flourishing,
and it ought to decay when the Yang decays. Then there would be no explanation of the
differences (which we observe) between (the light of) the sun and the moon.[f]

Summarising therefore, the Chinese physical universe in ancient and medieval times
was a perfectly continuous whole. *Chhi* condensed in palpable matter was not particulate in any important sense, but individual objects acted and reacted with all other
objects in the world. Such mutual influences could be effective over very great
distances, and operated in a wave-like or vibratory manner dependent in the last
resort on the rhythmic alternation at all levels of the two fundamental forces, the

---

[a] Cf. Vol. 2, p. 276.   [b] Cf. Vol. 2, p. 304.

[c] It is good to have the exact words: *chhu shih erh tzhu chhu chê, shui chhi chih thung yeh.*[7] Cf. Vitruvius,
De Archit. v, 3, vi ff. See also p. 203 below.

[d] Again: *hsiang hsiang erh hsiang chi, wu yuan pu chih, wu ko nêng sai chê.*[8]

[e] This is a most remarkable statement, approaching as it does so nearly a formulation of the law
of the conservation of energy—of course in terms appropriate to its time.

[f] *CSHK* (Chin sect.), ch. 39, pp. 5 a ff.; tr. auct. Already quoted in part, Vol. 3, p. 415.

[1] 劉智   [2] 論天   [3] 應   [4] 無門而通   [5] 雖遠相應
[6] 理   [7] 觸石而次出者水氣之通也
[8] 相響而相及無遠不至無隔能塞者

Yin and the Yang. Individual objects thus had their intrinsic rhythms. And these were integrated like the sounds of individual instruments in an orchestra, but spontaneously, into the general pattern of the harmony of the world.

We can express this prototypic wave-theory in our own terms by drawing two sine curves of opposite rhythm, as in the right-hand part of Fig. 277. As the Yang passes through its maximum the Yin begins to take over, then there comes a moment of equal balance after which the Yin rises to its maximum only to cede in its turn to the rising Yang. We can thus understand the intimate connection between the old Chinese

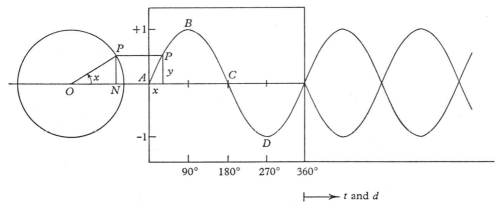

Fig. 277. Diagram to illustrate the relationship of cyclical and wave conceptions.

ideas of wave-motion and those cyclical formulations which were also profoundly characteristic of Chinese natural philosophy. In the apocryphal books of the Han, which were so much occupied with speculations about the two forces and the five elements, one can find passages like the following:[a]

For inferring from the numbers of things (the system is that they all) start from the cyclical sign *hai*[1] (N. 30° W.). This is the point which fixes the position of heaven and earth. The Yin and the Yang revolve in a cycle, ever returning to their points of departure; the myriad things die and pass away, yet ever again revive (in new forms) and once more come into being. This is how the Great Cycle (*ta thung*[2])[b] begins (and ends).

Now the forms of wave and circle are related together in a very simple mathematical way. As may be seen in any elementary introduction to trigonometry, and as every teaching museum of the history of science shows somewhere among its exhibits, the sine curve is derived from the perambulation of a point round the circumference of a circle when its excursion is plotted against the angle made by its radius with the diameter. Hence the curve in the left-hand part of Fig. 277 represents the function $y = \sin x$, for the ordinate measures the excursion (PN or y) of the point P, and the

[a] *Shih Wei Fan Li Shu*[3] (Apocryphal Treatise on the Book of Odes; the Pivot of the Infinite Calendar), in *YHSF*, ch. 54, p. 3*a*, tr. auct.
[b] The *thung* was one of the resonance periods; its value was 1539 years. See Vol. 3, pp. 406 ff.

[1] 亥   [2] 大統   [3] 詩緯汜曆樞

abscissa the magnitude of the angle x in degrees. But when we pass from the world of mathematics to that of physical phenomena, the continuous parameter becomes time or distance, and when these are expressed on the abscissa the wave-motion implicit in cyclical formulations can be seen. The schematic static circle changes into real temporal periodicity and recurrence. Excursion becomes empirical amplitude, and the degrees of the circle are transformed into empirical phase. As the point N passes back and forth along the diameter in simple linear harmonic motion, the curve of harmonic variation is generated. In Nature curves of this type are found whenever the forces tending to restore the changing matter to its intermediate position are exactly proportional to its displacement from that position, and smaller beats or pulses may of course be superimposed upon the main curve of vibration. Though the old Chinese naturalists never formulated the question in such terms, they were visualising clearly in their way those periodic phenomena in Nature due to resistances which act on matter in a state of change so as to retard the change and finally reverse it, in other words those multitudinous effects which we now attribute to wave-motion. On this is built one half of modern physics. One cannot but recognise the presence of similar conceptions in words such as those used by the writer of the *Pai Hu Thung Tê Lun*[1] about +80, when he said:[a]

> When the Tao of the Yang has reached its summit (*chi*[2]) then the Tao of the Yin takes over the task; when the Tao of the Yin has reached its summit, that of the Yang in turn takes over. Thus it is clear that two Yin in succession, or two Yang, would be absolutely impossible as a continuous process.[b]

No justification is necessary for the statement that Chinese natural philosophers tended to think in terms of cyclical recurrences. The fact has been noted by many observers.[c] In its simplest and oldest form it had to do with little more than the rhythm of the seasons and the rise and fall of the individual lives of men, yet Chuang Chou in the −4th century welded it with poetic fire into his Taoist philosophy of the ataraxic acceptance of the Tao of Nature. This 'doctrine of cyclically recurring change' (*hsün huan i pien lun*[3]) has been duly discussed in its place.[d] In the words of the *Tao Tê Ching*, 'returning is the characteristic movement of the Tao' (*fan chê Tao chih tung*[4]).[e]

Later generations developed more precise, if more symbolical, cyclical representations of natural phenomena. Already in the *I Ching* (Book of Changes),[f] as Wu Shih-Chhang (1) shows, one can find a four-membered series. Stability (*chiu*[5]) leads after a time to impasse (*chhiung*[6]), or antithesis, then there follows decisive change (*pien*[7]) with all its effects, ending in penetration (*thung*[8]), after which a new period of stability,

---

[a] 'Comprehensive Discussions at the White Tiger Lodge', ch. 27, p. 8*b*, tr. auct. adjuv. Tsêng Chu-Sên (1), vol. 2, p. 553.   [b] *Ming erh Yin erh Yang pu nêng hsiang chi yeh.*[9]
[c] E.g. Chatley (26); Wu Shih-Chhang (1); Huard & Huang Kuang-Ming (1).
[d] Vol. 2, pp. 75ff.   [e] Ch. 40.
[f] See Vol. 2, pp. 304ff.

[1] 白虎通德論   [2] 極   [3] 循環異變論   [4] 反者道之動   [5] 久
[6] 窮   [7] 變   [8] 通   [9] 明二陰二陽不能相繼也

a synthesis, supervenes. These stages may be represented by the successive points *A, B, C, D* in Fig. 277. It may very well be that this cycle arose from the contemplation of human historical, even dynastic, changes, but the naturalists of the Han would certainly not have hesitated to apply it to events in the non-human world also.

After the arrival of Buddhism, the Indian lore of *kalpas*, very long time-periods with alternating phases of creation, destruction and re-creation, marked by recurrent world conflagrations or dissolutions, became in the Thang a commonplace.[a] The four phases were termed stagnation (*chu*[1]), destruction (*huai*[2]), chaos or emptiness (*khung*[3]) and finally redifferentiation (*chhêng*[4]). One can insert them on the sine curve in just the same way. The Neo-Confucians of the Sung adopted the theory in their turn, laying particular emphasis on the course of events in its successive evolutionary phases,[b] and using a parallel set of technical terms, *yuan*[5] (beginning of spring), *hêng*[6] (beginning of summer), *li*[7] (beginning of autumn) and *chêng*[8] (beginning of winter). Chatley proposed for these terms the series, 'inception, climax, balance and anti-climax'. The four intervening periods he called positive increase, positive decrease, negative increase and negative decrease successively, no doubt with the sine curve in mind. And indeed it must have been in an attempt to visualise these four phases qualitatively in days long before the widening of man's imagination by coordinate geometry that throughout the centuries the Yin and Yang were divided into four— the Lesser and the Greater Yin (*shao Yin*,[9] *thai Yin*[10]), the Lesser and the Greater Yang (*shao Yang*,[11] *thai Yang*[12]). A text such as the *Pai Hu Thung Tê Lun*, which discusses[c] the 'ascendancy and decline of Yin and Yang' (*Yin Yang shêng shuai*[13]), may give only an arid account of the properties of each form, according to the system of symbolic correlations,[d] but the existence of four members in the set is in perfect agreement with the four parts of the sine curve.

Of course, the cyclical formulation was imposed from the start by the very subject-matter of certain sciences, such as calendar-making in astronomy, with its sixty-day and year counts,[e] or the recognition of the water-cycle in meteorology.[f] But ideas of circulation were also prominent in physiology and medicine, where a movement of *chhi* and pulsing blood was supposed to take place around the body daily.[g]

At an earlier stage we had occasion to remark upon the strange similarities between classical Chinese organic philosophy and the philosophy of the Stoics in mediterranean antiquity.[h] 'Ovum ovo non erit similius, quam Stoica sunt Sinensibus.' It is therefore particularly interesting to find that if anywhere in the ancient West it is

---

[a] See Vol. 2, p. 420; Vol. 3, p. 602.

[b] See Vol. 2, p. 486.

[c] Ch. 9, p. 11*b*; tr. Tsêng Chu-Sên (1), vol. 2, p. 433.

[d] See Vol. 2, pp. 253 ff.   [e] See Vol. 1, p. 79; Vol. 3, p. 396.

[f] Cf. Vol. 3, p. 467.

[g] The classical statement of this is in the *Huang Ti Nei Ching, Ling Shu*, ch. 15, which must be considered a Han text. The nature of these circulations, and the extent to which they prefigured modern knowledge of the circulation of the blood, will be fully discussed in Sect. 43 on physiology.

[h] Cf. Vol. 2, p. 476.

[1] 住   [2] 壞   [3] 空   [4] 成   [5] 元   [6] 亨   [7] 利
[8] 貞   [9] 少陰   [10] 太陰   [11] 少陽   [12] 太陽   [13] 陰陽盛衰

among the Stoics that adumbrations of wave-theory can be found.[a] In times corresponding to the Chhin and Han it was they who laid stress on propagation in a continuum of two or three dimensions using the analogy of water waves and ripples.[b] They also found it necessary that this continuum should be under tension—Wang Chhung's contemporary the astronomer Cleomedes said: 'Without one binding tension and without the all-permeating *pneuma* we would not be able to see and hear, for the sense-perceptions would be impeded by the intervening empty spaces.'[c] This amounted, as Sambursky so well points out, to the hypothesis of an elastic medium subjected to stresses. For tensional motion or vibration, probably actually observed in the form of 'standing waves' on a bounded water surface, the Stoics used a special term, *tonikē kinēsis* (τονικὴ κίνησις). These conceptions, which found application in fields as far apart as the physiology of Galen and the theology of Philo, clearly parallel the forms of thought which came to birth in China. What is particularly interesting and significant about them is their context of Stoic organicism, universal *pneuma* (with no void), influences ('sympathies') of things on other things even at great distances,[d] physical 'fields of force'[e] within the hierarchy of organisms. Indeed it was no coincidence that both the Stoics and the Chinese discovered the true cause of the tides.[f] How the continuum was visualised in ancient Chinese thought will shortly appear.[g]

Of course the wave-conceptions of medieval Chinese thought were never applied specifically and systematically to the interpretation of physical phenomena.[h] But the Greeks and Latins made no such attempt either. The full understanding of what the experimental method implied was necessary first, and this did not come until the time of Hooke and Huygens in +17th-century Europe. Hooke discoursed much on 'vehement vibratory motions' of small amplitude, applying wave-theories to light, heat and sound;[i] Huygens later built a whole structure of optics upon them.[j] The contest between the wave and the corpuscle theories of light in the +18th century is a well-known story. At first sight it would seem most improbable that the contact with China which Europe had then had for nearly two centuries provided any stimulus; no doubt the development of wave ideas was essentially a consequence of the study of vibration itself, as in springs. Yet we should not overlook the personal acquaintance which Robert Hooke himself had with Chinese visitors in London,[k] one of whom might well have dropped a hint about the great importance accorded in China to the long-term vibrations of the Yin and Yang.

[a] See especially Sambursky (1), pp. 138ff.; and more fully in Sambursky (2).
[b] As reported by Aetius (+2nd century), IV, 19 (in von Arnim).
[c] *De Motu Circ. Doctrina*, I, 1, 4 (−1st century).
[d] We shall return to this subject in Sect. 33. Meanwhile cf. Zeller (1); von Lippmann (1), p. 146.
[e] Cf. Vol. 2, p. 293.   [f] See Vol. 3, pp. 493ff.   [g] Pp. 28ff. below.
[h] With the exception, perhaps, of certain ideas which will be found in the Section on acoustics pp. 202ff. below.   [i] Cf. Andrade (1).
[j] Pledge (1), pp. 68ff.; Ronchi (1), pp. 196ff.
[k] See Gunther (1), vols. 6, 7, pp. 681, 694; vol. 10, pp. 258, 263. The dates in the diary are for July 1693. Hooke himself wrote on the Chinese language in the *Phil. Trans.* for 1686. Perhaps his chief informant was Shen Fu-Tsung[1] who came to Europe in 1683 and later worked with Thomas Hyde at Oxford (see Duyvendak, 13).

[1] 沈福宗

This is not the place to embark upon a history of atomic theories—excellent accounts are available, such as the monograph of Partington (2) or the book of J. C. Gregory (1).[a] The atomic theory was one of the greatest currents of Greek thought; founded by Leucippus and Democritus[b] in the −5th century, attacked by the Peripatetics and all other schools, reaffirmed by the Epicureans and immortalised in Lucretius' −1st-century poem, it was finally exiled throughout the early Christian centuries. The story of its resurrection in the +17th century, with Gassendi, Descartes and Boyle, is part of common knowledge and one of the cardinal features of the growth of modern science. But the atomism of the Indians and the Arabs has also to be considered, though its relation with European atomism remains rather obscure. Of the former something has already been said;[c] it figured in Brahminical systems such as the Vaiśeṣika and Nyāya philosophies, as also in heterodox Jain and Ājīvika schools.[d] Some roots of these may well be as old as the Greek beginnings of the theory. Arabic atomism flourished much later, mostly in the +9th and +10th centuries, when it was propounded by al-Ash'arī and al-Rāzī.[e] As Lasswitz (1) and Pines (1) have shown, it derived almost wholly from Indian and not from Greek versions; indeed it is an outstanding example of Indian influence on Arabic scientific thought, and furnishes another instance of the failure of Asian scientific achievements to diffuse to Europe, owing to the selectivity of the translators from Arabic into Latin.[f]

Now it is a striking, and perhaps significant, fact that the languages of all those civilisations which developed atomic theories were alphabetic.[g] Just as an almost infinite variety of words may be formed by different combinations of the relatively small number of letters in an alphabet, so the idea was natural enough that a large number of bodies with different properties might be composed by the association in different ways of a very small number of constituent elementary particles.[h] The trick of the anagram, as Gregory has put it, was part of the expository power of atomism. Lucretius, following Democritus, says so in so many words:[i]

> quin etiam passim nostris in versibus ipsis
> multa elementa vides multis communia verbis,
> cum tamen inter se versus ac verba necesse est
> confiteare alia ex aliis constare elementis.

[a] Lucidly written but undocumented.   [b] Almost an exact contemporary of Mo Ti.
[c] Vol. 1, p. 154; Vol. 2, p. 408.
[d] Brief account with good bibliography in Partington (2), to which add the recent book of Glasenapp (1).
[e] Mieli (1), pp. 54, 97, 139.   [f] Cf. Vol. 1, pp. 220 ff.
[g] In view of the great antiquity of the Phoenician alphabet, it is a little suspicious that Greek tradition referred to a Phoenician, Moschus, the first origin of atomism before Leucippus.
[h] This idea must have occurred to many; Dr Ehrensvaard (Stockholm) suggested it to me without knowing that it had already occurred to us.
[i] *De Rer. Nat.* II, 688. '...Nay, here in these our lines,
Elements many, common to many words,
Thou seest, though yet 'tis needful to confess
The words and verses differ, each from each,
Compounded out of different elements—', etc. (tr. Leonard).
The same argument is reproduced by Aristotle, *Metaphysics*, I, 4 (985 b), as Dr S. Sambursky (Jerusalem) reminded me.

On the other hand, the Chinese written character is an organic whole, a Gestalt, and minds accustomed to an ideographic language would perhaps hardly have been so open to the idea of an atomic constitution of matter. Nevertheless, the argument is weakened by the fact that the 214 radicals into which the Chinese lexicographers eventually reduced what they considered the fundamental elements of the written characters were essentially atomic,[a] and an immense variety of words ('molecules') were formed by their combinations. Moreover, the combinations of the components of the Symbolic Correlation groups of five[b] were understood from very early times to produce all natural phenomena. Thus we read in the *Sun Tzu Ping Fa*[1] (Master Sun's Art of War), ch. 5, written perhaps about −345:

There are not more than five musical notes, yet the combinations of these five give rise to more melodies than can ever be heard. There are not more than five primary colours, yet in combination they produce more hues than can ever be seen. There are not more than five cardinal tastes, yet combinations of them yield more flavours than can ever be tasted. In battle there are not more than two methods of attack, the direct (*chêng*[2]) and the indirect (*chhi*[3]), yet these two in combination give rise to an endless series of manœuvres.[c]

Again, it was the permutations and combinations of the broken and unbroken lines of the *kua* in the Book of Changes which were supposed to produce all the symbols which in their totality exhausted the possible states or situations in Nature. We may say, therefore, that while there is a certain plausibility in the correlation between alphabetism and atomism, the argument cannot be pressed too strongly.

As for the great debate between continuity and discontinuity as such, Chinese organic philosophy was bound to be on the side of continuity.[d] This might even be seen in the realm of mathematics.[e] The Greeks were so far from the idea of continuity that they could not imagine 'irrational' numbers like $\sqrt{2}$ to be true numbers at all. But, as we have seen,[f] the Chinese, even if they realised the special nature of these, were neither puzzled nor interested by them. Their universe was a continuous medium or matrix within which interactions of things took place, not by the clash of atoms, but by radiating influences. It was a wave world, not a particle world. And thus to the Chinese, as also to the Stoics, one of the great halves of modern 'classical' physics is owing.

[a] Cf. Vol. 1, pp. 31 ff.   [b] Cf. Vol. 2, pp. 261 ff.
[c] Tr. L. Giles (11), p. 36.
[d] Cf. Vol. 2, p. 281, and the accompanying discussions.
[e] As was pointed out to me by Dr W. W. Flexner.
[f] Vol. 3, p. 90.

[1] 孫子兵法   [2] 正   [3] 奇

(c) MASS, MENSURATION, STATICS AND HYDROSTATICS

Literary sinologists have often been inclined to suppose that nothing of any consequence could be found about these subjects in ancient and medieval Chinese writings. But such an impression was much too pessimistic. This subsection will deal first with basic knowledge about the properties of levers and with the history of the balance in Chinese culture, then with strength of materials and the philosophical ideas which such problems raised. After this will follow an account of the old Chinese theory and practice concerning the physical properties of liquids—hydrostatics, specific gravity, buoyancy, density, etc. We shall begin and end by discussions of mensuration itself, first introducing some further citations from the Mohist Canon by giving three propositions of metrological interest, and lastly by studying briefly two remarkable incidents in the pre-history of the metric system.

As a prelude to this, however, it may be of interest to read a remarkable passage in the *Huai Nan Tzu* book on quantitative measurement.[a] It is a lyrical exposition of measure and rule as seen in the operations of Nature no less than in those of man, and all the more striking in that it belongs to a civilisation which produced neither Greek deductive geometry nor the achievements of the world of Galileo. It runs thus:

As the great rules (*chih*[1]) for regulating and measuring the Yin and the Yang there are six measures (*tu*[2]). Heaven corresponds to the plumb-line (*shêng*[3]), earth to the water-level (*chun*[4]), spring to the compasses (*kuei*[5]), summer to the steelyard (*hêng*[6]), autumn to the carpenter's square (*chü*[7]), and winter to the balance (*chhüan*[8]).[b] The plumb-line serves to align the ten thousand things, the water-level to level them, the compasses to round them, the steelyard to equalise them, the square to square them and the balance to weigh them.[c]

The plumb-line as a measure is erect and unswerving. Draw it out as you will, it has no end. Use it as long as you wish, it will never wear out. Set it as far away as you can, it does not disappear. In its virtue it accords with Heaven, and in its brilliance with the spirits.[d] What is desired it obtains, what is disliked it destroys. From antiquity until today its straightness has remained unchangeable. Vast and profound is its virtue, broad and great so that it can encompass (all things). This is why the Rulers of Old (*shang ti*[9]) used it as the prime standard of things.[e]

[a] Written therefore somewhat before −120. The passage occurs in ch. 5, pp. 18*b* ff. Entitled Shih Tsê Hsün[10] (Teachings on the Rules for the Seasons), this chapter is for the most part textually identical with the *Yüeh Ling* (Monthly Ordinances), a much older work on which see Vol. 3, p. 195. But it appends some other material, including the present passage. Tr. auct. adjuv. Bodde (18).

[b] Although there are six of these measures, not five, we should probably take the meaning in the spirit of the symbolic correlations discussed already in Vol. 2, pp. 261 ff. Indeed the grammatical construction is typical of this.

[c] From this point onwards the text is rhymed.

[d] An echo from the *I Ching*; see Wilhelm (2), Baynes tr., vol. 2, p. 15. The paragraph seems to us to refer, almost in riddle form, to the calendrically indispensable gnomon (cf. Sect. 20*g*).

[e] These were the sentences which attracted Bodde's special interest to this passage, for he translated the phrase Shang Ti in its ancient sense as the 'Ruler Above', and was thus led to interpret the whole as a depiction of the activities of a personal divine engineer, if not exactly a divine creator. The passage

[1] 制   [2] 度   [3] 繩   [4] 準   [5] 規   [6] 衡   [7] 矩

[8] 權   [9] 上帝   [10] 時則訓

The water-level as a measure is even and without slope, flat and without declivity. Broad and great it is, so that it can encompass (all things), wide and all-embracing so that it can harmonise (them). It is soft and not hard, blunt and not sharp,[a] flowing and not stagnant, shifting and not choked. It issues forth and penetrates (everywhere) according to a (particular) principle (*chi*[1]). It is widespread and profound without being dissipated, and levels and equalises without any error. Thus the ten thousand things are all kept in equilibrium, the people form no dangerous plots, and hatreds or resentments never arise.[b] This is why the Rulers of Old used it as the equaliser of things.[c]

The compasses as a measure revolve without check, make circular lines without ends, and accommodate without permitting undue latitude. Broad and great they are, so that they can take (all things into their) embrace. In their action and reaction they follow a pattern (*li*[2]); in their issuing forth and (universal) penetration they have a (particular) principle (*chi*[1]). So accommodating, so indulgent, are they, that the hundred grudges make no appearance. They measure their curves without any error, and bring the *chhi* into its manifold patterns (*li*[2]) (of life).[d]

The steelyard as a measure moves with deliberation but not too slowly. It equalises without inducing resentment, extends benefits (for right-doing) without (ostentatious) virtue, and expresses sorrow (for wrong-doing) without (ostentatious) reproof. It takes care to equalise the people's means, prolonging thereby (the lives of those who would otherwise suffer) want. Glorious and majestic it is, in its operations never unvirtuous. It nourishes, gives growth, transforms and develops, so that the ten thousand things flourish exceedingly, the five grains come to fruition, and the fields and fiefs bring forth their produce. Its administration is without error, so that sky and earth are brightened thereby.[e]

The carpenter's square as a measure is severe but not perverse, hard but not obdurate. It collects without evoking resentment, and gathers in without harm. Though awe-inspiring,

thus has some relevance to the question discussed in Sect. 18 (Vol. 2), whether or not the conception of laws of Nature ever arose spontaneously in Chinese culture. We believe, however, that by the Han the original conception of a unitary sky-god (if it ever really existed) had long been lost, and that the meaning we here adopt was the most common one (striking examples may be found in the *Huang Ti Nei Ching Su Wên*, ch. 9, p. 6*b*; ch. 13, p. 4*a, b*). The same chapter of *Huai Nan Tzu*, it is true, like the *Yüeh Ling*, refers twice previously (pp. 8*a*, 15*b*) to sacrifices to a Huang Thien Shang Ti[3] (whether singular or plural is never clear); but if, as seems highly probable (cf. Vol. 2, pp. 580ff.) the 'Ruler Above' was a transcendentalised version of the 'Rulers of Old', i.e. the founding fathers and ancestors, the line between the two becomes very tenuous. Once again we see how misleading it can be to view Chinese thought through the spectacles of occidental monotheism; indeed Christian missionaries only adopted Shang Ti in despair for want of anything better. The whole passage is a poetical exposition of the measuring activities both of Nature and of man, and the earthly social significances are only just under the surface. Yet there is something to be said on both sides, and the debate may be followed fully in Bodde (18). A comparison of our versions illustrates instructively how different the nuance can be in a passage such as this according to the basic assumptions of the translators (cf. Forster, 1).

[a] The obvious echo from the *Tao Tê Ching*, chs. 4 and 56, authorises an inversion made by Bodde here.

[b] A sociological undertone now evidences itself, and grows clearer and clearer as the passage proceeds. The genuinely sublime aspects of metrology are being seen from the viewpoint of the feudal lord or the feudal-bureaucratic official. His function was to exact the maximum yield from the people in material rent or tax without giving rise to resentments and rebellions. Very relevant here is our discussion of weights and measures, devices and inventions, in the ancient Chinese feudal economy in Sect. 10 (Vol. 2, pp. 124ff.). This paragraph seems to us to involve a reference to the importance of irrigation works (cf. Sect. 28*f* below), for which the water-level was the indispensable survey instrument.

[c] See note *e*, p. 15.

[d] The thought here is of the burgeoning and rounding out of growing things in spring.

[e] The thought is of summer's brilliance and yet of the bounds set by Nature to the growth and development of animals and plants.

[1] 紀   [2] 理   [3] 皇天上帝

it does not terrify, nor can there be any denial of what it says. Deadly and effective are its attacks, bringing into subjection any opposing forces. It measures its squares correctly and without error, so that all those doomed to perish meekly submit.[a]

The balance as a measure acts quickly but not excessively, killing but not afflicting. It (presides over) the full, the real, the solid, wide and profound yet not dispersed. It destroys things without diminishing (the totality of the world) and executes the guilty without (possibility of) pardon. Its sincerity and trustworthiness give certainty, its firmness and genuineness give reliability. In its sweeping away of hidden evils it cannot but be straightforward. Thus when the administration of winter is about to take charge, it must weaken in order to make strong, and soften that which is to be made hard.[b] It weighs correctly and without error, so that the ten thousand things return into the treasuries of the earth.[c]

The regulations of the Ming Thang[d] are to take the water-level as the standard of quiescence, and the plumb-line as the standard of activity. In spring government follows the compasses, in autumn the square, in winter the balance, in summer the steelyard. Then drought and wet, cold and heat, arrive at the proper juncture, while sweet rains and enriching dews descend at their proper seasons.[e]

Thus was faith affirmed in a world of order, precise, clear, numerical, unvarying and repeatable, not vague and chaotic, not wholly composed of excess and defect. If there were no laws in Nature like the laws of men, there was no disorder either, but pattern above pattern, recognisable and measurable, level beyond level, from wisps of *chhi* to stars. And the measures of the world and the principles of its measurement did not change, before even the Rulers of Old, they were. Another writer of Liu An's invisible college had said so, in a passage which we cannot forbear from quoting again:[f]

At the present time the balance and the steelyard, the square and the compasses, are fixed in a uniform and unvarying manner. Neither (the people of) Chhin nor Chhu (can) change their specific properties—neither the northern Hu barbarians nor the men of Yüeh in the south (can) modify their appearance. These things are for ever the same and swerve not, they follow a straight path and do not meander. A single day formed them, ten thousand generations propagate them. And the action of their forming was non-action.

(1) THE MOHISTS AND METROLOGY

By way of background let us recall previous references to this subject. The mathematics Section showed how far back in Chinese history it is possible to trace the use of powers of ten for units of measurement,[g] the astronomical Section revealed an early attempt to develop something analogous to the standard platinum metre of

---

[a] The thought here is of autumn when vegetation and all life dies down, but the problem of the repartition of the harvest between lord and lad comes through clearly enough.

[b] Echo of the *Tao Tê Ching*, ch. 36, but a typical Taoist idea.

[c] In hibernation, in seeds, in ores, and so on.

[d] The imperial cosmological temple; cf. Vol. 2, p. 287; Vol. 3, p. 189.

[e] In our interpretation this is a statement of phenomenalism (cf. Vol. 2, pp. 378ff.), i.e. the doctrine universally held in the Han, that Nature was *en rapport* with human ethical behaviour, and reacted in accordance with its rightness or wrongness. Bodde, however, takes *chih* as a verb rather than a noun, and makes the instruments or their principles the actual methods of a divine government of the world at the successive seasons. But, as the Arabs say, Allah knoweth best.

[f] Ch. 9, p. 5a; tr. Escarra & Germain (1), p. 23; eng. auct. mod.   [g] Cf. Vol. 3, pp. 82ff.

modern times.[a] Presently the acoustics subsection will show us a 'statistical' method of fixing volumes.[b] Here now are the Mohist propositions, dating from the late −4th century, which have to do with standardisation in metrology.[c]

Ch 80/—/78.72. *Standardisation of length measurement*

C   A thing can be 'very' or 'not very' (*shen pu shen*[1]). The reason is given under 'following a standard' (*thi*[2]).

CS (People in different places) use very long or very short standards, but the 'very long' and the 'very short' should not be longer or shorter than the long and the short standard respectively. A 'standard's' standardisation may be true or false. All (individual) standards should conform to the (accepted) standard. (auct.)

>   With regard to 'very long' and 'very short' we may remember the system of technical terms adopted by the Chou and Han astronomers for stating different fractions of a degree of arc (Vol. 3, p. 268). The Mohist writer was doubtless thinking of the confusion of standards in different parts of the country, each feudal State having at least one set of measures of its own.

Ch 81/—/80.73. *Choice of units appropriately small*

C   Starting out with the lower, search out the upper. The reason is given under 'valley' (*tsê*[3]).

CS If one has to choose between higher and lower for a standard, it is not a question of 'mountain and valley'. It is better to live in lower rather than in higher places. The lower may really turn out to be the higher. (auct.)

>   Here there is an obvious echo of the *Tao Tê Ching*, ch. 6 (Vol. 2, p. 58 above), 66, etc. What the writer seems to be saying is that in choosing units of mensuration, the technician should not be misled by Confucian delusions of grandeur into making them too spacious and large, nor by Taoist humility into making them too small. Nevertheless, he inclines to the Taoist side, for smaller units are likely to be more convenient.

Ch 82/—/8.174. *Arbitrariness of standards*

C   The 'non-standard' can become the same as the 'standard'. The reason is given under 'no difference' (*pu chou*[4]).

CS If a standard has been standardised, then it must be a true standard. But nowadays standards which are supposed to be true are actually not so; therefore people call the non-standard the standard. Their standards are (false) standards, and do not agree with the true one, though people use them. So the standard of today is the same as what used to be a non-standard. (auct.)

---

[a] Cf. Vol. 3, pp. 286 ff.
[b] See pp. 199 ff. below.
[c] Interpretations, as given here, of this notoriously corrupted and difficult text, must be taken with all reserve. The key to the identification symbols used will be found in Sect. 11 (Vol. 2, p. 172).

[1] 甚不甚   [2] 題   [3] 澤   [4] 不州

(2) THE MOHISTS, THE LEVER AND THE BALANCE

The next group of excerpts from the *Mo Ching* give us an idea of the notions which were held by the Mohists about Aristotle's time (the end of the −4th century) on force and weight. It should again be emphasised that we have only surviving fragments and rather garbled texts to work on, so that one can hardly judge the extent of the physics of the Warring States period without much guesswork.

Cs 21/—/41.21. *Force and weight*

C Force (*li*[1]) is that which causes shaped things (*hsing*[2]) (i.e. solid bodies) to move (*fên*[3]).[a]

CS Weight (heaviness) (*chung*[4]) is a force. The fall of a thing, or the lifting of something else, is motion due to heaviness. (auct.)

Ch 26/—/48.20. *Balance of forces; consideration of pulley and balance*

C A suspending force acts in the opposite direction to the force which pulls (downwards). The reason is given under 'beating against' (*po*[5]).

CS For suspension (*chhieh*[6]) force is necessary, but free fall (*ying*[7]) occurs without the application of force (by us). A suspending force is not necessarily confined to the actual point at which it is applied (as in the case of a beam or a bridge). (Note how a cord is used for drilling.[b] (Consider a beam suspended on a rope. The side where the distance from the point of suspension (the fulcrum) is greater, and/or the weight heavier, will go down; (the side where) the distance from the point of suspension is shorter, and/or the weight lighter, will go up; so that the more the upper side gains the more the lower side will lose. When the rope is at a right angle to the beam, the weights are the same (on both sides) and a mutual balance is struck.

(Consider two weights suspended by a rope over a pulley.) The more loss (*sang*[8]) from the 'upper' side (by reduction of the amount of weight hanging), the more gain there will be on the 'lower' side. If the 'upper' weight is entirely taken away, the 'lower' (side) will fall altogether. (auct.)

In the last example it is assumed that at the beginning of the experiment the weight which is going to be changed is hanging at a point higher than the constant weight on the other side.

Here is an ancestor of Atwood's machine (+1780) for studying the relations between force, acceleration and mass. Cf. Wu Nan-Hsün (*1*), pp. 92 ff.

Ch 11/—/20.11. *Combination of forces*

C A force made up of (*ho*[9]) several forces, can act against one force. Sometimes there is a reaction (*fu*[10]) and sometimes not. The reason is given under 'parallelism' (*chü*[11]). (auct.)

---

[a] The word *fên* is of particular interest here, since it connotes rushing or accelerated movement, and originally meant the taking-off of a bird from the field in flight. If the Mohist writer had not had a vague idea of acceleration at the back of his mind, he would have used obvious words such as *hsing*,[12] *i*,[13] or *tung*.[14] In modern Chinese physical terminology, *fên li*[15] means impulse. Cf. Wu Nan-Hsün (*1*), p. 84.

[b] This is a reference to the rotary bow- or pump-drill. See Sect. 27 below.

[1] 力	[2] 形	[3] 奮	[4] 重	[5] 薄	[6] 挈	[7] 引
[8] 喪	[9] 合	[10] 復	[11] 矩	[12] 行	[13] 移	[14] 動
[15] 奮力						

CS (missing).

The brevity of this proposition and the loss of the commentary makes certainty difficult as to what it concerned. If *chü*[1] is the right character, then the Mohists were making some attempt at the resolution of vector forces, the parallelogram of forces, etc. Whether or not there would be a 'reaction' would depend upon whether the structure under consideration was in equilibrium or not. There is no way of telling, however, whether the Mohists had any methods of calculating the force distributions beforehand, and if so, what they were. Cf. Mikami (13) on the parallelogram of forces in post-Renaissance East Asian science.

Than Chieh-Fu, taking *chü*[2] ('pushing') to be the character intended, believed that this proposition concerned the transmission of energy by impact from one suspended ball to another through a series of intermediate suspended balls which remain unmoved; but this seems less likely than the former interpretation.

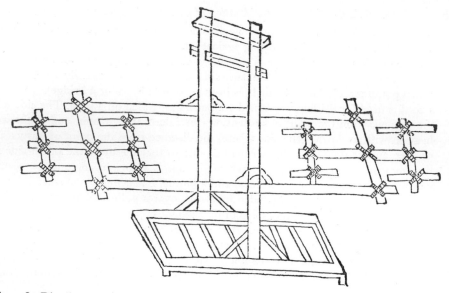

Fig. 278. Distribution of load: the diagram in the *San Tshai Thu Hui* encyclopaedia (+1609).

In practical engineering in ancient and medieval China there must have been many occasions for empirical knowledge of the combination of forces, though their theoretical resolution could not be undertaken. Not until the +17th century in Europe, indeed, was this fully understood. As Chhen Wên-Thao (1) has pointed out, the early aerodynamic invention of the kite in China was an application of the principle (cf. Sect. 27m below). An oustanding instance of the division of force was the ancient method used in China for carrying heavy loads—the *chhê ni*[3] or chariot-yoke principle, in which the four handles of the sledge on which the heavy weight rests are attached to a fanning-out series of carrying-poles, so that the weight resting on the shoulders of each individual carrier is not intolerable. This has often been described (cf. Fig. 279); e.g. by Esterer (1),

¹ 矩   ² 拒   ³ 車輗

PLATE XCVI

Fig. 279. Distribution of load among many porters, an ancient Chinese method still commonly used (photo. Gordon Sanders, near Chungking, 1943).

p. 143; G. L. Staunton (1), vol. 2, p. 113; and in *San Tshai Thu Hui* (I chih sect.),
ch. 7, p. 9*a*, *b* (Fig. 279). I myself have often seen heavy loads, such as dynamos
or transformers, still carried in this way. Similar branched linkage systems may
be seen today in aircraft testing establishments, where a single measurement
based on numerous points of attachment records the properties of a wing subjected
to various strains. As for the carrying-pole (*pien tan*[1]) itself, so universally
employed by Chinese workers and peasants, it must go back to very ancient times
(as, e.g. in Egypt, Wilkinson (1), vol. 2, p. 108). An Etruscan cinerary casket of
the −3rd century preserved at Orvieto shows a naked hierophant carrying two
baskets by a pole over one shoulder exactly like a Chinese porter.

Ch 27/—/?.19. *Mechanics of mobile scaling-ladders*

C   Suspension...(the rest missing).

CS The *chuan chhê*[2] (mobile scaling-ladder) is built (on a frame) with two of its wheels
higher (i.e. of larger diameter) than the other two (presumably for facility of steering).

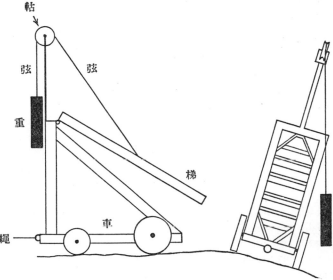

Fig. 280. Diagram of counterweighted scaling-ladder, to illustrate *Mo Ching*, Ch. 27.

A (counter-)weight is borne at the front end, which also has cords (lit. bowstrings,
hsien[3]) attached to it. (When it is ready) the cord is carried to the front and over the
pulley-wheel (*ku*[4]) (at the top), and the (counter-)weight hangs down from it in front.
As for the ladder, it moves when (the counterweight) is pulled or pushed. When
the weight is not pulled from above or pushed from below (the ladder will not move).
If the counterweight is not interfered with (by being allowed to get out of alignment)
sideways, it should hang straight down. When it slants (in relation to the whole
machine) some (irregularity in the ground) is spoiling it. Those who are unskilful
with scaling-ladders do not understand how to arrange that the weight hangs
perpendicularly. If it will not come down, (it is because it has) lost its side support.
The rope (*shêng*[5]) (for hauling the whole machine forward) is attached to the

¹ 扁擔   ² 輪車   ³ 弦   ⁴ 帖   ⁵ 繩

pulley-frame (at the front), just as a tow-rope is attached to the cross-piece at the bow of a boat. (auct.)

Here is a practical application of the pulley and counterweight mentioned already. The construction of the machine can readily be understood by the diagram, Fig. 280. Cf. Werner (3). The usual name for this machine is, of course, *yün thi*.[1] The discussion clearly exemplifies the Mohist interest in military technology. European parallels are in Uccelli (1), p. 218. It is remarkable that engineers such as Ramelli in +1588 were still interesting themselves in essentially the same devices.

Luan Tiao-Fu (1), p. 89, reconstructs the canon.

Ch 25/—/23?.—. *Lever and balance*

C   A balance (*hêng*[2]) can lose its equilibrium (*chêng*[3]). The reason is given under 'gaining' (*tê*[4]).

CS   If a weight is added to one side of a balance, that side will fall.

As for the steelyard (*chhüan*[5]) (balance with beam of unequal arms), let a quantity of material and a weight be balanced (*chung hsiang jo yeh*[6]), the distance between the fulcrum and the point where the material is suspended (*pên*[7]) being shorter (*tuan*[8]) than the distance between the fulcrum and the point where the weight is suspended (*piao*[9]). This will then be the longer (*chhang*[10]). If now to both sides the same weight is added, the weight must go down (*piao pi hsia*[11]) (because the distance between the fulcrum and the point where the weight hangs is greater than that between the fulcrum and the material). (auct.)

First a note concerning the technical terms. *Hêng* (K748*h,i*) essentially means the beam of any balance (whether of equal or unequal arms) and acquired in time the broad significance of balances in general. The character derives from a graph which depicts a carrying-pole or vehicle pole passing a cross-roads (a natural place for a primitive market). *Chhüan* (K158*o*), which combines the wood radical with a phonetic deriving from a picture of a heron, means essentially the weight of a steelyard, but occasionally by implication the steelyard itself, and later more commonly came to be used as a verb, to weigh.

K 748*i*

Now in this passage there is an exact parallel between the Mohist writer, who has plunged into the statics of the balance, and the propositions of Archimedes half a century later (*De Aequiponderantibus*, tr. Peyrard (1) and van Eecke (1); cf. Dugas (1), p. 24). The first statement of the commentary is identical with Archimedes' postulate no. 3, and presupposes 1 and 4. The second statement is the same as Archimedes' proposition III. The special case then follows of doubling the weight on the long arm, but only increasing that on the short arm by a small amount; if no change is to take place in the distances from the fulcrum, the result must be as stated. The whole entry gives useful technical terms for the −4th century.[a]

It should be remembered that this work of Archimedes was not available to the European Middle Ages (Dugas (1), p. 38).

---

[a] Cf. Wu Nan-Hsün (*1*), pp. 91 ff.

[1] 雲梯   [2] 衡   [3] 正   [4] 得   [5] 權   [6] 重相若也
[7] 本   [8] 短   [9] 標   [10] 長   [11] 標必下

The most important thing about this excerpt on the lever and balance is that it shows that the Mohists must have been essentially in possession of the whole theory of equilibria as stated by Archimedes. That it was widely understood in the Han emerges from a passage in the *Huai Nan Tzu*:[a]

Therefore if one has the benefit of 'position' (*shih li*[1]),[b] a very small grasp can support a very large thing. That which is small but essential can control that which is wide and broad. So a beam only 10 *wei* long can support a house 1000 *chün* in weight; a hinge only 5 inches in length can control the opening and closing of a large gate. It does not matter whether the material is large or small. What matters is its exact position (*so chü yao yeh*[2]).[c]

And obviously there were numerous other examples of the lever principle in practical life—one thinks of the swape (*shadūf*),[d] referred to usually as the *chieh kao*[3] (in *Li Chi*[e] as the 'bridge-balance', *chhiao hêng*[4]), and of the steering-oar.[f]

Naturally the balance was often referred to illustratively in texts. Two may be quoted. Mencius says (−4th century):[g]

By weighing we know what things are light and what heavy. By measuring we know what things are long and what short. The relations of all things may thus be determined. But it is of great importance to measure the (motions of the) mind. I beg your Majesty to measure *them*.

And in the *Shen Tzu* book (between the +2nd and +8th centuries) some Taoist writer lauds the quantitative as follows:[h]

Those who navigate on the sea can come to Yüeh by boat; those who go overland come to Chhin in carriages; Chhin and Yüeh, though far apart, may be reached just by sitting peacefully. This is all due to machines and appliances (*chiai*[5]) made of wood. As to the measurement (by such a machine) of weights such as the *chün*[6] and the *tan*[7]; even a person as clever as Yü the Great could not distinguish (merely by inspection) weights as small as the *tzu*[8] and the *chu*.[9] But when they are placed on the balance (*chhüan hêng*[10]), the difference of even a *li*[11] or a 'hair' (*fa*[12]) (the smallest weights) cannot be overlooked. For this one does not need the wisdom of Yü the Great. The knowledge of ordinary people is sufficient to attain this.

Shen Tzu would have agreed with Socrates, as Plato records him.[i]

---

[a] Ch. 9, p. 17*a*; tr. auct.
[b] Note the use of the Legalist technical term for 'princely influence'.
[c] Cf. 'With a long enough lever, one could move the world.' Cf. Duhem (2). The *Huai Nan Tzu* passage is echoed in *Chin Lou Tzu* (+6th century), ch. 4, p. 19*a*.
[d] Cf. below, Sect. 27*g*. And of its military derivative, the trebuchet.
[e] Ch. 1, p. 14*b* (Legge (7), vol. 1, p. 73). Mazaheri (3) suggests that the steelyard may have been derived from it.   [f] Cf. below, Sect. 29 in Vol. 4, pt. 2.
[g] I (1), vii, 13; tr. Legge (3), p. 20.   [h] P. 7*a*, tr. auct.
[i] *Republic*, x. 'Immersion in water makes the straight seem bent, but reason, thus confused by false appearances, is beautifully restored by measuring, numbering and weighing; these drive vague notions of greater or less or more or heavier right out of the minds of the surveyor, the computer, and the clerk of the scales. Surely it is the better part of thought that relies on measurement and calculation.'

| [1] 勢利 | [2] 所居要也 | [3] 桔槹 | [4] 橋衡 | [5] 械 | [6] 鈞 |
| [7] 石 | [8] 錙 | [9] 銖 | [10] 權衡 | [11] 絫 | [12] 髮 |

Unfortunately the comparative history of the balance has not yet been put together.[a] From authorities such as Ducros (1) and Glanville (1) we know that the balance was in use in ancient Egypt from at least the beginning of the -3rd millennium, and probably from pre-dynastic times. Nearly all the representations of such balances which have come down to us show beams of equal arms supported centrally by a pillar to which a plumb-line is attached.[b] The Greek world presumably knew the balance from its earliest origins, but pictures of it are very scarce. Some of them show a balance suspended from the central point, not supported on a column; as for example the celebrated black-figured dish treasured in Paris, dating from the middle of the -6th century, which depicts the weighing of silphium before Arcesilaus of Cyrene.[c]

It is generally agreed[d] that the suspended balance with unequal arms (the steel-yard)[e] is of much later date in the ancient occident, and its invention or adoption has sometimes been placed (though unconvincingly) in Campania about -200. It certainly needed a more sophisticated knowledge of the properties of the lever than that which had sufficed for the equal-armed balance. Named in Latin *statera*, it has often been called the 'Roman' balance, and was closely described by Vitruvius.[f] Parts of instruments from that time, and even whole examples, still exist,[g] as also contemporary representations of their use, e.g. the Gallo-Roman carving from Neumagen (Fig. 281) now at Trier. But this device never became the dominant type in Europe. It is curious that in China the opposite development occurred,[h] and the steelyard (*chhêng*[1,2]),[i] though of unknown origin,[j] seems to have been much more prevalent, at least from the Han onwards, than the simple balance with equal arms. This is called *thien phing*,[3] but the term is a late one, and since *têng* has the semantic significance of equality we can probably recognise the equal-armed balance in ancient and medieval texts under the name *têng tzu*.[4] But this can also mean the steelyard, if of small size, as in jewellers' and apothecaries' shops.

[a] There is only the fifty-year-old, though still meritorious, dissertation of Ibel (1). See also Mach-abey (1); Skinner (1); and Sanders (1) on the evolution of the pivot in weighing instruments. The knife-edge was not, so far as I know, developed in China. Moody & Clagett (1) and Clagett (2) have now published texts and translations of a number of ancient and medieval European works on the balance or lever, and on the forces produced by masses moving on inclined planes (*Scientia de Ponderibus*). These range from Hellenistic treatises bearing the names of Euclid and Archimedes, to the book of Jordanus Nemorarius (c. $+1225$) and that of Blasius of Parma, written about $+1400$.

[b] See, e.g. Klebs (3), fig. 76, p. 107 and fig. 116, p. 182.

[c] Cf. Neuburger (1), p. 206; Testut (1), p. 22.   [d] See, e.g. Feldhaus (1, 2); Testut (1).

[e] The steelyard has nothing to do with steel; its name derives from the word *stalhof*, the court in London where samples were shown by Hanseatic merchants to prospective customers. The first use in English seems to be as late as $+1531$.

[f] x, iii, 4. It is mentioned also by Varro, Cicero, Pliny and others.

[g] Cf. A. H. Smith (1), p. 162; Daremberg & Saglio (1), vol. 3, pp. 1226ff.

[h] One cannot help wondering whether this could have had anything to do with the practice, seemingly so widespread in all ages in China, of transporting loads on carrying-poles (*pien tan*[5]) borne on the shoulder. But this technique was common also in ancient Egypt, cf. Wilkinson (1), vol. 2, p. 185.

[i] The character (K 894g) combines the grain radical with what was originally a pictogram of a hand lifting something up (K 894b).

[j] The first form of the word occurs already in the *Sun Tzu Ping Fa* of c. -345 (Giles (11), p. 31), the second in the writings of Chuko Liang (c. $+210$).

K 894b

¹ 稱   ² 秤   ³ 天平   ⁴ 等子   ⁵ 扁擔

PLATE XCVII

Fig. 281. Steelyard depicted in a Gallo-Roman carving from Neumagen
(photo. Landesmuseum, Trier).

PLATE XCVIII

Fig. 282. Painted grass basket with a pair of scales and weights for them, together with a wooden comb, writing tablets of wood, and other things. From a tomb of the State of Chhu at Tso-chia-kung Shan, near Chhang-sha, dating from the −4th or −3rd century. The balance, one pan of which (on the left) still bears its suspension-cords, is the oldest Chinese example of weighing apparatus so far known. The set of weights takes the form of a series of thick rings of different sizes (photo. CPCRA and BCFA).

In the best-known form of the steelyard the determination of the unknown quantity is accomplished by moving the standard weight along the long arm and reading off the result on the scale provided. A diagram of the steelyard (*chu chhêng*[1]) was given about +1050 in the *Huang-Yu Hsin Yo Thu Chi*[2] (New Illustrated Record of Musical Matters of the Huang-Yu reign-period) by Juan I,[3] as quoted in Wang Chhi's[4] Ming-dynasty *Pai Shih Hui Pien*[5] (Informal History).[a] Fig. 283 reproduces the illustration in the *Thu Shu Chi Chhêng* encyclopaedia.[b] The beam (*kan*[6]) was divided into twenty-four divisions (*chu*[7]) each corresponding to 1 oz. (*liang*[8]), with every tenth

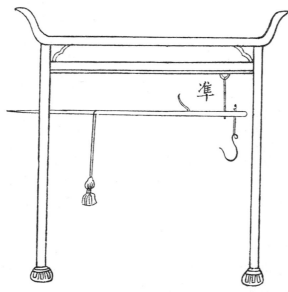

Fig. 283. Steelyard illustrated in the *Thu Shu Chi Chhêng* encyclopaedia.

division more prominently marked by a stud (*hsing*[9]). The standard weight (*chui*[10]) was moved along until equilibrium was reached. As in the West, the number of weights which have survived is far larger than that of the balances on which they were used. Chhin weights already have rings cast on them for attachment to the hook of the balance.[c] Obviously the Chhin and Han people must have known how to calibrate their steelyards, taking the weight of the beam itself into consideration. In Europe the steelyard of medieval and later times was also called 'bismar', a word of Scandinavian origin. But the Roman and Scandinavian-Slavonic steelyards differed, however, because in the former the divisions were equal, while in the latter they formed

---

[a] Juan I's account would correspond with the *Liber Charastonis*, a Latin version of an Arabic text of Thābit ibn Qurra (+836 to +901). It is not known whether the word meant the steelyard in Arabic, or referred to a legendary inventor Chariston. See Dugas (1), p. 37; Wiedemann (12); Mazaheri (3).

[b] *Khao kung tien*, ch. 13, *hui khao* 2, p. 17*b*.

[c] See, e.g. Wu Chhêng-Lo (2), pp. 148–53, new ed. pp. 70ff.

[1] 銖秤　[2] 皇祐新樂圖記　[3] 阮逸　[4] 王圻　[5] 稗史彙編
[6] 幹　[7] 銖　[8] 兩　[9] 星　[10] 錘

a harmonic series. According to Benton (1), China is the only country in Asia to have the 'Roman' type.[a] We find steelyards also in Gandhāra sculptures[b] and on Maurya coins.[c] An adequate history of these important instruments in all civilisations would be very illuminating.[d]

The use of more than one fulcrum would obviously permit of weighing in a series of different ranges, and Chinese steelyards were indeed (and often are) equipped with arrangements for more than one point of suspension. Going further in this direction the steelyard could be provided with a fixed weight and a shifting fulcrum, and this was in fact done in various parts of the world, forming a second type of unequal-armed balance which Sökeland (1) considered should be termed the 'desemer'. The weight was often a club-like expansion of the beam, but in Mediterranean antiquity could be formed into a lion head or other ornament. This balance was much used in Germany and western Russia, as also in Assam and Bhutan, but the best examples, oscillating well without upsetting, come from Tibet.[e] If known and used in China, it was certainly never common there. Some early Roman desemers have bridges along which the suspension slides, thus placing the centre of gravity well below the point of support; this helps oscillation without continual overturning but reduces sensitivity.[f]

For accurate weighing in China the equal-armed balance held its own, at least in the smaller kinds. Probably the oldest Chinese example of any weighing machine is that recently excavated from a tomb of the Chhu State ($-$4th century) and preserved in the Peking National Museum. The weights included rings of different sizes. It is a striking thought that we now have some pieces of simple physical apparatus contemporary with the Mohists (Fig. 282). Examples from the Han have long been known, some of the oldest being from the time of Wang Mang ($c.$ $+$10).[g] The equal-armed balance is frequently depicted in the frescoes of the cave-temples at Tunhuang.[h] In

---

[a] It is found, however, in the Shan States (Annandale, Meerwarth & Graves, 1).

[b] Bruhl & Lévi (1), fig. 26.

[c] Belaiev (5).

[d] Very recently Mazaheri (3) has made a determined effort to prove that the steelyard of Roman Europe was derived from that of China. Unfortunately, his argument is based on two unacceptables: (a) the belief that the text of the *Chou Li* is of the $-$10th century instead of about the $-$3rd, and (b) that the words *chhüan* and *hêng* which occur in it necessarily refer to the weights and beams of steelyards and not to scales and weights in general. Although, as we have seen, there is Chinese textual evidence for the balance of unequal arms in the $-$4th century, one of the Roman specimens has been dated as early as the $-$3rd. A coincidental converse of Mazaheri's unduly courteous bow from West to East appears in the learned memoir of Yamazaki (1), who makes the Chinese abacus derive from the Roman chiefly because he believes that it goes back to the $-$4th millennium in Egypt, a view for which we know no justification whatever (cf. Vol. 3, p. 79). Both these kind estimates of antiquity are greatly exaggerated, and as yet we simply do not know where the steelyard and the abacus originated. This is a pity, in view of their great importance in the history of commercial activity and intercourse.

[e] Cf. too the study of Annandale, Meerwarth & Graves (1) on the desemers of the markets of the Shan States. In some of these a scale-pan moves back and forth instead of the fulcrum.

[f] One of these steelyards is illustrated in Neuburger (1), p. 205, as well as in Sökeland's paper. See also Daremberg & Saglio (1), vol. 3, figs. 4474, 4475.

[g] See Wu Chhêng-Lo (2), p. 165, new ed. p. 78.

[h] Generally it hangs from a bar supported on two posts forming a stand like those used for bells and chime-stones. A bird is often perching on the bar; this is the dove waiting for the flesh donated by Śivi, one of the previous incarnations of the Buddha, and the flesh is being weighed. I have noted this in caves nos. 138 (late Thang), 98 (Wu Tai, *c.* $+$950), and 61 and 146 (early Sung, before $+$1000). Cf. too the $+$11th-century stele described by Shan Chhing-Lin (1).

his *Shih Yu Than Chi*[1] (Records of Discussions with my Teachers and Friends), Li Chih[2] wrote, towards the end of the +11th century, that it was agreed[a] that the steelyard would not do for goldsmith's work, and that it was necessary to pile up very small weights (*têng*[3]) on the *têng tzu*.[4] It is not quite clear whether there was any difference between this balance and the *thien phing*;[5] the latter seems first to be referred to in +1451 when an imperial edict ordered the construction of all three kinds of balances.[b] Weights used for it were known as *thung fa tzu*[6] (edict of +1506) or *fa ma*[7] (edict of +1529).

An ancient ordinance, preserved in the *Lü Shih Chhun Chhiu*, and therefore dating from the Chou period, prescribed the checking of all weights and measures at the autumn equinox.[c] No reason is given, but the temperature would then be neither too hot nor too cold. And indeed in the early +2nd century, Chang Hêng,[8] in his poetic essay on the Eastern Capital (*Tung Ching Fu*[9]) mentions the equalisation of measures in relation to contraction and expansion in cold and heat (*liang chhi chi shu yü han nuan*[10]).[d]

It is needless to emphasise the enormous importance of the balance in all branches of science. What we have here seen of an appreciation of the quantitative in −4th-century China may be paralleled, as Tasch (1) has shown, by the writers of the slightly earlier Hippocratic Corpus in Greece. Although it was not until the 17th and 18th centuries that modern chemistry arose out of the weighings of Rey and Lavoisier, the use of the balance in earlier ages (both East and West) must not be underestimated, and later on evidence will be brought forward to show the care with which some of the Sung alchemists and metallurgists made use of the balance.[e] We shall also trace it in pharmacy from the Han onwards.[f]

(3) TENSION, FRACTURE AND CONTINUITY

Let us begin with two more Mohist propositions.

Ch 24/—/—.—. *Strength of materials*

    C (Suppose a weight) to be supported (by a beam) which does not break. The reason is given under 'bearing' (*shêng*[11]).

  [a] He mentioned experts of the time such as Chhin Shao-Yu[12] and Hsing Ho-Shu[13].

  [b] The term *thien phing* is used in China today for weighing-machines or counter trip-scales, i.e. for instruments in which the scale-pans are placed above the beam and maintained constantly in horizontal position, giving accurate readings irrespective of the position of the load and the weights. All these depend upon combinations of linked load-carrying levers, and derive from the famous 'énigme statique' of de Roberval (+1670), the theory of which was not worked out until Desaguliers (+1740). It is very improbable that any device of this kind was known as early as the +15th century in China or elsewhere; cf. Testut (1), p. 72.

  [c] Ch. 36, tr. R. Wilhelm (3), p. 93.

  [d] I have to thank the late Dr E. R. Hughes for sending me this reference (*Wên Hsüan*, ch. 3, p. 17a).

  [e] See Sects. 30d and 33.   [f] Sect. 45.

[1] 師友談記　　　[2] 李廌　　　[3] 等　　　[4] 等子　　　[5] 天平　　　[6] 銅法子

[7] 法碼　　　[8] 張衡　　　[9] 東京賦　　　[10] 量齊急舒於寒煖　　　[11] 勝

[12] 秦少游　　　[13] 刑和叔

CS From a horizontal piece of wood a weight may be suspended, but the wood will not be broken because the centre (*chi*[1]) of the wood can bear (*shêng*[2]) the weight. But (in the like circumstances) a hand-twisted rope may break, if its centre is not able to bear the weight. (auct.)[a]

Ch 52/—/23.44. *Tension, breakage, and continuity*

C (It is upon) evenness, or continuity, that breaking or not-breaking depend. The reason is given under 'evenness, or continuity' (*chün*[3]).

CS Let a small weight hang on a hair. Even if it is very light, the hair will break. This is because the hair is not (truly) even, or continuous (*chün*[3]). If it were, it would not break. (auct.)

The first of these is an early example of attention to those problems of strain and load, bending and fracture, which must have occupied the minds of all ancient empirical engineers, architects and practical physicists,[b] but which were not tackled theoretically until the Renaissance with Leonardo, Galileo, Mariotte and Hooke.[c]

The second is of even wider interest. A parallel passage exists in the *Lieh Tzu* book, where although the *Mo Ching* is textually quoted, the purpose of the argument is exactly the opposite. In the Canon, the Mohist writer seems to be maintaining that the reason why a fibre breaks under tension is that it is formed of elements unequally strong, or unequally cohesive, so that a breaking-plane must occur somewhere.[d] This is an essentially atomic or particulate point of view, and fits in with the Mohist definitions of geometrical points[e] and indivisible instants of time. But the writer of the *Lieh Tzu* passage, on the contrary, is supporting continuity—as one would expect him to do, in view of what we have seen in the discussion of waves and particles.[f] Continuity (*chün*[3]) is, he says:[g]

the greatest principle (*li*[4]) in the world. The connected-togetherness (*lien*[5]) of all shapes and things (in the world) is due to continuity. Now a hair (might be thought to have) continuity. But 'let a small weight hang on a hair. Even if it is very light, the hair will break. This is because the hair is not (truly) even, or continuous.' But with real continuity there cannot be any fracture, or any separation. Many people do not believe this, but I will prove it by examples.

[a] Luan Tiao-Fu (*1*), p. 88, has a different reconstruction. Cf. Wu Nan-Hsün (*1*), pp. 88 ff.

[b] Later on we shall come across a remarkable example of strength-of-materials testing in relation to medieval stone beam bridges (Sect. 28 *e*).

[c] For some points in the later history of this important subject see Meyer (*1*); Frémont (*1*); Straub (*1*), pp. 62, 74, 79, 119; and the book of Timoshenko (*1*).

[d] This may have been also the point in the first proposition above, where the wooden beam is compared with the hand-twisted rope, the latter being more 'uneven'. Similar arguments made their appearance in European scholastic philosophy, e.g. the thesis retailed by Cyrano de Bergerac that a thread can carry an unlimited load if it is perfectly even, since there is then no reason why it should break at one place more than at any other. It is also possible, of course, that the second proposition above refers to the difference of tensile strength between the long axis and the transverse axis of a fibre, but this is improbable.

[e] Cf. Vol. 3, p. 91.

[f] Texts in *Lieh Tzu* cannot be exactly dated, but for this one the −2nd century would be very plausible.

[g] Ch. 5, p. 15 *a*, tr. auct. adjuv. Wieger (*7*), p. 139.

[1] 極   [2] 勝   [3] 均   [4] 理   [5] 連

Our Taoist is therefore in the great tradition. It is not in his power to prove his point by arguments which might have influenced Aristotle, but he embarks upon a series of parables and legends in the Taoist manner which show us clearly what his conception of the physical world was. Remember Chan Ho,[1] he says, who could bring up enormous fishes out of the abyss with a fishing-line made of a single silk fibre, such was his mental concentration and projection.[a] His teacher had been Phu Chü Tzu[2] the archer,[b] who brought down cranes from the clouds with a similar thread attached to his arrow.[c] There follows the story[d] of the exchange of hearts between two men[e] effected by the physician Pien Chhio,[3] tending to prove that the heart is the organ of continuity between the individual and the family. Other stories teach that music assures continuity between human beings, and between man and the rest of Nature. This is highly significant, for in the present context music means acoustics, and acoustics means wave transmission in a continuous medium. The legends are 'Orphic' in character—when Phao Pa[4] played his lute,[f] birds danced and fish leapt up, and when Shih Wên[5] sounded the pipe-notes appropriate to a particular season according to the symbolic correlations,[g] weather, grains, plants and animals responded like magic whatever the time of year. His teacher, Shih Hsiang,[6] far surpassed, swore that he equalled the masters of old,[h] Shih Khuang[7] and Tsou Yen.[8] This is particularly interesting, for, as we saw at an earlier stage,[i] Tsou Yen was the greatest of the School of Naturalists (Yin Yang chia[9]), indeed its founder about the beginning of the −3rd century. 'Tsou Yen blowing on his pipes' in this text is glossed as follows by Chang Chan:[10]

In the north there was a valley of good earth but so cold always that the five grains would not grow there. Tsou Yen blew on his pipe however and (permanently) warmed its climate, so that grain and millet could be raised abundantly.[j]

As Chang was making his commentary in the +4th century this legend is not necessarily very old, but the interest of it is that it should have become attached to the greatest systematiser of Yin–Yang theory, with all that that implied for wave rather

---

[a] Often cited later, e.g. *Po Wu Chih*, ch. 3, p. 8*b*.

[b] Further stories of archers follow later in the same chapter (Kan Ying[11] and Fei Wei[12]) as also of chariot-drivers (the famous Tsao Fu[13] and his teacher Thai Tou[14]), but these go over into the field of 'knack'-expertise (see Vol. 2, pp. 121 ff.) and therefore out of our present range of interest.

[c] We shall meet again with this technique in Sect. 28*e*. Phu Chü Tzu was a semi-legendary military technician, the nominal author of a book in the *Chhien Han Shu* bibliography on the use of fowling arrows with strings for recovery attached to them.

[d] Cf. Vol. 2, p. 54 above, and, on the man himself, Sect. 44 below.

[e] Kung Hu[15] of Lu, and Chhi Ying[16] of Chao.

[f] As the word *phao* means a bottle-gourd, this personage was probably originally a tutelary deity of sound-box instruments.

[g] This system has been explained in Vol. 2, pp. 261 ff.

[h] Shih Khuang is ascribed to the −6th century; he alarmed Duke Phing of Chin with his pipe-playing, which raised clouds, rain and violent winds.

[i] In Sect. 13*c*, Vol. 2, pp. 232 ff.

[j] *Lieh Tzu*, ch. 5, p. 18*b*, tr. auct. Cf. Hou Wai-Lu, Chao Chi-Pin *et al.* (1), vol. 1, p. 646; Forke (13), p. 504.

[1] 詹何	[2] 蒲且子	[3] 扁鵲	[4] 瓠巴	[5] 師文	[6] 師襄
[7] 師曠	[8] 騶衍	[9] 陰陽家	[10] 張湛	[11] 甘蠅	[12] 飛衛
[13] 造父	[14] 泰豆	[15] 公扈	[16] 齊嬰		

than particle conceptions. The *Lieh Tzu* writer pursues his theme[a] into regions of folk-lore where we need not follow—Han Ê[1] singing at an inn and the beams giving forth his song for three days afterwards, the thought behind Po Ya's[2] music being invariably understood by his friend Chung Tzu-Chhi,[3] etc. The essential teaching of the chapter is that of the universe as a continuum, in which all such phenomena as ripples on water, acoustic resonance,[b] invisible links between human organisms endowed with memories and emotions, or the action of the moon on the tides, all find a perfectly natural place.

The theme recurs in powerful form in many other books of the Warring States and Chhin and Han periods. I came across it quite early in my studies of Chinese thought, when reading the *Kuan Tzu* book with the late Gustav Haloun more than twenty years ago, that extraordinary work in which inspired and prophetic passages of nature philosophy alternate with diet and hygienics, magic and meditation-techniques. We had before us the following passage:[c]

Are you able to unite? Are you able to unify?[d] Then without the tortoise and the milfoil you will foreknow good and evil fortune. Are you able to stop?[e] Are you able to cease?[f] Are you able to refrain from asking others, and yourself get it from yourself? Thus was it said of old, meditate upon it, meditate upon it; if still you do not get it, the gods and spirits will teach it. But that will not be because of their show of force (in omens), but by (your) sending forth your essence (*ching*[4]) and your *pneuma* (*chhi*[5]) to the utmost degree (of human possibility, to enter into communication with them). What unifies the *chhi* so that it can change (external things) is called the essence; what unifies (human) affairs so that they can undergo change is called wisdom. Collecting and selecting is the way to grade matters. Changing to the utmost is the means by which to respond to things.[g] If one collects and selects there will be no disorder, if one changes to the utmost there will be no disappointment. The *chün-tzu*[h] holding fast to (the conception of) the unity[i] (of the universe) can alone perform this. Holding fast to the unity and not losing (sight of) it, he can command the myriad things. (Then) he shines in an equality with the sun and moon, and partakes of the same pattern-principle (*li*[6]) as heaven and earth. The sage commands things, and is not commanded by things.[j]

[a] His series includes the story of the automata made by Yen Shih[7] which so amazed King Mu of Chou (see Vol. 2, p. 53). According to Wieger (7), p. 145, it was incorporated here to hint that their constructor operated them by projected 'will-power' or mental concentration, though the text does not actually say this. Such a conception of remote control would have been curiously modern. It is strange to think that effects of this kind have been brought about in our own time precisely by following out to the end the same ideas of radiant energy travelling in a continuous medium.

[b] On this see further below, pp. 130, 185.   [c] *Kuan Tzu*, ch. 37, p. 7*b*, tr. Haloun (2).

[d] In Vol. 2, p. 46, the 'idea of the One' was granted mystical religious significance as well as proto-scientific meaning. I doubt now if its scientific aspect was there sufficiently stressed. 'Holding fast to the One' was also holding fast to the one Continuum.

[e] At the right interpretation, true hypothesis, or correct action, not going on beyond it by sophistical arguments. Cf. Vol. 2, p. 566.

[f] Cf. Vol. 2, p. 283.

[g] We have already quoted this in other contexts, Vol. 2, p. 60.

[h] See Vol. 2, p. 6.   [i] And hence the continuity, so important here.

[j] Many parallel passages can be found, e.g. in *Shih Tzu, Han Fei Tzu, Shen Tzu*.

[1] 韓娥   [2] 伯牙   [3] 鍾子期   [4] 精   [5] 氣   [6] 理
[7] 偃師

He commands things because he is aware of, and can use to the full, the fundamental interconnectedness (*lien*[1]) and non-isolation (*chü*[2]) of all things in the universe. With this remarkable passage, adumbrating many of the basic attitudes of mind characteristic of natural science, Haloun compared a particular chapter of the *Lü Shih Chhun Chhiu*, significantly entitled Ching Thung,[3] the 'Universal Permeation of Essences'. This text[a] is indeed concerned with demonstrating the action of things on each other even at considerable distances by means of 'sympathetic' influences radiating through the continuum. It is distinctly more scientific in the strict sense than either the *Lieh Tzu* or the *Kuan Tzu*, but it does not disdain to have recourse to folk-lore and legend in the usual Taoist style. Since it must date from the close neighbourhood of −240 it will be a century or so later than the *Kuan Tzu* passage. It is worth while to list its eleven arguments.

(1) The dodder,[b] (*thu-ssu*[4]) seems to have no root, but in fact it has one, i.e. the fungus called tuckahoe or Indian bread[c] (*fu-ling*[5]). These two plants are quite separated, with no connection between them, yet their relations are those of plant and root.

This was a misunderstanding, for in fact the two parasitic plants have nothing to do with one another.

(2) The lodestone draws to itself iron particles.[d]

(3) 'When the sage sits on his throne facing the south, thinking of nothing but loving and benefiting the people, the orders have hardly gone out of his mouth ere the people are stretching out their necks and standing on tiptoe to obey. His essential spirit has permeated to them.'

(4) The converse. Victims of a planned attack become uncomfortable, as if a spirit had told them.

(5) If a person is in Chhin (State) and his beloved (far away) in Chhi; then if one should die, will not the essential spirit of the other be restless?

(6) 'The virtue of a ruler is what all the people obey, just as the moon is the root and fount of all Yin things. So at full moon, shellfish (*pang ko*[6])[e] are fleshy, and all that is Yin abounds. When the moon has waned, the shellfish are empty and Yin things weak. When the moon appears in the heavens all Yin things are influenced right down to the depths of the sea. So the sage lets virtue flow forth from himself, and the four outer wildernesses rejoice in his benevolent love.'

This is the famous passage which we have met with before.[f] Though the case is not so clear for molluscs, one of the oldest of biological observations is that of the lunar periodicity of the reproductive system of echinoderms, especially sea-urchins. This was clearly stated by

---

[a] Ch. 45; tr. R. Wilhelm (3), pp. 114ff.

[b] *Cuscuta sinensis* (R156), a parasitic Phanerogam belonging to the Convolvulaceae which sucks the sap of its host (e.g. the willow) by means of special organs (haustoria). *PTKM*, ch. 18A, pp. 3a ff.

[c] *Pachyma*, the sclerotial condition of *Polyporus cocos* (R838), long used in pharmacy, cf. Burkill (1), vol. 2, p. 1618. *PTKM*, ch. 37, pp. 3aff.

[d] No misunderstanding here. Cf. Sect. 26i below.

[e] In modern times this term has come to be applied to the Unionidae lamellibranchs in general, mussels, but in the −3rd century it may easily have meant, or included, echinoderms.

[f] Vol. 1, p. 150. In the appropriate place below, Sect. 39, we shall deal in all fullness with this and similar texts.

[1] 連　　[2] 屬　　[3] 精通　　[4] 菟絲　　[5] 伏苓　　[6] 蚌蛤

Aristotle,[a] whose fishermen informants showed him how some kinds of sea-urchins were fat and good to eat at the full moon, and it has been amply confirmed by modern biological research.[b] The reference to the sea may partially conceal a reference to the action of the moon on the tides.

(7) Yang Yu-Chi[1] shot at night what he thought was a wild ox, but his arrow pierced a rock right to the feathering; this was because of the intensity of his belief that it was an animal.

(8) Po Lo[2] concentrated so much on the physiognomy of horses that he ended by being incapable of seeing anything else.

(9) Ting phao jen[3] the butcher concentrated so much on the carcases of oxen that his cutting and carving was almost miraculous. The *locus classicus* for this story in *Chuang Tzu* has already been given.[c]

(10) Chung Tzu-Chhi[4] knew all the story of a sad chime-stone player without need of any words. Here the parallel is in *Lieh Tzu*, as we have just seen.

(11) Shen Hsi[5] knew his mother when she came and sang sadly as an old beggar before the house, unaware that it was inhabited by the family which she had lost.

The chapter continues with general remarks on the invisible ties and responses of human relationships.

Thus the scholars of Lü Pu-Wei in this chapter assembled, to demonstrate the universal continuum, three observations in the natural sciences (two of which were perfectly correct), three examples of human relationships, three instances of mental concentration, and two incidents depending on the interpretation of acoustic (musical) phenomena.[d]

Some may feel that we have strayed far from physics. But it is not really so. In the ancient Chinese conceptions of the physical world, where sometimes *ching*[6] can almost be translated 'radiant energy', continuity, waves and cycles were supreme. There was no room for discontinuity and atomic particles. And so it was throughout the centuries of indigenous Chinese scientific thought.[e] By the time that modern physics found a home in China, the monopoly of the atom as a world explainer had long ceased.

A good deal of useful study could be devoted to defining how the world continuum was visualised by the ancient and medieval Chinese naturalists. So far as I know, the antithesis between wave-motion in a continuous medium of *chhi* and action at a distance in the strict sense across vacuous spaces was never decisively faced in old Chinese thought.[f] But so coherent and interrelated was the whole universe for the

---

[a] *De Part. Anim.* IV, 5 (680 a 31); *Hist. Anim.* 544 a 16.
[b] Cf. H. M. Fox (1).   [c] See Vol. 2, p. 45. Text in ch. 3.
[d] Among other texts of a similar kind cf. *Huai Nan Tzu*, ch. 6, pp. 2 b ff.

[e] It would be of great interest to collect examples of this physical world-outlook from the medieval centuries, and to see how the philosophers and the artisans were affected by it, but this would require extended research which is not at present available to us.

[f] Besides the many references in previous volumes to the concept of *chhi*, two Japanese studies must be mentioned, those of Hiraoka Teikichi (1) and Kuroda Genji (1). On action at a distance see numerous references in this and previous volumes. Cf. also those in Dampier-Whetham (1), and the special studies of Hesse (1, 3).

[1] 養由基   [2] 伯樂   [3] 丁庖人   [4] 鍾子期   [5] 申喜   [6] 精

Chinese that they would probably never have wanted to insist on the universality of a material medium if there had been any good reason for doubting its existence in particular places. Normally it was assumed.[a] In another of Martin's prophetic papers (6), written more than sixty years ago, he compared Neo-Confucian insistence on the universal presence of matter in some form or other, even if only as the most tenuous ethereal *chhi*, with the theories of the luminiferous aether in modern 'classical' physics.[b] Martin was able to cite some telling passages from Chang Tsai's[1] work[c] of +1076, the *Chêng Mêng*.[2] For example:[d]

In the great void *chhi* is alternately condensed and dissipated, just as ice is formed or dissolves in water. When one knows that the great void is full of *chhi*, one realises that there is no such thing as nothingness....How shallow were the disputes of the philosophers of old about the difference between existence and non-existence; they were far from comprehending the great science of pattern-principles (*li*[3]).

But in the upsurge of modern natural science in Europe the old scholastic axiom that 'matter cannot act where it is not' began to be questioned, and three physical models arose, not only the two ancient ones of Epicureans and Stoics.[e] The experimental and mathematical investigation of action by impact led to the Newtonian laws embodied in the *Principia*. The similar study of wave-motion gave rise to the hydrodynamics of Newton and the Bernoullis. But thirdly, true action at a distance presented itself in the phenomena of falling bodies, the solar system, and electric and magnetic attractions, attaining mathematical formulation in Newton's theory of central forces, including gravitation. Eventually all these models, irreconcilable among themselves, were subsumed into modern relativity theory and mathematical physics, an air which no 'everyday-life' analogies can breathe. By that time there had long ceased to be any distinction in science between men and minds—'neither Jew nor Greek', neither Chinese nor European, only human and universal.[f]

But perhaps we can find a few late echoes of Lü Pu-Wei and Lieh Yü-Khou. In the middle of the nineteenth century certain Japanese scholars engaged in a losing battle against the rise of modern science in their country. Thus in his *Hekija Shōgen*[4] (False Science Exposed), Ōhashi Totsuan[5] upheld about 1854 a transcendental interpretation of Neo-Confucian philosophy and emphasised the cultivation of self rather than the study of Nature. Among the things which, he said, the Western scientists and their friends do not understand, is *kakki*[6] (*huo chi*[6]), a certain vital force or energy in man continuous with that in the non-human world and capable of

[a] Cf. the heavens as piled-up *chhi* in the story of Chhang Lu-Tzu (Vol. 2, p. 41).
[b] On the history of these see Whittaker (1).
[c] Cf. Vol. 2, pp. 458, 562.
[d] Ch. 1, in *Chang Tzu Chhüan Shu*, ch. 2, p. 3a, or *Sung Ssu Tzu Chhao Shih*, ch. 1, p. 4a; tr. auct.
[e] See the interesting discussion in Hesse (2).
[f] Cf. the celebrated work of Li Tsung-Tao and Yang Chen-Ning in 1956 on the parity theory in nuclear physics.

[1] 張載   [2] 正蒙   [3] 理   [4] 闢邪小言   [5] 大橋訥菴
[6] 活機

development and utilisation with astonishing results.[a] Perhaps Ōhashi was partly reacting against atomic and mechanical materialism. At any rate his words recall the psycho-physical continuum of the archers and musicians of the *Kuan Tzu* book and the 'Spring and Autumn Annals of Master Lü'.[b]

(4) CENTRES OF GRAVITY AND THE 'ADVISORY VESSELS'

There seems to have been no theoretical treatment of the centre of gravity in China corresponding to the work of Heron of Alexandria on suspended objects of irregular shapes, in which the idea of moment was contained.[c] However, some empirical principles must have been followed, notably in the suspension of the chime-stones (*chhing*[1]), of which we have a description in the *Chou Li*.[d] These were L-shaped pieces of flat stone, the angle between the shorter limb (*ku*[2]) and the longer one (*ku*[3]) being obtuse.[e] The Chhing mathematicians, Chhêng Yao-Thien and Tsou Po-Chhi, occupied themselves with working out a reconstruction of the methods probably used by the Han technicians.[f]

A remarkable example of the application of knowledge about centres of gravity consisted in the famous hydrostatic 'trick' vessels which altered their position in accordance with the amount of water which they contained. For us, it is very easy to imagine the building in to a bronze vessel of a number of compartments with overflow channels into one another, so arranged as to give a variety of effects (Fig. 284). In ancient China, however, it was regarded as a great marvel, and evidently went back to a respectable antiquity. The oldest description of

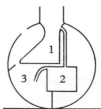

Fig. 284. Tentative reconstruction of an 'advisory vessel'.

such a vessel is in the *Hsün Tzu* book, ch. 28 of which is entitled *Yu Tso*,[4] i.e. 'The Advisory (Overturning Jars) Placed on the Right (of the Throne)'. If, therefore, the invention did not come from Confucius' time, it was certainly known in the − 3rd century. The passage runs:[g]

Confucius inspected the temple of Duke Huan of Lu State, and saw an inclining vessel (*i chhi*[5]). He asked the guardian of the temple what it was, and the guardian replied 'This is the Advisory Vessel which stands at the right hand side of the throne'. Confucius said: 'Ah, I have heard of these Advisory Vessels. If they are empty they lean over to one side, if they are half full they stand up straight, while if they are full they fall over altogether.' Confucius asked his disciples to pour water into one of the vessels; they did so, and its behaviour was just as he said.

---

[a] Meiji Bunka Zenshū ed., vol. 15, p. 111 (the second half of ch. 3).

[b] We are much indebted to Dr Carmen Blacker, who is engaged on a study of this thinker, for bringing him to our knowledge. Japanese scholars were liable to be intensely conservative, as we shall see in the case of the monk Entsū (Sect. 27*j* below), who adapted orrery clockwork to the most archaic of cosmologies.

[c] Dugas (1), p. 32.   [d] Ch. 12, p. 5*a* (ch. 42), tr. Biot (1), vol. 2, p. 531.

[e] Cf. the Section on acoustics, pp. 144 ff. below.

[f] Chhen Wên-Thao (1), pp. 67 ff.

[g] Ch. 28, p. 1*a*. Copied verbatim in *Khung Tzu Chia Yü*, ch. 2, p. 15*a*. Tr. auct.

¹ 磬   ² 股   ³ 鼓   ⁴ 宥坐   ⁵ 欹器

And they go off into a discussion of the morality of moderation in all things, of which principle the vessels were supposed to be a permanent reminder to princes.

These vessels persisted as a court wonder for more than a thousand years, with the alternative name of *chhi chhi*.[1] Chou examples seem to have lasted down to the end of the Han, but disappeared during the disturbances of the Three Kingdoms. About +260 Tu Yü[2] made a new set,[a] and at the same time the mathematician Liu Hui[3] wrote a *Lu Shih Chhi Chhi Thu*[4] (Diagrams of the Inclining Vessels of the Lu Officiants) which, though it did not survive, suggests that he and his contemporaries were in possession of some theoretical principles about centres of gravity. Some two centuries later Tsu Chhung-Chih[5] made more overturning jars,[b] and thereafter astronomers and mathematicians were always presenting them to emperors. In the middle of the +6th century (+538) elaborate ones were made by Hsüeh Chhêng[6],[c] and another illustrated book produced by Hsintu Fang.[7] At the beginning of the next Kêng Hsün[8] again had them made (+605)[d] and Lin Hsiao-Kung[9] wrote about them. A Thang prince, Li Kao[10] (Tshao Wang Kao[11]), whom we shall meet again in a moment,[e] interested himself in them; his were made of lacquered wood about +790 and probably turned out in quantity. One of the last references we have is to a presentation of them by the lexicographer Ting Tu[12] in +1052.[f]

Meanwhile these devices had aroused the keen interest of the Arabs, who greatly developed their possibilities, as may be seen by the *Kitāb fī ’l-Ḥiyal* of the Banū Mūsā,[g] the three sons of Mūsā ibn Shākir (+803 to +873), available to us in the translation of Hauser (1).

Another kind of trick vessel may be mentioned here, though not concerned with centres of gravity. Many of those whose lot it was to live in Chungking during the Second World War visited the park of Pei-wên-chhüan[13] in the Chialing gorge, and there saw a bronze vessel, preserved in a former temple, which had the curious property of sending jets of spray into the air from four directions when subjected to rubbing.[h] The date of this pan or bowl (now in the Chungking Museum) is unknown, but it may perhaps be of Thang or Sung origin,[i] and other bowls of somewhat similar shape and decoration are authentically Chou and Han (Fig. 285). The name for these vessels, of

---

[a] *Chin Shu*, ch. 34, p. 9b.
[b] Between +483 and +493; *Nan Shih*, ch. 72, p. 11b.
[c] *Chou Shu*, ch. 38, p. 10a.
[d] *Sui Shu*, ch. 19, p. 27b; ch. 78, p. 7bff; cf. Vol. 3, pp. 327, 329.
[e] And in Sect. 27g.
[f] Of course there are many late references to them; see, for instance, Hang Shih-Chün's *Yung Chhêng Shih Hua* (+1732), ch. 1, p. 11b.   [g] Mieli (1), p. 71.
[h] Having often been reminded of this in intervening years by my friend Dr Huxley Thomas, I was fortunate enough to be able to study it closely in the Chungking Municipal Museum in 1958, fifteen years later. Particular thanks are due to the director, Dr Têng Shao-Chhin, and to his assistant Miss Phan Pi-Ching, for their kindness.
[i] The bronze is so yellow that a brassy alloy is suspected, in which case the vessel will not be very early.

[1] 攲器   [2] 杜預   [3] 劉徽   [4] 魯史攲器圖   [5] 祖沖之
[6] 薛憕   [7] 信都芳   [8] 耿詢   [9] 臨孝恭   [10] 李皋
[11] 曹王皋   [12] 丁度   [13] 北溫泉

which other examples are known,[a] is *phên shui thung phên*[1] (bronze water-spouting bowls). The Szechuanese vessel is cylindrical with a slight taper, about 1 ft. 6 in. in diameter at the mouth, flat-bottomed and fairly shallow. Upon the bottom there appear four fishes in relief surrounding a central design, their open mouths terminating in radiating ridges which run up the sides of the bowl as far as the flanged rim.[b] When the two handles thereon are rubbed rather slowly and rhythmically with the wet palm of the hand, the bronze vibrates like a bell, and fountains of spray as high as the handles (about 3 in.) spring up from the sides of the vessel at the four places where the fish ridges end.[c] It is said that if the knack is learnt the spray may be raised as high as 3 ft. into the air.[d] At full blast the water-surface is covered with a very complex pattern of standing waves. The maximum disturbance is evidently produced at places corresponding to the nodes and antinodes of a struck bell, and the stationary waves formed when a wine-glass is rubbed may provide a much less impressive parallel,[e] but there must be some remarkable peculiarity in the shape of the vessel and its walls to give such an impulsion to the water. Perhaps the walls are under strain in some way, as in the case of the 'magic mirrors' presently to be discussed;[f] in any case the phenomenon clearly merits the attention of Chinese physicists historically minded.

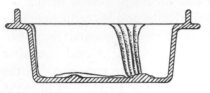

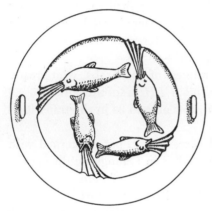

Fig. 285. Approximate cross-section and plan of the bronze water-spouting bowl conserved in the Chungking Municipal Museum.

This unexpectedly ancient type of standing fountain, seemingly unknown to any of the Alexandrian writers on pneumatics and hydrostatics, but very pertinent to the Sino-Stoic interest in wave-motion, finds a strange analogy in the effects of non-uniform electric fields (cf. Pohl, 1). If a point electrode surrounded by a looped electrode is immersed in a dish containing an organic liquid, and some 10,000 volts applied, the liquid is violently agitated and leaps up from the dish, individual drops remaining suspended in the air or describing spiral orbits around the lead-in wire. With higher

[a] One for instance is recorded as having been seen in April 1956 by a party of English jurists at a temple or house beside the lake at Hangchow (Gower (1), p. 114).

[b] A photograph was obtained by one of us (W.L.) in 1944.

[c] There is an optimum water level—the bowl should not have too much or too little. This fact recalls the admonitory purpose of the trick vessels we have just been discussing.

[d] At the time of my visit in 1958 only the museum attendant could do it successfully, but I found it easy enough to get the beginnings of the vibration build-up.

[e] Cf. what was said above, p. 12, on the observations of the Stoics on stationary waves.

[f] Cf. pp. 94ff. below.

[1] 噴水銅盆

voltages (though still very low amperages) a spray of liquid or powder may be thrown four to six feet high at the rate of a gallon a minute, forming thus a simple pump with no moving parts. That the standing fountain of medieval China was never thought of in this way illumines once again the difference between industrial and pre-industrial civilisations. So also, as we shall later see (p. 235), the ancient legend of 'Mahomet's coffin' has come true in the modern world, for samples of metal (e.g. Ti, Zn, V, Ta, Mo) weighing more than a pound may not only be suspended in mid-air by a powerful magnetic field, but actually melted in that position by induced high-frequency currents of high amperage.

Related to these questions are some early experiments concerning the pressure of the air. Siphons have already been discussed in connection with clepsydras,[a] and will receive further mention in Section 27b on engineering. As was noted above,[b] the earliest meaning of the word clepsydra was a pipette used for wine or oil. Since ancient Egyptian representations of pipes for sucking up liquids have come down to us,[c] they must have been a familiar device throughout Chinese history, and indeed the custom so common among the south-western tribal peoples of drinking wine ceremonially through long bamboo tubes is one which links them directly with the ancient peoples of the Fertile Crescent.[d] But the sucking-pipe does not become a pipette until it is realised that occlusion of the top will conserve the liquid in it so that it can be carried about from place to place.[e] The *I Chao Liao Tsa Chi* of +1200 has a discussion[f] on the meaning of the old terms *chhih i*[1] and *ku chi*,[2] both used in the Han dynasty. The former seems to have meant a leather wine vessel, and the latter, which had a 'belly as big as a pot', was the pipette for withdrawing the wine; but one cannot be sure that both terms did not designate the pipette, or indeed also the siphon.[g] A number of Thang references, however, are certain. The principle of the pipette is discussed in the +8th-century Taoist book, *Kuan Yin Tzu*:[h]

Take a bottle with two holes (one above and one below), and fill it with water. If you turn it upside down, the water will flow out, but if you close the upper orifice, the water will now not leave the lower one. This is because if something is not raised up, something else will not come down. A well may be eight thousand feet deep, but if you pull, water will come up. This is again because if something does not go down, nothing will come up. Similarly, the sage does not consider himself superior to creatures, but humbles himself before them.

On this the +13th-century commentator Chhen Hsien-Wei says that the *chhi* has to go up before the water will come down, and that without pressure (*pho*[3]) things will not move. This point of view must have been at least as advanced as anything which could be adduced from his contemporaries in Europe. Meanwhile, in the Thang, the

---

[a] Vol. 3, pp. 320ff.   [b] Vol. 3, p. 314.   [c] See e.g. Neuburger (1), p. 226.
[d] Earlier we saw (Vol. 3, p. 314) their elaboration of having a float valve between two of the bamboo nodes so as to prevent sucking too quickly or too slowly.
[e] Empedocles studied this (Diels–Freeman (1), p. 62), but does not seem to have understood the true reason.   [f] Ch. 2, p. 47a.   [g] See Sect. 27b below.
[h] *Wên Shih Chen Ching*, ch. 3, p. 11b; tr. auct.

[1] 鴟夷   [2] 滑稽   [3] 迫

wine pipette, with a bulbous body and various kinds of handle, was generally known as *chu tzu*,[1] later also as the 'sideways lifter' (*phien thi*[2]).[a] It seems to have been particularly popular from the beginning of the +9th century.[b]

A contemporary of the writer of the *Kuan Yin Tzu* book, Li Kao,[3] prince of the Thang, made experiments with bowls and plates fitting so well that no air could enter and displace liquids contained in them.[c] The *Thang Yü Lin* says:[d]

Kao, prince of Tshao, was very ingenious, and good at making vessels and other useful devices. When he was governor of Chingchow, two soldiers visited him with two barbarian drum bowls. Upon seeing them, (Li) Kao said 'What valuable objects!' and pointed out the extreme smoothness of their edges, but his guests could not understand his enthusiasm. So he said that he would demonstrate what he meant, and after selecting a plate as smooth as possible, and trying the fit of the plate and bowl together, he caused oil to be poured into one of the bowls. Then (covering the bowl with the plate, he turned it upside down) but nothing whatever came out. This showed the perfect cohesion (*wên ho wu chi*[4]) between bowl and lid.

A similar story is told of a palace official Li Yuan,[5] who used iron bowls ground very smooth at the edges, and of Jen Shih-Chün,[6] about +780.[e] These bowls were, as we shall see,[f] connected with tuning practices in acoustics and closed-vessel reactions in alchemy. The technological importance of an ability to produce smooth lapped edges of iron bowls should not be overlooked—the bowls were probably ground down to give the note desired.[g] The subject has also an obvious relation to the medical practice of 'cupping'; though this was not characteristic of Chinese medicine, we read occasionally of experiments with cups made to adhere by a partial vacuum within.[h] On tight seals in general, Horwitz (7) has some interesting things to say, e.g. the lacquering of coffin joints in China, and the preservation of salt in Japan by filling earthenware jars before firing.

---

[a] *Shih Wu Chi Yuan* (*c.* +1085), ch. 41, p. 14*a*, quoting *Hsü Shih Shih* (*c.* +960); as also *Shuo Fu*, ch. 10, p. 52*a*.

[b] The use of small bamboo pipettes in contemporary traditional Chinese technology is described by Hommel (1), p. 10.

[c] Cf. Fig. 286.

[d] Ch. 6, p. 5*b*, tr. auct.

[e] *Thang Yü Lin*, ch. 5, p. 26*a*.

[f] In Sect. 26*h*, pp. 192 ff. below, and in Sect. 33. Arising probably out of the latter use, they became the first gunpowder bombs; see Sect. 30.

[g] Cf. the use of abrasive sands in mineral grinding, Vol. 3, p. 667.

[h] *Tu Hsing Tsa Chih* (+1176), ch. 5, p. 2*b*. In the West, cupping goes back at least to the Hippocratic corpus, and was widely known and used in the Middle Ages.

[1] 注子　　　[2] 偏提　　　[3] 李臯　　　[4] 脗合無際　　　[5] 李琬
[6] 任使君

(5) Specific Gravity, Buoyancy and Density

The general idea of specific gravity must have existed from time immemorial. Mencius (−4th century) remarked[a] that gold was heavier than feathers, otherwise how could it be said that a hook of gold was heavier than a cartload of feathers? But there was nothing equivalent (so far as we know) to the treatise of Archimedes on floating bodies.[b] Empirical use, of course, was made of his principle, as in the floating of arrows[c] and vehicle wheels[d] in water by the Chou and Han technicians, in order to determine their equilibrium and add or remove material accordingly. By the Ming period tables of specific gravity were in common use.[e]

The classical appearance of the so-called principle of Archimedes in China is no doubt the well-known incident of the weighing of the elephant in the San Kuo period.[f] But it simply involved observation of displacement and buoyancy, analogous to Archimedes' Prop. v, and did not concern specific gravity as such. The text[g] is as follows:

The son of Tshao Tshao, Chhung,[1] was in his youth clever and observant. When only five or six years old his understanding was that of a grown man. Once Sun Chhüan[h] had an elephant, and Tshao Tshao wanted to know its weight. He asked all his courtiers and officials, but no one could work the thing out. Chhung, however, said, 'Put the elephant on a large boat and mark the water level; then weigh a number of heavy things and put them on the boat in their turn (till it sinks to the same level)—compare the two and you will have it'. Tshao Tshao was very pleased and ordered this to be done forthwith.

This would be just after +200.

It has, however, been suggested that the technicians of the Han were already familiar with the principle embodied in the famous story of the determination of the proportions of gold and silver in Hieron's crown. This depends on the interpretation of an important passage in the *Chou Li*[i] (Khao Kung Chi section). It concerns the makers of weights and measures. It says:

The workers called Li[2,j] make measures of capacity (*liang*[3]). They purify (separately) by successive heatings samples of metal (presumably copper) and tin, until there is no further loss of weight. Then they weigh them.

---

[a] *Mêng Tzu*, VI (2), i, 6.
[b] Dugas (1), p. 24 ff.; Thurot (1); v. Lippmann (3b). Arabic developments in Wiedemann (11).
[c] *Chou Li*, ch. 12, p. 6a (ch. 42); Biot (1), vol. 2, p. 534.
[d] *Chou Li*, ch. 11, p. 12b (ch. 40); Biot (1), vol. 2, p. 474. See especially Lu, Salaman & Needham (1). The words are: *shui chih i chen chhi phing shen chih chün yeh.*[4]
[e] See e.g. *Suan Fa Thung Tsung*, ch. 1, p. 4a (+1592). The *Ā'īn-i Akbarī* of +1590 gives a contemporary Indian parallel (Blochmann (1), pp. 41 ff.). See also Vol. 3, p. 33, on the *Sun Tzu Suan Ching*.
[f] Attention was drawn to this by Ardsheal and others.
[g] *San Kuo Chih*, ch. 20, p. 2a; tr. auct.
[h] Subsequently emperor of Wu; already met with in the geographical Section (Vol. 3, p. 538 above).
[i] Ch. 11, p. 25b (ch. 41).   [j] An old form of *li*.[5]

[1] 曹沖   [2] 奧   [3] 量   [4] 水 之 以 眂 其 平 沈 之 均 也   [5] 栗

Thus Biot.[a] The text goes on to say *chhüan chih, jan hou chun chih; chun chih jan hou liang chih*,[1] which Biot rendered 'After they have weighed them, they proportionalise (or equalise) them, and after that they measure them'. The sense of this is far from obvious. Chiang Yung, however, had already suggested[b] at the end of the 18th century that *chun* had meant in the Han 'weighing in water', while the ordinary word *chhüan* here meant weighing in air. The water radical in *chun* would thus be very significant, and though the word generally means equalising or levelling, one of its subsidiary meanings is certainly weighing. What the Han technicians were doing, therefore, was essentially what Archimedes did, namely to ascertain the proportions of an alloy by weighing in water as well as in air. Such an interpretation makes much more sense than Biot's.

General notions of buoyancy must have been widespread among sailors in China, as elsewhere, from early times. The *Shen Tzu* says:[c]

> Though a thing may be as heavy as the cauldron of Yen, or as much as 1000 *chün* in weight; if it is placed upon a boat of Wu it will be transportable; this is the principle of floating (*fou tao*[2]).

This would have been no news to the founders of China's canal transport system. But, as will later appear, Chinese nautical technology made use of the water-tight compartment much earlier than Europe.[d] And there was the monk Hui-Yuan,[3] in the Thang, with his floating and sinking water-clock bowls.[e]

Another monk, Huai-Ping,[4] was responsible for a method of raising heavy objects from the bottom of a river by the use of buoyancy, analogous to the pontoons filled alternately with water and air used in modern salvage operations.[f] The *Liang Chhi Man Chih* of +1192, after relating the story of Tshao Chhung and the elephant, goes on:[g]

> Another instance of remarkable ingenuity was the following. In Ho-chung Fu there was a floating bridge, fastened to the bank by means of eight iron oxen each one of which weighed several thousand catties. In the Chih-Phing reign period (+1064 to +1067) the bridge was broken during a sudden flood, and the iron oxen were swept away and buried under the water. Public proclamation was made to find someone capable of recovering them. It was then that the monk Huai-Ping from Chen-ting Fu suggested a method. He used two huge boats filled with earth, cables from them being made fast to the oxen in the river-bed (by divers). Hooks and a huge counterweighted lever were also used. Then the earth in the boats was gradually taken away so that the boats floated much higher and the oxen were lifted off the river bottom (and dragged up the river-bank in shallower water).
>
> Huai-Ping's success was reported to the emperor, who bestowed on him a purple robe of honour as recompense. He certainly followed the same principle as Tshao Chhung.

a (1), vol. 2, p. 503.
b *Chou Li I I Chü Yao*, ch. 6, p. 17b.
d Below, Sect. 29c.
f Cf. Masters (1).

c P. 8b; tr. auct.
e See Vol. 3, p. 315 above.
g Ch. 8, p. 11b; tr. auct.

1 權之然後準之準之然後量之   2 浮道   3 惠遠   4 懷丙

Presumably the role of the counterweighted lever was to help the divers fix the cables.[a]

It so happens that we are well informed about the origin of these iron oxen which secured the cables for one of the most important floating bridges across the Yellow River. The name of the place was Phu-chin[1] near Phuchow, a short distance north of the great bend at Thungkuan. An essay on the construction of the bridge in +724 was written by the imperial librarian Chang Yüeh[2] (+667 to +730). His *Chang Yen Kung Chi*[3] contains[b] it under the title *Phu-chin Chhiao Tsan*.[4] After various preliminaries we read that in this year:

There were thus collected together the most famous artisans, all eager to demonstrate their art. A tax of 'blown-with-bellows' iron[c] was imposed, as in the Chin State of old;[d] and they followed the classical metallurgical procedures (lit. the six alloy proportions) of the service of the Chou Empire.[e] The fans of the bellows[f] flew back and forth, and the furnaces furiously blazed. Some smelted (*lien*[5]), and others refined (*phêng*[6]) (the iron to wrought iron); some were filing (*tsho*[7]) while others were forging (*tuan*[8]) and beating with hammers. Thus they connected links together to form a great chain, and they cast (*jung*[9]) (iron)[g] into the shape of recumbent oxen, images which stood on both banks of the river connecting east (and west) amidst sandy beaches. The chain secured the lashed boats, and the oxen made fast the thick cables so that the bridge was safe against injury from objects floating downstream. Thus the boats with the pretty birds painted on their bows, all fixed firmly together (supported the road deck above).

Such were the iron anchors in the shape of oxen which Huai-Ping recovered from the river bottom nearly 350 years later.

As for the specific gravity or density of liquids, the question arose particularly in connection with the assessment of strengths of brine, and as salt was, at least from the Han onwards, a government monopoly or source of revenue, the procedures got a mention in literary works. From time immemorial salt-workers have used the

---

[a] Huai-Ping's method was proposed, or used, later in the West by Hieronymus Cardanus (of the 'Cardan' suspension, +1501 to +1576). His *De Subtilitate* contains an illustration showing the raising of a sunken vessel by means of barges progressively denuded of stones; this is reproduced in Ore (1), p. 16. In our own time Huai-Ping's method has become standard practice, as may be seen from the procedure for raising the Italian liner *Andrea Doria* sunk in 225 ft. of water 50 miles off Nantucket. The plan announced early in 1958 envisaged first the righting of the vessel by pumped-in compressed air, then raising her by cables from ore vessels progressively emptied of their burden of water. After towing into shallower water, the procedure would be repeated as often as necessary.

[b] Ch. 8, pp. 1*b*ff.; tr. auct.

[c] This is one interpretation of a famous passage in the *Tso Chuan* (Duke Chao, 29th year, i.e. −512).

[d] See Sect. 30*d* below.

[e] This is a reference to the classical passage on bronze alloy proportions in the *Chou Li* (Khao Kung Chi), cf. Biot (1), vol. 2, p. 491. We shall discuss it fully in Sect. 36 below. It was not very relevant here, as the matter concerned iron technology, but Chang Yüeh could not resist the literary allusion.

[f] On this see especially Sects. 27*b*, *f* and 30*d* below.

[g] Much will be said of iron-casting in Sect. 30*d*, but here it may be pointed out that no such casting of iron could have been performed in Europe for six further centuries after this date; cf. Needham (31). It is noteworthy that the wrought iron for the great chains was being produced from cast iron by refining or puddling, not from bloomery furnaces.

[1] 蒲津   [2] 張說   [3] 張燕公集   [4] 蒲津橋贊   [5] 鍊
[6] 烹   [7] 錯   [8] 鍛   [9] 鎔

swimming or sinking of eggs as a test of brine density, and this is mentioned in Galen.[a] But the favourite test object in China was the lotus-seed (*lien tzu*[1]). Late in the +11th century Yao Khuan[2] wrote:[b]

> When I was an official in Thaichow I tried to check corruption among salt merchants. Every day I tested the brine with lotus seeds. The heavier ones were selected for use. If brine can float 3 or 4 such seeds (out of 10) it is considered strong brine. If it floats 5 (out of 10) it is the strongest. Those seeds which float perpendicularly are preferred. If only 2 seeds float perpendicularly, or 1 perpendicularly and 1 horizontally, then the brine is considered thin and poor. If the seeds all sink to the bottom one will hardly succeed in obtaining any salt when such a liquid is evaporated. In Min (Fukien), however, they make these tests with eggs and peach-kernels. If the brine is strong both will float upright at the surface, and if it is half brine and half water, both will sink. The method is similar.[c]

Another writer of about the same time, Chiang Lin-Chi,[3] in his *Chia-Yu Tsa Chih*,[4] confirms this,[d] but the numbers of seeds differ; however, the statistical treatment is of interest in all these descriptions. Brine-testing methods (*yen yen fa*[5]) of this kind are often afterwards referred to, with variations, as in the mid+12th-century *Nêng Kai Chai Man Lu*[6,e] of Wu Tshêng.[7,f] And they have continued in use down to our own time. There does not seem to have been anything similar to the graduated floating 'hygroscopion' which Synesius of Ptolemais invented about +400.

(6) CHINA AND THE METRIC SYSTEM

In current common speech and thought the metric system is primarily associated with the decimal ordering of coinage, weights and measures. Though adopted in so many countries, this is still far estranged from the metrological chaos in our own. But decimalisation is not of the essence of the metric system; the real significance of this is that it was the first great attempt to define terrestrial units of measure in terms of an unvarying astronomical or geodetic constant. The metre was in fact defined as one ten millionth part of one quarter of the earth's circumference at sea-level.[g] Scientific lexicographers say that the metric system took its origin in the need imposed by the development of scientific thought for immutable and at the same time conveniently related units of physical measure; and they imply that this need was not satisfied until the last decade of the +18th century. This may be true enough for Europe, but as we shall shortly see, an approach was made to such an immutable unit in China in

---

[a] Feldhaus (1), col. 28; v. Lippmann (3c). Cf. *Chhi Min Yao Shu*, ch. 60 (p. 95).
[b] *Hsi Chhi Tshung Hua*, p. 44b.
[c] Tr. auct.
[d] 'Miscellaneous Records of the Chia-Yu reign-period', p. 38b.
[e] 'Miscellaneous Records of the Nêng Kai Studio', p. 22b.
[f] Also in the *Li Sao Tshao Mu Su*[8] (On the Trees and Plants mentioned in the *Li Sao*) by Wu Jen-Chieh[9] (+1197) with reference to lotus seeds (ch. 1, p. 4a).
[g] Weight measurement is of course derivative, the gram being a secondary standard based on the centimetre and the density of water.

[1] 蓮子	[2] 姚寬	[3] 江鄰幾	[4] 嘉祐雜志	[5] 驗鹽法
[6] 能改齋漫錄		[7] 吳曾	[8] 離騷草木疏	[9] 吳仁傑

the first decade of that century. Moreover, as in the case of so many post-Renaissance scientific developments, there is an earlier pre-history of this celestial–terrestrial bond, and we can find already in the +8th century in China a large-scale attempt to establish it.

Though decimalisation is not the main issue here, it is worth recalling that at an earlier stage we found a remarkable predilection for decimal metrology on the part of the ancient Chinese.[a] This goes back well into the Chou period, as foot-rules dating from the −6th century remain to witness, and was adopted on a still more considerable scale in the reforms of the first emperor, Chhin Shih Huang Ti, in −221.[b] In no other part of the world was the decimalisation of weights and measures so early and so consistent.[c] This went side by side with a remarkably advanced design of measuring instruments, as we shall find later on when we come to consider the sliding calipers used in the imperial workshops of the Han.[d]

But the real progress of metrology depended on the fixation of convenient length measures to comparatively unvarying natural reference standards far beyond the range of all those whims which might from time to time affect the princely givers of positive law. We shall see in due course how in ancient China acoustic measures were made to depend upon the volumes occupied by known numbers of standard cereal grains, those deviating widely from mean size being rejected.[e] This was one of the ways of defining the dimensions of the standard pitch-pipes.

A great step forward was taken when the idea arose in China of fixing terrestrial length measures in terms of astronomical units. That this could have occurred to the scholars at all was due to the fact that the sun's shadow thrown by an 8-ft. gnomon at summer solstice was a very convenient length at the latitude of Yang-chhêng (the 'centre of the Central Land')—about 1·5 ft. In the previous volume we gave an account[f] of the 'gnomon shadow template' (*thu kuei*[1]), a standard rule of pottery, terra-cotta or jade equivalent in length to the solstitial shadow and used for the determination of the exact date of the solstice each year. One does not hear of any Chinese system which based all length measures upon this standard length, though it might quite easily have developed; the relation with metrology came about in another way. As we saw in the astronomical Section,[g] there was a long-standing idea that the shadow length increased 1 in. for every thousand *li* north of the 'earth's centre' at

---

[a] Vol. 3, pp. 82ff. (Sect. 19). Cf. Fig. 287.

[b] Cf. Vol. 2, p. 210, where references are cited. For general information on the history of length measures in China, see Wu Chhêng-Lo (2); Lo Fu-I (1) and Yang Khuan (4).

[c] For the general background reference may again be made to the monographs of Wu Chhêng-Lo (2) and Yang Khuan (4). See also the studies of Ma Hêng (1), and Hsü Chung-Shu (6), tr. Sun & de Francis (1), pp. 7ff. On comparative aspects cf. Davidson (2) and the informative though idiosyncratic treatise of Berriman (1). The pre-Stevinian system of decimal weights developed by the assayer Ciriacus Schreittmann of Weissenberg about +1555 is reported by C. S. Smith (3) as very advanced for its time, as indeed in Europe it was.

[d] See Sect. 27a below, in Vol. 4, pt. 2.

[e] Cf. pp. 200ff. below.

[f] Vol. 3, pp. 286ff. (Sect. 20).

[g] Vol. 3, p. 292.

[1] 土圭

Yang-chhêng,[1] and decreased in the same proportion as one went south.[a] After the end of the Han, measurements made as far south as Indo-China soon disproved this numerical relation, but it was not until the Thang that a systematic effort was made to determine a great range of latitudes. This had the object of correlating the lengths of terrestrial and celestial measures by finding the number of *li* which corresponded to 1° of polar altitude (i.e. terrestrial latitude), and thus in effect fixing the length of the *li* precisely in terms of the earth's circumference. The meridian line so set up takes its place in history between the lines of Eratosthenes (*c.* −200),[b] and those of the astronomers of the Caliph al-Ma'mūn (*c.* +827).[c] It is worth examining quite closely.[d]

Uneasiness about the relation of gnomon shadows and latitude manifested itself as soon as the empire was reunified under the Sui. In +604 Liu Chhuo,[2] an eminent mathematician, realising the fallacy of the statement that a change of 1 in. in shadow length corresponded to 1000 *li* in north–south distance, memorialised the emperor as follows:[e]

We beg your imperial Majesty to appoint water-mechanics and mathematicians (*shui kung ping chieh suan shu shih*[3]) to select a piece of flat country in Honan and Hopei where measurements can be made over a few hundred *li*, to choose a true north–south meridian line, to determine the time with clepsydras, to (set up gnomons) on flat places (adjusting them with) plumb-lines, to follow seasons, solstices and equinoxes, and to measure the shadow (at different places) on the same day. From the differences in these shadow lengths the distances in *li* can be known. Thus the Heavens and the Earth will not be able to conceal their form, and the celestial bodies will be obliged to yield up to us their measurements. We shall excel the glorious sages of old and resolve our remaining doubts (about the universe). We beg your Majesty not to give credence to the worn-out theories of former times, and not to use them.

But the emperor Yang Ti succeeded Wên Ti in the following year and possibly because of this no action was taken on Liu Chhuo's proposal.

In the years +723 to +726, however, important expeditions were organised under the direction of an Astronomer-Royal, Nankung Yüeh,[4] and a Tantric Buddhist monk, I-Hsing,[5] one of the most outstanding mathematicians and astronomers of his age.[f] According to the sources, which are quite extensive and differ only on minor points,[g] at least eleven stations were established (including Yang-chhêng) with polar

---

[a] See, e.g. the *Hun Thien Hsiang Shuo* (Discourse on Uranographic Models), by Wang Fan, *c.* +260, cit. *Chin Shu*, ch. 11, p. 6*b*. Also *Hsü Po Wu Chih*, ch. 1, p. 5*b*, and many other places. Perhaps the earliest statement is that in the *Shang Shu Wei Khao Ling Yao*, then Chang Hêng in the *Ling Hsien* of +118, and later in the +2nd century Chêng Hsüan commenting on the *Chou Li*, all quoted in *Chou Pei Suan Ching*, comm. ch. 1, p. 10*a*.

[b] Alexandria to Syene, *c.* 795 km.

[c] Palmyra to Rakka, 187 km.; the plains of Sinjar, 109 km. (1°); Baghdad to Kufa, 146 km. On both the Greek and Arabic series of measurements see K. Miller (3). The chief source for the latter is al-Mas'ūdī's *Kitāb al-Tanbīh wa'l-Ishrāf* tr. Carra de Vaux (4). Cf. further Sarton (1), vol. 1, p. 558; Wolf (3), vol. 2, p. 125; Mieli (1), pp. 79 ff.; Bychawski (1).

[d] A brief account has already been given (Vol. 3, pp. 292 ff.).

[e] *Sui Shu*, ch. 19, pp. 20*a*, *b*; tr. auct. incl. E. Pulleyblank. Cf. *Chhou Jen Chuan*, ch. 12 (pp. 150 ff.).

[f] Cf. Vol. 3, p. 202.

[g] E.g. in some of the numerical values. *Chiu Thang Shu*, ch. 35, pp. 6*a*ff., abridged in *TCKM*, ch. 43, pp. 51*a*ff.; cf. *TH*, p. 1407. The text must have originated only some thirty years after the survey. Cf. also *Hsin Thang Shu*, ch. 31, pp. 5*b*ff.; *Thang Hui Yao*, ch. 42 (p. 755).

---

[1] 陽城   [2] 劉焯   [3] 水工并解算術士   [4] 南宮說   [5] 一行

PLATE XCIX

Fig. 286. Taoist immortal examining the trueness of a worked flat piece of jade. From the frescoes of the Yung-Lo Kung temple at Yung-lo-chen in Shansi, painted between +1325 and +1358. After Têng Pai (*1*); cf. Chêng Chen-To (*2*), pls. 21, 23.

PLATE C

k

j

i

h

g

f

e

d

c

b

a

Fig. 287. Explanation opposite.

altitudes ranging from 17·4° at Lin-i[1] (Indrapura in Champa, capital of the Lin-I State, not far from modern Hué in Annam), to 40° at Weichow[2] (an old city near modern Ling-chhiu, near the Great Wall in northern Shansi, and almost on the same latitude as Peking). The stations, which were not strictly on a north–south line but nearly so, were most numerous on the great plains north and south of the Yellow River. One only was on the northern border of China proper, and two were in the far south (Indo-China); the locations of all of them are given in Table 42. Along this meridian line of 7973 *li*, i.e. just over 2500 km. (more than three times as long as that of Eratosthenes), simultaneous measurements of summer and winter solstice shadows were made with standard 8-ft. gnomons.[a] The difference in shadow-length was found to be very close to 4 in. for each thousand *li*, or four times the amount accepted by the 'scholars of former times'. On account of its scale alone this whole operation must surely be regarded as the most remarkable example of organised field research carried out anywhere in the early Middle Ages, and in spite of the recognition of some 18th-century Western authors[b] it has remained one of the least known. The two chief observers who seem to have been responsible for it were Ta-Hsiang[3] and Yuan-Thai,[4] from their names seemingly monks, and probably trained by I-Hsing himself. These men were also in charge of the special expedition which at this time pushed down into the southern seas to observe and chart the constellations to within 20° of the south celestial pole.[c]

It is interesting to have the exact words of the *Chiu Thang Shu*:[d]

In the 12th year of the Khai-Yuan reign-period (+724) the Astronomer-Royal was ordered to make observations at Chiao-chow of the sun shadow at summer solstice, and it was found

---

[a] The texts note with interest that the +5th-century value for Chiao-chow, i.e. Hanoi in Tongking, was confirmed; the shadow falling *southwards* 3·3 in.

[b] D'Anville (1, 2) and of course Gaubil (2), pp. 76ff.

[c] See Vol. 3, p. 274 (Sect. 20).

[d] Ch. 35, p. 6a, tr. auct. incl. E. Pulleyblank.

[1] 林邑　　　[2] 蔚州　　　[3] 大相　　　[4] 元太

Fig. 287. A collection of standard length measures, wooden copies of bronze or ivory originals, made in Peking at the National Historical Museum for the late Professor W. P. Yetts. The dates and inscriptions are as follows (from below upwards): (a) Ming; 'Building foot measure of the Ministry of Works'. (b) Sung; 'Cloth and silk foot measure of the Finance Commission of the Sung (Dynasty)'. (c) Thang; 'Standard foot measure of the Khai-Yuan reign-period (+713 to +741) of the Thang (Dynasty)', actually decreed in +731. (d) Han; 'Standard foot measure of the Chien-Chhu reign-period (+76 to +83) of the (Later) Han (Dynasty)'. (e) Han; 'Bronze foot measure of Lü-chih hsien (city), made on the 15th day of the 8th month of the 6th year of the Chien-Chhu reign-period (i.e. +81)'. (f) Chin; 'Foot measure of the Chin (Dynasty), the same as that of the Chou (Dynasty), and the bronze one of Liu Hsin in the (Former) Han (Dynasty), and like also to the bronze one made in the Chien-Wu reign-period (+25 to +55) of the Later Han (Dynasty)'. (g) Chou; 'Foot measure standardised by the Huang-chung bell and the Pitch-pipes in the Chou (Dynasty)'. Reconstructed by Wu Ta-Chhêng. (h) Chou; 'Foot measure standardised by the Imperial Sceptre in the Chou (Dynasty)'. Reconstructed by Wu Ta-Chhêng. (i) Shu kingdom in the Three Kingdoms Period; 'Foot measure for making crossbow triggers, of the Chien-Hsing reign-period (+223 to +237) of the Han (Dynasty) continued in Shu'. Reconstructed by Wu Ta-Chhêng. (j) Chou; 'Foot measure for swords in the Chou (Dynasty)'. Reconstructed by Wu Ta-Chhêng, cf. Yang Khuan (4), p. 46. (k) Hsin; Sliding calipers and 6-in. measure 'Made on a *kuei-yu* day at new moon of the 1st month of the 1st year of the Shih-Chien-Kuo reign-period (i.e. +9)'. With the exception of the last, which will be discussed more fully in Sect. 27a below, all the standard measures are divided into 10 inches.

to be 0·33 ft. to the south of the gnomon. This was in good agreement with the observation made in the Yuan-Chia reign-period (+424 to +453). This being so, if one went south from Yang-chhêng along a road as straight as a bowstring to the point directly below the sun, it would not be as much as 5000 *li*. The Commissioners for Shadow Measurement (Tshê Ying Shih[1]) Ta-Hsiang and Yuan-Thai say that at Chiao-chow if one observes the pole it is elevated above the earth's surface only by a little more than 20°. Looking south in the 8th month from out at sea Lao-jen (Canopus) is remarkably high in the sky. The stars in the heavens below it are very brilliant, and there are many large and bright ones which are not recorded on the charts and the names of which are not known....

In the 13th year of the same reign-period (+725) Nankung Yüeh the Astronomer-Royal selected a region of level ground in Honan, and using water-levels and plumb-lines set up 8-ft. gnomons with which he made measurements....

Further light on the survey is given in the *Thang Hui Yao*:[a]

In the 12th year of the Khai-Yuan reign-period...a command was issued to the Astronomer-Royal, Nankung Yüeh, and the officials of the Bureau of Astronomy Ta-Hsiang and Yuan-Thai, to proceed by the post-station routes to An-nan, Lang-chow, Tshai-chow, Wei-chow, etc., and to measure the lengths of the sun shadows, a report to be made on the day of their return. For several years they took observations and when they came back to the capital they compared them in conference with I-Hsing....At Lang-chow, Hsiang-chow, Tshai-chow, Hsü-chow, Honan-fu, Hua-chow, Thai-yuan, etc. there were also in each case Commissioners, and all brought back various results. Then on the basis of the northern and southern sun-shadows I-Hsing made comparisons and estimates, using the 'right-angle triangle' method to calculate them....

Actually the work had begun at least as early as +723, for it was in that year that Nankung Yüeh set up an 8-ft. gnomon at Yang-chhêng which still exists there (Fig. 288).[b] This has an inscription on its south side saying 'Chou Kung's Tower for the Measurement of Sun Shadows'. The design is such that at the summer solstice of that time the shadow just extended to the top of the pyramidal base, the slope of the north side of which corresponded exactly with the edge of the shadow.

The results of the pioneer geodetic survey of I-Hsing and Nankung Yüeh are shown in Table 42 and Fig. 289. They form an impressive body of data, justifying Gaubil's charming comment:[c] 'Quand le Bonze Y-Hang n'auroit fait autre chôse que de procurer tant d'obsèrvations de la hauteur du Pôle, et de détèrminer la grandeur du Ly en le rapportant aux degrés de latitude, on lui auroit toûjours une obligation infinie.' But as it has turned out, there is much more in this set of figures than meets the eye. When subjected to a penetrating analysis by Beer *et al.*[d] a series of rather unexpected findings emerged.

[a] Ch. 42 (p. 755), tr. auct. incl. E. Pulleyblank.
[b] See Tung Tso-Pin *et al.* (1), pp. 38, 39, 40, 94.
[c] (2), p. 78.
[d] Beer, Ho Ping-Yü, Lu Gwei-Djen, Needham, Pulleyblank & Thompson (1).

[1] 測 影 使

First, it could be shown that the ground distances between the more distant stations were not measured but assessed by extrapolation on the basis of the result for the short central line of stations (3, 4, 6 and 7). Secondly, it appeared that all the winter solstice shadow-lengths and the great majority of the equinoctial shadow-lengths recorded were not measurements but values calculated from the summer solstice shadow series.

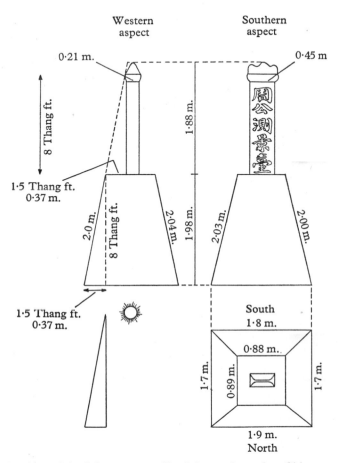

Fig. 288. Scale drawings of the 8-ft. gnomon still existing at the ancient Chinese central observatory of Yang-chhêng (now Kao-chhêng). It was set up by the Astronomer-Royal Nankung Yüeh in connection with the geodetic survey commissioned in +723. After Tung Tso-Pin *et al.*

Thirdly, it could be demonstrated that even the recorded polar altitudes were also not observations but values computed from the summer solstice shadow data. All these computations are accurate to nearly one part in a thousand.[a] If graphic construction methods had been employed, it would have been necessary to set up a plane

[a] It is possible to deduce from them the value employed for the obliquity of the ecliptic. This was 23° 40', almost exactly equivalent to 24 Chinese degrees (*tu*). The *Chiu Thang Shu* says (ch. 35, p. 5*b*) that this figure was in fact regarded as correct in I-Hsing's time, but it had been established half a century earlier by his great predecessor Li Shun-Fêng (see Vol. 3, p. 289).

table just over 100 ft. in diameter, together with stretched wires and reading devices of adequate delicacy. One may doubt whether this was a technical possibility at the time; in any case nothing which could be a reference to it has been met with in these or relevant astronomical texts. The alternative of course is the use by I-Hsing and his colleagues of tables of trigonometrical functions. They would have needed tables of tangents, or their equivalent in sine tables, and these would have had to be accurate to 1 part in 500, with intervals of the order of 5 minutes of arc (a tenth of a *tu*). Of the two possibilities this is much the more likely one, but the existence of such accuracy in

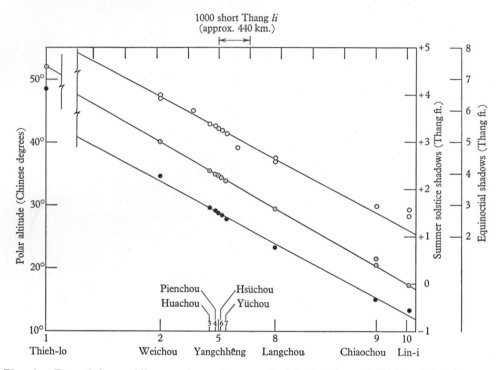

Fig. 289. Data of the meridian arc observations organised by I-Hsing and Nankung Yüeh (+724 to +726), in graphical form. Key: ⊙, polar altitude; ●, summer solstice shadow-length; ○, equinoctial shadow-length.

the +8th century is an unexpected conclusion, and quite different from what the data would at first sight seem to imply.

Trigonometry in Thang China is not in itself a surprise. Already we have seen in detail[a] how after the tabulation of chords by the Greek astronomers Hipparchus and Ptolemy, and the basic work in spherical trigonometry by Menelaus, this branch of mathematics was brought into its modern form by the Indians. The notion of sines and versed sines appears for the first time in the *Pauliśa Siddhānta* shortly after +400. Āryabhaṭa (c. +510) was the first to give a special name to the function, and to draw up a table of sines for each degree. His contemporary Varāha-Mihira, in the *Pañca*

[a] Cf. Vol. 3, pp. 108 ff., 202 ff.

The Meridian Line of +725. Cf. Gaubil (2), p. 76

No.	Station / Place	Summer solstice shadow of 8-ft. gnomon (Ch. ft.)	Equinoctial shadow (Ch. ft.)	Polar altitude (Ch. degrees and tenths of a degree)	Approximate latitudes of stations from modern maps (Ch. degrees and tenths)	Distances in li and (in brackets) pu
1	Thieh-lê 鐵勒 (in the country of the Tölös horde of Turkic nomads beside Lake Baikal)	+4·13	+9·87	52·0°	c. 52·76°	
2	Wei-chou (Hêng-yeh Chün 橫野軍) 蔚州 (northern Shansi)	+2·29	+6·44	40·0	40·38	
2a	Thai-yuan 太原	—	+6·0	—	38·22	
3	Hua-chou (Pai-ma hsien 白馬縣) 滑州	+1·57	+5·56	35·3	36·07	
4	Pien-chou (Ku-thai piao 古臺表 浚儀) near Chün-i (Khaifêng) 汴州	+1·53	+5·5	34·8	35·31	
5	Yang-chhêng 陽城 (the central astronomical observatory)	+1·48	+5·43	34·7	34·93	
6	Hsü-chou (Fu-kou piao 扶溝表) 許州	+1·44	+5·37	34·3	34·65	
7	Yü-chou (Wu-chin piao 武津表) near Shang-tshai 上蔡 豫州	+1·36	+5·28	33·8	34·09	
7a	Hsiang-chou 襄州	—	+4·8	—	32·46	
8	Lang-chou (Wu-ling 武陵) 郎州	+0·77	+4·37	29·5	29·42	
9	Chiao-chou (Hu-fu 護府) 交州 (capital of An-nan)	−0·33	+2·93	{ 21·6 / 20·4 }	21·03	
10	Lin-i 林邑 (capital of Lin-I; near Hué)	{ −0·57 / −0·91 }	+2·85	17·4	17·50	

Distances in li and (in brackets) pu:

6900

6112

5023

1861 (214)   1826 (196)

526 (270)

167 (281)

198 (179)   160 (110)

NOTES

(a) For the deviant values in some of the sources, as in Gaubil (2), see Beer *et al.* (1). On the Lin-i values cf. Stein (1), pp. 43 ff., 76 ff.

(b) It has been supposed that the meridian line was as long as 13,000 *li* (c. 3800 km.) and included the most northerly station (no. 1). But the texts seem to indicate that the figures for this place were extrapolated and not derived from observations made at this time. There is reason for thinking that the distance 1–5 was an independent estimate dating from the Chên-Kuan reign-period (+627 to +649).

(c) D'Anville (2) thought that the figure for the distance between stations nos. 3 and 4 was a mistake of Gaubil's for 168, but this is not so; all texts have 198. Besides, the addition of the whole set of three values for the central stations is unmistakably recorded in them. There was, however, one mistake in Gaubil's table, the reduction of the polar altitude of Yüchow.

Siddhāntikā (c. +505), gave formulae which in modern terms would comprise both sines and cosines. Then on the one hand the Indian work was taken over by the Arabs and transmitted to Europe, while in the other direction Indian monks or lay mathematicians who took service with the Chinese imperial Bureau of Astronomy spread the new development farther east.[a] Books on Indian astronomical methods were circulating in Chinese well before +600. Then one of these men, for instance, Chiayeh (Kāśyapa) Hsiao-Wei, assisted Li Shun-Fêng in his calendrical calculations of +665. By far their greatest representative was the astronomer and mathematician Chhüthan Hsi-Ta (Gautama Siddhārtha), at the height of his powers just when I-Hsing and Nankung Yüeh were busied with their meridian arc. In +729 he finished his *Khai-Yuan Chan Ching* (Treatise of the Khai-Yuan reign-period on Astronomy and Astrology), a work which still today constitutes our greatest surviving thesaurus of ancient Chinese astronomical quotations and fragments. Some parts of this had been ready, however, as early as +718, notably chapter 104, which consists of a translation into Chinese under the name *Chiu Chih* of the +6th-century *Navagraha* (Nine Upholders) calendar system of Varāha-Mihira. This was the first occasion in which a zero symbol appeared in a Chinese text,[b] but more relevant is the fact that the chapter also contains a table of sines.[c] It is typically Indian in that it tabulates at intervals of 3° 45′, a value derived by continuous bisection of the angle of 60°, the cosine of which was known to be $\frac{1}{2}$. But since this was half a dozen years before the great geodetic survey there is no reason why the clerks of the Bureau of Astronomy should not have been occupied in computing tables with much smaller intervals in preparation for the observational data which the Shadow-Measuring Commissioners would later on be bringing in.

For of real observations there was certainly no lack. From the fact that the figures in Table 42 were mostly computed and not directly measured values, it would be possible to conclude that in the course of the survey very few actual observations of any kind were made, and that theory took an overwhelming precedence of practice in +8th-century China. But such a judgment would be superficial and misleading. Our texts expressly indicate that the expeditions were real and thorough. In +723 the gnomon was established and used at Yang-chhêng. In +724 Commissioners were appointed and despatched to Chiao-chow in modern Indo-China and at least eleven other places up to the latitude of the Great Wall and covering a total arc of some 2500 km. In +725 the central line of stations on the Honan plain was set up and measurements collected. Only after two or three years did the Commissioners return to the capital and confer with I-Hsing on their various results.[d]

It may seem strange that the final set of figures presented by I-Hsing and Nankung Yüeh to the imperial court, and hence destined for transmission to us, should have been mainly computed and not observational. But surely what this means is that at that time it seemed more elegant to offer up a set of 'ideal values' computed by the

---

[a] Cf. Vol. 3, pp. 202ff.
[b] Cf. Vol. 3, pp. 10ff.
[c] This has been reproduced and studied by Yabuuchi (1).
[d] The period of their labours is circumscribed by the fact that I-Hsing died in +727.

most up-to-date mathematical methods. A predilection for trigonometric tables, then new and interesting as well as precise, would have been quite natural.[a] Sines and cosines could clearly achieve a computation not possible practically by any other means. Besides, today our minds are so accustomed to the statistical point of view that it is hard for us to imagine a time when this was not understood. In all probability I-Hsing thought it very undesirable to admit to his final tabulation a mass of raw data showing considerable scatter, and not being able to assess it statistically he used it only to satisfy himself that his calculated values came about where they should—indeed he probably believed that they were much more reliable than most of the observations. Except always the summer solstice figures, valued no doubt on account of the shortness of the shadow-length and a presumed greater accuracy of measurement,[b] so that they appear in the table as the primary, in fact the only, observational data. From these, and knowing the body of results from all the different places for other shadow-lengths and latitudes, he 'constructed a curve to fit'. In sum, there is no reason for doubting that a quantity of actual measurements of shadow-lengths by gnomon scale, and most probably also of polar altitudes by armillary sphere quadrants, came to rest in the archives of the Bureau of Astronomy. Unfortunately, as we possess only I-Hsing's final report, they have not come down to later generations.

The main result of the field survey was that the difference in shadow-length was found to be very close to 4 in. for each 1000 *li* north and south, and that the terrestrial distance corresponding to 1° of polar altitude was estimated at 351 *li*, 80 *pu*. There can be no doubt that this was derived only from the polar altitudes of the four central stations (3, 4, 6 and 7), the total distance of the line not exceeding 250 km., perhaps only 150 km. On this account alone it was bound at that time to be very inaccurate. But there were other sources of error also. Comparing the data given with the places as identified on modern maps a mean difference of −0·31° is found, so that nearly all the stations appear to be on the average some 28 km. south of their true positions. The fact that these deviations have the same sign implies some systematic error in the methods of observation, probably an incorrect convention in defining the edge of the shadow and penumbra.[c] Because of these uncertainties[d] the total measured arc may have extended to as much as 2·3° instead of 1·5°, and this would have given a result

[a] Thus a certain modification, perhaps, will be needed in the opinion expressed in Vol. 3, p. 203, commenting on the remarkably little influence exercised by Indian scientific thought on the course of Chinese astronomy. 'The equatorial mansions remained as before, the circle continued to have 365¼°, Indian trigonometry was not taken up, the zero symbol slumbered for another four centuries, and naturally enough the Greek zodiac remained buried in bizarre transliterations.' Nevertheless, the statement is still broadly true.

[b] I-Hsing's percentage error would have been much less if he had taken the long winter shadows.

[c] The mean deviation corresponds to only 0·4 in. in shadow-length. Five hundred years later these difficulties were well appreciated, and Kuo Shou-Ching in +1279 developed a special pin-hole device, the 'shadow-definer', for focusing the image of the cross-bar at the top of his 40-ft. gnomons. These were the observations which Laplace considered as among the most accurate ever made of solstitial sun shadows. Kuo Shou-Ching also caused a series of polar altitude observations to be made at this time, much more extensive than those of I-Hsing's survey. The data are recorded in *Yuan Shih*, ch. 48, pp. 12*b*ff., and it would be interesting to subject them to a similar study. On the shadow-measurements of Kuo Shou-Ching see Vol. 3, p. 299.

[d] The mean deviation irrespective of sign is 0·33°.

of 230 *li* per degree. Such a value would have been fairly correct for the normal short Thang *li*,[a] which (in spite of some doubts)[b] it seems tolerably sure that I-Hsing intended to use. An edict of +721 had laid down that the 'short measure' was to be used in astronomy, medicine and imperial paraphernalia, while the 'long measure' was to be applied to everything else whether official or private.[c] Thus while the work of I-Hsing and Nankung Yüeh fully achieved its aim of fixing a terrestrial civil unit in terms of 'the dimensions of Heaven and Earth' (1/351 of a degree), the level of their accuracy was not at all impressive. Yet their work takes an outstanding place in the pre-history of the metric system, not only on account of the spaciousness and amplitude of its plan and organisation, unmatched elsewhere throughout the Middle Ages, but also because of the advanced mathematical methods used to compute the ideal set of values.

In spite of the inaccuracy of the *li*/degree relation, it seems to have served the purposes of cartographers for a long time afterwards. Examination of the grid scale of the famous *Yü Chi Thu* map[d] carved on stone in +1137 suggests that I-Hsing's value was used for marking off long distances. On the other hand slightly better approximations were reached during the Sung, for example, in +1001 an arc of 3° latitude was taken as equivalent to 1000 *li*, giving a value of 333 *li* to the degree.[e] Towards the end of the dynasty, in +1221, the work of I-Hsing's survey was extended by the Taoist adept Chhiu Chhang-Chhun and his party, who made gnomon observations at the summer solstice on the bank of the Kerulen River in northern Mongolia (about 48° N. lat.) when travelling on a visit to Chingiz Khan at Samarqand.[f]

Whether or not I-Hsing attempted to derive from his measurements a value for the circumference of a spherical earth it is impossible to say.[g] Although no indication has so far been found in the texts to suggest that he did so, certain Chinese cosmological schools had from antiquity onwards assumed its sphericity.[h] This must have been well known to him, especially since just at the same time Chhüthan Hsi-Ta was engaged in

---

[a] The true figure would be 250 *li* per degree as against 208 for the long Thang *li*. The long Thang measure had a *li* of 532 m., giving 1·88 to the km. and 208 to the degree of latitude. The short Thang measure had a *li* of 442 m., giving 2·26 to the km. and 250 to the degree of latitude. Both had 300 double-paces (*pu*) to the *li*, and 6 ft. (*chhih*) to the double-pace, the length of the foot varying accordingly. These values agree with those derived by Mori Shikazo from extant Thang foot-rules. Values for the *li* as low as 193 m. seem to have been used in the Thang and earlier by Buddhist pilgrims, with some relation to the Indian *yōjana*; cf. Vost (1), Fleet (1), Weller (1), and others. That I-Hsing's *li* had 300 *pu* (double-paces) is clearly implicit in the figures for the distances given in the texts.

[b] See Beer *et al.* (1).

[c] See Wu Chhêng-Lo (2), 1st ed. p. 387, 2nd ed. p. 253.

[d] See Vol. 3, p. 547.

[e] Gaubil (2), p. 97. The reference is to *Sung Shih*, ch. 68, p. 2*a*, *b*. The value was accepted in three calendars, that of Wang Chhu-No in +962 (Ying Thien), that of Shih Hsü in +980 (I Thien), and that of Wu Chao-Su in +981 (Chhien Yuan). On account of the similarity of the figures d'Anville (2) supposed that 351 *li* must have been a mistake of Gaubil's for 331 *li*, but this is not so—all the texts read 351.

[f] The incident is recorded by Li Chih-Chhang in his *Chhang-Chhun Chen Jen Hsi Yu Chi*, ch. 1, p. 10*a*; tr. Waley (10), p. 66.

[g] This was the question asked by my friend Professor J. D. Bernal a number of years ago. Although it can probably never be answered, it led to the investigation of Beer *et al.* which made the present subsection possible.

[h] See Vol. 3, pp. 216 ff., 498 ff. Cf. Wang Yung (2), pp. 73 ff.; Wei Chü-Hsien (4).

compiling his great collection of ancient and early medieval astronomical writings. Moreover, I-Hsing's acquaintance with Indian and even indirectly with Hellenistic astronomy may well have informed him of the previous estimates of the earth's circumference.[a] There is thus no reason why he should have hesitated to use his data in this way, and indeed it is hard to see how he could have given a constant li/degree value if he had not had at least some *arrière-pensée* of a curved earth's surface. Yet after the account of the survey in the *Chiu Thang Shu* there follows a cosmological discussion[b] which attempts to reconcile the findings with the hypothesis of a flat earth. Using I-Hsing's figures, calculations of the size of the universe on the basis of some of the ancient cosmological ideas, and in rather archaic style, are set forth. In the end the writer is reduced to saying: 'But looking at it in this way, of what value for the cultivation of human morality are the hair-splitting discussions of Wang Chhung and Ko Hung?'[c]

Perhaps if I-Hsing had any conception of a spherical earth he kept it to himself or to his immediate circle of computers and observers. Though contrary to 'common sense' it would not have been seriously unorthodox for the scholars of the age, which was hardly one of Confucian ascendancy. It is noteworthy that the discrediting of the beliefs of the 'scholars of former times' (*hsien ju*[1]) by the new measurements was in no way glossed over in the texts, but rather emphasised. In fact one must say that the scholars of the Sui and Thang showed a relatively enlightened conception that age-old beliefs about the universe must bow to improved scientific observations. At the same time we may have to take account of some differences of opinion between the two chief figures, I-Hsing and Nankung Yüeh.[d] For after the death of the former in $+727$ Nankung supported a contention of Chhüthan Chuan that the monk's *Ta Yen* calendar was nothing but a plagiarism of the *Chiu Chih* calendar translated by Chhüthan Hsi-Ta ten years before. Though this could not be sustained, the fact indicates some disagreement between the Astronomer-Royal and the imperially favoured monastic amateur. Possibly the doctrine of the flat earth was still accepted in court circles and not susceptible to criticism by monks, however brilliant and talented they might be.

Whatever the truth of it, our two protagonists deserve a lasting fame for their organisation of the most outstanding piece of field research in any medieval civilisation. Perhaps it would only have been possible at that time in the empire of China and the lands neighbouring thereunto. Relating the civil distance measure, the li, to cosmic distances celestial and terrestrial, indeed defining it in terms of them, was a vital act in the pre-history of the metric system. I-Hsing and Nankung Yüeh were among the truest forefathers of the men of the Second International Geophysical Year, towards the end of which these words are written.

---

[a] On these see K. Miller (3); Fleet (2).
[b] Pp. 8$b$ff.
[c] P. 9b. For these debates of the $+$1st and $+$4th centuries see Vol. 3, pp. 218, 221, 226, and many other places.
[d] Cf. Vol. 3, p. 203.

[1] 先 儒

A whole millennium passes. We find ourselves in the last decade of the seventeenth century. Relations between East and West have long been greatly intensified, and the science of the Renaissance has become well known in China because of the activities of the Jesuit mission. Many of the missionary clergy have greatly distinguished themselves in furthering the mathematical, astronomical and geographical sciences.[a] And now it falls to one of them to take up a suggestion of the Khang-Hsi emperor and to anticipate the French Revolution by nearly a century in adjusting earthly measurements of distance to a celestial standard.[b]

Antoine Thomas (+ 1644 to + 1709) was a Belgian from Namur. While in Portugal he wrote a treatise on physics for Jesuit use, the *Synopsis Mathematica*, and then on the way out made astronomical observations at the Siamese capital. After his arrival in China in + 1685 An To[1] became Vice-Director and for a time Acting Director of the Bureau of Astronomy. In the last years of the century planning began for that great cartographic survey of the Empire[c] which was finished in + 1717, and one of the pre-requirements for this was a decision about the precise length of the *li*. Since estimates varied from less than 200 to more than 250 *li* to the degree, the idea grew up between Khang-Hsi and An To (Thomas) of pinning the *li* specifically to the degree of terrestrial latitude, i.e. to the earth's circumference. The crucial interview took place on 8 December 1698.

Measurement of the meridian line began in December 1702, the imperial commission having been given in the previous April. The plains near Pa-shih south-east of Peking were chosen for the purpose, and all the procedures, which were carried out with remarkable accuracy, have been described by Bosmans (3), who translated the subsequent account of Thomas in full. Thomas had as chief collaborator Yin-Chih,[2] the third son of the Khang-Hsi emperor, a prince who had become (in Thomas' words) 'a very clever observer familiar with apparatus, and a quick and accurate computer', as the result of training by the emperor himself. He contributed very actively to the success of the operations. Gnomons having been erected at both ends of the line, it was ascertained that 1° of terrestrial latitude corresponded to 195 *li* 6 *pu* of the standard in use at the time,[d] a 5-ft. bar of iron kept in the imperial palace

[a] Cf. Vol. 3, pp. 437 ff. above.

[b] The precise determination of the length of the meridian degree had presented itself as one of the most urgent cosmographic tasks to the scientific societies of 17th-century Europe, and the French Academy of Sciences, enjoying better State support than the Royal Society, achieved it under the direction of Picard by + 1670 (cf. Taylor (8), pp. 236 ff.). The role of societies in science, so often emphasised, may have been prefigured a thousand years earlier, for I-Hsing was a member of the Thang court's Chi Hsien Yuan (College of All Sages), as we shall see further below (Sect. 27*j*).

[c] Cf. Vol. 3, p. 585. A letter from J. B. Régis (Lei Hsiao-Ssu[3]) of + 1708 describes how the Jesuit geographers had official access to the *hsien chih*[4] (local topographies, cf. Vol. 3, p. 517) everywhere they went. They used the standard length measure fixed by A. Thomas and his collaborators. It was in + 1710 that they recognised the ellipsoidal nature of the earth's figure from observations made on the Manchurian plains by Régis and Jartoux (Tu Tê-Mei[5]); cf. Bosmans (3), p. 157. These were amongst the earliest observations of the kind, though Picard in + 1671 had already voiced the suspicion that the earth was no perfect sphere.

[d] It will be seen that this *li* was very nearly equivalent to the normal long Thang *li* (p. 52). D'Anville adopted 194 to the degree in his map of Asia.

[1] 安多　　　[2] 胤祉　　　[3] 雷孝思　　　[4] 縣志　　　[5] 杜德美

and representing the 'geometrical pace'. As the Khang-Hsi emperor wished to adjust the *li* to the degree in round numbers, it was decided to adopt the figure of 200 *li*, thus involving a diminution of the standard foot by a factor of $\frac{39}{40}$. Accordingly there were 72,000 standard paces to the degree, 1200 to the minute and 20 to the second.[a]

Thus the *li* was fixed astronomically some ninety years before the kilometre. For it was not until 1791 that the committee of the French Academy made their famous report in which it was suggested that the metre should be taken as one ten-millionth part of one quarter of the earth's circumference at sea level.[b] Once again social conditions in China had permitted large-scale scientific observation and rational public action based thereon some time before Europe was ready to follow suit. Yet only in Europe arose that modern phase of science which is now capable of defining the metre in terms of the wavelength of krypton 86.

(d) THE STUDY OF MOTION (DYNAMICS)

If there has been not a little to say about statics and hydrostatics, we shall now see that the study of motion (kinetics and cinematics) seems to have been, on the whole, conspicuously absent from Chinese physical thinking. Nevertheless, there are statements on the subject in the *Mo Ching*; indeed, some remarkable anticipations.[c]

Ch 63/—/45.56. *Movement in space (frames of reference)*

C   When an object is moving in space, we cannot say (in an absolute sense) whether it is coming nearer or going further away. The reason is given under 'spreading' (*fu*[1]) (i.e. setting up coordinates by pacing).

CS Talking about space, one cannot have in mind only some special district (*chhü*[2]). It is merely within a certain district that one can say that the first step (of a pacer) is nearer and his later steps further away. (The idea of space is like the idea of) duration (*chiu*[3]). (You can select a certain point in time or space as the beginning, and reckon from it within a certain period or region, so that in this sense) it has boundaries, (but time and space are alike) without boundaries. (auct.)

> Ssuma Piao's +3rd-century commentary on *Chuang Tzu*, ch. 33, says, 'The distance between Yen and Yüeh is limited, but that between the south and the north is infinite. Observing the limited from the viewpoint of the infinite, we find that Yen is not really separated from Yüeh. Space has no directions except that the place where you are yourself is the centre. Similarly, circulating time (the course of the seasons) has no end and no beginning; you can make any specific time start from whenever you like, according to what you are doing.' Many centuries later Nicholas of Cusa and Giordano Bruno spoke in the same way using almost the same words (see D. Singer, 1). Cf. Wu Nan-Hsün (1), p. 23.

---

[a] At 0·555 km. this *li* was closely similar to the *li* used in our own time.

[b] See e.g. Lloyd Brown (1), pp. 286 ff. As usual, there arises the question of transmission. The Jesuit work was, of course, well known in France, where it was popularised in writings such as those of d'Anville (1, 2). But both he and Gaubil misunderstood the work of Thomas, and supposed that the Khang-Hsi emperor had hit upon 200 *li* arbitrarily without the intention of making an astronomical definition.

[c] Though we shall not discuss it again, the Mo-Ming paradox about the flight of arrows (cf. Vol. 2, pp. 191 ff.) may be borne in mind.

[1] 數   [2] 區   [3] 久

Ch 64/271/47.57. *Movement and duration*

C Movement in space requires duration. The reason is given under 'earlier and later' (*hsien hou*[1]).

CS In movement, the motion must first be from what is nearer, and afterwards to what is further. The near and the far constitute space. The earlier and the later constitute duration. A person who moves in space requires duration. (FYL/B, mod.)

Cs 49/—/100.46. *Motion*

C Motion is due to a kind of looseness (*huo tsung*[2]) (i.e. to the absence of an opposing force).

CS There is motion (if a force is) allowed to work at the edge, just as a door-pivot is free when the bolt is not fixed. (auct.)

  This may be the remains of an attempt to discuss circular motion.

Cs 50/—/2.47. *Forces and motion*

C The cessation of motion (*chih*[3]), is due to the (opposing force) of a 'supporting pillar' (*chu*[4]).

CS If there is no (opposing force) of a 'supporting pillar' the motion will never stop. This is as true as that an ox is not a horse. Like an arrow passing through between two pillars (*ying*[5]) (without anything standing in its way, and following a linear motion without changing its direction).

  If there is (some kind of) 'supporting pillar' (some other force interfering with the motion), and nevertheless the motion does not stop (it may still be called motion but it will not be linear motion because there will have been a deflection). This is a case of something being 'a horse and yet not a horse'. It is like people passing over a bridge (i.e. they have to climb up to the top of the arch and down again, though they continue in motion). (auct.)[a]

Ch 28/—/53.19. *Inclined planes*

C Inclined planes (*i*[6]) should not be put horizontal. The reason is given under 'easier' (*ti*[7]).

CS Actions like carrying, pushing, pulling, and sliding down (lit. shooting) require an inclined position and not an upright (or horizontal) one. (auct.)[b]

Ch 62/—/43.55. *Unstable equilibrium of a spherical object*

C A perfect sphere cannot resist a force. The reason is given under 'revolving' (*chuan*[8]).

CS As for the sphere, no matter where it moves, it always keeps its centre. The course of its motion is like an object being whirled around on the end of a string. (auct.)

  Cf. the Centrifugal Cosmogony (Vol. 2, p. 371 above). Cf. also the statements in the *Chou Li* (Khao Kung Chi)[c] about the minimum friction as perfect circularity of wheels is approached.

---

[a] For a different reconstruction of this and the preceding proposition, see Luan Tiao-Fu (1), p. 70.
[b] This interpretation follows Than Chieh-Fu, but the entry may possibly refer rather to centres of gravity, the title being translated simply 'Leaning'.
[c] Ch. 40; see Biot (1), vol. 2, p. 466, also Lu, Salaman & Needham (1).

[1] 先後   [2] 或縱   [3] 止   [4] 柱   [5] 楹   [6] 倚   [7] 弟
[8] 轉

In order to place these Mohist propositions, as well as later Chinese thoughts, in their correct perspective, it is necessary to glance for a moment at the development of the study of motion of bodies in Europe.[a] As Dugas has said, this is a field in which one cannot speak either of 'the Greek miracle' or the 'night of the Middle Ages', for Greek mechanics was fallacious, and medieval progress rather striking. The dynamics of Aristotle was quite different from anything that we encounter in China. Ruling out motion in a vacuum *a priori*, Aristotle considered only real motion in a resistant medium. Everything for him had its 'natural place', and 'local' (i.e. terrestrial) motion took place in a linear manner towards the 'natural place' of the object concerned. Celestial motions, on the contrary, were 'naturally' circular. Aristotle therefore distinguished between 'natural' motion when an object was seeking its natural place, and 'violent' motion when it was compelled by some external force to move in some other manner. The velocity[b] of the object was thus proportional to the motive power and inversely related to the resistance of the medium. Nevertheless, Aristotle regarded the medium as necessary for the continuation of the motion, and supposed an 'antiperistasis' (ἀντιπερίστασις) in which the air rushed in owing to 'horror vacui' behind the projectile and sent it further on its way. As is well known, by the +15th century, in Leonardo's time, the path of a projectile had come to be imagined as first of all a violent motion in a straight line (*modus violentus*), which after a time, since 'nullum violentum potest esse perpetuum', gave way to a composite motion (*modus mixtus*), and finally to a vertical free fall in accordance with gravity (*modus naturalis*).[c] The discovery of parabolic trajectory by Galileo and the gunners was one of the great advances of the early phase of modern science.

It is also now well known that from the +6th century onwards the theory of antiperistasis was subjected to strong criticism. In the first place Simplicius,[d] commenting on Aristotle about +540, approved a rather old idea of Hipparchus that a special quality, impetus or inertia, was present in the moving body from the outset. Joannes Philoponus, a Byzantine (*d. c.* +560),[e] also considered that the disturbed medium could not possibly be the cause of the projectile's flight; it must have a motive virtue which travelled with it. These beginnings led to a whole movement of thought regarding dynamics among the philosophers of the Middle Ages. A Spanish Muslim, Ibn Bājjah (*fl.* +1118 to +1138),[f] perpetuated the views of Philoponus. John Buridan (*fl.* +1327; *d.* +1358) considered that impetus was something which gradually fades out, and the heavier the body the more impetus it could receive. He also said that it would last in perpetuity if it were not diminished by the resistance of the medium. In the same century the Oxford school, with men such as William Heytesbury (*fl.* +1330 to +1371), Thomas Bradwardine (*fl.* +1328),[g] and Wm. Collingwood,

[a] These paragraphs rely on the books of Dugas (1, 2) and Maier (1–7), an interesting review by Moody (1), the illuminating introductory article of Boyer (4), and Hall's work (1) on ballistics in the seventeenth century. Cf. also Clagett (1) and his source-book (2).
[b] N.B. not the acceleration.
[c] This whole conception probably derived from a simple optical illusion (Ronchi, 3).
[d] Sarton (1), vol. 1, p. 422.   [e] Sarton (1), vol. 1, p. 421.
[f] The Avempace of the Latins, see Hitti (1), p. 581; Mieli (1), p. 188; Moody (1); cf. Pines (2).
[g] His *Tractatus de Proportionibus* has been translated by Crosby (1).

perfected the concept of acceleration. Nicolas d'Oresme (*fl.* +1348; *d.* +1382) adopted impetus doctrine, and anticipated Descartes in the invention of coordinates; everything, he held, could be regarded as a continuously varying quantity. His coordinates were 'longitudo' and 'altitudo' or 'intensio'; an ascending straight line on a graph (as we should say) was 'uniformiter difformis', a curve was 'difformiter difformis'. The transmitters of these ideas to the Renaissance physicists, such as Blasius of Parma (*fl.* +1374; *d.* +1416), knew velocity as 'latitudo motus', acceleration as 'latitudo intensionis motus', and uniformly accelerated motion as 'latitudo intensionis motus uniformiter difformis'. So also for retardation, but none of them applied these conceptions to the free fall of bodies; this was left for Galileo.[a]

For trajectories Galileo assumed that horizontal motion remains uniform except as influenced by friction or air resistance. He disregarded the 'violent', 'mixed', and 'natural' phases. Having previously shown that points whose vertical and horizontal displacements were in a quadratic ratio lay on a parabola, Galileo concluded that the trajectory of a projectile must follow such a curve, gravity beginning to act as soon as the path is commenced. All ballistics followed from this.

Newton was not the first to state his First Law of Motion, that 'every body continues in its state of rest, or of uniform motion in a right line, unless it is compelled to change that state by forces impressed upon it'. Huygens and Descartes had said practically the same thing[b] and Galileo had used the principle though not applying it apart from projectiles. But it is also claimed that Ibn al-Haitham (+965 to +1039)[c] stated the law too. And now we find something at any rate extremely like it in the *Mo Ching* of the −4th or −3rd century (Cs 49), where motion is said to be due to the absence of an opposing force. The Mohist technical term 'supporting pillar' is to be understood only as that force which in Newton's first law changes the otherwise permanent state of motion of the moving body (Cs 50). The exposition distinctly states that if there is no such force the motion will never stop. Its writer seems to be trying, too, to describe non-linear or deflected motion as 'motion which is not quite motion in the fullest sense'. What remains in these brief fragments is so striking that we may be allowed to believe that if more of the physics of the Mohist school had been preserved, we should have found in it some discussion of trajectories, the effect of gravity, and so on. If the Mohists had no technical term corresponding to impetus, at least they did not suffer from the concept of 'natural place' or the awkward idea of antiperistasis.

As for the other Mohist propositions given above, the first two simply illustrate the relativistic and dialectical quality of the school. The fifth (Ch 28) reminds us that Pappus[d] was the only Western author (+3rd and +4th centuries) to treat of inclined planes, and that little progress was made until the time of Galileo. The sixth (Ch 62) seems to show the Mohists moving in the direction of the 'gravitas secundum situm'

[a] Exactly how much Galileo and his successors were indebted to the men of the +14th and +15th centuries is still one of the most contested problems in the history of science. The reader is referred to the writings of Koyré on the whole subject.
[b] L. W. Taylor (1), p. 130.   [c] Mieli (1), p. 105; Winter (3).
[d] Sarton (1), vol. 1, p. 337.

of Jordanus Nemorarius (+13th) and Leonardo, who both considered spheres or circles moving on inclined planes.

After these brilliant insights of the Mohist school, it seems almost incredible that through the subsequent millennia of Chinese history there are no recorded discussions of the motions of bodies, whether impelled or freely falling.[a] Though fully aware of the danger of negative evidence, we cannot but feel that if such discussions existed we should have come across them. There seem to be nothing but a few odds and ends. In the *Chou Li*[b] (Khao Kung Chi section) there are (as we have just seen) statements about vehicle-wheels in which it is emphasised that the periphery must touch the ground to the minimum extent. Inclined planes (gradual slopes; *ling i*[1]) are mentioned in passing from time to time (*Han Shih Wai Chuan*[2];[c] *Hsün Tzu*;[d] *Yen Thieh Lun*[3];[e] *Shuo Yuan*[f]), but never theoretically treated. In one of the odes of Chia I[4] it is said:[g]

If water is swept along (*chi*[5]) it becomes fiercely torrential (*han*[6]). If an arrow is shot forth it travels far. Yet everything has a force reacting back (*wan wu hui po*[7]) and vibrates in mutual opposition (*chen thang hsiang chuan*[8])....Such is the Tao of Nature.[h]

This is reminiscent of Newton's Third Law. The Mo-Ming paradoxes about motion occasionally received some discussion, in later times, as by Liu Hsiao-Piao[9] in the Liang.[i] But little more has been found.

In casting about for some explanation of the poverty of Chinese discussions on dynamics, we may first remember that it seems to have had no inhibitory effect at all upon practical technology in the eotechnic phase. So far as vehicles, projectiles, and engines of all kinds were concerned, Chinese mechanical practice was ahead of European, not retarded, down to the very time when the scholastics of the +14th century were preparing the way for Galileo, and even later.[j] Yet if the absence of Galileo needs no explanation, how was it that China had no thinkers corresponding to Philoponus, Buridan or d'Oresme? I have been tempted to wonder whether it was not because, as shown above (pp. 3 ff.), Chinese thinking was so averse from atomic or

---

[a] Much could be said, of course, concerning the failure of Mohist physics to develop. The extremely laconic nature of its propositions, which give the impression of having been notes accompanying lecture demonstrations to small groups of students, may well have been due to difficulties of writing, e.g. carving on bamboo tablets. Compare also the failure of the Han mathematicians to *explain* their techniques. But the particular point at issue here is why a science such as dynamics was so inhibited, while the studies of light and magnetism were not.

[b] Ch. 11, p. 6a (ch. 40).   [c] Ch. 3.   [d] Ch. 28.
[e] Ch. 58.   [f] Ch. 7.
[g] *Shih Chi*, ch. 84, p. 6a. About −170.   [h] Tr. auct.
[i] *Shih Shuo Hsin Yü*, ch. 1B, p. 13b. See Fêng Yu-Lan (1), vol. 2, pp. 176ff.
[j] For example, in +1400, though Europe had adopted during the preceding centuries many East Asian techniques, many were still to come—canal locks, iron-chain suspension-bridges, wagon-mills, sailing-carriages, rotary winnowing machines, square-pallet chain-pumps, and movable-type printing—to mention only a few. Moreover, we shall later show (Sect. 30h) that between the +5th and the +13th centuries at least, Chinese heavy ballistic weapons were greatly superior to anything known in the West.

---

[1] 陵夷   [2] 韓詩外傳   [3] 鹽鐵論   [4] 賈誼   [5] 激   [6] 悍
[7] 萬物回薄   [8] 振蕩相轉   [9] 劉孝標

particulate conceptions. Aristotle's dynamics may have been directed partly against the random motion of the atomists, but he participated in the same European tradition as they, and was prepared to think in terms of what happened to individual portions of matter in motion. The mechanical view of impulsions, impetus, fortuitous assemblies of flying particles, 'like one that presseth another in a strait', the motion and fall of bodies, never ceased to dominate European thinking from the time of Democritus onwards. But for a wave-oriented civilisation, the chief thing would be the alternation of phases in a continuum, and individual bodies in motion would tend not to be visualised so clearly as large manageable particles, 'macro-atoms', separated in thought, at least, from all the rest of the surrounding world of matter. And here we note with special interest that it was precisely the Mohists, and only they, with their distinct leanings towards atomism (cf. their definitions of points and instants, and their notion of imperfect continuity as the reason why fibres broke in tension), who conceived of the motion of bodies in an abstract way.

The absence of explicit dynamics from Chinese physics is the more extraordinary, however, because of two other considerations. If it be true that the thinking of the Chinese was always in terms of continuity, and not particulate,[a] it follows that the idea of action at a distance can never have been difficult for them. We have seen this already in their early appreciation of the true cause of the tides, and of such phenomena as acoustic resonance.[b] Still more striking was the Chinese familiarity with action at a distance in magnetic fields, since they knew the directive, as well as the attractive, property of the lodestone and magnet much earlier than Europe.[c] Yet the inability to conceive of action at a distance was one of the factors which, in the West, long delayed an appreciation of the true nature of gravity. Among the scholastics, only William of Ockham and Thomas Aquinas admitted the possibility of action at a distance; all the rest regarded gravity as quite different from magnetic attraction, in that it did not, as they thought, die away at a certain distance. Duhem (1) has seen in what he called the 'philosophie aimantiste' a necessary preparation for Newton.[d] For example, Fracastorius about +1545 thought that when two parts of one whole are separated, each emits a 'species' which fills the intervening space.[e] Then in the following century doctrines of 'effluvia', in such writers as Francis Bacon, Walter Charleton, and the van Helmonts, paved the way for the vast generalisations at the end of the century on the gravitational field. Conversely, Chinese thought on continuity, so much older, should have been favourable to the discovery of the laws of motion and of gravity, if it had been possible in China to consider the motion of individual bodies at all. But apparently it was only in Europe that the Chinese knowledge of the field physics of the magnet could exert its immense revolutionary influence.

The second consideration concerns the relative valuation of rest and motion. The concept of inertial movement, as Koyré (3) has said, is self-evident to us, yet for the

---

[a] Martin (5) is worth reading on this.
[b] Vol. 1, p. 233; Vol. 3, pp. 584ff.; also, pp. 130, 185ff. below.
[c] Cf. the Section on fundamental ideas, Vol. 2, p. 293 above; and below, pp. 332 ff.
[d] See Butterfield (1), pp. 56, 79, 126 ff.; Hall & Hall (1); Boas & Hall (1).
[e] *De Sympathia et Antipathia Rerum.* Cf. Hesse (1, 3), and the general argument in Ho & Needham (2).

Greeks and the people of the occidental Middle Ages, it was an absurdity.[a] He suggests that this was because the Greeks thought rest intrinsically 'superior' to motion. They could not think of both motion and rest as states on the same ontological level. Being, pure existence, pure contemplation, the state of the Deus Otiosus, was naturally superior to motion. Of course this may be said to have been a valuation of human social classes elevated into a philosophical principle. But in any case the Chinese did not participate in this valuation. No doubt in their culture also the lords and scholars sat still, while the slaves and the people moved about at their work—yet certain healthier factors, hard to define with precision,[b] influenced Chinese thought in such a way that motion was regarded as, if anything, superior to rest.[c] The sage, like Heaven, should never rest. At the very opening of the 'Book of Changes' we read[d] *Thien hsing chien, chün-tzu i tzu chhiang pu hsi*[1] (The movement of Heaven is full of power; so also the *chün-tzu* is strong and unresting). Confucius, like Heraclitus, often stood beside a running stream; 'How it flows on,' he said, 'never ceasing by day or night'.[e] But the Confucians drew from this no pessimistic philosophical conclusions; they simply taught that it symbolised the unresting efforts of the sages to achieve perfection. So also the first attribute of the cosmological principle of dynamic fittingness, *chhêng*[2] (discussed in Sect. 16*d* above), was unrestingness (*wu hsi*[3]).[f]

Then in the Section on Taoism we saw in a multitude of quotations from the great masters of the school, such as Chuang Chou, their conviction that the Tao, the Order of Nature, was a principle of ceaseless motion, change and return. It did not conflict with the great principle[g] of *wu wei*,[4] since, properly interpreted, this was a concept not of non-action, but of no action contrary to Nature. Wang Pi, it is true, spoke of the 'impossibility of motion controlling motion',[h] and used phraseology reminiscent of the Aristotelian 'unmoved mover', but he also said[i] 'The cessation of activity always means quiescence, but this is not something opposed to activity'. Wang Pi thought of the Tao as the ground or root of activity not in a metaphysical sense, but as a kind of field of force including all fields of force with the motions generated by

[a] Even medieval 'impetus' was thought of as a force immanent in the moving object. The necessity of this was just what the Renaissance physicists denied.

[b] One cannot help making a correlation here between this mentality and the outstanding industriousness of the Chinese people, itself in its turn surely connected with the garden-character of their agriculture.

[c] The Yang was always equated with movement and the Yin with quiescence; this, as Fêng Yu-Lan (1), vol. 2, p. 96, points out, vitiates any analogy between these two concepts and the Pythagorean 'Limited' and 'Unlimited'.

[d] *I Ching*; the Hsiang (App. II) to the first *kua*, Chhien (R. Wilhelm (2), vol. 3, p. 4).

[e] *Lun Yü*, IX, xvi. Cf. *Mêng Tzu*, IV (2), xviii. As Chou I-Chhing has said: 'Toute la sagesse chinoise est marquée par cette tendance; elle n'est pas, à l'image de la pensée occidentale, philosophie de l'Être, mais de l'Agir', (1), p. 78.

[f] *Chung Yung*, ch. 26.

[g] Vol. 2, pp. 68 ff.

[h] A quotation from him on this subject has been given above, Vol. 2, p. 322.

[i] Commentary on the *I Ching*, *re* the 24th kua, *fu*,[5] reversion.

[1] 天行健君子以自彊不息　　[2] 誠　　[3] 無息　　[4] 無爲　　[5] 復

them. This was a silent and supreme nothingness. Sometimes he called it 'the mind of Heaven and Earth'.[a]

To all this we might add a famous story about the Chin general Thao Khan[1] (+259 to +334) told in the *Chin Shu*.[b] In one of his official posts, 'He often found himself with no business at all to be done. So every morning he carried out a hundred large earthenware jars from his study, and in the evening brought them in again. When people asked him why he did this, he said that he was devoted to the guard of the district assigned to him, and that he was afraid of sinking into inactivity. This was his method of strengthening his will.'

Such was the unrestingness of the sage, seen through the eyes of a simple-minded Chinese officer. Once more, if other factors had permitted such considerations, would not this element in Chinese thought have been favourable to the idea of the effortless unrestingness of material particles in motion?

(e) SURFACE PHENOMENA

The transition from the study of motion to that of heat should have come about through the study of friction. But in spite of the relatively high level of practical technique, Chinese literature seems to lack theoretical disquisitions on the subject. The production of fire by rubbing pieces of wood was of course anciently known,[c] and is referred to from time to time in books such as the *Huai Nan Tzu*; there is also the note on the friction of wheels in the *Chou Li*. But as in Europe, the physics of rubbing solids had to await post-Renaissance times.

Above (p. 38) we saw some interest taken, in the Thang period, in the goodness of fit of smooth surfaces, by the prince Li Kao and others. Two further indications may be given of the study of liquid surfaces. In the +13th century Chou Mi noted some properties of what we should now call monomolecular films.[d]

'Bear alum' (*hsiung fan*[2]) (presumably bear-fat mixed with alum) can disperse dust. It can be tested as follows. A vessel of clean water is scattered over with dust. When a grain of this alum is then put in, the collected dust opens (becomes clear). The preparation has proved successful against cloudiness of the eyes.

Here the fat was forming a film and acting as what would now be called piston-oil.[e] This was an anticipatory hint of the work of Pockels (1891) on which most of our knowledge of films is founded. It was Rayleigh who took up her work and first

---

[a] The discussion of Fêng Yu-Lan (1), vol. 2, pp. 180, 181, should be read in this connection, but as usual (in my opinion) he interpreted Wang Pi in too metaphysical a sense.

[b] Ch. 66, p. 7a; tr. auct.

[c] Cf. the reference in the fragments of Tsou Yen (Vol. 2, p. 236 above).

[d] *Chhi Tung Yeh Yü*, ch. 4, p. 14b. Tr. auct.

[e] There is a reference to the use of grease for the floating of a needle on water in the *Huai Nan Wan Pi Shu* (see on, p. 277), which may well date from the −2nd century.

[1] 陶侃   [2] 熊礬

suggested that the films were truly monomolecular.[a] In another Sung book, the *Yü Huan Chi Wên*[1] (Things Seen and Heard on my Official Travels), Chang Shih-Nan[2] says:[b]

> The following is the method of testing the quality of lacquer. Good lacquer is as clear as a mirror. A hanging thread of it looks like a fish-hook. On shaking, its colour is seen to be like amber. If stirred, it makes floating bubbles. For testing the quality of lacquer or tung-oil, they make a ring of a small sliver of bamboo, and dip it into the liquid. If upon being withdrawn a film is formed across the ring, the lacquer or tung-oil is good; if they have been adulterated no ring-film will be formed.

This ring-test must have been used qualitatively for centuries. In 1878 Sondhauss introduced it as a quantitative method for determining surface-tension by finding the force necessary to break the surface. The ancestor of the apparatus of du Nouy was thus already empirically known and used around +1230.

(f) HEAT AND COMBUSTION

In the West the understanding of heat was one of the last achievements in the elaboration of the general structure of physics as we know it. In antiquity and the Middle Ages the necessary conceptions and definitions were, as Heller says,[c] completely lacking; in spite of the rich store of practical information about expansion and contraction, change of state, vaporisation and solidification, with many similar such processes, accumulated in the trades and industries. One has only to read the detailed story of the invention of the thermometer,[d] or consider the perplexities about heat (the 'caloric' fluid) which persisted into the nineteenth century,[e] to realise that thermal science, based on the kinetic theory,[f] is one of the relatively youngest parts of modern physics. In these circumstances it is perhaps not surprising that little can be found on heat as such in ancient and medieval Chinese writings.[g] But like their occidental counterparts[h] they yield many interesting examples of the observation and use of thermal phenomena in the technological field.

Theoretical arguments about heat, dryness, etc. in old China were quite in the style of the Aristotelians or the technicians like Leonardo,[i] who would have been at home

---

[a] Cf. N. K. Adam (1).
[b] Ch. 2, p. 6*b*; tr. auct.
[c] (1), vol. 1, pp. 153, 393. Cf. Gerland (1); Gerland & Traumüller (1), pp. 312ff.
[d] Sherwood Taylor (1); Gerhard & Traumüller (1), p. 166.
[e] Lilley (2).
[f] Cf. Dampier-Whetham (1), pp. 248ff.
[g] Wu Nan-Hsün (*1*) has discussed (pp. 153ff.) the conceptions of the nature of heat entertained by the great alchemist Ko Hung early in the +4th century.
[h] In this field Chinese and Western conceptions were more on a par than in that of dynamics, where, as we saw, the European Middle Ages made substantial advances.
[i] Cf. Vol. 3, p. 160.

[1] 游宦紀聞 　　[2] 張世南

with them. In Han texts (*c.* −2nd century) such as the cosmological chapters of the *Huang Ti Nei Ching Su Wên*[a] we may read as follows:

Huang Ti said, 'What about the floating (*phing*[1]) of the earth?'
> (Comm.[b] It was said that in the great emptiness there was no barrier to the earth's body; how then could it float where it does?)

Chhi-Po replied, 'The great *chhi* (*pneuma*) keeps it raised aloft. Dryness (*sao*[2]) hardens it, heat (*shu*[3]) steams (evaporates) it, wind moves it, damp (*shih*[4]) soaks it, cold hardens it, and fire warms it. Thus the wind and the cold are below, the dryness and the heat are above, and the damp *chhi* is in the middle, while fire wanders and moves between. Thus there are six 'entries' (*ju*[5])[c] which bring things into visibility out of the emptiness and make them undergo change. When dryness dominates the earth becomes hard, when heat is in the ascendant it becomes hot, when wind arises it is made to move, when dampness floods forth it is turned to mud, when cold reigns it is split and cracked, and when fire governs all it becomes solid and compacted.'[d]

All this is pre-Socratic enough, but such vague ideas were not decisively superseded until the time of the Renaissance in Europe, and modern (or relatively modern) ideas of thermal phenomena were therefore among the more valuable gifts brought by the Jesuits to China early in the 17th century. We have already seen the picture of the first thermometer there.[e]

Just before this time the traditional ideas on fire and flame were conveniently summarised by the great naturalist Li Shih-Chen in his *Pên Tshao Kang Mu*.[f] By +1596 generally held opinion had crystallised as follows. Fire (*huo*[6]) has *chhi*[7] (*pneuma*) but no *chih*[8] (corporeal matter). Unlike all the other elements, which are unitary, fire is of two kinds, Yin and Yang.[g] Its classification follows three 'net-cords' (*kang*[9]), i.e. heaven, earth and man; and twelve 'net-meshes' (*wang*[10]), i.e. four sorts of heavenly fire, five sorts of earthly fire, and three kinds of human fire. There were in addition four varieties of incandescence or luminescence which Li Shih-Chen could not classify.[h] His system is explained in Table 43.

When we remember that even at the beginning of the 19th century in Europe it was by no means generally conceded that heat was a form of motion of the particles of matter, and indeed more widely thought an imponderable self-repellent fluid *sui*

---

[a] The greatest of the medical classics; cf. Sect. 44 below.

[b] By Wang Ping of the Thang (+8th century).

[c] Or six *chhi*, according to the commentator. This technical term no doubt derived from the conception that the visible differentiation of the mundane world was brought about by the incoming of various forms of cosmic *chhi*. If this idea paralleled to some extent medieval Western notions of starry and planetary influences, it was also not so far removed from present-day conceptions of the earth's receipt of radiant energy from the sun and from cosmic space.

[d] *Pu Chu* ed., ch. 67, pp. 1aff., tr. auct.   [e] See Vol. 3, p. 466.

[f] Ch. 6, studied especially by de Visser (1).

[g] Reminiscent, as always, of positive and negative electricity (cf. Vol. 2, p. 278). But this particular theory about fire differed from most of the older views. See e.g. Vol. 2, pp. 263, 463.

[h] He says they were not really 'burning' at all.

1 馮   2 燥   3 暑   4 溼   5 入   6 火   7 氣
8 質   9 綱   10 网

Table 43. *Li Shih-Chen's classification of the varieties of fire*

	FIRE	
	YANG	YIN
Heavenly	(1) *Thai yang chen huo*[1] (heat of the sun) (2) *Hsing ching fei huo*[2] (light of stars and meteors)	(1) *Lung huo*[3] (dragon fire) (2) *Lei huo*[4] (lightning)
Earthly	(1) *Tsuan mu chih huo*[5] (fire and heat produced by drilling wood) (2) *Chi shih chih huo*[6] (sparks produced by striking stone) (3) *Ka chin chih huo*[7] (sparks produced by tapping metal)	(1) *Shih yü chih huo*[8] (fire from burning petroleum) (2) *Shui chung chih huo*[9] (ignis fatuus; burning methane?)
Human	(1) *Ping ting chün huo*[10] (general metabolic heat)[a]	(1) *Ming mên hsiang huo*[11] (heat of the viscera and generative organs)[a] (2) *San mei chih huo*[12] ('samādhi' or meditation heat)
Unclassified	(1) *Hsiao chhiu huo*[13] (= *han huo*;[14] cold heat; flame of natural gas) (2) *Tsê chung chih yang yen*[15] (phenomena like ignis fatuus; burning methane?) (3) *yeh wai chih kuei lin* (see p. 72 below) (4) *chin yin chih ching chhi*[16] (glitter of gold, silver and gems)	

generis, namely caloric; then the efforts of Li Shih-Chen to classify different forms of flame and combustion do not seem so archaic as might appear at first sight. Let us give them a brief commentary.

To begin with the physiological, it may be that Li Shih-Chen meant to make a distinction between the essential heat-production of the body and the heat connected with muscular movement, foreshadowing our categories of basal metabolic rate and total heat-output.[b] The 'meditation'-fire was a notion probably derived from the

[a] For discussion of these terms see Sect. 43 below.
[b] Cf. 17th-century distinctions between 'natural' and 'adventitious' heat, as in Sanctorius.

[1] 太陽眞火　　[2] 星精飛火　　[3] 龍火　　[4] 雷火　　[5] 鑽木之火
[6] 擊石之火　　[7] 戛金之火　　[8] 石油之火　　　　　　[9] 水中之火
[10] 丙丁君火　　[11] 命門相火　　[12] 三昧之火　　　　　　[13] 蕭丘火
[14] 寒火　　[15] 澤中之陽燄　　[16] 金銀之精氣

yogistic feat, well enough attested, of maintaining a high body-temperature under continued exposure to cold.[a] What Li Shih-Chen meant by 'dragon'-fire is not very obvious, perhaps lightning which may strike and melt without burning or singeing anything.[b] In his time there was little point in separating the different forms of heat and light produced by friction and impact; the fundamental study of the former was something reserved for later generations. It is curious that he did not know where to classify the 'cool' flames of natural gas,[c] a source of heat which the Chinese perhaps of all peoples were the first to make use of on an industrial scale. We must not here anticipate what will be said of this subject in relation to brine-fields and the deep drilling of boreholes,[d] but it is most probable that the systematic use of natural gas for evaporating salt started in the Chhin and early Han. From the +2nd century onwards the references are numerous and informative,[e] but since other texts describe brine wells and boreholes from the −4th century onwards, mainly in Szechuan, and since it is unlikely that the striking of natural gas there would have been much delayed, the −2nd century may be considered a safe estimate. Lastly, Li Shih-Chen did not realise that oils are 'inextinguishable' because they continue to burn when floating. So he thought that this kind of flame must have some especially Yin quality[f] which enabled it to overcome the normally quenching effects of water. But he knew the phenomenon of the flash-point, for he says that 'if strong alcohol (*nung chiu ching*[1]), or oil, is heated, it will burst into flame of itself'.

Spontaneous combustion of substances at lower external temperatures was a matter of interest in China for many centuries. Chang Hua says in his *Po Wu Chih*, towards the end of the +3rd century:[g]

If a full hundred catties of oil are accumulated in store, the oil will ignite itself spontaneously. The calamitous fire which occurred in the arsenal in the time of the emperor Wu[h] (of the Chin dynasty) in the Thai-Shih reign-period (+265 to +274) was caused by the stored oil.

This explanation could not be correct quite as it stands. More light is thrown upon the matter, however, by an interesting story reported in the +13th century by the jurist Kuei Wan-Jung[2] in his *Thang Yin Pi Shih*,[3] a work the title of which might be translated 'Parallel Cases solved by Eminent Judges'.[i]

---

[a] Cf. Vol. 2, p. 144 above.

[b] Cf. the careful description of lightning effects by Shen Kua and others; Vol. 3, p. 482.

[c] Cf. the observations of Ko Hung; Vol. 2, p. 438.

[d] Sect. 37 below.

[e] For example, first the *Po Wu Chi* (+190) (*YHSF*, ch. 73, p. 4a) quoted in *Hou Han Shu* (+450), ch. 33, p. 3a (*TPYL*, ch. 869, p. 6a); then two mentions in the *Po Wu Chih* (+290), ch. 2, p. 7a (*TPYL*, ch. 869, p. 6a); then *Hua Yang Kuo Chih* (+347), ch. 3, p. 7b; also *Ku Chin Chu* and *Shih I Chi*, both of the +4th century.

[f] Like all the other categories on the right-hand side of Table 43.

[g] Ch. 4, p. 3a, tr. auct. adjuv. van Gulik (6). Cit. *TPYL*, ch. 869, p. 6a, hence Pfizmaier (98), p. 17.

[h] Reigned +265 to +290.

[i] See the full translation with elaborate notes by van Gulik (6).

1 濃酒精   2 桂萬榮   3 棠陰比事

When the Director of the Sacrifices Department (of the Ministry of Rites) Chhiang Chih[1] was serving as Intendant of the Imperial Guards at the palace at Khaifêng, oiled curtains[a] had been left piled up in the open air, and one night they caught fire. According to the law the men responsible for looking after them all incurred the penalty of death. But at the preliminary hearing of the case Chhiang Chih conceived doubts about the cause of the fire, so he summoned the workmen who had made the curtains and questioned them. These artisans said that during the manufacture of the curtains a certain chemical was added (to the oil), and that if they were left for a long time piled up, then on getting damp they might start burning. When Chhiang Chih reported this to the emperor Jen Tsung[b] the emperor was suddenly struck with an idea and said, 'The fire which recently occurred in the mausoleum of the emperor Chen Tsung[c] started in oiled garments. So that was the cause!' The guards were let off with a lighter punishment.

Chang Hua thought that the fire which occurred formerly during the (Western) Chin dynasty in the arsenal originated from the oil which was stored there, but in fact it must have been from the same cause as mentioned here (the spontaneous ignition of oiled cloth).[d]

Chhiang Chih was abundantly right in believing that the combustion was spontaneous, but the workmen may or may not have been right about the cause of it. Certain kinds of oiled cloths are liable to ignite spontaneously when packed in layers, without the action of any additional chemical. Processes of autoxidation, desaturation, aldehyde formation, etc. may give out so much heat as to bring the surrounding oil to ignition point, after which the cellulose of the textile fibres affords ample fuel to burn in the air available between the layers. This was studied in the 18th century, notably by Duhamel (1). For example, in +1757 (the year of his paper) canvas sails treated with ochre in oil and quickly dried in a July sun were found to be burning at the centre of a stack a few hours after storage. These same sail-lofts at Rochefort had several times before experienced such conflagrations. In +1725 piles of serges and woollen cloth stacked before de-greasing and fulling had ignited themselves in the same way. Twenty years after Duhamel's publication, says Thomson,[e] two accidental spontaneous combustions in Russia were ascribed to treason, but the empress Catherine II suspected their true origin and had experiments made which fully confirmed the findings of Duhamel. This property of oiled cloth has now long been well known in technical circles, and instructions against piling it up are issued to all army and navy storekeepers.

On the other hand, the workmen may have been right that some chemical had been added to the oil as applied to the curtains. If so, the only substance probable would have been calcium oxide (quicklime), perhaps used as a whitening agent, perhaps even substituted sometimes for other white powders. Here we have to do with nothing else than one of the compositions formerly believed to have been that of 'Greek Fire'. Although the feasibility of igniting oils such as petroleum or naphtha by the heat

---

[a] Probably tents.
[b] Reigned +1023 to +1063. The incident would have been somewhere about +1050.
[c] Reigned +998 to +1022.
[d] Ch. 1, no. 7B, tr. van Gulik (6), p. 84.   [e] (1), vol. 1, p. 293.

[1] 強至

generated by quicklime on contact with water has often been denied, it seems, never-theless, from a number of investigations[a] that the process is quite possible under favourable conditions. But it would not satisfy the extant descriptions of Greek Fire, and it is much more probable that this consisted of distilled light petroleum fractions ejected from flamethrowers by pumps.[b] In any case, the presence of quicklime in the curtains with the oil would be quite a plausible interpretation of the workmen's claim that dampness would lead to spontaneous ignition. No one in the +11th century except perhaps a few Taoist alchemists would have known enough about the behaviour of chemical substances to be able to distinguish between two phenomena so apparently similar. As for the arsenal fire in the +3rd century, Kuei Wan-Jung (or his com-mentator) was right enough in explaining it as probably due to oiled cloth. Even the drying oils should not give rise to spontaneous combustion if stored in such a way as to protect them from accidental ignition. Or perhaps somebody in the Chin arsenal was 'looking for a gas leak with a lighted candle', as modern idiom might say.

The *Po Wu Chih* has another interesting passage on oil which deals with the pheno-mena which occur when it is heated; this is contained in a small section entitled Wu Li,[1] 'On the natures of things'.[c]

When hempseed oil is heated, after the steam has been given off, no more smoke is to be seen, and it no longer boils. (It seems to have) returned to a cold condition. Indeed, it can be stirred with a finger. (But afterwards) if water be added, the oil bursts into flame, and this cannot be put out even if it is scattered. These things have been tested and proved to be true.

These observations are not bad. In the first phase all the water has been driven off as steam and the boiling-point of the oil not yet reached. If sufficiently wetted the finger would not come into contact with the oil, protection being given by a layer of steam. Then when the oil is near its boiling-point, added water will cause violent disturbance and lead to the ignition of drops thrown out. But the observation of ebullition phenomena in old China and Japan may have led to something much more subtle, namely modern knowledge of the successive phases of boiling in all fluids.

It is now known that these are three in number.[d] In nuclear boiling large bubbles form at active points on the hot metal surface. When the temperature difference rises slightly beyond a critical value, a dramatic change occurs, the boiling becomes notice-ably louder and the rate of heat-transfer falls. The hot surface is now covered with a thin layer of vapour which acts as an insulating blanket, and within which innumer-able small explosions occur. This is known as transition boiling. Eventually the vapour layer becomes so thick that it smothers all these explosions, and the third phase is

[a] E.g. Richardson (1).

[b] Such is the conclusion of Professor J. R. Partington, to whom we are greatly indebted for much of the information contained in this and the preceding paragraph. The whole question of incendiary weapons and other precursors of gunpowder both in China and the West will be fully dealt with in Sect. 30 below.   [c] Ch. 4, p. 2b, tr. auct.

[d] See Westwater (1), and more extensively Jakob (1) or McAdams (1).

[1] 物理

reached, stable film boiling; now heat transfer and vapour generation fall to their lowest levels. A continuous drumming noise rumbles from the liquid. When a red-hot metal is quenched in a liquid all three types of boiling occur in the reverse order as the metal cools first slowly, then rapidly, then slowly again. The most rapid phase of heat loss corresponds with nuclear boiling, after which the last cooling is by natural convection.

The discovery and elucidation of the three phases began with work done in 1934 by Nukiyama Shirō in Japan. He it was who found the paradoxical and challenging effect which occurs when platinum wire is heated in water; it will cause boiling at less than 150° and above 300°, but will not accept a temperature between these limits. Drews (1) pointed out that the rise of such studies in Japan was probably no coincidence, for in all East Asian culture during many centuries great attention was paid to the water used for tea-making, and nowhere more than in the Japanese tea-ceremony. For this, says Brinkley (1):

a brisk fire should be used. The water gives the first indication of heat by a low intermittent singing and by the appearance of large slowly rising bubbles known as 'fish eyes'. The next stage is marked by agitation like the seething of a hot spring, accompanied by a constant succession of rapidly ascending bubbles. In the third stage waves appear upon the surface, and these finally subsiding, all appearance of steam is lost. The water has now attained the condition of maturity; it is 'aged hot water'. If the fire is good and well sustained, all these stages can be distinctly noted.[a]

Thus the tea-kettle which accompanies the Taoist sage or naturalist in so many pictures by Chinese artists acquires a new meaning for us.

A curious passage about steam occurs in the *Huai Nan Wan Pi Shu*[1] (The Ten Thousand Infallible Arts of the Prince of Huai-Nan). This no longer exists as a separate book, but some 116 semi-magical recipes from it remain as fragments in e.g. the *Thai-Phing Yü Lan*.[b] While we have no certain evidence that it goes back to the time of Liu An himself (−2nd century), its archaic character compels its acceptance as a text of the Han period. The eighth recipe runs as follows:

To make a sound like thunder in a copper vessel (*thung yung*[2]). Put boiling water into such a vessel and then sink it into a well. It will make a noise which can be heard several tens of *li* away.[c]

[a] It would be interesting to trace the previous history of such observations in Chinese culture itself. Fortune (1), vol. 2, p. 230, quoted a so-far unidentified passage from Su Tung-Pho: 'When tea is to be made, water should be taken from a running stream and boiled over a lively fire. This is an old custom. Water from springs in the hills is said to be the best, then river water, while well water is the worst. A lively fire means a clear and bright charcoal fire. The water should not be boiled too hastily. First it begins to sparkle like crabs' eyes, then bubbles somewhat like fish eyes appear, and finally it boils up like pearls innumerable, springing and waving about. This is the way to boil the water.'
[b] Chs. 736, 758, 988, 993, etc.
[c] Ch. 736, p. 8b, tr. auct. Ch. 758, p. 3b, under *yung*, repeats the passage, adding that the orifice of the vessel has to be closed extremely tightly.

[1] 淮南萬畢術   [2] 銅甕

If it is assumed that the vessel was full of steam when the lid was tightly closed, then the procedure would seem to be an anticipation of the principle of causing a vacuum by suddenly cooling a vessel full of steam and thereby condensing it, as in the earliest forms of the steam-engine.[a] If the vessel were thin-walled there could be an implosion, and the noise of the crumpling of the container would be intensified by the echoing of the sides of the well. It is astonishing that a manifestation of that power which became available in 17th-century Europe for raising water and for so many other purposes should first have been described as far back as the −2nd century in China though only for military or thaumaturgical purposes.[b] Presently we shall show that the steam-jet also was known in that culture-area as well as in Alexandria, though used for a different purpose, and we shall tell further of the coming to China of the steam-turbine in the +17th and the steam-engine proper in the 19th centuries.[c]

Here a word or two on fire-making itself may not be out of place.[d] Ancient Chinese literature naturally contains references to the prehistoric and primitive methods of igniting tinder by the heat of pieces of wood rubbed together—we have seen an example among the fragments of Tsou Yen (−4th century).[e] For many centuries, down indeed to recent times, the flint and steel replaced the fire-drill, just as among our own ancestors.[f] Later on at the appropriate place there will be something to say of the fire-piston, never common in China proper though perhaps of much importance for Chinese technology.[g] But even less known than any of these is the priority which China seems to have in the invention of the sulphur match. In his *Chhing I Lu*[1] (Records of the Unworldly and the Strange), written about +950, Thao Ku[2] tells us about matches.[h]

If there occurs an emergency at night it may take some time to make a light to light a lamp. But an ingenious man devised the system of impregnating little sticks of pinewood with sulphur and storing them ready for use. At the slightest touch of fire they burst into flame. One gets a little flame like an ear of corn. This marvellous thing was formerly called a 'light-bringing slave' (*yin kuang nu*[3]), but afterwards when it became an article of commerce its name was changed to 'fire inch-stick' (*huo tshun*[4]).

[a] It will be remembered that the condensation vacuum was first applied to the raising of water by the Marquis of Worcester (+1630 to +1670) (Dircks, 1) and Thomas Savery (+1698). These machines contained no pistons. The piston was the introduction of Thomas Newcomen (+1712); James Watt's essential improvement later was the introduction of a separate chamber for the condensation, and the admission of the steam on both sides of the piston alternately, exactly as air had been in the Chinese double-acting piston bellows (see Sect. 27*b* below). On the whole subject of the development of the steam-engine cf. Usher (1), 2nd ed. pp. 342 ff.; Dickinson (4) and Triewald (1).

[b] For my first introduction to this interesting passage some twenty years ago I am indebted to Professor Huang Tzu-Chhing. Wu Nan-Hsün also discusses it, (1), p. 201.

[c] Sect. 27 below, in Vol. 4, pt. 2.

[d] Among many general discussions see, e.g. Mason (2), pp. 88 ff.; Hough (1), pp. 84 ff.

[e] Above, Vol. 2, p. 236. Cf. *TPYL*, ch. 869, p. 6*a*, quoting *Po Wu Chih*.

[f] Yet another method involved the use of burning-mirrors and burning-glasses. On these much will be said below in Sect. 26*g*, pp. 87 ff.

[g] See below, Sect. 27*b*, in Vol. 4, pt. 2.

[h] Ch. 2, p. 28*b*, tr. auct. We are much indebted to Dr Werner Eichhorn for telling us of this passage, and indeed for bringing the whole subject to our notice.

[1] 清異錄   [2] 陶穀   [3] 引光奴   [4] 火寸

An enlarged form of the same note occurs in the *Cho Kêng Lu*[1] of Thao Tsung-I,[2] written in +1366, who adds the tradition that the invention was not made by the people of Hangchow, as generally assumed, but by the impoverished court ladies of the Northern Chhi in +577 at the time of the conquest of the empire by the Sui.[a] However, sulphur matches were certainly sold in the markets of Hangchow when Marco Polo was there as they are mentioned in a list of wares in the *Wu Lin Chiu Shih*[3] (c. +1270).[b] Here they are called *fa chu*[4] or *tshui erh*.[5]

Information of this kind can never be placed in its proper perspective without a knowledge of comparative developments in other parts of the world. It is thus striking to find that no positive evidence of sulphur matches in Europe exists before +1530.[c] They continued in use throughout the 18th century, but after about +1780 phosphorus devices in various forms began to replace them. For instance, the 'philosophical bottles' were small vessels containing partly oxidised yellow phosphorus and kept tightly corked; when a light was wanted a sulphur match was pushed in, turned round and quickly drawn out, igniting on contact with the air. In 1805 Chancel introduced the plan of dipping sticks impregnated with potassium chlorate and sugar into phials of strong sulphuric acid, thus giving an automatic ignition, but the most important advance was made just after 1830 when Sauria in France and Kammerer in Germany[d] employed a mixture of yellow phosphorus, sulphur and potassium chlorate. Sulphur as a constituent of matches has thus continued, through many vicissitudes, into modern times, and it is interesting that the first location of its use seems to have been in +6th-century China.

(i) DIGRESSION ON LUMINESCENCE

The study of thermal changes, with their accompanying phenomena of incandescence, was for many centuries bedevilled by a crowd of miscellaneous observations most of which had nothing to do with heat at all. The effects seen are classified today as the various types of luminescence.[e] The emission of light from substances only during the time when they are exposed to different kinds of radiation[f] is known as fluorescence; if it persists after the exciting radiation is cut off we may call it phosphorescence.[g] A delayed phosphorescence may appear on heating (thermoluminescence). Lights accompanying electric discharges are termed electroluminescence;

---

[a] Cf. Vol. 1, p. 122.
[b] By Chou Mi.[6] The reference is to the Ming edition, ch. 8, p. 25b.
[c] Cf. Sherwood Taylor (4), pp. 203 ff.
[d] Cf. Niemann (1).
[e] The current classification stems from the work of Wiedemann (16) in 1888. Within the past few years we have been able to profit by an elaborate treatise on the history of man's knowledge of luminescence phenomena, by Newton Harvey (1).
[f] For instance, visible light, ultraviolet light, X-rays, gamma rays, cathode rays (electrons), and many other types of particles.
[g] Materials exhibiting phosphorescence are known as phosphors; see p. 76 below.

[1] 輟耕錄　　[2] 陶宗義　　[3] 武林舊事　　[4] 發燭　　[5] 焠兒
[6] 周密

they may appear in the air as well as in vacuum tubes, and the aurora borealis itself[a] has often been placed in this category. The nature of triboluminescence and piezo-luminescence is not perfectly understood; these are emissions of light brought about by friction, especially the crushing or rubbing of crystals, and may be either an electron bombardment of surfaces or electroluminescence resulting from their forcible separation. Light given forth during certain chemical reactions, especially notable where the element phosphorus is involved,[b] is naturally known as chemiluminescence; and all the phenomena of 'cold' biological light,[c] visible radiations produced by living organisms, though called bioluminescence, are really a variety of chemiluminescence. What did Chinese writers[d] have to say of these phenomena?

Let us return to the classification of Li Shih-Chen in +1596. He was baffled by the marsh-lights or ignis fatuus, but so are the most modern investigators,[e] and we are still not sure whether to ascribe the famous 'will-o'-the-wisp' to burning methane,[f] spontaneously inflammable organo-metallic compounds,[g] phosphine (phosphorus trihydride) or alkyl phosphides,[h] or to electrical discharges. Li Shih-Chen called it the 'Yang flames of marshes', but he also had another name, the 'devil-lights of the outer wildernesses' (*yeh wai chih kuei lin*[1]). There are many references in Chinese literature to the ignis fatuus, e.g. the 'water lanterns' (*shui têng*[2]) mentioned in the *Kuei Hsin Tsa Chih* of Chou Mi (late +13th century), and other lights on water described in the *Shan Thang Ssu Khao*[3] (Books Seen in the Mountain Hall Library) by Phêng Ta-I[4] (+1595). Collections of texts have been made,[i] and it is needless to recapitulate them here.

At the same time ancient observations had accumulated on the bioluminescence of putrefying substances, due, as we now know, to luminous bacteria and fungi.[j] As the living origin of this light was not known, its manifestations were naturally confused with the marsh-lights, and the idea grew up that the ignis fatuus was derived from old blood. The *Chuang Tzu* text at one time contained the statement that 'the blood of horses becomes marsh-light (*lin*[5]) and that of men becomes wildfire (*yeh huo*[6])'.[k]

[a] See Sect. 21*h*, Vol. 3, p. 482.

[b] Here, of course, the great advance was made by Robert Boyle in his work (1680) on the 'Aerial' and the 'Icy' Noctiluca. Cf. Wolf (1), p. 349; Newton Harvey (1), pp. 427ff.

[c] See Newton Harvey (2).

[d] For a parallel study of Arabic knowledge see Wiedemann (5).

[e] See Newton Harvey (1), pp. 263ff.

[f] As suggested in a well-known letter from Volta to Priestley (10 December 1776), cf. Newton Harvey (1), p. 265.

[g] R. E. D. Clark (1).

[h] Karrer (1), p. 125. Many of the alkyl phosphides inflame spontaneously on contact with air and oxidise to the corresponding phosphonic acids. But neither they nor phosphine itself have yet been shown to arise as a product of organic decomposition.

[i] E.g. de Groot (2), vol. 4, pp. 80ff.; de Visser (1), pp. 162ff.

[j] In the European medieval and later periods there were many reports of shining fish, flesh and wood (Newton Harvey (1), pp. 461ff.). The true cause was not revealed until the work of the Coopers in 1838 and of J. F. Heller in 1843.

[k] So *TPYL*, ch. 869, p. 2*b*, but the Harvard-Yenching index does not locate it.

[1] 野外之鬼燐   [2] 水灯   [3] 山堂肆考   [4] 彭大翼   [5] 燐
[6] 野火

The *Huai Nan Tzu* says[a] that 'old pieces of *Sophora* wood (*huai*[1]) shine like fire, and old blood turns into marsh-light (*lin*[2])'. Chang Shih,[3] one of the Neo-Confucian philosophers,[b] says that he saw an old battlefield studded with lights,[c] and Lu Yu[4] relates similar personal experiences.[d] The idea that the ignis fatuus derived from old decaying blood arose partly no doubt from the thought that 'the life is in the blood', a notion common to Sung philosophers[e] such as Huang Kan[5] or Chen Tê-Hsiu[6] and to men of the Renaissance such as William Harvey.[f] For example, it was a commonplace to say that 'the essence is (in) the blood' (*hsüeh ching yeh*[7]), or 'the blood is the abode of the spirit' (*hsüeh shen chih shê*[8]). As the *shen*, or upward-striving soul (*hun*[9]) was a Yang thing, it would naturally turn into fire, while the complementary *kuei*[10] and *pho*[11] souls would eventually descend to water, but this would be invisible.[g]

An account of the glowing of cast insect or reptile skins, probably due to luminous bacteria, came to my notice in the *Chhun Chu Chi Wên*[12] (Record of Things heard at Spring Island), a book containing many things in natural history and alchemy written by Ho Wei[13] (*fl. c.* +1095). He says:[h]

When Chang Tsê[14] of Hêng-hai, the governor of Chhing-chhih, was returning one night from his office to (his residence at) Tung-chhêng near Yün-chow, there was no moon and it was so dark that one could not easily see one's way, and might easily wander off the path. Suddenly he saw a bright light shining like a candle among the branches of the trees. When he reached home the walls (of his garden) prevented him from seeing it further. But next morning he went back to the place and found between the branches the exuviae of a 'dragon' (*lung shui*[15]), (in appearance) very like the shell of a newly emerged cicada (*chhan*[16]). The head, horns, claws and tail were all complete, empty within but hard outside, and when tapped made a tinkling noise like gems. It shone with a dazzling luminosity, giving out light in a darkened room as if a candle was there. So it was kept as a precious curiosity in the family.

Afterwards Shen Chung-Lao,[17] who was fond of investigating miscellaneous matters, said that in the Shao-Shêng reign-period (+1094 to +1097) when he was attending upon his elder brother in the capacity of private secretary at Chhing-chow, they found a cast skin of shape and form similar in every way to that of Mr Chang, on grape-vine trellises in front of

---

[a] Ch. 13, p. 20*a*. The statement is repeated in *Lun Hêng*, ch. 20, and (partly) in the *Huai Nan Wan Pi Shu* (*TPYL*, ch. 736, p. 8*a*). A parallel reference, poetically expressed, is in the Ai Sui ode of the Chiu Ssu sequence, written by Wang I about +120 (see *Chhu Tzhu Pu Chu*, ch. 17, p. 11*b*; tr. Hawkes (1), p. 180).

[b] The closest friend of Chu Hsi. He lived from +1133 to +1180.

[c] *Hsing Li Ta Chhüan Shu*, ch. 28.

[d] *Lao Hsüeh An Pi Chi*, ch. 4, p. 2*a*. As he was about ten years old at the time, the date would be *c.* +1135.

[e] Huang Kan lived from +1152 to +1221; Chen Tê-Hsiu from +1178 to +1235. Both have statements of this kind in *Hsing Li Ta Chhüan Shu*, ch. 28.

[f] Cf. Bayon (1), and especially Pagel (4). See *De Gen. Animalium*, Eng. ed. 1653, p. 459, e.g.

[g] See on this subject Vol. 2, p. 490.

[h] Ch. 2, p. 7*a*, tr. auct.

---

1 槐	2 燐	3 張栻	4 陸游	5 黃幹	6 眞德秀
7 血精也		8 血神之舍	9 魂	10 鬼	11 魄
12 春渚紀聞		13 何薳	14 張澤	15 龍蛻	16 蟬
17 沈中老					

the court when it was under repair—only it did not give out light. (He said that) magical dragons transform themselves into these things, so one could not expect them to be of a definite size large or small, but he did not know why some should be luminous and others not.

Doubt may perhaps remain in the mind as to whether the exuviae concerned were not those of some small lizard rather than of a snake or insect, but the description is not unclear. Whatever they were they seem to have been much too large for any of the truly bioluminescent insects, even the Chinese candle-fly *Fulgora candelaria*.[a]

The 'five-element' chapters in the dynastic histories[b] contain many records of observations of this kind. For example, in the *Sung Shu* we find the following entry:[c]

In the time of the emperor Ming of the (Liu) Sung, on a *ping-wu* day in the fifth month of the second year of the Thai-Shih reign-period (+466), a Taoist, Shêng-Tao,[1] from a mountain temple near Huang-chhêng in Lin-i district in southern Lang-hsieh, (reported that) a pillar in an apartment next to the Hall of Salvation was spontaneously shining brightly in the dark. Thus wood had lost its natural properties. Some people said that when wood goes rotten it shines of itself.

Here the chief point of interest is that the association with rotting was clearly understood though the portent was inserted officially along with others connected with aberrations of the element Wood. Elsewhere in the same history[d] we have records of flames burning in the middle of lakes in +359 and +360.

Not only were there observations on marsh-lights and the glowing of luminous bacteria and fungi, but sparks of static electricity were also noted, and the electroluminescence known as ignis lambens.[e] The *Po Wu Chih* contains an interesting passage[f] on this, dating from about +290.

In places where fights have been fought and people have been slain, the blood of men and horses changes after a number of years into will-o'-the-wisps (*lin*[2]). These lights stick to the ground and to shrubs and trees like dew. As a rule they are invisible, but wayfarers come into contact with them sometimes; then they cling to their bodies and become luminous. When wiped away with the hand, they divide into innumerable other lights, giving out a soft crackling noise, as of peas being roasted. If the person stands still a good while, they disappear, but he may then suddenly become bewildered as if he had lost his reason, and not recover before the next day.

Nowadays it happens that when people are combing their hair, or when dressing and undressing, such lights follow the comb, or appear at the buttons when they are done up or undone, accompanied likewise by a crackling sound.

Similar descriptions occur in later books.[g]

[a] On which see Newton Harvey (1), p. 561.   [b] Cf. Vol. 2, p. 380.
[c] Ch. 30, p. 5b, tr. auct. Cf. Pfizmaier (58), p. 366.
[d] Ch. 33, p. 30a; cf. Pfizmaier (58), p. 439.   [e] Newton Harvey (1), pp. 263, 266.
[f] Ch. 9, p. 3a, tr. de Visser (1), mod.
[g] E.g. the *Tai Tsui Phien*[3] (On Substitutes for Getting Drunk), mentioned by de Visser (1), p. 191; a book about which one would like to know more.

[1] 盛道   [2] 燐   [3] 代醉編

The 'phosphorescence' of sea-water, due, as we now know, to the bioluminescence of Protozoa and minute Metazoa, was noticed early in Chinese literature.[a] The *Hai Nei Shih Chou Chi*, an account of marvellous sea islands attributed to Tungfang Shuo (−2nd century) but probably of the +4th or +5th century, says that 'if one travels on the sea, one may see fiery sparks when the water is stirred'.[b] In +1371 innumerable lights were seen on or in the sea, and people thought that they were the spirits or decaying blood of drowned men, but those acquainted with the sea said that it was often thus when the sea was agitated by wind or rain.[c] Similar effects were described by Phêng Tsung-Mêng[1] in his *Hai Yen Hsien Thu Ching*[2] (Illustrated Historical Geography of Sea-Salt City) about +1528. Not until 1830 was the problem solved by Michaelis's discovery of the luminescent dinoflagellates.

Bioluminescence arising quite obviously from larger animals had of course long been known in China as in most other cultures; namely the light of fireflies and glowworms (*chhung huo*[3]). Anciently they were called 'night-travellers', as in the *Shih Ching* (Book of Odes)[d] dating from the early centuries of the −1st millennium; 'flashing go the night-travellers' (*i yao hsiao hsing*[4]) says the text of one of these old folk-songs. The modern names appear in the Chhin and Han, when the *Erh Ya* encyclopaedia speaks[e] of the *ying huo*[5] and the *chi chao*.[6] The +4th-century commentator, Kuo Pho, says that they fly about in the air and have lights underneath their bodies. But the term *ying*[7] may go back to the −7th century, if the *Yüeh Ling* (Monthly Ordinances of the Chou Dynasty) may be dated so early.[f] There, among the events of the third month of summer, it is said that 'decaying grass becomes fireflies' (*fu tshao wei ying*[8]).[g] Li Shih-Chen in +1596 distinguished three main kinds of luminous insects, the common firefly (*Luciola* spp. and many other genera), the glow-

---

[a] But not perhaps as early as in Greek writings, for a report in Diels attributes to Anaximenes (*c.* −545) the view that 'lightning is due to the cleaving of the clouds by the force of the winds; the effect is the same as that which we observe when the sea flashes as the oar-blade cleaves it' (Freeman (2), p. 72). Almost the same remark is made by Aristotle in the *Meteorologica* (II, 9 (370a), Loeb ed. p. 229), who attributes it to a −5th-century philosopher, Cleidemus (Freeman (2), p. 278), and disagrees with it. Aristotle denies any intrinsic light emanating from the water, and attributes the appearances to reflections.

[b] Quoted by Chhen Tshang-Chhi[9] in his *Pên Tshao Shih I*, *c.* +725, preserved in *PTKM*, ch. 5, p. 18b.

[c] So Shen Chieh-Fu[10] in his *Chi Lu Hui Pien*.[11] And indeed it is the case that storms are often preceded by unusually brilliant displays of sea light, for a long spell of hot and calm weather favours the growth of the dinoflagellates (cf. Newton Harvey (1), p. 533).

[d] Odes of Pin, no. 3; Mao no. 156, tr. Karlgren (14), p. 101; Waley (1), p. 116.

[e] Ch. 15, p. 17a.

[f] On the dating of this ancient calendar, which arguments from astronomy may help to fix, see Vol. 3, p. 195. The reference to the text in the *Li Chi* is ch. 6, p. 66b; tr. Legge (7), vol. 1, p. 277.

[g] This is another instance of those strange transformations or spontaneous generations of living things so numerous in proto-scientific botany and zoology. We have already had occasion to notice such beliefs, in Vol. 2, pp. 79, 421, etc., and they will be dealt with faithfully in the zoological section below (Sect. 39). The statement is also found in Li Shun-Fêng's[12] *Kan Ying Ching*[13] (On Stimulus and Response), a +7th-century tractate (*Shuo Fu*, ch. 9, p. 1a), on which see Ho & Needham (2).

---

[1] 彭宗孟	[2] 海鹽縣圖經	[3] 蟲火	[4] 熠燿宵行	[5] 螢火
[6] 即炤	[7] 螢	[8] 腐草爲螢	[9] 陳藏器	[10] 沈節甫
[11] 紀錄彙編		[12] 李淳風	[13] 感應經	

worm with the luminous part towards the tail (*Lampyris noctiluca*, etc.), and luminescent flies living in or near water (probably midges or mayflies infected with luminous bacteria).[a] The collection of fireflies is a pastime of great antiquity in China and Japan.[b]

Turning to the luminescence of inorganic materials, we may remember that at a much earlier stage[c] there was talk of the 'night-shining jewel' (*yeh kuang pi*[1]), a mineral of some kind brought from Roman Syria to China in Han and Chin times. Reasons were given for thinking that this gem was probably chlorophane, a variety of fluorspar (calcium fluoride), which lights up on being heated or scratched.[d] This would be tribo- or piezoluminescence, but there are many other possibilities, including various natural or artificial phosphors.[e] Perhaps the most famous of these was the 'Bononian stone' or lapis solaris which so much interested the scientists of the seventeenth century.[f] This mineral was a heavy spar, a native barium sulphate rich in sulphur, found by Vincenzo Cascariolo about +1603 to glow in the dark after calcination. The presence of baryta deposits in China[g] may have afforded some indigenous 'night-shining jewels' of this kind. Laufer[h] has assembled a number of Chinese texts from various periods dealing with this subject, but most of them are rather fabulous and it is not easy to be sure in any particular case what exactly was the phenomenon under observation.[i]

McGowan (5) drew attention long ago to a remarkable story indicating that the preparation of artificial phosphors may have been known in the Sung. It occurs in a book of miscellaneous notes by the monk Wên-Jung[2] entitled *Hsiang Shan Yeh Lu*[3] (Rustic Notes from Hsiang-shan) and written in the +11th century.[j]

The Provincial Legate Hsü Chih-O[4] was fond of collecting curios. He paid 50,000 mace (coins) to a barbarian merchant for a stuffed bird's head, very brightly coloured, which he used as a pillow. He also got hold of an extraordinary painting which he presented to Li Hou

---

[a] *PTKM*, ch. 41, pp. 21aff. (R67). A point of particular interest in Li Shih-Chen's account is that he quotes Thao Hung-Ching, the distinguished physician of the late +5th century, as saying that the magical technicians (*fang shu chia*[5]) place fireflies in wine to kill them. This must be a remarkably early instance of the use of alcohol as a preservative, antedating the knowledge of alcohol distillation. The use of alcohol in strong solution for keeping small animal objects was introduced by Robert Boyle in the 17th century: cf. Needham (2), 2nd ed. p. 158.

[b] Here the *locus classicus* is *Chin Shu*, ch. 83, p. 9a, where we read the story of the poor but diligent student Chhê Yin[6] (d. c. +397), who collected a bagful of fireflies each night in order to read his books, their light replacing the oil his family could not afford to buy. In Japan in later ages firefly collecting became an important annual festival.

[c] Vol. 1, p. 199.

[d] On luminous gems in Greek and Roman literature see Newton Harvey (1), p. 33.

[e] Cf. the introductory article of Prener & Sullenger (1).

[f] See Bromehead (3) and Newton Harvey (1), pp. 306ff.

[g] Cf. p. 102 below.

[h] (12), pp. 67ff.

[i] Cf. *TSCC*, *Chhien hsiang tien*, ch. 95.

[j] Ch. 3, p. 20b; tr. auct. adjuv. Yang Lien-Shêng in Newton Harvey (1), p. 18. Rupp (1), p. 147, knew of the story from McGowan, but attributed it wrongly to the Japanese.

¹ 夜光璧   ² 文瑩   ³ 湘山野錄   ⁴ 徐知諤   ⁵ 方術家
⁶ 車胤

Chu.[a] This ruler (upon the extinction of the Southern Thang) passed it on as tribute to (the second Sung emperor) Thai Tsung.[b] Chang Hou-Wan[1] showed it to the court. On the painting there was an ox which during the day appeared to be eating grass outside a pen, but at night seemed to be lying down inside it. None of the officials could offer any explanation of this phenomenon. Only the monk Lu Tsan-Ning,[2] however, said that (he understood it). According to him, the 'southern dwarf island-barbarians' (Nan Wo jen[3]), when the tide is out and there remains only a little dampness on the shore, collect certain remaining drops of liquid from a special kind of oysters (chu pang thai chung[4]) and use this to mix a coloured paint or ink which appears only at night and not by day. At another place there are burning mountains (volcanoes) where sometimes violent winds blow, whereupon certain rocks fall down to the shore; these are ground and used to mix a coloured paint or ink which appears only in the daytime and not at night. All the scholars maintained that there was no basis for this story, but Tsan-Ning said that it could be found in a book called the *Hai Wai I Chi*[5] written by Chang Chhien[6] (the famous envoy sent to the western regions by Han Wu Ti in the −2nd century).[c] Afterwards Tu Hao[7] examined the collections in the imperial library and found the reference in a manuscript dating from the Liu Chhao period (+3rd to +6th centuries).

Now the interest of the matter is that in 1768 John Canton did in fact describe a phosphor made from oyster shells—an impure calcium sulphide made by calcining the carbonate with sulphur.[d] This became known as Canton's phosphorus. By adding the sulphides of arsenic, antimony or mercury, phosphors with blue or green luminescence can be obtained, as Osann showed in 1825. Our opinion is that Newton Harvey[e] is a little too sceptical in estimating the probability of the preparation of such luminescent substances by the alchemists of the early Sung. Tsan-Ning was a learned naturalist much respected by his contemporaries,[f] and the book of Wên-Jung contains many other relations about the effects of different kinds of light and heat. For example:[g]

Conferring with Tsan-Ning, Liu Chung-Thu[8] said that in his gardens after rain, there formed blue flames (chhing yen[9]) like sunset glow, but when one approached them they disappeared. He wondered what kind of untoward thing this could be. Tsan-Ning replied

---

[a] Li Yü,[10] the third and last emperor of the Southern Thang. He surrendered to the Sung in +975, which may date the transfer of the painting at about +977. His name is connected with the origin of the foot-binding practice.

[b] Reigned from +976 to +997.

[c] No such book by Chang Chhien now exists, but we have already encountered one with a very similar title which was ascribed to him (Vol. 1, p. 176). See Bretschneider (1), vol. 1, p. 25. Some discrepancy lies in the fact that the great envoy travelled to the far west (Bactria and Sogdiana), while the southern 'Japanese' barbarians would seem to be Liu-chhiu, Formosa, or Hainan people.

[d] Newton Harvey (1), pp. 329, 346.   [e] (1), p. 19.

[f] He was a friend of Wang Yuan-Chih,[11] of Wang Chhu-No[12] the astronomer, and of one of the greatest prose writers of the early Sung, Hsü Hsüan[13] (d. +991). He was a botanist who wrote a monograph on bamboos, and the author of a well-known biographical history, the *Sung Kao Sêng Chuan*.[14]

[g] Ch. 3, pp. 5a, b; tr. auct.

[1] 張後苑   [2] 錄贊寧   [3] 南倭人   [4] 諸蚌胎中   [5] 海外異記
[6] 張騫   [7] 杜鎬   [8] 柳仲塗   [9] 青爓   [10] 李煜
[11] 王元之   [12] 王處訥   [13] 徐鉉   [14] 宋高曾傳

that this was the *lin huo*.[1] Wherever there had been battles with much bloodshed, and the blood of oxen or horses got into the earth and coagulated there, these phenomena appeared, and even after a thousand years would not quite disperse. Afterwards Liu had the place dug up, and pieces of old weapons were found there.

Thus we are back again at the marsh-lights. But it is now time that we left them and turned to those steadier lights which lighted the development of more assured knowledge in the Chinese Middle Ages.

(g) LIGHT (OPTICS)

If Chinese optics never equalled the highest level attained by the Islamic students of light such as Ibn al-Haitham,[a] who benefited by the accessibility of Greek geometry, it nevertheless began at least as early as the optics of the Greeks. It has proved so easy to find traces of optical thought and experimentation in Chinese literature that an extended investigation would be likely to find much more, though a great deal (even apart from the Mohist material) is likely to be irretrievably lost. In this subject we are not quite so devoid of previous historical work as before; mention may be made of the discussions of Laufer (14) and Forke[b] on mirrors and lenses, the writings of de Visser (1) on fire and light, etc.[c] Only Chinese historians of science, however, have so far taken into account the most important school of Chinese optics, namely the Mohists; as in the excellent monographs by Than Chieh-Fu (1) and Chhien Lin-Chao (1). The Taoists might talk about the wonders and beauties of Nature, the Naturalists might bring forward their generalised explanations of her phenomena, the Logicians might argue about the proper way of discussing, but only the Mohists actually took mirrors and light-sources and carefully looked to see what happened.

Before speaking of light-sources in the abstract, a few words should be said of the lights themselves.

Presumably the walls of the feudal princes of the Chou flared with torches (*chü*[2]) made of miscellaneous combustibles (bamboo, pine resin, etc.). For smaller rooms throughout historical times oil was burnt with wicks in lamps or cruses (*têng*[3]), at first of pottery and bronze.[d] From the beginning of history China probably never lacked vegetable oils, but we gave one text above[e] which described the use of whale or seal oil in lamps at the court of a coastal prince of Yen in −308. Oil from seals (*jen yü*[4]) was used for the lamps in the tomb of Chhin Shih Huang Ti in −210, according to Ssuma Chhien.[f] Of the lamps and perfume-burners in gimbals made by Ting Huan about +180 we shall have more to say below (Section 27*d*). The hundred and twenty

---

[a] Cf. Winter (4); Ronchi (2).   [b] (4), vol. 2, p. 496.

[c] Ordinary histories of optics such as Papanastasiou (1), or Ronchi (1) excellent though it is, never mention Chinese material. We know of only one article on this, by Wang Chin-Kuang (1), and it has not been available to us. There is now also a brief communication by Chhien Lin-Chao (1).

[d] The various types may be studied in accounts such as those of Hough (2), or M. R. Allen (1).
[e] Vol. 3, p. 657.

[f] *Shih Chi*, ch. 6, p. 31*a*; cf. *TPYL*, ch. 870, p. 1*b*, quoting a *San Chhin Chi*,[5] and ch. 938, pp. 2*a*, 7*a*, quoting *Kuang Chih*[6] as in *YHSF*, ch. 74, p. 48*b*.

[1] 燐火   [2] 炬   [3] 燈   [4] 人魚   [5] 三秦記   [6] 廣志

iron lamps in the assembly-hall of Shih Hu, emperor of the Later Chao (*c*. +340) were long famous.[a]

The small and simple flat pan or domical cruse which served so far and wide in ancient civilisations to hold oil and wick, and which persisted so long that I myself have supped and read in the depths of the Chinese countryside by its feeble but mellow light, was given an ingenious development in medieval times. Someone realised that much oil was being lost by evaporation without efficient combustion at the wick, so saucer-shaped cruses were made with a reservoir underneath to hold cold water—something like the hot-water chafing dishes which were familiar on Victorian breakfast-tables, but for the opposite purpose. These lamps, which came from Chhiung-lai Hsien in Szechuan, were called *Chhiung Yao shêng yu têng*,[1] and were made of *chhing tzhu*[2] pottery. This place was famous for its glazed wares during the Sui and Thang periods. Some good specimens of these 'economical lamps' are in the Chungking Museum, where I had the pleasure of seeing them in 1958. The *locus classicus* for their literary description is in the *Lao Hsüeh An Pi Chi* of *c*. +1190.[b] There Lu Yu says:

> In the collected works of Sung Wên An Kung there is a poem on 'economic lamps'. One can find these things in Han-chia; they are actually made of two layers. At one side there is a small hole into which you put cold water, changing it every evening. The flame of an ordinary lamp as it burns quickly dries up the oil, but these lamps are different for they save half the oil. When the Venerable Shao Chi-Mu was at Han-chia he sent several of them to the scholars and high officials at court. According to Wên An one can also use dew. Han-chia has been producing these for more than three hundred years.

At this estimate the industry must have started in the Thang, perhaps early in the +9th century. It was an interesting anticipation of the water-jacketing of the chemical condenser in distillation, and of the steam and water circulatory systems of all modern technology.[c] Was not this an essential element also in the development of the internal combustion engine?

Another word, *chu*,[3] which originally also meant a torch, came early to be applied to candles, i.e. solid cylinders of fatty substances (*kao*[4]) surrounding wicks of fibre or bamboo. The problem is, exactly when. The word *chu* occurs many times in the Chhin and Han rituals (*Li Chi*, *I Li*, etc.), mostly quoted in *Thai-Phing Yü Lan*,[d] where it often clearly means lights borne by liturgical assistants. In the *Chou Li* the Ssu Hsüan Shih[5] official was responsible at court ceremonies for the *chu*, but these seem to have been hempseed oil-lamps rather than torches or candles.[e] *Chu* are often mentioned in books of the Warring States, Chhin and Han periods, for example, *Chuang Tzu* and *Mo Tzu* (two references each), the *Chan Kuo Tshê* and the *Huai Nan Tzu* (seven references). The oldest mention of candles made specifically of beeswax may be the *mi chu*[6] of

---

[a] *Yeh Chung Chi*, cit. *TPYL*, ch. 870, p. 1*b*.　　　　　　[b] Ch. 10, p. 9*a*, tr. auct.

[c] As we shall see in Section 33, the device was greatly elaborated by the medieval Chinese alchemists; meanwhile see Ho & Needham (3).

[d] Ch. 870, pp. 3*b*ff. (tr. Pfizmaier, 98).

[e] Ch. 10, pp. 4*b*, 5*a* (ch. 38), tr. Biot (1), vol. 2, p. 381.

[1] 邛窰省油燈　　　[2] 青瓷　　　[3] 燭　　　[4] 膏　　　[5] 司烜氏　　　[6] 蜜燭

the *Chi Chu Phien* dictionary[a] of about −40. The oldest occurrence of the surer term *la chu*[1] seems, however, to be in the *Chin Shu*, with reference to the statesman Chou I[2] who died in +322.[b] It is quite likely that beeswax candles were made already during the Warring States period, so that the Mohist natural philosophers may well have used them in their experiments.

Hommel gives an account[c] of traditional candle-making in China—always by dipping, never moulding. Vegetable tallow (fat from *Stillingia sebifera*, R 332) forms the core, and insect wax of higher melting-point (from *Ericerus pela*, R 11) the coat. This coccid wax (cf. Sect. 39) was replacing beeswax in the Thang and Sung, while the tallow-tree wax came into general use only from the Yuan period. The wick, always projecting below, is either a hollow reed which can go on a pricket, or a solid bamboo stick to fit a socketed holder. Everyone who has lived in the Chinese countryside remembers with affection the characteristic tapering red candles, with their light flickering through lattice windows.

In the Sung period mineral wax seems to have been used as well as that from vegetable and insect sources. In the *Lao Hsüeh An Pi Chi* (c. +1190), Lu Yu says:[d]

Sung's poem on the 'White Stone Candles' (*pai shih chu*[3]) has the verse: 'You only like those that are bright like wax—why despise those that are black in colour?' In fact these latter come from Yen-an, and when I was in southern Chêng I often saw them. They are as hard as stone, burning extremely brightly, and they gutter like wax. But the smoke is so thick that it can be used for smoking (curing) things. As it spoils curtains and clothes (when used domestically) the Westerners do not like it either.

Lu Yu is here presumably referring to Persians and Arabs. Since, as is well known, the natural petroleum of Yümên is rich in waxy fractions, it seems very likely that seepages at various places in west China may from time to time have yielded natural deposits of dark wax which could have served as the basis for local rough candle industries. Some would be darker than others, thus accounting for the two kinds, white and black.[e]

Bronze objects which may well have been candlesticks have been found in the Chhu tombs of the Warring States period (c. −4th century) excavated at Chhangsha.[f] They take the form of small pans with elegant handles on three short legs and having a central pricket. I saw one of these at the Archaeological Institute of Academia Sinica in Peking in 1958. At the Loyang Museum there is also an earthenware bowl about 8 in. diam. with a socket hollowed as if for a candle sticking up inside it. This is of Chin date (+4th century). The director, Mr Chiang Jo-Shih, told me that companion pieces had traces of wax in them when found.[g]

a Ch. 4, p. 34a.   b Ch. 69, p. 10b (cit. *TPYL*, ch. 870, p. 4b). On him, G 417.
c (1), pp. 318 ff.   d Ch. 5, p. 9b, tr. auct.
e Cf. Vol. 3, p. 609.   f See Anon. (11), p. 115 and pls. 65, 66.
g On lanterns see Shan Shih-Yuan (1). Some further notes on this subject will be found in Wu Nan-Hsün (1), pp. 63 ff. For an introduction to the comparative history of lighting in other civilisations, see Forbes (15), pp. 119 ff.; O'Dea (1, 2); Robins (3).

¹ 蠟燭   ² 周顗   ³ 白石燭

(1) MOHIST OPTICS

Let us now examine the propositions on optics contained in the *Mo Ching*, truncated and fragmentary though they are.

Ch 16/271/—.—. *Shadow formation*

C   A shadow (*ying*[1]) never moves (of itself). (If it does move, it is owing to the moving of the source of light or of the object which casts the shadow.) The reason is given under 'changing action' (*kai wei*[2]).[a]

CS When light arrives, the shadow disappears. But if it were not interfered with, it would last for ever. (FYL/B, mod.)

This is Chhien's interpretation. Others, including Than, have regarded the proposition as identical with the famous paradox of the Logicians (PC/25 above, Vol. 2, p. 191) 'The shadow of a flying bird never moves'. But this seems less likely, partly because the shadow-casting object is here not in motion, and partly because the opening proposition of a series on optics would be quite likely to state the fixity of a shadow if light-source and object are both fixed. Cf. Luan Tiao-Fu (*1*), p. 80; Wu Nan-Hsün (*1*), p. 98.

Ch 17/—/38.16. *Umbra and penumbra*

C   When there are two shadows (it is because there are two sources of light). The reason is given under 'doubling' (*chhung*[3]).

CS Two (rays of) light grip (*chia*[4]) (i.e. converge) to one light-point. And so you get one shadow from each light-point. (auct.)

This clearly indicates that the Mohists appreciated the linearity of light-rays. Cf. the famous passage in *Chuang Tzu*, ch. 2 (Vol. 2, p. 51 above) where Umbra (*ying*[1]) has a discussion with Penumbra (*wang liang*[5]). This is closely analogous to the present propositions in that the motion of the shadows is affirmed to depend solely on the motion of the light-sources or shadow-casting objects. Chuang Chou was about contemporary with the Mohists. Wang-liang was a kind of demon (see Chiang Shao-Yuan (*1*), pp. 168 ff.); I do not know why the name was used for penumbra.[b]

The above is Than's interpretation; Chhien, Wu and Luan point out that the proposition may refer to two shadows cast by two entirely different light-sources, rather than to a penumbra formed because the light was not a point-source. But in any case the linearity of light-rays was understood.

Ch 20/—/44.17. *Size of shadow dependent on position of object and of source of light*

C   As for the size of the shadow—the reason is given under 'whether slanting like a steering-oar (*tho*[6]) (i.e. not perpendicular to the direction of the light-rays), or upright (*chêng*[7]) (i.e. perpendicular to the direction of the light-rays); whether far (*yuan*[8]) or near (*chin*[9]).'

[a] Note that the word *kai* is one which we have never met in the Taoist writings, where so much play is made with *pien*[10] and *hua*[11] (cf. Vol. 2, p. 74, above). This is because its ancient scientific meaning must have been change of mutual position only.

[b] This use occurs again in an account of a Thang Taoist, Kuo Tshai-Chen,[12] who had special names for the successive penumbrae resulting from increasing the number of light-sources (*Yu-Yang Tsa Tsu*, ch. 11, p. 6*a*).

[1] 影   [2] 改爲   [3] 重   [4] 夾   [5] 罔兩   [6] 杝
[7] 正   [8] 遠   [9] 近   [10] 變   [11] 化   [12] 郭采眞

CS If the post is slanting, like a steering-oar (not perpendicular to the rays of the sun or other light-source), its shadow is short and intense. If the post is upright (perpendicular to the rays of the sun or other light-source), its shadow is long and weak. If the source of light is smaller than the post, the shadow will be larger than the post. But if the source of light is larger than the post, the shadow will still be larger than the post. The further (from the source of light) the post, the shorter and darker will be its shadow; the nearer (to the source of light) the post, the longer and lighter will be its shadow. (auct.)

> Here the experimentalist must have had a fixed light-source and a fixed screen, with the post able to move back and forth between them. Cf. Wu Nan-Hsün (*1*), pp. 101 ff.

Cs 48/—/98.45. *Pinhole*

C The 'collecting-place' (*khu*[1]) (or; the 'wall' (*chang*[2])) is the place where the 'change' (*i*[3]) (i.e. the inversion of the image) starts.

CS It is an empty (round) hole (*hsü hsüeh*[4]), like the sun and moon depicted on the imperial flags. (auct.)

> We do not know which is the right word in the first part of this proposition. *Khu*[1] might refer to the whole enclosed space of the camera obscura.[a]

Ch 18/—/40.17. *Definition of focal point and inversion of image*

C The image is inverted (*tao*[5]) because of the intersection (*wu*[6]). The intersecting place is a point (*tuan*[7]). This affects the size of the image (*ying*[8]). The reason is given under 'point' (*tuan*[7]).

CS An illuminated person shines as if he was shooting forth (rays). The bottom part of the man becomes the top part (of the image) and the top part of the man becomes the bottom part (of the image). The foot of the man (sends out, as it were) light (rays, some of which are) hidden below (i.e. strike below the pinhole) (but others of which) form its image at the top. The head of the man (sends out, as it were) light (rays, some of which are) hidden above (i.e. strike above the pinhole) (but others of which) form its image at the bottom. At a position farther or nearer (from the source of light, reflecting body, or image) there is a point (*tuan*[7]) (the pinhole) which collects (*yü*[9]) the (rays of) light, so that the image is formed (only from what is permitted to come through the collecting-place (*khu*[1] or *chang*[2]). (auct.)

> Note here the way in which the man whose image is to be studied in the camera obscura is supposed to 'shoot forth' light. But it is 'an illuminated person', not a seeing person, so the conception is quite different from the Greek theory of the emission of rays from the eye in vision. Moreover, in CS the Mohists distinctly said 'as if', showing that they knew he was giving off reflected light.
>
> The technical terms are interesting. *Chang*,[10] meaning a dyke or embankment, was at all times combined with *ai*[11] as *chang-ai*, to mean barrier or obstacle. Hence the significance of the fact that we shall find Shen Kua in the +11th century using *ai*[11] as a technical term for focal point. It was Pi Yuan who suggested emending *khu*[1] to *chang*.[2] The fact that the Mohists worked with pinhole and camera obscura in the −4th century is of great interest, as this is usually

---

[a] Luan Tiao-Fu (*1*), pp. 72 ff. interprets also Cs 61, discussed above in Vol. 3, p. 91, as concerned with rays of light coming through small apertures.

[1] 庫   [2] 牆   [3] 易   [4] 虛穴   [5] 倒   [6] 午   [7] 端
[8] 影   [9] 與   [10] 障   [11] 礙

placed much later by historians of physics (Arabic, early + 11th century). Later, we shall see further work on this by the Chinese (p. 97). Cf. Wu Nan-Hsün (*1*), p. 100.

Ch 19/—/42.—. *Plane mirror*

C   A shadow can be formed by the reflected (*ying*[1]) (rays of) the sun. The reason is given under 'turning' (*chuan*[2]).

CS   If the light(-rays) from the sun are reflected (from a plane mirror perpendicular to the ground) on to a person, the shadow (of that person) is formed (on the ground) between that person and the sun. (auct.)

> Than thought that this implied a knowledge of the law of equality of the angle of incidence with the angle of reflection, but it does not seem necessarily to do so. Cf. Luan Tiao-Fu (*1*), p. 83; Wu Nan-Hsün (*1*), p. 103.

Ch. 21/—/26.—. *Combination of plane mirrors*

C   Standing on a plane mirror (*chien*[3]) and looking downwards, one finds that one's image is inverted (*tao*[4]). (If two mirrors are used) the larger (the angle formed by the mirrors within the limit of 180°) the fewer (the images). The reason is given under 'smaller district' (*kua chhü*[5]) (i.e. the distance between the free edges of the mirrors, hence the angle).

CS   A plane mirror (*chêng chien*[6]) has only one image. Its shape, bearing, colour, white or black; distance, far or near; and position, slanting or upright—all depend upon the (position of the) (object or the) source of light. If now two plane mirrors are placed at an angle (*tang*[7]), there will be two images. If the two mirrors are closed or opened (as if on a hinge), the two images will reflect each other. The reflected images are all on the opposite side (from where the eye is). A person reflected in the mirror (*chien chê*[8]) (shoots his light-rays) at certain mirror-targets (*nieh*[9]), and wherever he stands (within the angle of the two plane mirrors as long as it is less than 180°) the image is never unreflected. The image-targets are numerous (i.e. there are many images) but (the angle between the two mirrors) must be less than when they were originally in the same line (i.e. 180°). The reflected images are formed from the two mirrors separately. (auct.)

> Other later examples of experiments with multiple reflection will be found below (p. 92). Again we have the idea of light-rays being shot forth from objects, which is reminiscent of Democritean vision theory and Epicurean *idola*. And Brewster's kaleidoscope (1817) is foreshadowed. Cf. Wu Nan-Hsün (*1*), p. 104.

Ch 56/—/57.21. *Refractive index*

C   The (apparent) size of a thorn (*ching*[10]) (in water) is such that the sunken part (*shen*[11]) seems shallow. The reason is given under 'appearance' (*hsien*[12]).

CS   The sunken part is (only) the appearance of the thorn, hence the shallowness of the sunken part is not the shallowness of the thorn itself. If you compare it (you find that the difference between the real and apparent depth is) one part in five. (auct./WNH)

> It would be surprising if the Mohists had not noted and studied the very obvious phenomenon of refraction, but the true nature of this proposition was not recognised until the work of Wu Nan-Hsün (*1*), p. 111. Earlier investigators such as Than Chieh-Fu (*1*) and Forke (*3*), misled by corruptions of three important

---

[1] 迎   [2] 轉   [3] 鑑   [4] 倒   [5] 寡區   [6] 正鑑   [7] 當
[8] 鑑者   [9] 臬   [10] 荆   [11] 沈   [12] 見

words, had interpreted it (with great difficulty) as having something to do with economics and the value of money. Translating 'one part in five' gives a refractive index of 1·25, that of water being of course 1·33. Wu Nan-Hsün, however, takes *tshan*[1] in its technical meaning of 'one in three' instead of 'compare', and so reads 'one five in three fives', obtaining 1·5. In any case it would perhaps hardly be reasonable to ask for better values in the −4th century.

Ch 22/—/28.17. *Concave mirror*

C   With a concave mirror (*wa chien*[2]), the image may be smaller and inverted (*hsiao erh i*[3]), or larger and upright (*ta erh chêng*[4]).

> Here is the distinction between what we now call the real and inverted image, and the erect, magnified, or virtual image.

The reason is given under 'outwards from the centre area' (*chung chih wai*[5]) (i.e. away from the centre of curvature); and 'inwards from the centre area' (*chung chih nei*[6]) (i.e. from the focal point to the surface of the mirror).

CS (Take first) (an object in) the region between the mirror and the focal point (*chung chih nei*[6]). The nearer the object is to the focal point (and therefore the farther away from the mirror), the weaker the intensity of light will be (if the object is a light-source), but the larger the image will be. The farther away the object is from the focal point (and therefore the nearer to the mirror), the stronger the intensity of light will be (if the object is a light-source), but the smaller the image will be. In both cases the image will be upright. From the very edge of the centre region (i.e. almost at the focal point) (*chhi yu chung yuan*[7]), and going towards the mirror, all the images will be larger than the object, and upright.

(Take next) (an object in) the region outside the centre of curvature and away from the mirror (*chung chih wai*[5]). The nearer the object is to the centre of curvature the stronger the intensity of light will be (if the object is a light-source), and the larger the image will be. The farther away the object is from the centre of curvature, the weaker the intensity of light will be (if the object is a light-source), and the smaller the image will be. In both cases the image will be inverted.

(Take lastly) (an object in) the region at the centre (*ho yü chung*[8]) (i.e. the region between the focal point and the centre of curvature). Here the image is larger than the object (and inverted). (auct.)

> This is a striking series of observations. To get the last one right must have involved very careful procedure. The Mohists do not seem to have had special technical terms for the focal point and the centre of curvature of the mirror, and Chhien doubted whether they understood the difference. We think that they must have done, otherwise what they say about the two regions 'outside' and 'inside' would, in one case or the other, be incorrect. Other references to concave mirrors will be discussed below (p. 87).

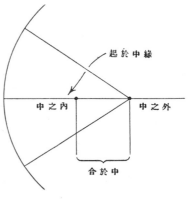

[1] 參   [2] 窪鑑   [3] 小而易   [4] 大而正   [5] 中之外
[6] 中之內   [7] 起於中緣   [8] 合於中

This work with light-sources makes it probable that the use of mirrors as lamp-reflectors arose early in China, and not late, as was thought by Horwitz (7). Cf. the discussion of Luan Tiao-Fu (*1*), p. 85; Wu Nan-Hsün (*1*), pp. 105 ff.

Ch 23/—/—.—. *Convex mirror*

C   With a convex mirror (*thuan chien*[1]) there is only one kind of image. The reason is given under 'the size of the shape' (*hsing chih ta hsiao*[2]).

CS   The nearer the object is to the mirror the stronger the intensity of light will be (if the object is a light-source), and the larger the image will be. The farther away the object is the weaker the intensity of light will be (if the object is a light-source) and the smaller the image will be. But in both cases the image will be upright. An image given by an object too far away becomes indistinct. (auct.)[a]

Once again, in order to place these contributions in perspective, it is necessary to sketch as briefly as possible the parallel origins of the science in Greece. The oldest, and the most widely accepted, theory of light and vision was the Pythagorean belief that visual rays were emitted by the eye, ran in straight lines to the object, and by touching it gave the sensation of sight.[b] This error had the advantage that the progress of geometrical optics was not hindered, it being necessary only to reverse the direction of the light along its paths when the true situation was realised. Its propagation was generally considered instantaneous. Next most important was the Epicurean theory, according to which all objects gave off images or *idola* in all directions, some of which penetrated the eyes of seeing persons and gave rise to the sensation of sight. We need not here consider the theories of the Stoics and of Plato.

Four branches of the study of light were distinguished, (*a*) optics *sensu stricto*, or the study of vision, (*b*) catoptrics, or the properties of mirrors, (*c*) scenography or perspective, (*d*) dioptrics, or the measurement of angles by optical methods (sighting and surveying). The only work contemporary with the Mohists which has come down to us from Greek antiquity is the *Optics* of Euclid[c] (tr. Burton). This consists of 58 theorems, treated like the geometrical ones, and based on four definitions, that light rays travel in straight lines, that figures comprised by them are cones, that one can only see things upon which the light rays fall, and that the apparent size of things depends upon the angle of the light cone. It is evident that all these ideas were familiar to the Mohists. Euclid develops the basic laws of perspective, stating, for example, that a cylinder seen from one point always appears as slightly less than a hemicylinder. Whether he wrote on catoptrics, like the Mohists, is not known, since the work later attributed to him on this is probably due to Theon of Alexandria (a +4th-century contemporary of Ko Hung). Archimedes (c. −250) almost certainly did, but the treatise which survived (if it was really his) has been missing since the +15th century.[d]

---

[a] Wu Nan-Hsün (*1*), p. 109, reconstructs the text differently but the topic is not in doubt.

[b] On ancient optics in the West see the excellent account of Ronchi (*1*), as also B & M, pp. 817 ff. Empedocles propounded a double emission, from the object and from the eye. Aristotle, as so often, hedged a great deal, but ended by deciding that light was some form of motion between the eye and the object.

[c] Unfortunately much remodelled by Theon of Alexandria (c. +365), though the original text is now approximately known.   [d] On burning-mirrors, see below, pp. 87 ff.

[1] 團鑑   [2] 形之大小

The oldest extant Greek writing on mirrors is the *De Speculis* of Heron of Alexandria (*c.* +100);[a] in this there is treatment of plane, concave and convex mirrors, with a proof of the fact that the angle of reflection is equal to the angle of incidence. It will be seen, then, that the Mohist beginnings are rather earlier than anything we have from Greece.[b] But the *Optics* of Ptolemy, which must have been written early in the +2nd century (about the time of Chang Hêng),[c] goes much beyond any systematic exposition which has remained in Chinese literature; it extends the discussion to spherical and cylindrical mirrors, and above all deals thoroughly with refraction. Ptolemy gave tables of quantitative data for this phenomenon, worked out more or less experimentally,[d] and applied his knowledge to astronomical problems. Such is the general framework in which we have to consider Mohist optics. The Mohists must have made careful and extensive use of the experimental method, but they suffered from the lack of a developed geometry.

It is perhaps worth emphasising that ancient and early medieval optics in the West was dominated by the curious theory already mentioned that visual rays were emitted by the eye and ran in straight lines to the object of vision. This was by no means such a ridiculous idea as some modern historians of science have naturally supposed.[e] The comparatively recent discovery of 'animal radar', in which it has been shown that bats[f] (and probably many other flying animals, including birds)[g] emit bursts of short-wave radiation, to the echoes of which they then 'listen', so guiding their flight, demonstrates that living organisms can do just what most of the Greeks thought happened in sight. It has even been shown that many fishes make use of similar means of orientation in water.[h] Nevertheless, the Greek conception of light and vision was fundamentally wrong, and it was not until the brilliant revolutionary work of Ibn al-Haitham (Alhazen, +965 to +1039)[i] that the correct view prevailed. Now so far as we can find, the theory of visual ray emission was quite foreign to ancient Chinese thought.[j] The Mohists would have had more in common with the Epicurean minority since their light-rays reflected from seen objects entered the eye like the images or simulacra of Lucretius.[k]

---

[a] Diocles, to whom the invention of the parabolic mirror was attributed, may have been considerably earlier; he wrote on burning-mirrors (cf. Sarton (1), vol. 1, p. 183).

[b] Nor apparently can India compete. There is a little on light rays, mirrors, mirages, etc. in the *Nyāya Sūtra*, as Mallik has pointed out, but this may be as late as the +3rd century, and its true epoch is not known. It is said to contain statements similar to those of Empedocles.

[c] Lejeune (1).   [d] Lejeune (2).

[e] Indeed it is not impossible that the light-producing organs on the heads of abyssal fishes function in relation to their eyes, faintly illuminating regions which would otherwise lie hid in perpetual night.

[f] See Hartridge (1); Galambos & Griffin (1); Griffin (1, 2). There is stereophonic perception of the echoes of repeated supersonic cries. This 'sonar' is enormously more efficient than any echo-location system so far devised by man.   [g] See Griffin (3, 4).

[h] See Lissmann (1, 2); Lissmann & Machin (1). It seems that they emit a weak electrical discharge and then by special receptors perceive disturbances in the field so set up, caused by invisible objects with conductivities different from that of water.

[i] A good account of his contribution will be found in Ronchi (1), pp. 33 ff.

[j] Yet Demiéville, in his criticism of Ku Pao-Ku's translation of the *Kungsun Lung Tzu*, urged regarding ch. 5 (Ku (1), p. 62) that the text should be emended to conform with the idea that light was radiated from the eye.

[k] *De Rerum Nat.* IV, ll. 42 ff.

(2) MIRRORS AND BURNING-MIRRORS

A development parallel to, and probably much older than, the explicit catoptrics of the Mohists, was the use of burning-mirrors for igniting tinder from the sun's rays.[a] References to these are very common in ancient Chinese writings. They were called *fu sui*[1] or *yang sui*.[2,3] By the Chhin and Han, variants of the word *sui* had acquired the general meaning of concavity; thus we find that in the account of bell-making in the *Khao Kung Chi*[b] the concave part of the bell is called *sui*,[4] moreover, its 'depth' (*shen*,[5] i.e. radius or arc) is exactly standardised. Similarly, *sui tao*[6] meant an underground passage, especially leading into an imperial tumulus tomb, and this was perhaps its original meaning.[c]

Mirrors of bronze (*chien*[7]) must go back extremely far into the Chinese Bronze Age. One of the earliest literary references to them is doubtless that for −672 in the *Tso Chuan*.[d] The oldest existing dated mirrors, however, are from +6 and +10, and we have many from Later Han times. There is, of course, a great literature on ancient Chinese mirrors, which (on account of their ornamentation) have been so prominent in studies of archaeology and art-history; with such aspects we are not here concerned.[e]

Among Han writings, the *Chou Li* comes first to mind, since so much archaic material seems to be embedded in it. It mentions two kinds of officials, the 'Directors of Fire Ceremonies' (Ssu Kuan[8])[f] and the 'Directors of Sun Fire' (Ssu Hsüan Shih[9]).[g] As has been made clear[h] the duties of the former were concerned with obtaining the 'new fire'[i] by means of the fire-drill (*tsuan sui*[10]), for which a variety of different woods were used at five periods during the year.[j] It was the latter who had to make ceremonial use of the burning-mirror for the 'new fire'. The *Chou Li* says:

They have the duty of receiving, with the *fu sui*[11] mirror, brilliant fire from the sun; and of receiving with the (ordinary) mirror (*chien*[12]) brilliant water from the moon. They carry out these operations in order to prepare brilliant rice, brilliant torches for sacrifices, and brilliant water.[k]

[a] Wu Nan-Hsün (1), pp. 67ff., has also given an account of this.
[b] *Chou Li*, ch. 11, pp. 24a, 25a (ch. 41), tr. Biot (1), vol. 2, p. 499.
[c] *Tso Chuan*, Duke Yin, 1st year; Duke Hsi, 25th year. It would be the opposite of the convexity of the mound itself.
[d] Duke Chuang, 21st year (Couvreur (1), vol. 1, p. 176).
[e] A convenient account is that of Swallow (1). The monograph of Hirth (5) will be referred to again below. The review by Pelliot (23) of the well-known books of Lo Chen-Yü (2) and Tomioka (1) is instructive. Literary and social aspects are briefly dealt with by A. Ripley Hall (1). Cf. Bulling (8).
[f] Ch. 7, p. 24b (ch. 30), tr. Biot (1), vol. 2, p. 194.
[g] Ch. 10, p. 4b (ch. 37), tr. Biot (1), vol. 2, p. 381. Cf. p. 79 above.
[h] E.g. by de Visser (1).
[i] A numinous thing in so many antique civilisations and primitive cultures. And still we see it in the moving ceremonies of the Latin Church on Easter Saturday. On this see Frazer (1), one-vol. ed. p. 614. A typical Han reference will be found in the *Hou Han Shu*, ch. 15, p. 5a.
[j] Cf. Vol. 2, p. 236, as also Sect. 27a below, and *Chan Yuan Ching Yü*, ch. 1, p. 8b (+14th century).
[k] Tr. Biot (1), vol. 2, p. 381; eng. auct.

[1] 夫遂	[2] 陽遂	[3] 陽燧	[4] 隧	[5] 深	[6] 隧道
[7] 鑑	[8] 司爟	[9] 司烜氏	[10] 鑽燧	[11] 夫燧	[12] 鑑

For the moment we will neglect the Yin or lunar part of their work, to return to it later.[a] It comes again, of course, in the *locus classicus* in *Huai Nan Tzu*,[b] about −120:

> When the *yang sui*[1] sees the sun, there is a burning and fire is produced. When the *fang chu*[2] sees the moon, there is a dampness (or secretion) and water is produced.

On this Kao Yu,[3] of the Later Han, comments:

> The burning mirror is of metal. One takes a metal cup untarnished with verdigris and polishes it strongly, then it is heated by being made to face the sun at noon time; in this position cause it to play upon mugwort tinder and this will take fire. The *fang chu*[2] is the Yin mirror (*yin sui*[4]); it is like a large clam(-shell) (*ta ko*[5]). It is also polished and held under the moonlight at full moon; water collects upon it, which can be received in drops upon the bronze plate. So the statements of our ancient teacher are really true.[c]

Other Han references to burning-mirrors occur in the *Li Chi*,[d] the *Chhun Chhiu Fan Lu*,[e] and the *Lun Hêng*.[f] In the +3rd century Ko Hung gives us an eye-witness account—'I have often seen men getting fire from the morning sun and water from the evening moon',[g]—and in the +4th Tshui Pao also refers to it.[h] Among numerous other references it may be significant that burning-mirrors are associated[i] with the name of the +3rd-century mathematician Kaothang Lung;[6] perhaps he made some catoptric studies now lost. Another mention of burning-mirrors in the biography of the Thang general, Li Ching,[7] who used to carry one, suggests that they were used by soldiers for making fire in the field.[j] The philosophical significance of the two kinds of mirrors was ever present in the minds of Chinese scholars, as is shown, for instance, by the mention of them in Chu Hsi's commentary on Wei Po-Yang.[k]

The use of burning-mirrors among the Chinese, which may reasonably be regarded as going back to the time of Confucius, has many parallels in other civilisations. A famous passage in Plutarch's *Lives* reminds us of the antiquity of the tradition in Rome. He is speaking of the fire of the Vestal Virgins in his account of Numa Pompilius:

> If it (the fire) happens by any accident to be put out, as the sacred lamp is said to have been at Athens, under the tyranny of Aristion; at Delphi, when the temple was burnt by the

---

[a] On both, see the special study by Thang Po-Huang (*1*).

[b] Ch. 3, p. 2*a*. The burning-mirror is often mentioned elsewhere in the *Huai Nan Tzu* book, e.g. ch. 6, pp. 4*a*, 10*a*; ch. 17, p. 13*b*.   [c] Tr. auct. adjuv. de Visser (*1*), p. 117.

[d] Ch. 12, p. 52*a* (Legge (7), vol. 1, p. 449).

[e] Ch. 14, where the term *ching chin*[8] (neck metal) is used, either because the instrument was hung round the neck, or because necks are concave things.

[f] Ch. 32 (Forke (4), vol. 1, p. 272), ch. 74 (Forke (4), vol. 2, p. 412).

[g] *Pao Phu Tzu*, ch. 3 (Feifel (*1*), p. 200).

[h] *Ku Chin Chu*, ch. 7; cf. *Chung Hua Ku Chin Chu*, ch. 2, p. 8*b* (+10th century).

[i] In the *Wei Ming Chhen Tsou*[9] (Memorials written by Eminent Ministers of the Wei (San Kuo) Dynasty) by Chhen Shou,[10] the +3rd-century author of the *San Kuo Chih*; quoted in *TPYL*, ch. 717, p. 3*b*.

[j] *Hsin Thang Shu*, ch. 93, p. 4*b*. He died in +649.

[k] *Tshan Thung Chhi Khao I*, ch. 1, p. 19*a*. Many other references in *TSCC*, *Chhien hsiang tien*, ch. 95.

[1] 陽燧	[2] 方諸	[3] 高誘	[4] 陰燧	[5] 大蛤	[6] 高堂隆
[7] 李靖	[8] 頸金	[9] 魏名臣奏		[10] 陳壽	

Medes; and at Rome, in the Mithridatic war, as also in the civil war, when not only was the fire extinguished, but the altar overturned—it is not to be lighted again from another fire, but new fire is to be gained by drawing a pure and unpolluted flame from the sunbeams. They kindle it generally with concave vessels of brass, formed by hollowing out an isosceles rectangular triangle, whose lines from the circumference meet in one single point. This being placed against the sun, causes its rays to converge in the centre, which, by reflection, acquiring the force and activity of fire, rarefy the air, and immediately kindle such light and dry matter as they think fit to apply.[a]

This supposedly refers to the second, semi-legendary, Sabine king of Rome, c. −700.[b] In Chinese literature, however, there does not seem to be any parallel to the famous story of the burning-mirrors used by Archimedes in the defence of Syracuse.[c] But the contemporaneity of the first references in Chinese and European (Latin) literature[d] is rather striking, and probably indicates the spread in both directions of a technique originally Mesopotamian or Egyptian.

This is not the place to say much of the metallurgical composition of the burning-mirrors of the Chou and Han, a subject which will be appropriate in Section 36. There, in a later volume and a chemical context, we shall have to review a quite large literature on the alloys of the ancient Chinese metallurgists. Certain texts there are which tell us what they were doing, or at least what contemporary scholars thought they were doing; and these it has been possible to check against the results of modern analyses of their products. Here we need only say that while the famous table of alloy proportions in the *Chou Li*[e] (Khao Kung Chi) states that metal mirrors are composed of 50 per cent copper and 50 per cent tin, modern analyses show that in fact the tin never exceeded 31 per cent. Beyond about 32 per cent the alloy becomes excessively brittle, and increasing tin content brings no further advantages of any kind; this the Han metallurgists evidently knew. Indeed they knew much more, for they almost always added up to 9 per cent of lead, a constituent which greatly improved the casting properties. Han specular metal is truly white, reflects without tinning or silvering, resists scratching and corrosion well, and was admirably adapted for the purposes of its makers.

As for the moon-mirrors, mentioned in many other places,[f] it is clear that what they collected was condensing dew. That this was so much prized was a kind of philo-

[a] Tr. Langhorne (1), vol. 1, p. 195.

[b] Amerindian parallels in Spinden (1), pp. 173, 218; Frazer (1), one-vol. ed. p. 485.

[c] The story begins late, not being mentioned before Galen (see B & M, p. 359). It is not generally known that Buffon, in 1747, succeeded in producing, by the aid of a combination of 168 plane mirrors, effects similar to those ascribed to Archimedes; cf. J. T. Needham (2). See also J. Scott (1); Heath (6), vol. 2, p. 200. Modern technique has done much better, and the solar furnaces which have been developed in France should be capable of volatilising all known substances; they can already melt the most refractory oxides (e.g. thoria at 3000° C.). The reader is referred to the interesting review of Trombe (1).

[d] Later mentions are quite numerous in the West. Burning-mirrors are discussed under catoptrics by Theon (c. +350), and Diocles wrote a book on them in the −2nd century. So also did Anthemius of Tralles (d. +534).   [e] Ch. 11, p. 20b (ch. 41), tr. Biot (1), vol. 2, pp. 491 ff.

[f] *Huai Nan Wan Pi Shu*, quoted in *TPYL*, ch. 58, p. 7a. Also *Chhien Han Shu*, ch. 99 B, p. 29b, where a figure of a *hsien*[1] holds a *chhêng lu phan*[2] (plate for receiving dew).

[1] 仙   [2] 承露盤

sophical superstition.[a] It may be mentioned, however, that there was a good deal of confusion between this and another firm belief of the ancient Chinese, namely that there were certain marine animals which waxed and waned in correspondence with the moon. This we encountered first in connection with culture contacts,[b] and again in the discussion on the idea of continuity and action at a distance in the physical universe (p. 31 above). We shall do full justice to it in Section 39 on zoology; it is of considerable interest because based on facts which have been ascertained to be true. Possibly the confusion occurred because of the concavity of the shells of lamellibranchiate molluscs—we heard the *Huai Nan Tzu* commentator saying (above) that the moon-mirror was like a clam(-shell). Schlegel supposed[c] that the meaning of *fang chu*[1] was 'perfectly square', and if the moon-mirrors were so shaped it would have been quite in accord with theory, since roundness pertained to Yang and heavenly fiery things, while squareness pertained to Yin and earthly watery things. Umehara has, indeed, described (*1*) square mirrors believed be of Chhin date.

Social uses of mirrors and literary allusions to them have been collected by Hall (1) and Maspero (17). Chuang Chou often likens the mind to a mirror without dust; it is part of the Taoist emphasis on passive observational receptivity.[d] Later Buddhist sources strengthened such analogies, as Demiéville (1) has shown in an interesting paper. Tunhuang frescoes of Thang date often show people meditating in front of what appear to be mirrors, both concave and convex (Fig. 290).[e] This may connect with older divination practices,[f] or with Taoist tests of the efficacy of respiratory exercises in which mirrors remained undimmed.[g]

That some of the scientific accuracy of the Mohists still persisted in the Han may appear from a statement in *Huai Nan Tzu*:[h]

It is like collecting fire with a burning-mirror. If (the tinder is placed) too far away (*su*[2]), (the fire) cannot be obtained. If (the tinder is placed) too near (lit. prompt, *shuo*[3]), the centre point (*chung*[4]) will not (be hit either). It should be just exactly between 'too far away' and 'too near'. The direction of the light moves its position from early morning till late afternoon. (If you insist on placing the mirror in a position suitable for) the oblique rays, when the rays are falling (nearly) vertically, the experiment will fail.

[a] The same idea is met with in India, where the moon-mirror was called *chandrakānta*, in parallel with *sūryakānta* (Laufer (14), p. 222). Thang Po-Huang believes that this was a transmission from China to India. But there was also a cult of dew in ancient Egypt, as de Savignac (1) has described.

[b] Vol. 1, p. 150.

[c] (5), p. 612.

[d] *Chuang Tzu*, chs. 5, 7 and 13 (Legge (5), vol. 1, pp. 225, 266, 331).

[e] E.g. caves nos. 65 and 66. The frescoes in cave no. 156, dating *c.* +851, show two of these mirrors, one on the balustraded roof of a pavilion, the other on the top of the pedestal of a 'banner of victory'. All have characteristically 'crank-handled' supports.

[f] Cf. *Pao Phu Tzu*, in *TPYL*, ch. 717, p. 3*a*, on 'crystal-gazing'.

[g] Cf. *Shu Chü Tzu* (Ming), cf. Forke (9), p. 451.

[h] Ch. 17, p. 3*b*. Interpretation follows Kao Yu's commentary. Tr. auct.

[1] 方諸   [2] 疏   [3] 數   [4] 中

Here Liu An and his friends were aware of the necessity of finding the focal point. In another place[a] we find him observing mirrors formed by water surfaces:

> If you observe your reflection on the surface of water in a (large and shallow) tray, you will see a round image. But above the water (at the edge of) a (small) cup, you will see an elongated image (owing to the meniscus). The shape of the face has not changed; the reason for the difference lies in the difference of shape of the surface at which you are looking.

These ideas continue *Mo Ching* Ch 22. Another instance of combinations of plane mirrors (Ch 21) occurs in a fragment of the *Huai Nan Wan Pi Shu*:[b]

> A large mirror being hung up (above a large trough filled with water)[c], one can see, even though scated, four 'neighbours'.

All this, and still more all that follows, implies smooth and plane (or precisely curved) mirrors of bright finish and high reflectivity. Without the techniques of making these, China's medieval experimenters could not have made the observations they did, for image quality falls off rapidly with increasing number unless the surface is truly smooth. That high-tin bronze (specular metal) was used from Chou times onward is certain, and that it was sometimes coated with a layer of tin by heating above 230° C. is highly probable (see Sect. 36); this would give at least 80% reflectivity. Later the tin was deposited by means of a mercury amalgam, and the appearance of the Taoist patron saint of mirror-polishers, Fu Chü hsien-sêng[1] ('Mr Box-on-his-Back') in an entry in the *Lieh Hsien Chuan*,[d] would valuably attest the technique for the +4th century, if we had not the anterior evidence of the *Huai Nan Tzu* book[e] for the −2nd. Taoists used mirrors as demonifuge armour in their mountain excursions.[f] A relic of the Han preoccupation with mirrors is seen in that part of the *Hsi Ching Tsa Chi*[g] which deals

Fig. 290. Sketches of objects which seem to be mirrors, both concave and convex, from Thang frescoes on the walls of cave-temples at Chhien-fo-tung, Tunhuang (caves nos. 65 and 66). Perhaps connected with certain meditation techniques.

with the mysterious instruments in the treasury of Chhin Shih Huang Ti, found when the first Han emperor captured it. A rectangular mirror (4 ft. × 5 ft. 9 in.), gave

---

a Quoted *TPYL*, ch. 758, p. 7*a*. Tr. auct.

b Preserved in *TPYL*, ch. 717, p. 3*a*. A much earlier source is Chang Hua's[2] *Kan Ying Lei Tshung Chih*[3] (Record of the Mutual Resonances of Things), c. +295, in *Shuo Fu*, ch. 24, p. 20*a*. Also earlier is the quotation in the Thang encyclopaedia *I Lin*[4] (*Shuo Fu*, ch. 11, p. 32*b*). Tr. auct.

c This necessary amplification is in the commentary. Cf. Wu Nan-Hsün (*1*), p. 167.

d Ch. 63, tr. Kaltenmark (2), p. 174.

e Ch. 19, p. 7*b*. Cf. *Shih Lin Kuang Chi* (*Hsü Chi*), +1478 ed., ch. 9, p. 13*b*.

f *Pao Phu Tzu* (*Nei Phien*), ch. 17, p. 2*a*; cf. Needham (8).

g Ch. 3, p. 3*a*.

¹ 頁局先生　　　² 張華　　　³ 感應類從志　　　⁴ 意林

inverted images of people standing before it; moreover, it revealed the viscera and other opaque parts of men and women, which was useful in medical diagnosis and the investigation of 'dangerous thoughts' entertained by imperial concubines.[a] But Yuan Ti, the Liang emperor (*c.* +550), employed plane mirrors much more soberly for lighting the interiors of well shafts.[b]

In the +10th century, Than Chhiao[1] (or whoever it was who wrote the *Hua Shu*[2]) sought to illustrate his philosophical theories of subjective realism by experiments with mirrors. He considered the infinite regress of images of an object reflected in oppositely placed plane mirrors.[c] Each of these images (*ying*[3]) perfectly reproduces the form and colour of the object (*hsing*[4]). Since it can exist without them, it is not alone and in itself complete (*shih*[5]), but since they can copy correctly its form and colour they are not in themselves empty (*hsü*[6]). Putting Than Chhiao's thought in modern terms, the object is not entirely real, but the images are not entirely unreal; and things of that sort, he concludes, are not far from the Tao. Now this fascination with the effects produced by multiple mirrors was not new in Taoist circles about +940; it had been even more thoroughly investigated (if that is the right word) by the Buddhists a couple of centuries before. According to a famous metaphor of Indian origin, the 'Nodes of Indra's Net',[d] every object in the universe mirrors every other object, just as the moon gives rise to an infinite number of separate reflections on the wavelets of an expanse of water.[e] This theory was discussed, for example, in the *Chin Shih Tzu Chang*[7] (Essay on the Golden Lion) which the monk Fa-Tsang[8] wrote for an empress in +704.

As his students failed to understand it, he used a clever expedient. He took ten mirrors and arranged them so that one occupied each of the eight compass-points, with one above and one below, in such a way that they all faced one another, a little over ten feet apart. He then placed the figure of a Buddha at the centre, and illuminated it with a torch so that its images were reflected back and forth. Thus his students came to understand the theory of passing from (the world of) 'sea and land' into (the world of) infinity.[f]

An even earlier reference occurs in the Sui commentary of Lu Tê-Ming[9] (*fl.* +583) on *Chuang Tzu*, ch. 33, where the ideas of the philosophical schools are discussed.[g] We are already familiar with Hui Shih's paradox 'The South has at the same time

---

[a] For similar Western legends about mirrors, see Laufer (18). The Arab legend about the television mirror on the Alexandrian Pharos got into the *Chu Fan Chih* (Sung) and was embroidered in the *San Tshai Thu Hui* (Ming).   [b] *Chin Lou Tzu*, ch. 4, p. 19a.   [c] P. 2b.

[d] This was several times referred to in Vol. 2, e.g. p. 483.

[e] It will be remembered that this doctrine, or something very like it, was adopted by Leibniz in the *Monadology*. In our own time, Whitehead has spoken of 'a focal region where each thing is, but its influence streams away from it throughout the utmost recesses of space and time' (2), p. 202. However, in some presentations, including those of Fa-Tsang himself (+643 to +712), the doctrine of the nodes of Indra's net seems to have similarities with the 'homoeomereity' (ὁμοιομέρεια) of Anaxagoras; full 'lion-ness' was contained within each single hair of the lion's fur. See Cornford (7), Peck (3, 4).

[f] *Sung Kao Sêng Chuan*, ch. 5 (Taishō ed.), p. 731; tr. Bodde in Fêng Yu-Lan (1), vol. 2, p. 353, mod.

[g] *Chuang Tzu Pu Chêng*, ch. 10B, p. 19a; Wang Hsien-Chhien (2), p. 103.

[1] 譚峭   [2] 化書   [3] 影   [4] 形   [5] 實   [6] 虛
[7] 金師子章   [8] 法藏   [9] 陸德明

a limit and no limit'.[a] On this Lu Tê-Ming remarked: 'There is the mirror and the image, but there is also the image of the image; two mirrors reflect each other, and images may be multiplied without end. He spoke about the South, but he was only taking it as one example.'[b] Further research will doubtless bring to light more intermediate discussions and experiments between the time of Lu Tê-Ming and that of the author of the *Huai Nan Wan Pi Shu* and the Mohists.

Fa-Tsang and Lu Tê-Ming might have been rather astonished if they could have learnt that in future ages this method of setting up inward-facing mirrors would prove useful to mathematicians in their efforts to describe and list all the possible uniform polyhedral solids. Uniform polyhedra have regular faces meeting in the same manner at every vertex. Besides the five Platonic and the thirteen Archimedean solids, the four regular star-polyhedra of Kepler and Poinsot, and the infinite families of prisms and antiprisms, there are at least fifty-three more, of which forty-one were discovered in the 19th century, while the rest remained unknown till 1953. This final achievement was the work of Coxeter, Longuet-Higgins & Miller, who described how Möbius in 1849 had made use in such studies of a 'polyhedral kaleidoscope', i.e. an assembly of mirrors arranged so as to face inwards at known angles.[c] This led to Wythoff's kaleidoscopic construction in the mathematics of polyhedral solids, and so to the present systematisation, which is believed to be complete.

We shall shortly recur to the optical investigations of the +10th century Taoist, Than Chhiao, in connection with lenses; here it may be noted that he said:[d]

At a distance greater than 100 yards (*pu*[1]) the mirror can see man, but man cannot see the image in the mirror.

He seems thus to have perceived that although the image on the mirror may be so small as to be on the limit of visibility, the rays of light go straight to the mirror surface all the same. In the +11th century, Shen Kua occupied himself with mirrors, as with so many other things of scientific interest. He wrote:[e]

The ancients made mirrors according to the following methods. If the mirror was large, the surface was made flat (or concave); if the mirror was small, the surface was made convex. If the mirror is concave (*wa*[2]) it reflects a person's face larger; if the mirror is convex (*tieh*[3]) it reflects the face smaller. The whole of a person's face could not be seen in a small mirror, so that was why they made the surface convex. They increased or reduced the degree of convexity or concavity according to the size of the mirror, and could thus always make the mirror correspond to the face. The ingenious workmanship of the ancients has not been equalled by subsequent generations. Nowadays, when people get hold of ancient mirrors, they actually have the surfaces ground flat. So perishes not only ancient skill, but even the appreciation of ancient skill.

---

[a] HS/6, cf. Vol. 2, p. 191.   [b] Cf. Wu Nan-Hsün (*1*), p. 173.
[c] Cf. also Coxeter (*1*). Though Coxeter *et al.* referred to the study of certain tessellations by Abū'l-Wafā' in the +10th century (cf. F. Woepcke, *1*), they did not know of his contemporary Than Chhiao, still less of Fa-Tsang.   [d] *Hua Shu*, p. 4*a*; tr. auct.
[e] *Mêng Chhi Pi Than*, ch. 19, para. 9; cf. Hu Tao-Ching (*1*), vol. 2, pp. 630ff.; tr. auct. adjuv. Hirth (*5*). Cf. also Wu Nan-Hsün (*1*), pp. 184ff.

[1] 步   [2] 凹 or 窪   [3] 凸

But Shen Kua carried the matter no further. Perhaps the Mohists were in closer contact with the technicians of their time than Shen Kua was able to be with his. However, as we shall shortly see, Shen knew, from pinhole experiments, that the collection of rays at the hole was analogous with the focal point of a mirror and that this was the same as the point of concentration of heat-rays. He compared the cone of light-rays with the space swept out by an oar in its travel. Shen's remarks on the decay of technique in his time were perhaps too severe, for in another book, the *Hou Shan Than Tshung*[1] (Collected Discussions at Hou-Shan) by Chhen Shih-Tao,[2] written at about the same time, we read that mirrors of iron with specular surfaces were then being made in which the 'depth' (*hsien*[3]) was carefully regulated.[a]

(3) MIRRORS OF UNEQUAL CURVATURE

A number of years ago I read another entry of Shen Kua's, which I shall now reproduce. At the time we found it impossible to believe that he was not really talking of lenses. He says:[b]

There exist certain 'light-penetration mirrors' (*thou kuang chien*[4]), which have about twenty characters inscribed on them in an ancient style which cannot be interpreted. If such a mirror is exposed to the sunshine, although the characters are all on the back, they 'pass through' and are reflected on the wall of a house, where they can be read most distinctly.

Those who discuss the reason for this say that at the time the mirror was cast, the thinner part became cold first, while the (raised part of the) design on the back, being thicker, became cold later, so that the bronze formed (minute) wrinkles. Thus although the characters are on the back, the face has faint lines (*chi*[5]) (too faint to be seen with the naked eye).

From this experiment with light, we can see that the principle of vision may be really like this.[c]

I have three of these inscribed 'light-penetration mirrors' in my own family, and I have seen others treasured in other families, which are closely similar and very ancient; all of them 'let the light through'. But I do not understand why other mirrors, though extremely thin, do not 'let light through'. The ancients must indeed have had some special art.

In fact, Shen Kua was quite right; we know now that in his time and long before,[d] mirrors were made which had the property of reflecting from their polished faces the designs executed in relief on their backs. Fig. 291 shows such a mirror and its reflection.

---

[a] Chhien Lin-Chao (2) has recently described three extant concave bronze or specular burning-mirrors, two with handles, of Sung date. He gives detailed measurements.

[b] *Mêng Chhi Pi Than*, ch. 19, para. 12; tr. auct. adjuv. St Julien (3). Here, for once, Hu Tao-Ching (1) could not elucidate.

[c] The meaning of this remark is not obvious. Perhaps Shen Kua thought that just as the surface of the mirror could send out a pattern without having any visible pattern itself, so the retina could receive a pattern without showing one. For this, he would have to have recognised the retina as the receiving organ, which was not decided until the time of Leonardo in the West.

[d] It has not usually been thought that this technique antedated the +5th century, but the magazine *La Chine*, 1959 (no. 11), p. 31, figures a 'magic mirror' apparently of Han date, presented to the national museums in Peking by the historian Chin Kung, whose family had long possessed it.

[1] 後山談叢   [2] 陳師道   [3] 陷   [4] 透光鑒   [5] 跡

PLATE CI

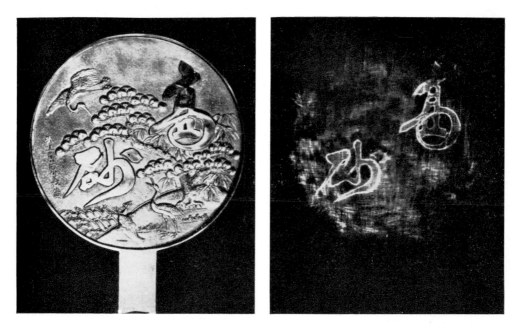

Fig. 291. A 'magic mirror' of Japanese provenance (from Dember, 1). Although the polished face (not here seen) appears to the eye perfectly smooth, the characters executed in relief on the back (left) are clearly visible in the reflection (right). They are read *Takasago*, the name of a *nō* play.

The interest of the story lies perhaps mainly in the fact that this technique, current already in the +11th century in China, exercised some of the best scientific minds of the 19th century before an explanation was reached. And when the matter was settled somewhat over fifty years ago, it turned out that Shen Kua's explanation had been essentially correct, though the invisible relief on the surface of the polished side is not simply brought about by differences of cooling rate.

Before describing the investigations of the 19th-century physicists, there are two other passages in Chinese writings which claim our attention. They show that the mirrors were attracting interest at the end of the +13th century. Chou Mi, in his *Kuei Hsin Tsa Chih Hsü Chi*, says:[a]

The principle of the 'light-penetration mirrors' is really inexplicable. Almost the only scholar who has discussed them is Shen Kua, but his theory was forced and trivial. I first saw one in the house of Hsienyü Po-Chi,[1] and later on another two when staying with Ho Chhing-Fu.[2] The strangest of all was one which Hu Tshun-Chai[3] had—when it was made to reflect sunlight, even the faintest lines in the design on the back could clearly be seen. It is really mysterious. Ma Chih-Chi[4] the poet wrote a famous essay on this. Most of these mirrors show only part of the design, or show it indistinctly. In the *Thai-Phing Kuang Chi*, ch. 230, it says that a Mr Hou[5] gave one of these magic mirrors to Wang Tu[6]...from which we can conclude that they were also rare in former times.

The last remark is interesting, for Wang Tu was a +5th-century official.[b] Chou Mi's contemporary, the great archaeologist Wuchhiu Yen,[7] was less hopeless of finding an explanation.[c] He said:

These mirrors have their effect owing to the use of two kinds of bronze of unequal density. If on the back of the mirror a dragon has been produced by the mould, then an exactly similar dragon is deeply engraved on the face of the disc, and the cuts filled up with denser (*cho*[8]) bronze. This is incorporated with the body of the mirror, made of finer (*chhing*[9]) bronze, by heating; then the face is planed and prepared, and a thin layer of lead or tin placed over it.[d]

He added that he had seen such a mirror broken in pieces, and had convinced himself of the truth of this explanation. But it was wrong.

Five centuries later, developed Renaissance science began to get to work. In 1832 Prinsep came across one of these 'magic mirrors' in Calcutta and described it (1) in the *Journal of the Asiatic Society*; shortly afterwards the English physicist Brewster

---

[a] Ch. 2, p. 27*b*; tr. auct.

[b] Biogr. in *Wei Shu*, ch. 30. Cf. *KCCY*, ch. 56, p. 3*a*, where Mr Hou's gift is placed in the Sui dynasty, on the authority of a book about which I have not been able to find information, the *Ku Ching Chi*[10] (Record of Ancient Mirrors).

[c] D.+1311. The quotation is from *Hsüeh Ku Phien*. Repeated in *Yü Tung Hsü Lu*[11] (Late Winter Talks) by Ho Mêng-Chhun[12] (Ming); cit. *KCCY*, ch. 56, p. 6*b*.

[d] Tr. St Julien (3), eng. Ayrton & Perry. This was perhaps a reference to the mercury amalgam used for polishing.

[1] 鮮于伯機   [2] 霍清夫   [3] 胡存齋   [4] 麻知幾   [5] 侯
[6] 王度   [7] 吾邱衍   [8] 濁   [9] 清   [10] 古鏡記
[11] 餘冬序錄   [12] 何孟春

examined the mirror, and without knowing anything of Wuchhiu Yen, decided (1) that the effect was due to density differences. In 1844 the famous French astronomer Arago presented one to the Académie des Sciences, and this started a series of contributions in which St Julien (3) adduced two of the passages from Chinese books given above. All the physicists agreed that the effects must be due to minute differences in the degree of curvature of the convex polished surface, except Séguier (1) who thought that they were produced by some kind of compression. Person (1) and Maillard (1) were proved to be right in the elaborate experiments of Ayrton & Perry (1) of 1878. This was preceded by a vigorous exchange of letters in *Nature* in the previous year.[a] Ayrton & Perry had the advantage of residence in Japan, where the mirrors were still being produced, so that they were able to add to their laboratory investigations a first-hand acquaintance with the technicians of the industry. Careful and extended optical experimentation demonstrated that the surfaces of 'magic' mirrors reproduced the designs on the backs because of very slight inequalities in curvature, the thicker portions being very slightly flatter than the thinner ones, and even sometimes actually concave.[b] Though all such mirrors are convex, the moulds used are quite flat, and the convexity is produced afterwards by a hand scraping-tool or 'distorting rod', with which a series of parallel scratches are made in different directions. The thinnest parts of the mirror buckle under the stress, and remain strained in the opposite direction, i.e. more convex, after the stress is removed. The polishing pressure also tended to make the thinner parts more convex than the thicker. Any minute cavities in the reflecting surface were filled up with small particles of bronze, and these hammered in and polished off; this is the probable explanation of what Wuchhiu Yen saw in his broken specimen. Subsequently, Ayrton & Perry (2) gave reasons for thinking that the polishing process exaggerated the minute increased convexity of the thinnest parts not only by pressure, but by means of the amalgam used, for they found that when a brass bar is amalgamated on one side it will show considerable expanding and bending. Since the polishing was done with an amalgam consisting of 69 % Sn, 30 % Hg and 0·64 % Pb, it would seem that this must have had a share in the effect.

In the meantime other interesting experiments had been performed in Italy and France. Govi (1) was able to make all mirrors with designs embossed on the back show the effect, by heating; the thinnest parts expanded most quickly and the design

[a] Most of the contributors thought that the invisible differences of relief on the mirror's surface could be produced by stamping from the back, and some showed that this could be done; cf. Atkinson (1); Highley (1); Darbishire (1); Sylvanus Thompson (3); and Parnell (1). But unfortunately all the Chinese and Japanese mirrors in question were cast and not stamped with dies. Cf. Berson (1).

[b] Fig. 292. Occasional references in Chinese literature speak of 'magic' mirrors which showed, on reflection, a design different to the design cast on their backs. One of these came into the hands of Dember (1) who found that it consisted of two quite separate plates attached by a common edging band; the back plate with the visible design having no reflecting surface, and the front reflecting plate having a hidden design behind it. Another similar 'double' mirror carried a hidden tuning fork and could be made to give forth musical notes. The dates of these were uncertain.

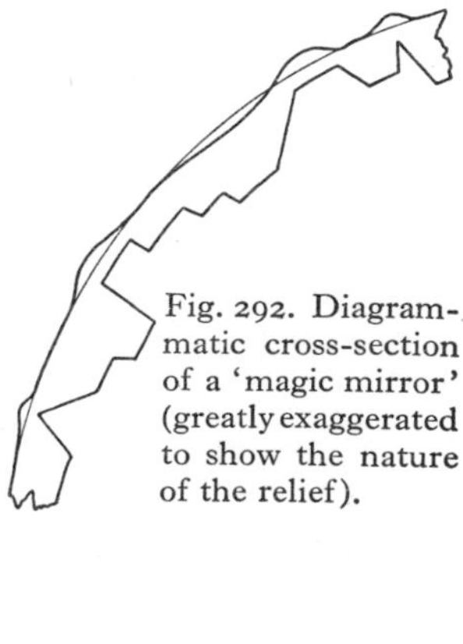

Fig. 292. Diagrammatic cross-section of a 'magic mirror' (greatly exaggerated to show the nature of the relief).

appeared. This was taken note of by astronomers interested in thermostatic control of telescope mirrors. Then Bertin & Duboscq (1) repeated the experiment using powerful pressure from behind the mirror, and succeeded in the same way. Finally, Muraoka (1–4) showed that mirrors of the properties required could be made from any metal.[a]

To conclude, mirrors of unequal curvature, reflecting the pattern on their backs, must have been first produced as an empirically discovered marvel at some time before the +5th century. That people were then studying the reflections of mirrors on screens is a point not to be overlooked. Shen Kua, with his 'traces' on the reflecting surfaces, gave an explanation substantially correct, though differences in cooling rate were not responsible. Wuchhiu Yen was further off the mark, though no more so than Brewster. The technicians of the +11th century would perhaps have been flattered if they could have known that it would take a hundred years' work before Sir William Bragg (1) could write a definitive explanation of 'magic mirrors' in 1932.[b] What they had essentially discovered was the magnifying effect of a long light path. It was perhaps the first step on the road to knowledge about the minute structure of metal surfaces as exemplified in such beautiful developments as multiple-beam interferometry.[c]

(4) CAMERA OBSCURA

It seems that in the Thang and Sung there was much interest in experiments with pinholes and darkened rooms. Shen Kua has an important passage on the subject in his *Mêng Chhi Pi Than* (+1086):[d]

The burning-mirror (*yang sui*[1]) reflects objects so as to form inverted images. This is because there is a focal point (*ai*[2]) in the middle (i.e. between the object and the mirror). The mathematicians call investigations about such things Ko Shu.[3] It is like the pattern made by an oar (*lu*,[4] for *lu*[5]) moved by someone on a boat against a rowlock (*nieh*[6]) (as fulcrum). We can see it happening in the following example. When a bird flies in the air, its shadow moves along the ground in the same direction. But if its image is collected (*shu*[7]) (like a belt being tightened) through a small hole in a window, then the shadow moves in the direction opposite to that of the bird. The bird moves to the east while the shadow moves to the west, and vice versa. Take another example. The image of a pagoda, passing through the hole or small window, is inverted after being 'collected'. This is the same principle as the burning-mirror. Such a mirror has a concave surface, and reflects a finger to give an upright image if the object is very near, but if the finger moves farther and farther away it reaches a point where the image disappears and after that the image appears inverted. Thus the point where the image disappears is like the pinhole of the window. So also the oar is fixed at the rowlock somewhere at its middle part, constituting, when it is moved, a sort of 'waist'

[a] General reviews of the subject by Rein (1) and Chamberlain (1); on the physical side the most interesting is that of Bertin (1). Cf. Hirth (5); Ayrton (1).

[b] His explanation is slightly different from that of Ayrton & Perry, adding another factor. 'In the scraping', he says, 'the thin parts give way and do not lose so much metal as the thick. After the tool has gone over them, they rebound and stand up as minute elevations too small to be seen by the naked eye. Only the magnifying effect of reflection makes them plain.'

[c] Cf. Tolansky (1).

[d] Ch. 3, para. 3. Cf. Hu Tao-Ching (1), vol. 1, pp. 111 ff.

[1] 陽燧   [2] 礙   [3] 格術   [4] 艣   [5] 櫓   [6] 臬   [7] 束

(*yao*[1]), and the handle of the oar is always in the position inverse to the end (which is in the water). One can easily see (under the proper conditions) that when one moves one's hand upwards the image moves downwards, and vice versa. [Since the surface of the burning-mirror is concave, when it faces the sun it collects all the light and brings it to a point one or two inches away from the mirror's surface, as small as a hempseed. It is when things are at this point that they catch fire. This is indeed the place where the 'waist' is smallest.]

Are not human beings also rather like this? There are few people whose thinking is not restricted in some way. How often they misunderstand everything and think that true benefit is harmful, and that right is wrong. In more serious cases they take the subjective for the objective, and vice versa. If such fixed ideas are not got rid of, it is really difficult to avoid seeing things upside down.

[The *Yu-Yang Tsa Tsu* said that the image of the pagoda is inverted because it is beside the sea, and that the sea has that effect. This is nonsense. It is a normal principle (*chhang li*[2]) that the image is inverted after passing through the small hole.][a]

This is undeniably interesting. The reference to the mathematicians recalls Kaothang Lung, mentioned above, and suggests that active study had been going on for a long time previously. Then the comparison of the two cones of light-rays with the two spatial segments made by an oar in motion around its fulcrum is a striking analogy. The identification of three separate radiation phenomena is excellent (pinhole, focal point and burning-point).

The mention of the passage in the *Yu-Yang Tsa Tsu*,[b] written by Tuan Chhêng-Shih early in the +9th century, brings up the question of the antiquity of the study of the pinhole and camera obscura. *Mo Ching*, Cs 48, has already shown us that the Mohists about −300 were familiar with the former; now we see that inverted pagodas were being looked at at least as early as about +840. Yet the earliest work with the camera obscura is usually attributed to Ibn al-Haitham, the great Arab physicist (+965 to +1039),[c] when not put down to 16th- and 17th-century people such as della Porta or Kircher.[d] Alhazen used the camera obscura particularly for observing solar eclipses, as indeed Aristotle is said to have done, and it seems that, like Shen Kua, he had predecessors in its study, since he did not claim it as any new finding of his own.[e] But his treatment of it was competently geometrical and quantitative for the first time. There can be little doubt that both the Chinese and the Arabs were interested in it from the +8th century onwards, and the full unravelling of the story must be left for further research.

Inverted images continued to attract interest later. The phenomena, and Shen Kua's explanations of them, were discussed, though not further advanced, by Lu

[a] The sentences in square brackets are printed in some editions as commentary, but seem to be statements of Shen Kua himself. Tr. auct.

[b] Ch. 4, p. 7*a*.

[c] Sarton (1), vol. 1, p. 721; Mieli (1), p. 106; Winter (3); Winter & Arafat (2); Feldhaus (1), col. 225; Wiedemann (3, 4); Würschmidt (1); Werner (1).

[d] Cf. Pauschmann (1).

[e] Private communication from Dr M. Nazif Bey, through Dr K. Borch.

[1] 腰   [2] 常理

Yu[1] in the +12th, Yang Yü[2] in the +14th,[a] and Ku Chhi-Yuan[3] in the +17th century.[b] A century later Yü Chao-Lung[4] made experiments[c] with screens (*chang*[5]), holes (*tou*[6]), and model pagodas, noting the way in which the rays diverged or 'breathed forth' (*hsi*[7]).

The genius of Shen Kua's insight into the relation of focal point and pinhole can better be appreciated when we read in Singer (9) that this was first understood in Europe by Leonardo da Vinci (+1452 to +1519), almost five hundred years later. A diagram showing the relation occurs in the *Codice Atlantico* (p. 216r). Leonardo thought that the lens of the eye reversed the pinhole effect, so that the image did not appear inverted on the retina; though in fact it does. Actually, the analogy of focal-point and pin-point must have been understood by Ibn al-Haitham, who died just about the time when Shen Kua was born.[d]

(5) LENSES AND BURNING-LENSES

(i) *Rock-crystal and glass*

It had originally been intended to treat of glass in connection with ceramics (with which it is of course related, through glazes and enamels), but since it is impossible to appraise the evidence about the Chinese use of lenses without a knowledge of the history of glass[e] we have to sketch this here. Until recently, great uncertainty has reigned about it, for the earliest literary references to what might be glass-making in Chinese writings are somewhat obscure, while the earliest clear statements about glass-making, though formerly accepted at their face-value, have in fact turned out to be long posterior to the period for glass-making in China now assured by the archaeological evidence. Laufer (14) conducted a veritable campaign to prove that China had neither rock-crystal lenses nor glass before the +5th century. As Pelliot used to say, there is now not much to be retained of all this argumentation, though of course, as always, a great debt is owing to Laufer for drawing attention to little-known sources.

Let us first recall the nature of rock-crystal and glass before glancing at the appearance of the former in Chinese mineralogy. We may then summarise present knowledge about the history of glass in China. At that point we shall be in a position to re-examine

[a] *Shan Chü Hsin Hua*, p. 36b (tr. H. Franke (2), no. 100). This passage explains the curious idea of the *Yu-Yang Tsa Tsu* that water had something to do with it. The inverted image of the dark room was confused with the inverted reflection of a pagoda in a pool or lake at its base. Cf. the contemporary book *Cho Kêng Lu* (+1366), ch. 15, on the three pagodas of Sungchiang. See also *Hua Man Chi*, ch. 7, p. 3b.

[b] In *Kho Tso Chui Yü*[8] (My Boring Discourses to My Guests), c. +1628. He called the pinhole *chhi*.[9]

[c] In *Thien Hsiang Lou Ou Tê*[10] (Occasional Discoveries of the Heavenly Fragrance Pavilion), before +1740.

[d] Cf. Winter & Arafat (3).

[e] See the bibliography of Duncan (1) and the lavishly illustrated account of Vávra (1). Cf. also Forbes (14); Harden (1).

[1] 陸游	[2] 楊瑀	[3] 顧起元	[4] 虞兆漋	[5] 障	[6] 竇
[7] 吸	[8] 客座贅語	[9] 隙	[10] 天香樓偶得		

the evidence as to the use of lenses by the Chinese. Lastly, the question of the invention of spectacles will come up for consideration.

Rock-crystal is pure transparent crystalline quartz (SiO_2), i.e. silica. Anciently it could only be worked like jade, for temperatures high enough to fuse it and make vessels of it have been attainable only in modern times. Li Shih-Chen knew it[a] as *shui ching*,[1] *shui ching*,[2] *shui yü*,[3] or *ying shih*.[4] The first of these names was that under which it was known in the first pharmacopoeia to include it, the *Pên Tshao Shih I* of +725. The third name comes from the *Shan Hai Ching*[b] and would therefore be Han, while the fourth is that which occurs in the +5th-century (Northern Wei) dictionary, the *Kuang Ya*. If Li Shih-Chen was wrong in these identifications[c] (and of course it is always difficult to be sure of exactly what was meant by such ancient technical terms), Laufer (14) might be right in his view that rock-crystal was not generally known till the Thang, but the balance of probability seems entirely the other way.[d] There was at that time, it is true, a marked influx of burning-lenses, probably of rock-crystal, from foreign countries,[e] but that is a separate question. Another Chinese name for rock-crystal seems to have been *huo ching*[5] ('fire-essence' instead of 'water-essence').[f]

Rock-crystal is found more often coloured or translucent than clear and transparent. The *Pên Tshao* writers mention many kinds of coloured silica, purple or yellow, and some kinds, the 'tea-crystal' (*chha ching*[6]) or 'black crystal' (*mo ching*[7]), were used in the Ming for making dark spectacles, just as we use dark sunglasses today. In the West, acquaintance with rock-crystal is of ancient date; a famous worked piece of Babylonian provenance (assessed at the −9th century) was described by Layard.[g] Many Greek and Roman authors, such as Aristophanes and Pliny, mention rock-crystal; and many rock-crystal balls, which may or may not have been used as lenses (Singer, 9), have been found at sites such as Pompeii, Nola, Mainz, etc.

Glass, in its commonest and earliest form, is essentially a silicate of sodium or potassium (or both) and calcium,[h] produced by the fusion of sand (silica), with limestone

---

[a] *PTKM*, ch. 8, p. 56a.

[b] And the *Hai Nei Shih Chou Chi*, relative to the Khun-Lun mountains.

[c] Kuo Pho, of the +4th century, says in his commentary on the *Shan Hai Ching* that *shui yü* = *shui ching*, and he ought to have known.

[d] It will be remembered that numerous texts assembled by Hirth (1), dealing with Ta-Chhin (Roman Syria), mention the pillars in palaces there, made of, or ornamented with, rock-crystal—*Hou Han Shu*, ch. 118; *San Kuo Chih*, ch. 30; *Chin Shu*, ch. 97; *Chiu Thang Shu*, ch. 198 (Hirth, pp. 40, 44, 51, 71).

[e] See below, p. 115.

[f] *Hsü Han Shu*,[8] by Hsieh Chhêng[9] (+3rd century), where it is said that it came from the country of the Ai-Lao[10] barbarians, between Yunnan and Burma. This term would add weight to the argument which will be presented below, that the Chinese were well aware of the use of burning-lenses in the late Han, if not before. It would be natural to regard rock-crystal as congealed fire, if one did not prefer to consider it congealed water. The passage is quoted in ch. 808 of the *Thai-Phing Yü Lan* twice. In the first (p. 4b) the *huo ching* is coupled with glass (*liu-li*, see immediately below) as imports from the Ai-Lao people. But in the second (p. 6a) the name is written *shui ching*, not *huo ching*. The two characters were of course easily confused by copyists.   [g] (1), p. 197.

[h] A formulation often given is $Na_2O.CaO.6SiO_2$. Before the +17th century, however, lime generally got in unintentionally (see Turner, 1, 3).

---

[1] 水精   [2] 水晶   [3] 水玉   [4] 英石   [5] 火精   [6] 茶晶
[7] 墨晶   [8] 續漢書   [9] 謝承   [10] 哀牢夷

(calcium carbonate), sodium or potassium carbonate, or other alkali ingredients.[a] Another form, developed at a later stage in history, is lead-silicate glass, in which the alkaline earth elements are to some extent replaced by lead. A famous story in Pliny[b] recounts that glass was discovered accidentally by Phoenician merchants, who used bags of the natron (sodium carbonate) which they were carrying, to prop up their cooking-pots when bivouacking at the sandy bar where the river Belus meets the sea near Mount Carmel. Some source of lime may also have been present, and they saw, to their astonishment, small rivulets of glass running from the intense heat of the fire. This story is now dismissed as a fable,[c] though the manufacture of glass is certainly very ancient in Phoenicia and the Lebanon, and it may well be that some especially pure sands there were used. The oldest piece of glass which can be positively dated was made somewhere in Mesopotamia in the -3rd millennium,[d] and the generally accepted view now is that glass manufacture may go back to the neighbourhood of -2900. By the middle of the -2nd millennium solid glass objects were becoming quite common both in Egypt and Babylonia.[e] In the Roman world the industry was a very great one, centred on Alexandria,[f] and some of its products certainly reached China.[g] Fig. 293 shows a bowl from Begram.

(ii) *Chinese glass technology*

Archaeological discoveries have revolutionised our knowledge of ancient Chinese glass-making. The earliest Chinese glass, in the form of opaque beads dating from the Chou period, has been thoroughly studied by Beck & Seligman (1). This glass is of the lead-silicate type, found only as rare and isolated examples in Western Asia from about -700 onwards.[h] But the Chinese glass exhibits another feature unknown among other ancient glasses, in that it contains large amounts of barium. One bead, with a specific gravity of 3·57, was found to contain no less than 19·2% of barium (as BaO).

---

[a] The alkali was often derived from the ash of marine plants, e.g. *Salsola kali* and *Salicornia herbacea*, or from forest potash—hence the name 'Waldglas'.

[b] *Hist. Nat.* XXXVI, 190ff. (tr. Bailey (1), vol. 2, pp. 147ff.), chs. 65–7.

[c] B & M, pp. 706ff. Interesting analyses of the sand at the mouth of the Belus in Haifa Bay have been made (Turner, 3). It would have been quite usable.

[d] Glazed faience beads go back to the -4th (cf. Stone & Thomas, 1)—but this is indeed the pre-history of glass.

[e] Babylonian glass- and glaze-makers early introduced the frit technique, in which the constituents of a glaze are fused to a glassy mass, then broken up and pulverised before being applied to the ware. Such a frit may be $CaO.CuO.4SiO_2$. There are cuneiform textual references both from the -17th and from the -7th centuries; see the useful general reviews of Turner (1–3, 6, 7).

[f] Cf. the histories of occidental glass, e.g. Kisa (1), Neuburg (1) and Honey (1).

[g] Sarton (1), vol. 1, p. 389. Although his account of Chinese glass technology is no longer acceptable, his views on the export of Roman glass to China were abundantly justified by the remarkable findings of Hackin, Hackin, Carl *et al.*, excavating the $+3$rd-century warehouse at Begram, a way-station in Gandhāra on the Old Silk Road.

[h] Notably a cake of sealing-wax-red glass from Nimrud, *c.* -700 (Turner, 4*a*, *b*, 5), a single bead of the -7th century from Rhodes, and three samples of red Egyptian -2nd-century glass (Neumann & Kotyga, 1). The Assyrian specimen had 22·8% lead oxide. There may be some doubt whether the cuneiform texts mention the use of lead in glass, but it is clearly stated in the technical treatises of the two Western monks Heraclius (late $+10$th century) and Theophilus (late $+11$th century). Cf. Turner (1, 7).

Other specimens contained no barium but very large amounts of lead, as much as 70 % (PbO) being found in one bead which had a specific gravity of 5·75.[a]

Many of the pieces of Chinese glass, particularly the beads of the type known as 'eye-beads', in which decorations in the form of eyes were made by drilling the original opaque bead and inserting layers of glass of different colours, were based on European designs. Eye-beads there became common from about −480 in the La Tène period, and by −300 at least, the time of Tsou Yen and Chuang Chou, these beads were reaching China and being copied there. It is worth noting that this export trade preceded the opening of the Old Silk Road by some two centuries.[b] The copying was by no means servile, and later extended to small lion amulets and glass vessels. But most of the Chinese glass was formed into objects of distinctively Chinese character such as dragons and cicadas, often colourless and opalescent, with 12·5 % barium and a specific gravity of 3·75. The discoidal pi[1] emblems,[c] of cosmological significance, have been found made of translucent grey or green glass, and even elaborate sword-furniture, purely Chinese in design, all intended for placing in graves. The number of glass objects which have now been found in tombs of the Warring States, Chhin and Han periods is so large as to suggest that they were then comparatively common.[d] It seems indeed that glass was used by the poorer families as a standard substitute or cheap imitation of jade for funerary purposes.

The extensive observations and analyses made by Seligman & Beck (1), aided by the spectrographic studies of Ritchie (1), have successfully distinguished the Chinese glass objects from the European importations. As a number of the latter were found in Chinese tombs a very careful study was necessary. The Chinese beads were almost invariably made of lead-silicate glass, while the European beads were of soda glass. The presence of barium in many of the beads, which has already been mentioned, enabled their Chinese origin to be established beyond question, for barium is not found in any occidental glass of this early period.[e] The compositions of a number of objects of purely Chinese shape, mostly of lead-silicate glass containing barium, confirmed the conclusions reached from a study of the eye-beads. Thus the high barium content (averaging 4 % BaO) differentiates Chinese glass sharply from European.

The consistent use of lead in early Chinese glass, and still more the use of barium, poses the question as to how far the addition of these two elements was intentional. Barium occurs in North China and Korea, often in conjunction with lead, in the form

---

[a] Seligman & Beck (1). The record density was that of a Buddhist amulet of pre-Thang date.
[b] Cf. Vol. 1, pp. 181 ff.
[c] See below, pp. 105, 112, 145 and Vol. 3, pp. 334 ff.
[d] See e.g. Yang Khuan (3).
[e] Graham & Dye (1) were among the first to show that some of the ancient Chinese glass had lead but not barium. It would be interesting to know the composition of the glass beads, believed to have come from China in ancient times, which have been found in some quantity in Indonesia and New Guinea; see van der Sande (1), vol. 3, pp. 218 ff.; von Heine-Geldern (1), p. 146. Chemical analyses might reveal whether these were European exports faring still further afield or the products of Chinese workshops. On lead glass in general, see Charleston (1).

[1] 璧

of barytes (the sulphate) and witherite (the carbonate). Some specimens of Chinese glass also contain strontium. The suggestion has been made[a] that barium was introduced simply as an impurity in the lead ores. This does not seem very likely in view of the fact that a number of beads with large amounts of lead contain no barium. Furthermore, the lead glazes of the Han period on pottery contain no barium either. Lead and barium possess the property of giving a high refractive index and low dispersion to the glass containing them; they have been used in modern times since 1884 for these purposes.[b] They also lower the melting-point of the glass and make it easier to work. During the Second World War barium was used as a substitute for lead in half proportions. It may be, therefore, that the ancient Chinese craftsmen were aware of the fact that the use of lead and barium gave a slight additional brilliance to the glass, resulting from the higher refractive index.

In passing from the Han to the Thang periods, Chinese glass tends to go over from a Pb-Ba-silicate type to a Pb-Na-Ca-silicate type, and finally to ordinary soft glass (Na-Ca-silicate).[c] This is a rough generalisation, with exceptions which have been already noted.[d] The abandonment of the lead-silicate type of glaze from the Thang period onwards seems also to have been complete.[e]

It is not yet possible to say when the earliest glass was made in China. The evidence of Seligman and Beck places the first manufacture no more precisely than the late Chou period, but Sarton (7), basing himself on the archaeological data of White, accepts the mid −6th century, the time of Confucius, for many beads and for a gilt bronze plaque, the centre inlaid with a convex disc of barium-containing glass, decorated with the eyes design.[f] Sarton acknowledged that this was evidence for glass-manufacture about a thousand years earlier than the formerly accepted date in China, and pointed out that the literary documents would have to be re-examined in the light of it.[g] Glass in plaques made for inlaying metal has been found fairly frequently in Han tombs, and the art may have been the ancestor of the cloisonné technique. Fig. 294 shows the Winthrop mirror, a piece in somewhat the same style as the plaque and disc of White and Sarton.

Particularly important from the optical point of view is the question as to how far it was possible in the Han to make glass of any reasonable degree of transparency.[h]

[a] E.g. by Professor C. G. Cullis in correspondence with Sir Harry Garner.

[b] Actually the first intentional introduction of barium was by J. W. Döbereiner for optical glass in 1829.

[c] See the paper of Seligman, Ritchie & Beck (1).

[d] For instance, the many pieces of ancient Chinese glass which have lead but no barium. Seligman (5) described a number of glass objects of this kind indubitably Han in date. One, a vivid lapis blue sleeve-weight mounted in a bronze ring, had been dated by the impression of a coin of Wang Mang in its base when still pasty; it was probably made therefore about +10. Many other small objects, such as tubes, rods, broken rings and a stylised lizard, certainly made in Han China and not of foreign origin, proved to be of similar composition.   [e] Cf. Hetherington (1).

[f] If the date of −550 is not accepted, the only alternative is −379, just before the birth of Mencius. This is venerable enough.

[g] All this evidence, as he said, cancelled the entry in Sarton (1), vol. 1, p. 389.

[h] The origin of the technique of glass-blowing is unknown, but it was used extensively in the Roman +1st century, with moulds of clay. The art centred on Sidon. It is said by Honey (1) that blown glass was not made in China until the +17th century, but this seems hard to believe. Beautiful blown glass

Many of the objects so far mentioned are translucent, and there are also curious white glass rods of hexagonal cross-section, like pieces of blackboard chalk, but of unknown use.[a] Transparent glass beads, yellow and red, have been found, and are ascribed to Han or Chin.[b] The most splendid piece of old Chinese glass is the so-called Buckley Bowl (Fig. 295), of ribbed spiral form, but opinions differ greatly as to its date, Seligman & Beck regarding it as Chin, and Honey (1) as Ming. In sum, there seems no reason for believing that the glass technicians could not have made objects of transparent colourless glass in the Han.[c]

We are now in a position to examine the literary references to glass, and to lenses made of rock-crystal or glass. The first difficulty which confronts us is the uncertainty of identification of the terms which may have meant glass in early writings. Broadly speaking one may consider *liu-li*[1] as meaning opaque glass or glaze[d] and *po-li*[2] as meaning more or less transparent glass. But the questions raised are very complex, and the last word has probably not yet been said on them.

The expression *liu-li*[3] first prominently appears in the *fu*[4] poetry of the early Han period. In one of these odes or poetical rhapsodies which can be dated −11, Yang Hsiung[5] wrote:[e] 'they break up with mallets the night-shining *liu-li*, and open the shells of oysters pregnant with moon-bright pearls'. Presumably in this place *liu-li* must be a thing. Demiéville (2) favoured some kind of blue precious stone, possibly lapis lazuli or turquoise; yet glass or frit glaze would meet the case just as well if not better. But in another ode by the same writer[f] the term *liu-li* simply means radiance— 'how attractive the radiance of the red decorations!' And in an older poetical essay[g] by Ssuma Hsiang-Ju[6] (*d.* −117) it has the even broader sense of dispersion, occurring as an epithet for birds flying forth in all directions. During the early Han therefore

vases and other vessels have been found in Korean tombs by Hamada & Umehara (1), apparently dating from the late Han period, and certainly pre-Thang. They may, of course, have been imports from the Mediterranean region, and the subject remains open. The *Wu Lin Chiu Shih* of about +1270, but referring to the events of the previous hundred years, has an account (ch. 2, p. 19*a*) of a technique for making 'frameless lamps' (*wu ku têng*[7]), i.e. glass globes, by pouring molten glass round a silk bag full of rice. This internal mould (lit. embryo, *thai*[8]) was afterwards removed, so that a glass bulb (*po-li chhiu*[9]) was obtained. The technique resembles the Chinese practice of moulding lacquer on a framework which is afterwards taken away.

[a] They recall the 'aggry' beads of +4th-century Palestine (Neuburg), but perhaps in China they were destined for mathematical counting-rods. Cf. Vol. 3, pp. 70ff.

[b] In the Canton Museum on Yüeh Hsiu Shan one may see three black glass bowls about $3\frac{1}{2}$ in. in diameter from the former Han, and a green glass robe hook (*tai kou*) from the Later Han.

[c] The author is greatly indebted to Sir Harry Garner for helpful discussions which resulted in the revised form of the preceding paragraphs.

[d] Hence later, by implication, glazed pottery. Note that in the first of these characters the phonetic is the same as that of the word 'to flow'. The semantic significance would thus be 'fusible jade'.

[e] *Yü Lieh Fu*[10] (Ode on the Hunt attended by Soldiers dressed in Feather Coats), in *Chhien Han Shu*, ch. 87A, p. 30*b* and *Wên Hsüan*, ch. 8, p. 15*a*.

[f] *Kan Chhüan Fu*[11] (Ode on the Sweetwater Springs), in *Chhien Han Shu*, ch. 87A, p. 13*b*, and *Wên Hsüan*, ch. 7, p. 4*b*.

[g] *Shang Lin Fu*[12] (Ode on the Imperial Hunting-Park), in *Wên Hsüan*, ch. 8, p. 6*b*. The explanation of the meaning of the phrase is given in the +3rd-century commentary by Chang I.[13]

[1] 琉璃	[2] 玻璃	[3] 流離	[4] 賦	[5] 揚雄
[6] 司馬相如	[7] 無骨燈	[8] 胎	[9] 玻璃毬	[10] 羽獵賦
[11] 甘泉賦	[12] 上林賦	[13] 張揖		

PLATE CII

Fig. 293. Italian gilt glass bowl from the +3rd-century warehouse or customs depot at Begram in Afghanistan, one of the way-stations on a loop of the Old Silk Road (after Hackin, Hackin *et al.*). Such wares were carried from Europe to China at this time.

Fig. 294. The Winthrop mirror (after Seligman, 5). The flat glass surface of the back of the bronze mirror includes as inlays most of the forms of the compound and stratified 'eye' designs found in Chinese glass beads of the late Chou period. This object is likely to be as old as the middle of the −6th century and can hardly be later than the Han (−2nd century).

PLATE CIII

Fig. 295. The Buckley bowl (after Honey, 1). A beautiful ribbed spiral design, but of very uncertain date, estimates varying from Chin to Ming.

liu-li must have come to mean a glassy radiance, and so stabilised as a noun for glassy substance. Indeed the characters were used in the transliterated form of Skr. *vaiḍūrya*, meaning a blue precious stone, as *pi-liu-li*.[1] Yang Hsiung's *liu-li* has been regarded as an abbreviation of this.[a]

The *Chhien Han Shu* has two references to *pi-liu-li*,[2] both of which have always been taken to mean glass. The first says[b] that it was imported from Chi-Pin[3] (Gandhāra, Greek North-west India);[c] the second[d] that the Han emperors sent official agents abroad by sea to southern countries to buy it. Both these statements apply to the time from about −115 onwards. Epigraphic evidence also exists, notably one of the inscriptions seen in the Wu Liang tomb-shrines (c. +147), which speaks of *pi-liu-li* with reference to a carving of a *pi*[4]—this would have been written not long after Pan Ku was composing his history.[e] As we have just seen, actual glass imitations of these jade discs (cosmological symbols)[f] have been recovered from tombs by archaeologists. Soon the term *liu-li* began to be written with the jade radical. In apocryphal treatises of the Chhan-Wei type, dating from the + or −1st century, we learn that the saliva of the gods turns into all kinds of precious things including glass mirrors (*liu-li ching*[5]),[g] or that jade-green glass (*pi-liu-li*[6]) becomes plentiful when there is good government.[h] These ideas may be Taoist, but all the late Han or San Kuo Buddhist texts write *liu-li* in the same way,[i] though generally referring no doubt to the gem-stone *vaiḍūrya*. Later on, Buddhist scholars of the Thang, such as Hsüan-Ying[7] in his *I Chhieh Ching Yin I*[8] (Explanation of Sounds and Technical Terms in the Vinaya), carefully distinguished between *liu-li* as a divine precious stone and the *liu-li* made artificially by man.

We thus have the puzzling situation that the earliest recognisable term for glass accepted by philologists was one of foreign origin,[j] although (as we have seen) the substance had been manufactured in China for centuries previously. Naturally the philologists could not foresee the progress of archaeological knowledge, still less of archaeology backed by chemical analyses. Perhaps the Han court and the scholars were more interested in rarities from abroad than in the imitations of jade produced

---

[a] Pelliot (43); Demiéville (2). But there remain some persistent doubts. China had no Buddhist missionaries nor Sanskrit scholars until the +2nd century (or at best the end of the +1st) yet the term was developing in the −1st. Also, anciently *pi* had a final guttural unsuitable for this transliteration. We may yet have to return to an indigenous evolution of the term 'green fusible glassily radiant substance'.

[b] Ch. 96A, p. 11a.   [c] Cf. Vol. 1, pp. 171, 191ff., and p. 101 above.

[d] Ch. 28B, p. 40a. Noted by Ferrand (3) and Pelliot (30); we shall give a full translation of this interesting passage in Sect. 29 below.

[e] See Chavannes (9), vol. 1, p. 170.

[f] See also extensively below, p. 112.

[g] *Hsiao Wei Yuan Shen Chhi*[9] (Apocryphal Treatise on the Filial Piety Classic; Cantraps for Salvation by the Spirits), in *TPYL*, ch. 808, p. 3b, and *Ku Wei Shu*, ch. 28, p. 10a.

[h] *Hsiao Wei Yuan Shen Chhi*, in *Khai-Yuan Chan Ching*, ch. 114, p. 3b.

[i] References in Demiéville (2).

[j] 'On sait', wrote Pelliot (44), 'que *verulia* est à la base de *pi-liu-li*, le premier nom que les Chinois, vers le début de notre ère, aient connu pour le verre.' This is the Prakrit form of the word.

[1] 璧流璃   [2] 璧流離   [3] 罽賓   [4] 璧   [5] 琉璃鏡   [6] 碧琉璃

[7] 玄應   [8] 一切經音義   [9] 孝緯援神契

for the popular market by Taoist artisans at home, so that the original term for glass was lost. Even if the manufacture was a family or temple secret there must have been some common name for the substance in the Warring States, Chhin and Han periods.[a] But what it was we do not know.

The term *pho-li* or *po-li*[1,2,3] was much studied by the older sinologists.[b] Appearing a good deal later than *liu-li*, it was current during the Thang, and there is mention[c] of red *po-li* glass being sent as tribute by the Bathrik (the Nestorian Patriarch)[d] of Antioch in +643. Hirth (1, 7) thought of deriving *po-li* from a word for glass, *bolor*, found in Mongol and other Central Asian languages, but the Chinese themselves, e.g. Hsüan-Ying in the work just mentioned (c. +649), believed that it originated from Skr. *sphaṭika* (transliterated *sê-pho-thi ka*[4] or *tsai-pho-chih-ka*[5]), meaning rock-crystal.[e] In the *Ta Thang Hsi Yü Chi* of Hsüan-Chuang (+646) it appears as *pho-thi*.[6] Thang naturalists confused it with rock-crystal, or perhaps with obsidian (volcanic glass), entertaining the idea that it was water or ice which had concreted after thousands of years in the earth.[f] As for *bolor*, Pelliot (43) concluded that the word came from a Turco-Persian form *bilūr*, itself certainly descended from *vaiḍūrya* through a Prakrit intermediary *verūlya*. It is thus a derivative parallel to *pi-liu-li*, not to *po-li*.

This pattern of the double origin of glass, indigenous and imported, continued throughout the centuries, and it is impossible to separate the story into distinguishable strands.[g] Let us therefore glance at the texts in a historical order. The oldest statement that may have something to do with glass-making is the legend of the female semi-creator, Nü-Kua,[7] who 'fused minerals of five colours in order to repair the blue heavens'.[h] One of the oldest sources of this is a fragment in the *Lieh Tzu* book,[i] but it is also mentioned in *Huai Nan Tzu*[j] and in the *Lun Hêng*.[k] Schlegel (8) thought that it was a reference to coal, and de Lacouperie (3) believed it had something to do,

[a] Chang Hung-Chao (*1*) wished to identify *liu-li* with a mysterious stone mentioned in the Yü Kung chapter of the *Shu Ching*, *chhiu-lin*,[8] and to regard both as having originally meant lapis lazuli. Demié-ville (2) considered that some kind of sky-blue jade was a much better solution for *chhiu-lin*. All the ancient references agree that whatever *chhiu*[9] (and its homophone *chhiu*[10]) were, their colour was caerulean. Perhaps they ought to be reconsidered as candidates for glass.

[b] E.g. Parker (1); Hirth & Rockhill (1), p. 228.

[c] *Hsin Thang Shu*, ch. 221 B, p. 11a; (tr. Hirth (1), pp. 60, 294).

[d] The Po-to-li.[11]

[e] The use of this new term might be paralleled by the French custom in modern times of referring to any 'very pure and limpid white glass' as 'cristal'.

[f] So Chhen Tshang-Chhi, who was the first to include *po-li* glass in a general natural history. This was his *Pên Tshao Shih I* of +725; the entry is preserved in *PTKM*, ch. 8, p. 55b. The idea of the transformation from ice seems to be of foreign, presumably Indian, origin.

[g] The approach of Laufer (10) is now untenable. He argued that the term *liu-li* must refer to glazed pottery and not glass on the ground that glass was not made in China in Han times, and that the Chinese hardly ever used glass for vessels. On the other hand he may have been at least partly right in his idea that some of the *liu-li* imported to China from Western Asia was frit for glazes rather than glass.

[h] *Lien wu sê shih i pu tshang thien.*[12]

[i] Ch. 5, p. 3b (tr. L. Giles (4), p. 85; Wieger (7), p. 131).

[j] Ch. 6, p. 7a.

[k] Ch. 31 (Forke (4), vol. 1, p. 250), ch. 46 (Forke (4), vol. 2, p. 347).

[1] 玻瓈	[2] 頗黎	[3] 玻璃	[4] 塞頗胝迦	[5] 窣坡致迦
[6] 頗胝	[7] 女媧	[8] 璆琳	[9] 璆	[10] 球
[11] 波多力	[12] 鍊五色石以補蒼天			

not with glass, but with clays of different colours (which were certainly called *wu sê shih*[1] in later times); while Laufer (14) dismissed the whole legend as mythological nonsense. In view of what we know now about glass-making in pre-Han times, however, it does seem likely that the legend was a garbled reference to glass. Ancient glass had a strong tendency to iridescence, and its broken edges would be likely to show the spectral colours. Interest in spectral colours arose quite early in China. Though the *Shih Tzu*[2] book, attributed to Shih Chiao,[3] is not, as we have it, likely to be genuinely Chou, or even Han, it may well be pre-Thang; in this we find:[a]

> The five colours of sunlight are the essence of the Yang and resemble the virtue of a prince. They (may be made to) stream forth with great brilliance.

And Joseph Priestley wrote, in his history of optics (+1772):[b] 'The colorific property, as I may call it, of prisms, makes them of great value in the East, as we learn from Fr. Trigautius, in his account of his mission to China, who says that 500 pieces of gold had been given for one of them, by a person who got it upon these terms with great difficulty, it being thought to be fit for sovereign princes only.[c] Fr. Kircher says the same in his *China Illustrata*.' Of course, Europeans in Han times, such as Seneca, had known the identity of rainbow colours and those formed at the edge of a piece of glass. Surely the Nü-Kua legend was the repairing of the transparent heavens by a rainbow art like that of the glass-makers.

The first really suggestive evidence of glass-making occurs in the Later Han, in the *Lun Hêng* of Wang Chhung, but as it has a specific connection with burning-lenses we may defer consideration of it for a page or two. From the +3rd century onwards references become much more frequent. About +264 the author of the *Wei Lüeh*[4] recorded in his account of Roman Syria (Ta-Chhin)[d] that among the products of that country were ten sorts of *liu-li* (or frit) of different colours.[e] Then about +300 Wan Chen, in his *Nan Chou I Wu Chih* (Strange Things of the South), wrote:[f]

> The basic substances of *liu-li* glass are minerals. In order to make vessels from them they must be worked by means of soda-ash (*tzu-jan hui*[5]). This material has the appearance of yellow ashes; it is found on the shores of the southern seas and can also be used for the washing of clothes.[g] There is no need to soak it (for a long time); it is just thrown into water

---

[a] Ch. 2, p. 1a. Tr. Forke (13), p. 528; eng. auct.   [b] P. 169.

[c] Gallagher tr., p. 318. Cf. Vol. 3, p. 438.   [d] See Vol. 1, pp. 174, 186.

[e] Preserved in *San Kuo Chih*, ch. 30, p. 33b; *TPYL*, ch. 808, p. 4a, etc. Tr. Hirth (1), p. 73.

[f] Cit. *TPYL*, ch. 808, p. 4b; tr. auct. adjuv. Laufer (10), p. 145.

[g] The connection here between the glass and soap industries is quite significant, for it shows that the product from the south was a crude sodium carbonate like the Levantine 'polverine' or the Spanish 'barilla' used by European medieval glass-makers (cf. Sherwood Taylor (4), pp. 69ff.). These 'glass-maker's salts' could also be mainly potassium carbonate if derived from land plants. When the lye was 'sharpened' by boiling with quicklime the caustic alkali hydroxides were formed, and with these, fats could be saponified. Chinese literature has many references to 'salt' obtained by burning a swamp-weed, the *shui po*[6] or *shui tshai*[7] (*Menyanthes trifoliata*, R171, B11, 398, III, 199; Stuart, p. 263). In +828 the process was for a time forbidden as infringing the State salt monopoly (*Tshê Fu Yuan Kuei*, ch. 494; *Thang Hui Yao*, ch. 88 (p. 1611); *Hsin Thang Shu*, ch. 54, p. 2b). I am indebted to Dr D. C. Twitchett for bringing this to our notice.

[1] 五色石   [2] 尸子   [3] 尸佼   [4] 魏略   [5] 自然灰   [6] 水柏
[7] 睡菜

(and dissolves quickly), giving a solution slippery to the touch[a] like moss-covered stones. Without these ashes (the other minerals) will not dissolve.

This passage does not suggest that glass was made only outside China but rather hints that crude carbonate was imported at this time. Glass vessels were now becoming quite widespread. In the *Chin Shu* we read[b] of glass wine-cups (*liu-li chung*[1]), and other sources speak of glass bowls (*liu-li wan*[2]).[c] Among Wan Chen's contemporaries was the great alchemist Ko Hung. In his *Pao Phu Tzu*, written also about +300, there is a striking passage to which Waley (14) first drew attention:

> The 'crystal' vessels, which are made outside China, are in fact prepared by compounding five sorts of (mineral) ashes. Today this method is being commonly practised in Chiao and Kuang (i.e. Annam and Kuangtung). Now if one tells this to ordinary people, they will certainly not believe it, saying that crystal is a natural product belonging to the class of rock-crystal.[d]

And he goes on to say that ordinary people are so stupid that they will not believe it is possible to make gold because natural gold already exists. Moreover, in the *Shen Hsien Chuan*, possibly of about the same date, it is said that 'the eight minerals can be liquefied until they resemble water'.[e]

It is not long after this that we come to the passages which were formerly considered the earliest relating to glass-manufacture. In the *Pei Shih* we read:[f]

> During the time of the emperor Thai Wu (of the Northern Wei dynasty) (r. +424 to +452), traders came to the capital of Wei from the Ta Yüeh-chih[3] country,[g] who said that by fusing (*chu*[4]) certain minerals they could make the five colours of *liu-li* glass. They then gathered (materials) and dug in the hills, and fused the minerals at the capital. When ready, the material so obtained was of even greater brilliancy than the *liu-li* glass imported from the West. An edict was issued that a movable palace should be made of this material, and when it was done it held more than a hundred people. It was bright and transparent so that all who beheld it were astonished and thought it was made by magical power. After this, articles made of glass became considerably cheaper in China than they had been before, and no one regarded it as particularly precious.

---

[a] A characteristic property of strong alkalis in solution.
[b] *Chin Shu*, ch. 45, p. 10b.
[c] *Yuan Chien Lei Han*, ch. 364, p. 31b.
[d] Ch. 2, p. 13a. Tr. Feifel (1), p. 179; Waley (14), p. 13; mod.
[e] Ch. 4 (Feifel (1), p. 197).
[f] Ch. 97, p. 19a; repeated verbally in *Wei Shu*, ch. 102, p. 19b. Tr. auct. adjuv. Hirth (1), p. 231; Hirth & Rockhill (1), p. 227. A similar passage is said to occur in the *Sung Shu*, to the effect that the king of Ta-Chhin (Roman Syria) sent glass-makers to the Sung emperor, about the same date as above. But I have not been able to find it, and suppose that Hirth could not either, as he gave no reference.
[g] Presumably the Kushān empire, or what remained of it, in Bactria, Gandhāra and North-west India. *TPYL*, quoting the *Wei Shu* text, makes the traders come from Thien-Chu[5] (India); ch. 808, p. 4a.

---

[1] 琉璃鍾   [2] 琉璃椀   [3] 大月氏   [4] 鑄   [5] 天竺

From the description, these glass technicians must have been making glass for some kind of screens. Elsewhere in the *Pei Shih* it is said that

By the time of the Sui (+581 to +618) there had been for a long time no more glass-makers, and no one dared to attempt it. But Ho Chhou[1] succeeded in making it like glaze on pottery, and it was the same as of old.[a]

The +6th century also had its imported glass. In +519 the first Liang emperor had received tribute of glass pitchers (*liu-li ying*[2]) from Khotan.[b] They must indeed have been well packed to survive their long journey across the Gobi over the Old Silk Road.

The persistent uncertainty of scholars about the distinction between artificial glass and obsidian or rock-crystal is well seen in the commentary of Yen Shih-Ku[3] (+579 to +645) on the *Chhien Han Shu* passage mentioned above concerning the glass from Chi-Pin. It follows a few words by Mêng Khang,[4] a San Kuo commentator (c. +180 to +260).[c]

Mêng Khang says: '*liu-li* glass is of a green colour like jade'.

Yen Shih-Ku says: 'according to the *Wei Lüeh*, the country of Ta-Chhin exports *liu-li* glass (or frit) of ten different colours, light red, white, black, yellow, blue-green, green, yellowish green, deep purple, dark red, and purple. The definition of Mêng Khang was much too narrow. This material is a natural product, variegated, glossy and brilliant, exceeding all hard stones (lit. jade), and most constant in colour. Nowadays the common people make (something of the kind) by melting certain minerals, to which they add a number of chemical substances, and so pouring (into moulds, make vessels).[d] The mass, however, is hollow, brittle and not evenly compact—it is not the genuine article.'

This is rather reminiscent of the famous story about Lord Curzon's inkstand.

There would be no point in assembling many quotations about glass from the Sung and later times.[e] By now the statements were becoming quite explicit if not always very accurate. An interesting note in the *Yün Lin Shih Phu* (Cloud Forest Lapidary) of +1133 says:[f]

At the western capital, in the Lo River, they find pieces of bluish white stone with spots of five colours in it. The whitest of these are compounded with lead, and mixed with other minerals, then after heating it is all changed into 'false jade' or *liu-li* glass for use.

If this was really the first occasion on which a scholarly writer mentioned the use of lead salts in glass-making, then it had taken some fifteen centuries for this information

[a] Ch. 90, p. 18a, tr. auct.

[b] *Liang Shu*, ch. 54, pp. 43a, b. Cf. *TPYL*, ch. 758, p. 4a, quoting the *I Yuan*.

[c] *Chhien Han Shu*, ch. 96A, p. 11a. Tr. auct. adjuv. Laufer (10), p. 145. Yen Shih-Ku's description of glass-making was quoted in later times, as in the *I Chao Liao Tsa Chi*[5] of c. +1200 (ch. 2, p. 54a).

[d] *Chin su so yung chieh hsiao shih chih, chia i chung yao, kuan erh wei chih.*[6]

[e] On later Chinese glass see Bushell (2), vol. 2, pp. 58ff.

[f] Ch. 2, p. 3b, tr. auct.

[1] 何稠　　　[2] 瑠璃罌　　　[3] 顏師古　　　[4] 孟康
[5] 猗覺察雜記　　　[6] 今俗所用皆消石汁加以衆藥灌而爲之

to pass from the artisans to the literati. One might almost think that it was easier for technical knowledge to travel the length and breadth of the Old World than to pass the social barriers within a single civilisation. The reference to the 'false jade' is also interesting, recalling as it does the production of the cheaper substitute for Han tombs.

Another book of about the same time[a] mentions baskets of raw materials for making glass which had apparently been presented as tribute from Arabia, in +1114. Glass objects themselves were brought in considerable amount by the embassy[b] from the Chola king Kulottanga I in +1077, but Gode (2) who has studied the matter believes that they were made not in Ceylon or on the Coromandel coast of South India, but in southern Arabia. Hence the interest of the following passages. The *Yen Fan Lu* of +1175 compares Chinese with foreign glass as follows:[c]

The *liu-li* which is made in China is rather different from that which comes from abroad. The Chinese variety is bright and sparkling, and the material is light but fragile. If you pour hot wine into it, it will immediately break. That which is brought by sea is rather rough and unrefined, and the colour is also slightly darker. But the strange thing is that even if hot water is poured into it a hundred times, it behaves like porcelain or silver and will never break.

The explanation of this is found when we read what Chao Ju-Kua says[d] in his *Chu Fan Chih* (Records of Foreign Peoples) about +1225.

Liu-li glass comes from several of the Arab (Ta-Shih[1]) countries. The method they follow in heating and melting it (*shao lien*[2]) is the same as that used in China; it is made by heating lead (carbonate), natron (*hsiao*[3])[e] and gypsum (*shih kao*[4]).[f] To these materials the Arabs add borax from the south (*nan phêng sha*[5]), which causes the glass to be elastic without brittleness, and indifferent to temperature, so that one may put it in (hot) water for a long time without spoiling it. Thus it is more valuable than the Chinese product.

Apparently, therefore, some of the glass exported to China by the Islamic countries in the +13th century was of the borosilicate or pyrex type. Finally, the idea that natron was one of the ingredients led to the use of yet another term for glass in Yuan and Ming books, *hsiao-tzu*[6] (offspring of natron).[g]

---

[a] *Thieh Wei Shan Tshung Than*, ch. 5, p. 19b.
[b] On this see Chao Ju-Kua's *Chu Fan Chih*, ch. 1, p. 18b; tr. Hirth & Rockill (1), p. 96. The Chinese name for the Chola kingdom was Chu-Lien.[7]
[c] Cit. by Chang Hung-Chao (1), p. 14.
[d] Ch. 2, p. 11b, tr. auct. adjuv. Hirth & Rockhill (1), p. 227.
[e] Strictly this word should mean saltpetre, potassium or sodium nitrate. But in all medieval cultures there was confusion between the nitrate and the carbonate; cf. Bailey (1), vol. 1, p. 169. The question is important in relation to the history of gunpowder (see Sect. 30 below), and any given reference to saltpetre can only be assured by noting the properties attributed to it. It is not impossible that saltpetre was really intended here; perhaps it was a way of making potash glass of flint or Bohemian type.
[f] Calcium sulphate, here perhaps a mistake for some form of the carbonate.
[g] Thus in the *Ko Ku Yao Lun*, ch. 6, p. 5a. It was the author of this book, the archaeologist Tshao Chao (+1387) who exploded the persistent belief in the origin of rock-crystal from ice. 'This is obviously false', he wrote, 'since green and red varieties of rock crystal occur in Japan.' Three centuries earlier Marbodus in Europe had been equally sceptical (Laufer (14), p. 190), as also in China Thang Shen-Wei (*Chêng Lei Pên Tshao*, ch. 3, p. 44a).

[1] 大食   [2] 燒煉   [3] 硝   [4] 石膏   [5] 甫鵬砂   [6] 硝子
[7] 注輦

Thus the general impression which is given by the evidence both archaeological and literary is that almost from the middle of the − 1st millennium onwards an indigenous Chinese glass industry, the roots of which lay doubtless in ancient Mesopotamia, ran parallel with a considerable trade in imported glass wares of particular kinds and some special raw materials.[a] As the imports aroused the attention of the educated scholars more than the home products we remain rather in the dark about the details of glass manufacture as the Chinese practised it. The art seems sometimes to have been recondite in character, and often distinctly localised, so that here and there it had to be revived from time to time. But the chief question which this discussion set out to answer, namely whether the Chinese could have had burning-glasses or other lenses from the Han onwards, seems to be settled clearly in the affirmative. We must now return to optical science, the proper subject of this Section, and consider lenses themselves.

(iii) *Burning-glasses and the optical properties of lenses*

With regard to lenses, it was easy for Laufer (14) to show that Schlegel[b] had been wrong in attributing a rock-crystal burning-lens (*shui ching ta chu*[1]) to *Huai Nan Tzu*, by mixing up the original − 2nd-century text with a passage about it written as late as the Ming (+ 1579).[c] He also set right other mistakes. But when he came to the *Lun Hêng* of + 83 his interpretation that only bronze mirrors were meant seems now open to question in the light of the archaeological findings.

Wang Chhung has three passages about preparing an instrument for bringing the rays of the sun to a focal point. In the first[d] he simply says, in connection with the *yang sui*,[2] that in the fifth month, at the height of summer, people 'liquefy and transform five minerals, casting therewith an instrument with which they can catch fire (from heaven)' (*hsiao lien wu shih...chu i wei chhi nai nêng tê huo*[3]). The second[e] adds to these words the information that the Chi Tao chih Chia[4] (Taoist technicians) do it, and that they do it on a *ping-wu*[5] day in the fifth month. They cast *yang sui*[2] (surely bronze mirrors) as well as the 'instrument'. Wang Chhung makes the interesting remark here that all this 'is not exactly a natural occurrence, but Heaven's response makes it a natural process' (*fei tzu-jan yeh, erh Thien jan chih yeh*[6]). But the third[f] is the most interesting of all, and requires quotation with fuller context than Laufer (14) gave it.

---

[a] In Sect. 35 below, on ceramic technology, we shall find an outstanding example of this in the cobalt ore used for the blue colouring of pottery and porcelain. Young & Garner (1) have shown that before the Ming all the cobalt pigment was free from manganese and therefore probably emanated from Persia as an imported raw material in the form of arsenical cobalt oxide. During the Chhing period the blue pigment is always rich in manganese, revealing the use of indigenous Chinese high-Mn ores. The Ming itself is a transition period, cobalt from both sources being used.

[b] (5), p. 142.

[c] The *Liu Chhing Jih Cha*[7] (Diary on Bamboo Tablets) of Thien I-Hêng.[8]

[d] Ch. 80 (Forke (4), vol. 2, p. 132).   [e] Ch. 47 (Forke (4), vol. 2, p. 350).

[f] Ch. 8 (Forke (4), vol. 1, pp. 377 ff.).

[1] 水精大珠   [2] 陽燧   [3] 消鍊五石鑄以爲器乃能得火   [4] 伎道之家

[5] 丙午   [6] 非自然也而天然之也   [7] 留青日札   [8] 田藝衡

In the Tao of Heaven there are genuine (*chen*[1]) things and counterfeit (*wei*[2]) things; the true things are firm in their correspondence with Heaven's naturalness; the artificial things are due to human knowledge and art—and the latter are often indistinguishable from the former.

The Tribute of Yü (chapter of the *Shu Ching*) speaks of bluish jade (*chhiu-lin*[3]) and *lang-kan*[4] (possibly agate, ruby, or coral). These were the products of the earth, and genuine like jade and pearls. But now the Taoists melt and fuse (*hsiao shuo*[5]) five kinds of minerals and make 'jade' of five colours out of them. The lustre of these is not at all different from that of true jade. Similarly, pearls from fishy oysters are like the bluish jade of the Tribute of Yü; all true and genuine (natural products). But by following proper timing[a] (i.e. when to begin heating and how long to go on) pearls can be made from chemicals (*yao*[6]), just as brilliant as genuine ones. This is the climax of Taoist learning and a triumph of their skill.

Now by means of the burning-mirror (*yang sui*[7]) one catches fire from heaven. Yet of five mineral substances liquefied and transmuted on a *ping-wu* day in the fifth month, an instrument (*chhi*[8]) is cast, which, when brightly polished and held up against the sun, brings down fire too, in precisely the same manner as when fire is caught in the proper way.

Indeed, people go so far now as to furbish up the curved blades of swords, so that when held against the sun they attract fire also. Though curved blades are not (strictly speaking) burning mirrors, they can catch fire because of the rubbing to which they have been subjected.

[And he goes off into an analogy with the effect of education in improving men.][b]

Can there be much doubt that this is an account of the making of glass burning-lenses? The whole passage is built of antitheses between 'genuine' and 'imitation' things.

---

[a] We are quite aware that these words, *sui hou*,[9] were translated by Forke (4) as 'the Marquis of Sui'. But as has already been seen (Vol. 2, pp. 330ff.), the *huo hou*[10] or 'fire-times', i.e. the times when heating should begin and end, were both ancient and important in Chinese alchemy. It almost looks as if we are here in the presence of an old alchemical pun. The legend about the Marquis of Sui was that having healed a wounded snake with drugs, it brought him afterwards in gratitude a (real) pearl (*Chuang Tzu*, ch. 28; tr. Legge (5), vol. 2, p. 154). This he used in a singular way, namely, for shooting at birds (cf. the proverbial reference in *Pao Phu Tzu*, ch. 1; tr. Feifel (1), p. 130). More important is the passage in *Huai Nan Tzu* (ch. 6, p. 3*b*) which reads as follows: 'As for the Tao, it has no private desires or preferences, whether to go or to come. If you know how to make use of it you will have abundance; if you are unskilful you will meet with poverty. Following it you will get benefits, going against it will bring misfortune. For example there was "the Marquis of Sui's pearl" and "Mr Ho's jade disc (*pi*[11])".' Getting (the Tao) means wealth and losing it means poverty.' Now the +3rd-century commentator Kao Yu took this *au pied de la lettre*, glossing it with the Marquis legend, and giving the famous story about Pien Ho[12] (*Shih Chi*, ch. 83, p. 10*b*, *Lun Hêng*, chs. 8, 15, 29, 30). Pien Ho was a man of the −7th century who brought a piece of uncut stone from Ching-shan (*Ching-shan pho*[13]) to Prince Hsiung (Wu Wang) of Chhu. When the jade-workers pronounced it to be worthless, the man was punished by having his feet cut off, yet afterwards it turned out to be excellent jade. But this eventual recognition could not restore 'Mr Ho's' lost feet; 'getting the Tao' had not been a benefit to him. Perhaps we may be permitted the alternative suggestion, in view of what is now known about ancient Chinese alchemy and glass-making, that Liu An was using two appellations which, to the initiated, referred to the opaque glass substitute for jade. This would no doubt have brought prosperity to its manufacturers. And the fact that in the story the Marquis of Sui gave forth drugs (chemical substances or minerals) and subsequently got a pearl, would have been just the kind of analogy which the Taoist alchemists would have delighted to use as a punning description of their processes for making artificial pearls and jade. Of course *sui hou*[14] (the Marquis of Sui) and *sui hou*[15] (following the (fire-)times) would have been very easily confused by copyists.

[b] Tr. auct. adjuv. Forke (4).

[1] 眞	[2] 僞	[3] 璓 琳	[4] 琅 玕	[5] 消 爍	[6] 藥
[7] 陽 燧	[8] 器	[9] 隨 侯	[10] 火 候	[11] 璧	[12] 卞和
[13] 荆 山 璞		[14] 隋 侯	[15] 隨 侯		

Elsewhere, Wang Chhung has said that there can be things not 'natural' by occurrence and production, yet 'natural' by function, though man-made. Now he contrasts first the opaque glass 'jade' and 'gems' made by the Taoists with the real thing; secondly, the artificially made 'pearls' with true pearls; and thirdly, the 'instrument' made by liquefying five different minerals, which can concentrate the sun's rays just as well as the traditional bronze mirror. Even more 'ersatz' are the hollow-ground sword-blades with which the soldiers light their fires. All these passages were considered by Laufer (14) to refer to the casting of bronze, but it is not obvious why five different minerals should be so carefully specified. Bronze would require only two ores, perhaps even only one, with possible addition of a flux. Glass needs silica, limestone, an alkaline carbonate, and perhaps litharge or the barium mineral, together with colouring matter. Of course, the text does not clearly tell us that lenses were being made; the instruments might have been simply glass mirrors imitating the bronze ones. Yet the fact that the passage on burning-glasses follows immediately on the discussion of pearls is very suspicious, since the term 'fire-pearl' (*huo chu*[1]) in later centuries undoubtedly meant a burning-lens. The whole quotation also fits in perfectly with what we have seen above concerning the substitution of glass for jade and bronze objects in tombs. In sum, therefore, we may be justified in concluding that in the +1st century, and probably as far back as the −3rd century, biconvex lenses of glass could be artificially made.[a]

As for the mysterious importance of the *ping-wu* day, Pelliot (23) pointed out that we find this mentioned in inscriptions on many bronze mirrors. Such a day was considered auspicious for casting operations since the first cyclical character was associated with the west and therefore with metal, while the second was associated with the south and therefore with fire. But there is no reason why this should be confined strictly to the casting of bronze; glass has always partaken of the nature of metal, as we see in the term 'pot-metal' still used by glass-makers. Correspondingly, the moon-mirrors were cast on *jen-tzu*[2] days in the twelfth month, these cyclical signs being associated with the complementary elements of water and wood.[b]

Next comes, from the end of the +3rd century, the curious story about the use of ice as burning-lens. In a section on conjuring tricks (*hsi shu*[3]), the *Po Wu Chih* says:[c]

A piece of ice is cut into the shape of a round ball and held facing the sun. Mugwort tinder is held to receive the bright beam from the ice, and thus fire is produced. There has been much talk about getting fire by the use of pearls, but this (ice) method has not been employed.

---

[a] The conclusions in this paragraph were reached without knowledge of the many finds of glass objects in tombs dating back as far as the Warring States period, made in recent years. They are naturally strengthened considerably by such evidence. We are also glad to find our interpretation of these texts accepted by contemporary historians such as Yang Khuan (3), pp. 231, 239.

[b] *Huai Nan Wan Pi Shu*, cit. *TPYL*, ch. 58, p. 7a.

[c] *Po Wu Chih*, ch. 4, p. 6a. Identical passage in the *Huai Nan Wan Pi Shu*, cit. *TPYL*, ch. 736, p. 9a, without the last sentence about the pearls. Tr. auct.

[1] 火珠   [2] 壬子   [3] 戲術

Although ice can be used in this way (experiments of the kind were made by Robert Hooke in the early days of the Royal Society),[a] it seems more likely that what Liu An and Chang Hua were really talking about was a lens of rock-crystal or glass.[b] We may remember that the original meaning of the Greek word *crystallos* (κρύσταλλος) was ice, and that, as already mentioned, there was a persistent theory in China, probably of Buddhist origin, that ice turned into rock-crystal after thousands of years.

About +520 envoys of Fu-Sang[1] are said[c] to have arrived in China, bringing with them a precious stone suitable for observation of the sun (*kuan jih yü*[2]) 'of the size of a mirror, measuring over a foot in circumference, and as transparent as glass (*liu-li*); looking through it in bright sunlight, the palace buildings[d] could be very clearly distinguished'. This comes from the *Liang Ssu Kung Chi* (Tales of the Four Lords of Liang),[e] written about +695. Schlegel[f] thought this was rock-crystal, but glass seems at least as possible. The same book also contains the following interesting account:[g]

A large junk of Fu-Nan (Cambodia) which had come from Western India arrived (in China) and (its merchants) offered for sale a mirror of a peculiar variety of pale green glass (*pi po-li ching*[3]), 1 ft. 4 in. across its surface, and 40 catties in weight. On the surface and in its substance it was pure white and transparent; and on its obverse it displayed many colours (presumably spectral). When held against the light and examined, its substance was not discernible. On enquiry about the price, it was given at a million strings of copper coins. The emperor ordered the officials to raise this sum, but the treasury did not hold enough.... Nobody in the empire understood what the traders said, or dared to pay their price.[h]

Of more direct optical interest is the consideration of 'fire-pearls' (*huo chu*[4]). Li Shih-Chen, in the *Pên Tshao Kang Mu* (end of +16th century) gave an account of them as an appendix to his entry on rock-crystal.[i] By a confusion which will be explained immediately below, they were sometimes thought to be of *huo chhi*,[5] a substance which must undoubtedly be identified with mica.[j] All the dynastic histories

---

  [a] Priestley (1), p. 170.

  [b] I find the style and content of this text archaic, but Laufer firmly denied its authenticity. The whole of his argument about the 'ice lenses' seems to me to show him at his most unconvincing. Having made up his mind that the Chinese had no knowledge of lenses before the importation of foreign rock-crystal ones (fire-pearls) in the Thang, there was no hope for Chang Hua. Chou Mohists and Thang Taoists would have been surprised to learn that 'the Chinese never cultivated natural observation or optical study' (Laufer (14), p. 224). He also ignored the *Huai Nan Wan Pi Shu* parallel.

  [c] The country is of uncertain identity. Schlegel (7) thought Sakhalin and Northern Japan. Cf. Sect. 29.

  [d] Texts vary here, as Schlegel (7a) noted long ago, and these words may refer to palace buildings in the sun, not upon earth, in which case the passage is a garbled account of the observation of sunspots. Cf. Vol. 3, p. 436.

  [e] Cit. *TPYL*, ch. 805, p. 8b.   [f] (7a), p. 138.

  [g] Cit. *TPYL*, ch. 808, p. 5b.

  [h] What the traders said was a garbled version of the legend about the collection of diamonds by exposing the flesh of large beasts. Cf. Laufer (12). Tr. auct. adjuv. Laufer (14), p. 200, (12), p. 19.

  [i] Ch. 8, p. 56b, tr. Laufer (14), p. 189.

  [j] The commonest name for this is *yün mu*.[6] Schafer (5) has given an interesting account of its place in medieval Chinese mineralogy and pharmacy.

[1] 扶桑   [2] 觀日玉   [3] 碧玻璨鏡   [4] 火珠   [5] 火齊
[6] 雲母

agree[a] that this mineral came from India (still one of the greatest world sources of mica); as also do several +3rd- and +4th-century accounts.[b] China, however, does not lack mica deposits. The *Shih I Chi*[c] speaks of a mica mirror which gave a sound echo as well as reflecting light, referring it to the −6th century; all this means is that Wang Chia had seen something of the kind himself in the +3rd. The use of fire-screens (very similar to one of the modern industrial uses of mica) gave rise, no doubt, to the name of 'fire-regulating substance'. But mica is not a substance of which lenses could ever have been made.

Apart from the indications we have already seen in texts such as the *Lun Hêng* and the *Po Wu Chih*, the first widespread mention of *huo-chu*[1] ('fire-pearls') comes in the Thang. The *Hsin Thang Shu* says,[d] in discussing the Lo-Chha[2] and Tan-Tan[3] kingdoms:[e]

Their country produces fire-pearls in great number, the biggest attaining the size of a hen's egg. They are round and white (transparent), and emit light to a distance of several feet. When held against the rays of the sun, mugwort and rush (tinder) will be ignited at once by fire springing from the pearl. The material looks like rock-crystal.[f]

These burning-lenses, which Laufer took to be rock-crystal (though glass does not seem at all excluded), were constantly being presented, as by a king of Lin-I in +630,[g] from Magadha and Kashmir in +641, etc.[h] In +607 Chhang Chün,[4] while on a mission to the Chhih-Thu[5] kingdom (Siam), went to Lo-Chha to buy them.[i] The Japanese scholar-monk Ennin[6] presented one to the Shintō gods in thanksgiving for his safe arrival in China in +839 after a hazardous sea voyage.[j] According to some editions of the *Chêng Lei Pên Tshao*, they were being made in China in the Sung.[k] As has been pointed out, *huo chu*[1] is a direct translation of Skr. *agnimaṇi*,[l] but the term was not new in Thang China, since it had earlier been applied to meteors. There can be little doubt that one of the applications of the lenses in Thang and Sung was in medicine for cauterisation. Li Shih-Chen says that in his time physicians ignited moxa with them so as not to hurt the patient.

[a] *Liang Shu*, ch. 54, p. 21*b*; *Nan Shih*, ch. 78, p. 14*b*, cit. *TPYL*, ch. 802, p. 11*b*; etc.
[b] *Nan Chou I Wu Chih*, cit. *TPYL*, ch. 809, p. 2*a*; *Wu Lu Ti Li Chih*, cit. *TPYL*, ch. 809, p. 1*b*.
[c] Ch. 3, p. 5*a*.   [d] Ch. 222c, p. 2*a*.
[e] Lo-Chha, though 'the country of the Rākṣas', is not Ceylon, but rather Pahang in Malaya; Tan-Tan is probably Kelantan; cf. Gerini (1), Purcell (3).
[f] The last sentence appears only in the *Chiu Thang Shu* version, ch. 197, p. 1*b*, cit. *TPYL*, ch. 803, p. 1*b*. Tr. Laufer (14); Pfizmaier (94), p. 630.
[g] *Hsin Thang Shu*, ch. 222c, p. 1*b*.
[h] *Ibid*. ch. 221A, p. 12*a*; ch. 221B, p. 8*a*, etc.
[i] Cf. *Sui Shu*, ch. 82, p. 4*a*; Hirth & Rockhill (1), p. 8; Pelliot (17).
[j] His account of the presentation of the *hitorutama* is in his diary, the *Nittō Guhō Junrei Gyōki*,[7] tr. Reischauer (2), p. 117.   [k] Ch. 3, p. 44*a*.
[l] 'New fire' was traditionally obtained by the use of burning-glasses (*sūryakānta*) in India; cf. Blochmann (1), p. 48, translating *Ā'īn-i Akbarī* (c. +1590). 'There is also a shining white stone, called *chandrakānta*, which upon being exposed to the beams of the moon, drips water'; cf. p. 90 above.

[1] 火珠   [2] 羅刹   [3] 丹丹   [4] 常駿   [5] 赤土   [6] 圓仁
[7] 入唐求法巡禮行記

In this same passage[a] Li Shih-Chen quoted the *Hsin Thang Shu* wrongly, writing *huo-chhi chu*[1] instead of *huo chu*. Certainly in his mind they were exactly the same thing. He stated that the *huo-chhi chu* had first been mentioned in the *Shuo Wên* (the great dictionary of +121), that these *huo-chhi* pearls were what the *Chhien Han Shu* had known as *mei-hui*,[2] and identical with the *huo ching*[3] referred to in the *Hsü Han Shu* (see p. 100 above). Li Shih-Chen might have been surprised at the warmth with which Laufer,[b] intent on proving the absence of burning-glasses in pre-Thang China, rejected his ideas. Laufer maintained that all these terms meant only mica,[c] and that this had constantly been confused with lenses of rock-crystal and glass. Chang Hung-Chao,[d] on the other hand, believed that *huo-chhi chu* should be translated 'burning-lenses brilliant and vitreous like mica'. Later, Demiéville (2), while adhering to the interpretation of *mei-hui* as mica, gave several further lexicographic references from the Han time onwards in which *huo-chhi chu* is always given as the explanation of *mei-hui*. But Chang Hung-Chao was equally firm (and justified) in his conviction that biconvex lenses never could have been made from the flat sheets of mica. It is therefore very difficult to tell where mica ended and lenticulate pieces of glass or rock-crystal began, but archaeological evidence accruing since these discussions took place has somewhat changed the balance of probability. Furthermore, in +725 Chhen Tshang-Chhi defined *liu-li* glass as the same thing as *huo-chhi chu*.[e] He quoted in support the +4th-century dictionary *Yün Chi*[4] of Lü Ching.[5] So even if we reserve judgement about the *huo-chhi chu* of the Han, it now seems very likely that many of the pre-Thang references imply lenses. Thus in +519 Fu-Nan (Cambodia) sent *huo-chhi* pearls as tribute.[f] In +528 the Tan-Tan country sent more of such pearls,[g] but seven years later its offering included simply *liu-li* glass—perhaps a significant juxtaposition.[h] There seems no doubt that we shall have to envisage a much more widespread use of glass and rock-crystal lenses in early medieval China than has generally been supposed. The question is, into what shapes were they cut?

It is at the period between the Thang and the Sung that we find one of the most interesting of all Chinese references to lenses. In an earlier Section (Vol. 2, pp. 444 ff.) a good deal was said about the *Hua Shu* (Book of Transformations), attributed to Than Chhiao, and datable at about +940. It has a passage concerning four optical instruments, which, because of the term used (*ching*[6]) have been supposed, by those few who have ever read it, such as Forke (12), to be mirrors. It has not been remembered, however, that while there are only three fundamental types of mirrors (plane, concave and convex), there are four fundamental types of lenses (plano-concave, biconcave, plano-convex and biconvex). Let us read this passage, therefore, on the hypothesis that Than Chhiao was really experimenting with the four types of lens.

[a] *Pên Tshao Kang Mu*, ch. 8, p. 56b.   [b] (14), p. 191.
[c] He excepted *huo ching*, which he considered a pink or red variety of rock-crystal.
[d] (1), pp. 55 ff.   [e] *PTKM*, ch. 8, p. 57a.
[f] *Liang Shu*, ch. 54, p. 11a.   [g] *Liang Shu*, ch. 54, p. 16a, b.   [h] *Liang Shu*, ch. 54, p. 16b.

[1] 火齊珠   [2] 玫瑰   [3] 火精   [4] 韻集   [5] 呂靜   [6] 鏡

I have always by me four lenses. The first is called *kuei*[1] (the 'sceptre', a diverging bi-concave lens). The second is called *chu*[2] (the 'pearl', biconvex). The third is called *chih*[3] (the 'whetstone', plano-concave). The fourth is called *yu*[4] (the 'bowl', plano-convex).

With *kuei* the object is larger (than the image).[a]

With *chu* the object is smaller (than the image).[b]

With *chih* the image appears upright.[c]

With *yu* the image appears inverted.[d]

When one looks at shapes or human forms through such instruments, one realises that there is no such thing as (absolute) largeness or smallness, beauty or ugliness....[e]

Here there can be no doubt about the identification of *chu*, which was the old biconvex burning-glass. *Chih* as plano-concave is strongly suggested by the shape of the traditional Chinese whetstone, which is not a wheel, but a plate of stone held upright in supports, its upper surface becoming concave by the continual grinding.[f] *Yu* could well be a solid glass hemisphere,[g] and *kuei* takes its position by exclusion, though it is not difficult to see why a biconcave lens should have been so named.[h] One wishes that a fuller account of Than Chhiao's experiments had come down to us.[i]

Little is known as yet about optical studies in later times. The interest of Shen Kua in mirrors, burning-mirrors, 'magic' mirrors and the camera obscura has already been discussed.[j] One would have expected a man such as Kuo Shou-Ching to be particularly interested in mirrors and lenses, and it is probable that a good deal of Yuan work on this subject may yet come to light. Kuo's invention of a 'shadow definer' we have already mentioned.[k] Presumably optics and catoptrics shared in the general decline of the physical sciences during the Ming period, but after the coming of the Jesuits interest was stimulated as in so many other scientific subjects. We took note at an earlier stage[l] of the first Chinese book on the telescope, the *Yuan Ching Shuo*[5] (Far-Seeing Optick Glass) of Adam Schall von Bell (Thang Jo-Wang[6]) in +1626. But it was not until the 19th century that the upsurge really began, with Chêng Fu-Kuang's[7] *Ching Ching Ling Chhih*[8] (Treatise on Optics by an Untalented Scholar), written about 1835, and Chang Fu-Hsi's[9] *Kuang Lun*[10] (Discourse on Optics) about

[a] We assume that he kept the object beyond twice the focal distance.

[b] A magnifying glass.   [c] Because this is a diverging lens.

[d] Because this is a converging lens.   [e] Tr. auct.

[f] Hommel (1), p. 257.

[g] Unless indeed it was a lens of concavo-convex type instead of simply planoconvex. But although glass cups were certainly available in Than Chhiao's time, it would seem very unlikely that any study was made of either the positive or negative meniscus forms of concavo-convex lenses at such an early date.

[h] Some of the ancient jade sceptres (*kuei*[1]) and ceremonial bronze hatchets from which they seem to have derived show a marked tendency towards biconcave sides (cf. Laufer (8), pp. 74 and 95).

[i] All the more so since in Europe there was no systematic treatment of the properties of the fundamental types of lenses until the *De Refractione* of G. B. della Porta, published in +1593. An interesting analysis of this book, the forerunner of Kepler's masterly treatise of +1604, is given by Ronchi (1), pp. 65 ff.

[j] Pp. 93, 94, 97 above.   [k] Vol. 3, p. 299.   [l] Vol. 3, p. 445.

[1] 圭   [2] 珠   [3] 砥   [4] 盂   [5] 遠鏡說   [6] 湯若望

鄭復光   [8] 鏡鏡詅癡   [9] 張福僖   [10] 光論

five years later. Both treat systematically of the properties of light, and of the different shapes of mirrors and lenses, and the former goes on to describe the making of telescopes and sextants.[a]

One cannot evaluate the story so far without a glance at parallel developments in Europe. As to the optics and catoptrics of the Greeks, something has already been said (pp. 85 ff. above). Though lenses were not considered by them, there is no doubt that 'perspicilia' and burning-glasses of rock-crystal or glass were known in the West from an early time.[b] A famous passage in Aristophanes[c] refers to a diaphanous stone (*hyalos*, ὕαλος) with which some legal documents were to be set on fire. Pliny mentions[d] the use of rock-crystal lenses for cauterising, and says[e] that glass balls filled with water will ignite textiles. Seneca wrote[f] that letters, however minute or indistinct, could be magnified and made readable by a water-filled ball of glass.

Though refraction as well as reflection was considered by Cleomedes and Ptolemy, it is agreed[g] that the first to examine the properties of lenses was Ibn al-Haitham (+965 to +1039), and even he dealt mainly with biconvex or spherical burning-glasses.[h] The Polish physicist Witelo (first half of the +13th century) knew something of lenses from al-Haitham, and drew up a table of refractions at surfaces of air, water, and glass. His contemporaries Robert Grosseteste (+1175 to 1253)[i] and Roger Bacon (+1214 to +1292)[j] were the first after Ibn al-Haitham to use plano-convex lenses, and still later John Peckham (*d.* 1292)[k] alluded to the possibility of concave and plano-concave lenses. It was about this time that the invention of spectacles was made. If, therefore, our interpretation of the work of Than Chhiao is correct, we see again, as we saw in the case of the pinhole and camera obscura, that the Chinese were advancing in parallel with the Arabs (indeed in this case in advance of them), though handicapped as always by the lack of deductive geometry.

(iv) *Eye-glasses and spectacles*

It has sometimes been stated that the invention of spectacles was Chinese. This may, in part, have derived from a paper by Laufer (16) containing many inconsistencies, which were afterwards cleared up by Chhiu Khai-Ming (2). If Laufer had been

---

[a] Furthermore, it contains an appendix called Huo Lun Chhuan Thu Shuo[1] which gives an illustrated account of the engines of steam paddle-boats. Cf. Sect. 27 *h.*

[b] Some maintain that the plano-convex rock-crystal object found by Layard (1), p. 197, at Babylon cannot have been intended for use as a lens (Singer, 9).

[c] *Clouds*, 768.

[d] *Hist. Nat.* XXXVII, 28 (ch. 10).

[e] *Hist. Nat.* XXXVI, 199 (ch. 67); tr. Bailey (1), vol. 2, p. 153.

[f] *Quaest. Nat.* I, 6, 5; tr. Clarke & Geikie (1), p. 29.

[g] Cf. Singer (9).

[h] Sarton (1), vol. 1, p. 721; Winter (4). Ibn al-Haitham also applied geometrical theory to the passage of light through cylindrical glass surfaces both convex and concave.

[i] Sarton (1), vol. 2, p. 584.   [j] Sarton (1), vol. 2, pp. 952, 957.

[k] Singer (9).

[1] 火 輪 船 圖 說

justified in accepting as authentic a mention of spectacles (*ai-tai*[1]) in the *Tung Thien Chhing Lu*[2] of Chao Hsi-Ku,[3] written not long after +1240, then the mention of these aids to better vision in China would have antedated European references by about half a century. The passage runs:[a]

Ai-tai resemble large coins, and their colour is like mica. When old people are dizzy and their sight tired, so that they cannot read fine print, they put *ai-tai* over their eyes. Then they are once more able to concentrate, and the strokes of the characters appear doubly clear. *Ai-tai* come from Malacca in the western regions.

But bibliographical study showed that this was not in the best and oldest versions of Chao's book, so that it must have been added by someone in the Ming. Besides, the mention of Malacca would have been an anachronism. In fact, the earliest books which refer to spectacles were written in Ming times, the *Chhi Hsiu Lei Kao*[4] of Lang Ying[5] (+1487 to +1566) and the *Fang-Chou Tsa Yen*[6] of Chang Ning[7] (*fl.* +1452).[b]

From their accounts it is clear that spectacles were known in China, though not very common there, during the early years of the Ming dynasty, i.e. the +15th century.[c] For example, Chang Ning saw a pair at the house of the army commander Hu Lung,[8] whose father had had them as a present from the emperor about +1430. He described them as like mica 'or what is commonly called *hsiao-tzu*[9]', i.e. glass,[d] and said that they were believed to come from Malacca. Lang Ying, writing nearly a century later, told of seeing a pair which his friend Ho Tzu-Chhi[10] had obtained from barbarians (presumably Arab or Persian merchants) in Kansu. All these Ming accounts say that the spectacles were monoculars which could be connected together at will. Duyvendak (19) pushed back the first known mention of spectacles in Chinese literature to the early years of the +15th century when he noticed that a record exists of the presentation as tribute of ten pairs of spectacles by the king of Malacca in +1410. This has come down to us in the *Hsi-Yang Chi*[11] (Story of...the Western Oceans), a novel written by Lo Mou-Têng[12] in the last decade of the +16th century.[e] Though full of fabulous material, it is based on very sober sources dealing with the famous voyages of Chêng Ho and Wang Ching-Hung, and preserves certain pieces of information (concerning, for example, lists of tribute and details of armament such as cannon) which were contained in sources since lost. There is, of course, no ground for assuming

---

[a] Ch. 1, p. 9*b*. Tr. Chhiu Khai-Ming (2), mod.

[b] Both cit. *KCCY*, ch. 58, p. 22*a*.

[c] There is a story in the literature that Marco Polo came across spectacles when he was in China, but the late Professor A. C. Moule confirmed to me the correctness of Hirschberg's view (1), vol. 13, p. 265, that Polo never mentions them. Werner (2), p. 280, even claimed spectacles in Thang China, but as the nearest titles of books to those given as authorities by him are Yuan or Ming, he must have made some mistake.

[d] This term will be found in the *Ko Ku Yao Lun* of +1387 (ch. 6, p. 5*a*), where a description of glass-making is given.

[e] Ch. 50, p. 36; ch. 99, p. 25. See further in Sect. 29.

[1] 靉靆	[2] 洞天清錄	[3] 趙希鵠	[4] 七修類稿	[5] 郎瑛
[6] 方洲雜言	[7] 張寧	[8] 胡𧩺龍	[9] 硝子	[10] 霍子麒
[11] 西洋記	[12] 羅懋登			

this to have been the first introduction, but whether the coming of spectacles really antedates the fall of the Yuan dynasty, as Chhiu Khai-Ming believes, remains unsure.[a] The mention of Kansu suggests that they came in, like cotton, both overland by the north and through maritime contacts in the south.

While the first term for them was *ai-tai*[1] (or *ai-tai*[2]), the modern term *yen-ching*[3] was already current at the beginning of the +15th century. Chhiu Khai-Ming believes, plausibly enough, that the term *ai-tai* must be a transliteration of some foreign, probably Arabic, word, and we find indeed *al-'uwaināt* ('little eyes') for spectacles in that language. Since *ai-na*[4] is one variant, the suggestion that Pers. *'ainak* (sing.) was another source is also convincing. But one cannot help wondering whether the scholars who first chose the name could have punned on the old word *ai* of Shen Kua, which meant a focal point.[b]

As to the origin of spectacles in Europe there has been much obscurity. The difficulty is that most of the evidence concerning the inventors has been lost. Roger Bacon certainly mentioned the possibility of improving sight, but the credit for the first actual making of spectacles has been ascribed (*inter alia*) to a Salvino degli Armati, who is said to have died about +1317, or to Alessandro Spina, a Pisan religious (*d.* +1313). Singer (9) summed up the circumstantial evidence by saying that they came into use at the end of the +13th century, first in Italy, beginning with convex lenses (*occhiali*) for presbyopia, and not developing concave lenses for myopia till as late as the mid +16th.[c] Although the recent and exhaustive researches of Rosen (2, 3) have not changed this general picture, they have unravelled an exceptionally tangled skein of evidence concerning the original invention.[d] It is therefore now possible to date the first making of spectacles shortly after +1286. The inventor was neither of the above-mentioned men, nor is his name known to us; he was probably a layman of Pisa, perhaps a worker in glass, who kept his method secret for trade reasons. The earliest report of it which has come down to us is contained in a sermon by the friar Giordano of Pisa at Florence in +1306, and it is clear that he had known the craftsman personally. The oldest mention of spectacles themselves occurs in a series of regulations[e] of the Venetian guild of workers in rock-crystal and glass, under the name of *roidi da ogli*, in +1300. The first appearance of spectacles in a picture is seen in a portrait of +1352 at Treviso.[f] It is therefore clear that their

---

[a] The editors of the Khang-Hsi dictionary, at any rate, thought so (entry under *ai*).

[b] Transliteration from the Arabic would not exclude a suitable Chinese meaning, and *ai-tai*[1] could be interpreted 'loving to reach through a cloudy screen'. Moreover, the light-rays lovingly assemble together at the focal point. Duyvendak (19) agrees that the characters were chosen learnedly.

[c] The standard histories of spectacle-making by Greeff (1, 2); Bock (1); Oliver (1, 2); etc. may be consulted, but all under radical correction by Rosen (2, 3) for the earliest developments. Lynn White (1) sets the invention in its historical context, with further references. Cf. E. C. Watson (1). For the relation to the history of ophthalmology see the elaborate volumes of Hirschberg (1).

[d] This work constitutes one of the most remarkable 'detective stories' of modern research in the history of science and technology. The phrase is particularly appropriate because it brought to light a whole series of deliberate falsifications (as well as innumerable inaccuracies) from the 17th century onwards.

[e] Supplementary to those of +1284 which do not mention eye-glasses. See Rosen (2), p. 211.

[f] Von Rohr (2); Rosen (2), p. 205.

[1] 靉靆   [2] 僾逮   [3] 眼鏡   [4] 矮納

transmission to China must have taken place comparatively fast, no doubt because of the fact that this was a period of intensified trade relations, first with Europe under the Mongol aegis, then with South Asia during the Ming navigations.

The Sung people, however, did have two techniques which may be considered introductory to spectacle lenses; one was the magnifying glass, and the other dark glasses as eye-protection. Regarding the former, Liu Chhi[1] wrote, in his *Hsia Jih Chi*[2] (Records of Leisure Hours)[a] some time before his death in +1117, that his contemporary, Shih Khang,[3] and other judges, used to use various magnifying lenses of rock-crystal (*shui ching*[4]) for deciphering illegible documents in criminal cases.[b] The judges also used to use dark glasses made of smoky quartz (e.g. *chha ching*[5]), not, as we do, for driving against the sun,[c] but so as to disguise from litigants their reactions to the evidence.[d] Indirect support for the use of magnifying glasses comes, moreover, from the old and favourite practice of inscribing very minute characters on art objects. For example, in +1360 Yang Yü wrote:[e]

It is said that the magistrate Wang I[6] owned a brush pen, which, though not much larger than an ordinary pen, was of greater diameter at both ends, about half an inch across. Between these swellings there was engraved a picture of an army, men, horses (even with their hair), pavilions, terraces, and distant waters, all on a scale of extreme minuteness. Each scene was accompanied by two verses of a poem. It hardly seemed the work of man. The lines of the drawing shone as if made with chalk, so that they could easily be seen in reflected light. It was said that the teeth of rats had been used as engraving tools. In the collection of Tshui Shen[7] there is an essay on this brush of Mr Wang, which shows how it was treasured.

I once heard that in the house of the Wang family, on Belltower Street in Peking, there is an archer's jade thumb-ring preserved, about the same size as the ring underneath a begging-bowl. Yet a whole chapter of the *Hsin Ching*[8] is engraved on it.[f]

Moreover, my late father, the State Counsellor, used to say that he once saw a bamboo tortoise, similar to that which I have in my own collection, but with an inscription in ivory inlaid in ebony. A whole chapter of the *Hsiao Ching* (Filial Piety Classic) was thus written. It was no bigger than the first finger of one's hand. When one compares these things with the brush of Mr Wang, they would seem even more skilfully made, and anyone would take them for the work of devils.

This technique has continued down to modern times.[g]

[a] In *Shuo Fu*, ch. 4, p. 39a.   [b] Later echoes of this are mentioned by Chhiu Khai-Ming (2).

[c] For protection against excessive light, however, eye-screens of yak-hair were worn in Tibet and the western provinces of China (Rockhill (2), pl. 30), while in the north Mongols used the Eskimo device of horn or bone eye-screens bearing a horizontal slit (Mason (1), pp. 281 ff.). There is some reason for thinking that metal ones were used in +6th century China.

[d] This piece of information, which I fully believe to be true, comes from a paper on fire-pearls and spectacles by Pi (1), which, though interesting, is full of serious and misleading mistakes. The same applies to the paper by Rakusen (1).

[e] *Shan Chü Hsin Hua*, p. 29a. Tr. H. Franke (2), no. 80.

[f] A popular name for the *Prajñā-pāramitā Sūtra*.

[g] Not long after I had written this I was shown by my friend H.E. Dr Fêng Hsüan a very minute piece of ivory upon which a poem by Chairman Mao Tsê-Tung had been inscribed.

[1] 劉跂   [2] 暇日記   [3] 史抗   [4] 水晶   [5] 茶晶   [6] 王倚
[7] 崔鋋   [8] 心經

Details about traditional Chinese spectacles have been collected by Rasmussen (1), and Hommel.[a] Davis (1) and Wells Williams[b] (1836–46) saw the use of dark glasses, and found rock-crystal more commonly used for spectacles than glass. The traditional spectacle-makers had an empirical system in which the concavity of lenses was estimated in a series of grades designated by the duodenary cyclical characters and running from −0·5 to −20 diopters.[c] This was for myopia (*chin kuang*[1]), but convex lenses were also in use for presbyopia (*lao kuang*[2]).

(6) Shadow-Play and Zoetrope

A celebrated incident in Chinese history concerns the evocation, by the magician Shao Ong,[3] of a moving image of one of the dead concubines of the emperor Wu of the Han, in −121. The story is told both in the *Shih Chi*[d] and the *Chhien Han Shu*;[e] I reproduce the fuller version of the latter.

(After the death of Li Fu Jen[4]) the emperor could not stop thinking of her. Shao Ong, a magician from Chhi, said that he could cause her spirit to appear. So after certain offerings of wine and meat had been set forth, and when certain lamps and candles had been disposed about a curtain, the emperor took his place behind another (diaphanous) curtain. After a time he saw indeed at a distance a beautiful girl sitting down and walking back and forth. But he could not approach her. Afterwards the emperor thought upon she all the more, and being sad, composed a poem somewhat to this effect: 'Was it really she or was it not? I could not help rising from my seat. How was it that she walked so gracefully, yet seemed to come towards me so slowly?...'

This was always remembered in later times.[f] Moreover, we have quite circumstantial accounts of the same kind of illusion being produced in the Thang. The *Pei Mêng So Yen*[5] (Dreams of the North and Trifling Talk), written by Sun Kuang-Hsien[6] about +930, ascribes[g] to a hermit, Chhen Hsiu-Fu,[7] exactly similar techniques. The *Thang Chhüeh Shih*[8] (Thang Memorabilia) by Kao Yen-Hsiu[9] records[h] another hermit accomplishing the same for a scholar's lost love.[i]

It is improbable that the Han procedure could have been anything else than one of those shadow-plays (*ying hsi*[10]) which have been traditional for centuries in many Asian countries.[j] Such also was the opinion of the Sung author of the *Shih Wu Chi*

---

[a] (1), p. 198.   [b] (1), vol. 2, p. 22.
[c] A diopter is the reciprocal of the distance in metres between the lens and its focal point.
[d] Ch. 28, p. 24b (tr. Chavannes (1), vol. 3, p. 470). In this version the name of the girl was Wang, not Li; and another, somewhat alchemical, character, the Spirit of the Furnace, also made its appearance.
[e] Ch. 97A, p. 14a, tr. auct.
[f] Cf. *Shih I Chi; Yeh Kho Tshung Shu*, ch. 11, p. 6b, etc.
[g] Ch. 8, p. 7a.   [h] Ch. 2, p. 15b.
[i] It is curious that quite similar conjurations were part of the magical stock-in-trade of adepts in +17th-century England, notably the astrologer William Lilly (+1602 to +1681), the friend of Ashmole and Pepys, and the predecessor of Salford and Case.
[j] Cf. Jacob & Jensen (1).

---

[1] 近光   [2] 老光   [3] 少翁   [4] 李夫人   [5] 北夢瑣言
[6] 孫光憲   [7] 陳休復   [8] 唐闕史   [9] 高彦休   [10] 影戲

Yuan.[a] In the Thang examples, having in mind what has already been related of Chinese optical knowledge, it is just possible that someone had the idea of placing one or more lenses at the pinhole of a closed chamber. This was, of course, the new element in the invention of the magic lantern by Cornelius Drebbel in +1630 or Athanasius Kircher in +1646.[b] But it seems much more probable that the Thang continued, perhaps with refinements, what Shao Ong had started.

Another ancestor of the cinematograph was a variety of zoetrope, which may well have originated in China, namely a light canopy hung over a lamp, and bearing vanes at the top so disposed that the ascending convection currents cause it to turn. On the sides of the cylinder[c] there would be thin panes of paper or mica, carrying painted pictures, which, if the canopy spun round fast enough, would give an impression of movement of animals or men. Such devices certainly embodied the principle of a rapid succession of images.[d] In its semi-fabulous account of Chhin Shih Huang Ti's treasury, already quoted, the *Hsi Ching Tsa Chi* speaks[e] of the sparkling of scales of turning dragons after a lamp was lit. It also describes what must have been a small windmill or air turbine, saying that

there was a jade tube 2 ft. 3 inches long, with 26 holes in it. If air was blown through it, one saw chariots, horses, mountains and forests appear in front of a screen, one after another, with a rumbling noise. When the blast stopped, all disappeared.

This was called *chao hua chih kuan*[1] (the pipe which makes fantasies appear). Rising currents of hot air were evidently used by Ting Huan[2] (c. +180) who made a 'nine-storied hill-censer'[f] (*chiu tshêng po shan hsiang lu*[3]), on which many strange birds and mysterious animals were attached. All these wonderful creatures moved quite naturally (*chieh tzu-jan yün tung*[4]), presumably as soon as the lamp was lit.[g] Similar

[a] Ch. 9, p. 34a. In his time they were very popular, and special guilds of operators existed; cf. *Mêng Liang Lu*, ch. 13, p. 12b; ch. 20, p. 13b; *Wu Lin Chiu Shih*, ch. 2, pp. 16a, 19b; ch. 6, p. 17a. From these references we know that in Marco Polo's time the name for the zoetrope was a development from this, *hsüan ying hsi*,[5] 'lathe-turned' or 'revolving' shadow-plays.

[b] Cf. Feldhaus (1), col. 824. The first use of a lens with the camera obscura was apparently due to Daniele Barbaro at Venice in +1568, followed by della Porta in +1589 (Feldhaus (1), col. 226). It is not generally known that the first lecturer to use lantern-slides was a Jesuit of the China mission, Martin Martini (1614–61). The reader will remember his important geographical work described in Vol. 3, pp. 554, 586. A pupil of Athanasius Kircher, he returned to Europe in +1651 and for four years travelled widely, meeting many scholars; cf. Bernard-Maître (15), Duyvendak (13), and Vol. 1, p. 38. His lectures illustrated by the new technique were given at Louvain in +1654; see Liesegang (1).

[c] A common modern name is 'umbrella lamp' (*san têng*[6]), no doubt from the form both of lamp and character (Liu Hsien-Chou (1), p. 77).

[d] Could some similar thought have been at the basis of the representations of gods and goddesses with attributes held by many limbs? China inherited these with Buddhism from India; Fig. 296 shows the Kuan-Yin of the Four Cardinal Points and the Thousand Arms in the Chieh-thung Ssu temple, Chiangsu. Though the arms may have been sometimes necessary to grasp all the god's symbols, their multiplication may also, perhaps, have been intended to signify intensity and vigour of rapid motion, frozen iconographically into a composite 'still'.

[e] Ch. 3, p. 3b, tr. auct. adjuv. Dubs (2), vol. 1, p. 57.

[f] Cf. the Section on geography, Vol. 3, p. 580 above, with relation to relief maps.

[g] *Hsi Ching Tsa Chi*, ch. 1, p. 8a. Cf. Laufer (3), pp. 180, 196.

[1] 昭華之琯   [2] 丁緩   [3] 九層博山香爐   [4] 皆自然運動
[5] 鏇影戲   [6] 傘燈

apparatus (*hsien yin chu*[1]), in which moving shapes were seen and tinkling noises heard, after the lighting of a candle or lamp, is mentioned in the +10th century *Chhing I Lu*.[a] A toy of this kind was presented to a temple by an emperor in memory of a young Thang princess who had died. It sounds exactly like the charming devices which still commonly appear in Scandinavian countries at Christmas time.

The +12th-century Sung scholars Fan Chhêng-Ta[2] and Chiang Khuei[3] wrote poems[b] describing how with the *ma chhi têng*[4] (horse-riding lamp) one can see the shadow-horses prancing round after the lamp is lit. A similar *tsou ma têng*[5] (pacing-horse lamp) is referred to in the *Mêng Liang Lu* (a description of Hangchow in +1275).[c] One of the Jesuit fathers, Gabriel Magalhaens, was delighted with the thing, and wrote, about the middle of the seventeenth century:[d]

...the Lamps and Candles, of which there are an infinit number in every Lanthorn, are intermix'd and plac'd within-side, so artificially and agreeably, that the Light adds beauty to the Painting; and the smoak gives life and spirit to the Figures in the Lanthorn, which Art has so contriv'd, that they seem to walk, turn about, ascend and descend. You shall see Horses run, draw Chariots and till the Earth; Vessels Sailing; Kings and Princes go in and out with large Trains; and great numbers of People both a-Foot and a-Horseback, Armies Marching, Comedies, Dances, and a thousand other Divertisements and Motions represented....

Modern writers have described[e] toys of this kind made in contemporary China, especially Peking, and the passage which Bodde translated from Tun Li-Chhen's[6] *Yen-ching Sui Shih Chi*[7] (Annual Customs and Festivals of Peking) is so interesting that it is worth quoting fully. Writing in 1900 Tun Li-Chhen said:

Pacing-horse lamps are wheels cut out of paper, so that when they are blown on by (the warm air rising from) a candle (fastened below the wheel), the carts and horses (painted on it) move and run round and round without stopping. When the candle goes out the whole thing stops. Though this is but a trifling thing, it contains in truth the whole underlying principle of completion and destruction, rise and decay, so that in the thousand ages from antiquity down to today, as recorded in the Twenty-four Histories, there is not one which is not like a 'pacing-horse lamp'.[f]

Besides this lamp, there are others of such things as carts, sheep, lions, and embroidered balls (so made that they remain lighted from within even when rolled over and over on the

---

[a] Ch. 2, p. 24b.

[b] Respectively, the Shang Yuan Chieh Wu Shih[8] (Poem on objects belonging to the Lantern Festival), and the Kuan Têng Shih[9] (Poem on Watching Lanterns); both cited in Wei Sung's encyclopaedia, ch. 17, p. 6a. I am much indebted to Professor Derk Bodde for these exact references.

[c] Ch. 13, p. 8a. This book also describes (ch. 1, p. 3b) illuminated dragons which seemed to fly and run. There may be some connection between these and hot-air balloons (see below, Sect. 27m).

[d] (1), p. 105.

[e] Bodde (12); Eder (1), p. 24. For Indo-China, Huard & Durand (1), p. 80.

[f] Mr Tun had found a striking metaphor for that element of cyclic recurrence which is present in the successive Chinese dynasties (cf. Sect. 48 below). Perhaps he was also thinking of Neo-Confucian ideas of recurrent change (cf. Vol. 2, p. 485 above). But perhaps also he was unduly influenced by the misconceptions of Westerners at that time, who found nothing but stasis in Chinese culture.

---

[1] 仙音燭   [2] 范成大   [3] 姜夔   [4] 馬騎燈   [5] 走馬燈
[6] 敦禮臣   [7] 燕京歲時記   [8] 上元節物詩   [9] 觀燈詩

PLATE CIV

Fig. 296. Image of Kuan-Yin of the Four Cardinal Points and the Thousand Arms, in the Chieh-thung Ssu temple, Chiangsu. Flashing action and infinite vigour symbolised by the multiplication of limbs and attributes (photo. Boerschmann, 3a).

ground).[a] Each year, at the coming of the 10th month (i.e. at the approach of the New Year), such places as Chhien Mên, Hou Mên, Tung Ssu Phai-lou and Hsi Tan Phai-lou, all have them, and for people who have leisure, it is a delightful thing to go leading a child to one of these places, joyfully buy some and so return.

The 'pacing-horse lamp', with its wheel which is controlled by a flame, and its mechanism revolved by that wheel, is in the same class with the steamships and railways of the present day. For if its (principle of operation) had been pushed and extended, so that from one abstruse principle there had been a searching further for the next abstruse principle, who knows but that during the last few hundred years there might not have been completed a mechanism of real utility? What a pity that China has so limited herself in the scope of her ingenuity, that for the creations of her brain, and the perfected essence of her inventors, she has nothing better to show than a children's toy! At the present day, when others make a step we too must make a step, when others move forward we too must move forward. If we are amazed at the wonderful powers (of Westerners), and remain content in our own stupidity, how can we then say in self-extenuation that the flow of genius produced by the universe should be widespread among them alone, and narrow only among us? Is it not indeed something for which we should be angry with ourselves?[b]

With typical occidental assurance, the Vice-President of the Royal Society, W. B. Carpenter, writing in 1868, supposed that the zoetrope had originated with Faraday in 1836.[c] Neither he nor Tun Li-Chhen made sufficient allowance for the ancient mechanicians of genius such as Ting Huan. Neither of them realised that a close relative of the pacing-horse lamp, the Chinese helicopter top, had already been the object which awakened the father of modern aerodynamics to his revolutionary labours;[d] or that another invention of Ting Huan's mentioned by Tun Li-Chhen, the suspension which the world still attributes to Jerome Cardan, would also be of basic importance to flight when it came to house the gyrostatic wheel.[e] There was too much self-assurance on one side and too much diffidence on the other; only the balance of history could ultimately show that Europe had as much leeway to make up in the 10th century as China had at the beginning of the 20th.

[a] This is the Cardan suspension—see below, Sect. 27d.
[b] Tr. Bodde (12). I have to thank Professor Derk Bodde for sending to me this excerpt when his book was not available in Cambridge.
[c] In fact it had been described by John Bate in +1634 (*Mysteryes of Nature and Art*, pp. 30ff.). The principle of persistence of vision, fundamental to cinematography, had also been described by Paris and Cruickshank in 1827.
[d] See Sect. 27m below.   [e] See Sect. 27d below.

(h) SOUND (ACOUSTICS)

(1) INTRODUCTION

The term acoustics may be defined either broadly to cover the nature of sound in general, as a branch of physics; or more narrowly to denote our knowledge of sound as applied to the properties of halls and buildings. It is here taken in the broad sense, and this Section will therefore aim at describing not only the positive achievements of the Chinese in this field,[a] but also their attitudes to acoustic phenomena in ancient and medieval times. The whole subject is one of particular interest from the point of view of the history of science because it was one of the earliest fields, both in East and West, where quantitative measurement was applied to natural phenomena.

While no one has essayed hitherto an evaluation of the development of Chinese ideas on acoustics as a science,[b] scholars who have written on Chinese music have naturally had to deal with the subject to some extent. The earliest authoritative exposition by a Westerner was that of the Jesuit father Jean Joseph-Marie Amiot (+1718 to +1793; Chhien Tê-Ming[1]) who did (1) for this field something of what Antoine Gaubil[c] did for astronomy. Two indispensable monographs on Chinese music are those of Courant (2) and Levis (1), works which largely superseded the older contributions.[d] More recently the brilliant syntheses of Picken (1, 2) have become available.[e] Particularly important is the translation by Chavannes[f] of the Yo Chi[2] (Record of Ritual Music and Dance),[g] a precious document of the late Chou period;[h] as also his study of the standard pitch-pipes.[i] In our opinion, Chao Yuan-Jen (2) was too modest in his estimate of the musical contributions of the Chinese; for he hardly did justice to their great sensitivity to timbre,[j] their achievement as a people who formulated the only theory of melodic composition in a tonal language,[k] and the

---

[a] On the relations between music and mathematics we may cite the stimulating lecture of Archibald (2). The book of Jeans (2) would be very useful for the reader of the following pages to have at hand.

[b] The unpublished work of Kuttner (2) may do something to fill this gap, and an important book by Wu Nan-Hsün is eagerly awaited.

[c] Cf. Vol. 3, pp. 182 ff.

[d] Such as those of Faber (1); Wagener (1); van Aalst (1); Dechevrens (1); Soulié de Morant (1), etc. Levis's book is considered rather idiosyncratic by some scholars.

[e] See also the dictionary articles of Robinson (3); Eckardt (1, 2) and Crossley-Holland (1). Some widespread misconceptions about Chinese music are considered by Robinson (5). Cf. Daniélou (1).

[f] (1), vol. 3, pp. 238–86.

[g] Contained in Li Chi, ch. 19 (tr. Legge (7), vol. 2, pp. 92 ff.), which constitutes a parallel text to that of Ssuma Chhien in Shih Chi, ch. 24. It will be remembered (Vol. 2, p. 4) that there once existed a Yo Ching[3] (Music Classic), but this was lost very early. Three of the Han apocryphal treatises (cf. Vol. 2, pp. 380, 382) connected with this work have been preserved (YHSF, ch. 54), but they have not yet been investigated from the point of view of the history of acoustics. Besides these, a large number of musical and acoustic fragments, from the Han onwards, exist in the collections of Ma Kuo-Han (YHSF, chs. 30, 31) and Yen Kho-Chün, offering a field of promise for further research. The term yo (music) of course included ritual miming in the Chou, Chhin and Han.

[h] Another book of the same title, attributed to Liu Hsiang, exists in fragmentary form in YHSF, ch. 30, pp. 68 a ff.

[i] (1), vol. 3, pp. 630 ff. (Appendix 2).   [j] Van Gulik (1).

[k] Levis (1).

[1] 錢德明   [2] 樂記   [3] 樂經

great wealth of their characteristic and distinctive melodic fund.[a] On Chinese musical instruments there is a substantial Western literature,[b] to which we shall refer from time to time as occasion arises, and useful books by Li Shun-I (*1*) and others.

An adequate treatment of Chinese musical literature throughout the ages would go far beyond our frame of reference, yet it is rather difficult to separate the primarily acoustic works from those primarily musical. The tradition of the late Chou *Yo Chi* gradually blossomed forth into encyclopaedic studies which included basic musical theory, tables of mode-keys, etc.,[c] systematic descriptions of instruments, orchestral arrangements, dances and costumes. None of these has come down to us from a time earlier than the Sung, but we still possess (if in incomplete form) the admirable *Yo Shu*[1] (Treatise on Acoustics and Music) written by Chhen Yang[2] late in the +11th century.[d] A similar work has been preserved in Korea. The *Akhak Kwebŏm*[3] (Standard Patterns in Musicology) was compiled in +1493 by Sŏng Hyŏn at the command of King Sŏngjong[4] (*r.* +1470 to +1494), to preserve the studies of court music which had been made by an outstanding musician, Pak Yŏn[5] (*fl.* +1419 to +1450).[e] This work, afterwards many times reprinted, is succinctly written, excellently arranged, and well illustrated; in all essentials it is of the Chinese tradition, with suitable Korean additions and modifications.[f] Apart from such encyclopaedias there is a great mass of information about acoustics and music in the chapters on these subjects in the successive dynastic histories, sources which have been utilised best by Courant (*2*) of Western scholars. In addition, much is to be found in ordinary encyclopaedias, such as the *Chhu Hsüeh Chi*[6] of Hsü Chien[7] (*c.* +700), in ethnological works, e.g. the *Fêng Su Thung I*[8] (The Meanings of Popular Traditions and Customs) by Ying Shao[9] in +175, and in miscellaneous books on scientific subjects like Shen Kua's[10] *Mêng Chhi Pi Than*[11] (Dream Pool Essays) of +1086, so constantly referred to throughout the present work.[g]

Besides all this there are numerous important studies by individual scholars. In due course we shall refer[h] to a special monograph on drums, written by Nan Cho[12] in +848, the *Chieh Ku Lu*,[13] while in a somewhat later generation Tuan An-Chieh[14] wrote

---

[a] Picken (*1, 2, 3*). An important Chinese secondary source which was not available to us until our work was nearly finished is the book of Wang Kuang-Hsi (*1*).

[b] For a general survey in relation to the musical instruments of other cultures, the books of Sachs (*1, 2*) and Schaeffner (*1*) are to be consulted. A brief survey will be found in Montandon (*1*), pp. 695 ff. Extensive studies specific to China are those of Fernald (*1*), Mahillon (*1*), Norlind (*1*), and above all Moule (*10*). An excellent album of illustrations of Chinese musical instruments has been produced recently by the Central School of Music's Research Institute at Shanghai; see Chhien Chün-Thao (*1*). The instruments probably used in the Shang period (*c.* − 14th century) have been discussed by Gibson (*1*).

[c] See pp. 161, 169, 215, 218 below.

[d] Cf. *Lü Lü Hsin Lun*, ch. 2, pp. 16 ff. Chhen Yang's book is not to be confused with another of the same title, written by Hsintu Fang (cf. p. 189) about +570.

[e] Though the chief editor spoke slightingly of them in the preface.

[f] Cf. Hazard *et al.* p. 143.

[g] Cf. Vol. 1, p. 135, and many subsequent mentions as indexed.

[h] P. 161 below.

[1] 樂書	[2] 陳暘	[3] 樂學軌範	[4] 成宗	[5] 朴堧
[6] 初學記	[7] 徐堅	[8] 風俗通義	[9] 應劭	[10] 沈括
[11] 夢溪筆談	[12] 南卓	[13] 羯鼓錄	[14] 段安節	

his *Yo Fu Tsa Lu*[1] (Miscellaneous Notes on the Bureau of Music) in the Wu Tai period (+10th century). This deals with musical instruments and their origins, songs, dances, and famous performers. But just as pharmaceutical science in China through the centuries waits for Li Shih-Chen, so also in acoustics and musicology it was the last decades of the +16th century which produced the greatest master of the subject, overshadowing all his predecessors—Chu Tsai-Yü.[2] Much will be said of him later on;[a] here we wish only to refer to the elaborate monographs which he prepared with mathematical precision and illustrated with some of the best line-drawings in any Chinese technical work. The earliest of these treatises, the *Lü Li Yung Thung*[3] (The Pitch-pipes and their Calendrical Concordances) appeared in +1581,[b] and three years later saw the conclusion of the *Lü Hsüeh Hsin Shuo*[4] (New Account of the Science of the Pitch-pipes). The 'Essential Meaning of the Standard Pitch-pipes' (*Lü Lü Ching I*[5]) was ready by +1596,[c] and the 'New Account of the Science of Calculation (in Acoustics and Music)', *Suan Hsüeh Hsin Shuo*,[6] by +1603.[d] As we shall see later, Chu Tsai-Yü was not influenced by the coming of the Jesuits though his own influence on Europe may have been very great; he represents the final climax of indigenous acoustic and musical theory. During the +18th century Chinese works have to stand comparison with what was going on in post-Renaissance Europe, but even so, the productions of a Chiang Yung,[7] whose *Lü Lü Hsin Lun*[8] appeared about +1740, or a Tai Chen,[9] who in +1746 devoted himself to an admirable archaeological reconstruction of the forms of ancient bells,[e] are still deserving of careful study today.

In acoustics, as in so many branches of science, the Chinese approach was rather different from the European. Where ancient Greece was analytic, ancient China was correlative. We might look vainly before the Thang for such questioning as that recorded by Plutarch[f] who enquires

why the narrower of two pipes[g] of the same length should speak (sharper and the wider) flatter? Why, if you raise the pipe, all its notes will be sharp; and flat again if you stoop it? And why, when clapt to another, it will sound the flatter; and sharper again, when taken from it?...And why, when one would have set up a copper Alexander for a Frontispiece to a Stage at Pella, the Architect advis'd to the contrary, because it would spoil the Actors Voices?[h]

[a] Pp. 139, 220ff. below.
[b] This was later included in his *Li Shu*[10] (Calendrical Opus), together with an 'Imperial Longevity Permanent Calendar' (*Shêng Shou Wan Nien Li*[11]), which has already been referred to in Vol. 3, p. 713.
[c] This was really only the first part of his *Lü Shu*[12] (Pitch-pipe Opus).
[d] This work, together with the *Lü Lü Ching I* and the *Lü Hsüeh Hsin Shuo*, was combined c. +1620 to form the *Yo Lü Chhüan Shu*[13] (Collected Works on Music and the Pitch-pipes).
[e] This was embodied in his *Khao Kung Chi Thu*[14] (Illustrations for the *Artificers' Record* (of the *Chou Li*), with a critical archaeological analysis), to which a special paper has been devoted by Kondō Mitsuo (*1*).   [f] Wang Chhung's contemporary in the +1st century.
[g] Baxter translated 'flute' here, yet the *aulos* was not a flute, but a pipe with a double reed. Among the Greeks true flutes were, if not absent, surprisingly rare.
[h] *Works*, 1096A, 'Pleasure not attainable according to Epicurus', tr. Baxter (*1*), vol. 2, pt. 4, p. 118, mod.

[1] 樂府雜錄	[2] 朱載堉	[3] 律厤融通	[4] 律學新說	
[5] 律呂精義	[6] 算學新說	[7] 江永	[8] 律呂新論	[9] 戴震
[10] 厤書	[11] 聖壽萬年厤	[12] 律書	[13] 樂律全書	
[14] 考工記圖				

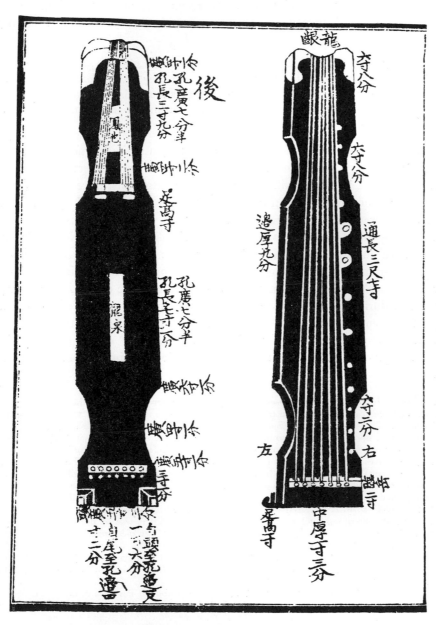

Fig. 297. The classical Chinese 'lute' (*ku chhin*), properly described as a seven-stringed half-tube zither. *Akhak Kwebŏm*, ch. 6, p. 21*b* (+1493).

Tung Chung-Shu, for example, in the −2nd century, when confronted by the much more striking phenomenon of sympathetic resonance, accepts it simply as being 'nothing miraculous', since it accords so well with the typically Chinese organic world-view.

> Try tuning musical instruments such as the *chhin*[1] or the *sê*.[2] The *kung* note or the *shang* note struck upon one lute[a] will be answered by the *kung* or the *shang* notes from other stringed instruments. They sound by themselves. This is nothing miraculous, but the Five Notes being in relation; they are what they are according to the Numbers (whereby the world is constructed).[b]

But in China we have to deal with two distinct currents, the literary tradition of the scholars, and the oral tradition of the craftsmen who were expert in acoustics and music. From what follows it will be seen that the latter must have done a great deal of experimentation, asking questions quite parallel to those asked by the Greeks— but the details were only rarely recorded.[c]

Tung Chung-Shu, indeed, was among the most scientific and philosophical minds of his age. In ancient and medieval times acoustic phenomena were often enough regarded as portents. Many strange sounds were recorded, but enquiry was concerned rather with what they could mean than with how they were caused. For example, it is noted that during the reign of the emperor Chhêng there occurred in −18 a case of a great rock emitting a noise like thunder.[d] The prognosticatory tradition was that such an event implied a disturbance of the element Metal,[e] due to unbridled love of war and conquest on the part of rulers. The people said simply that soldiers would come.

---

[a] Both the instruments named have commonly been considered lutes, but the term is literary and imprecise. The classical *chhin* is still in use today (Fig. 297)—an instrument of seven strings, correctly to be described as a half-tube zither (Sachs, 2), for it consists of a flat elongated board concave below and convex above, upon which are mounted the silk strings. A musician playing on a *chhin* may be seen on the back of a bronze mirror made in Tung Chung-Shu's own time (Bulling (8), pl. 31). The *sê* (Fig. 298) survives only in the form of a descendant called the *chêng*,[3] which has thirteen strings of brass wire but retains the integral board body. In true lutes the resonator box is distinct from the long or short neck upon which the strings are extended. Such instruments the Chinese had, but not during the seminal period of their acoustics on which much of our discussion will turn. All were variants of the celebrated *phi-pha*,[4] so often referred to in later poetry and literature. The history of this short lute has been carefully examined by Picken (6), who concludes that it was of non-Chinese provenance, introduced from some Central Asian people, probably Iranised Turco-Mongols, in the +2nd century. The most important of the earliest sources include Liu Hsi's[5] *Shih Ming*[6] (Explanation of Names) *c.* +200, and the *Phi-Pha Fu*[7] (Rhapsody on the Phi-Pha) by Fu Hsüan[8] (+217 to +278) in *CSHK* (Chin sect.), ch. 45, p. 6a. On musical interchange between East and West Asia see further Farmer (1, 2, 3).

[b] *Chhun Chhiu Fan Lu*, ch. 57, tr. auct., cf. Vol. 2, p. 281 above. Parallels in *Lü Shih Chhun Chhiu*, ch. 63, vol. 1, p. 122, tr. R. Wilhelm (3), p. 161; and *Chuang Tzu*, ch. 24, tr. Legge (5), vol. 2, p. 99; *Huai Nan Tzu*, ch. 11, p. 11a, cf. Wu Nan-Hsün (1), p. 167.

[c] One of the most stimulating comparisons of Greek with Chinese music and acoustics is that of Laloy (2). The work of Amiot and Chavannes in this context will be referred to below (p. 176) in connection with the 'Pythagorean controversy'.

[d] *Chhien Han Shu*, ch. 27A, p. 20a; *TSCC*, *Shu chêng tien*, ch. 158, p. 2b.

[e] Cf. Vol. 2, pp. 243ff. and Eberhard (6), p. 19.

---

[1] 琴   [2] 瑟   [3] 箏   [4] 琵琶   [5] 劉熙   [6] 釋名
[7] 琵琶賦   [8] 傅玄

Nevertheless, Chinese interest in sound, though it followed a different course from that of the Greeks, was by no means fruitless. Chinese invention enriched the world's civilisation in the sphere of acoustics and music no less than in other fields. The pages which follow will attempt to show first how the social life of the Chinese in pre-Han times brought them to focus attention on sound as a manifestation of Nature in equilibrium and disequilibrium. This entails a study of the concept of *chhi*,[1] subtle matter,

Fig. 298. The extinct 'great lute' (*sê*), a horizontal psaltery with twenty-five silk strings. *Hsiang Yin Shih Yo Phu*, ch. 1, p. 2*b*, in Chu Tsai-Yü's *Yo Lü Chhüan Shu* (+1620).

vital breath, or emanation. We shall then try to trace the advance towards acoustics as a science, with steadily improving systems of classifying sounds, and devices for measuring the pitch of musical notes. Finally, we shall describe some of the contributions which China has made to the world's understanding of sound, and of the nature of music.

(2) CORRELATION OF SOUND WITH FLAVOUR AND COLOUR

Few peoples ancient or modern have proved themselves more sensitive than the Chinese to the timbre of musical sounds. Van Gulik mentions[a] sixteen different 'touches' in playing on the silk strings of the classical 'lute' (*chhin*[2]) and lists yet

[a] (1), pp. 105, 125.

[1] 氣   [2] 琴

other manners of striking and pulling them. To take one example only, the vibrato termed *yin*:[1]

A finger of the left hand moves quickly up and down over the spot indicated. 'A cold cicada bemoans the coming of autumn.' The plaintive, rocking drone of the cicadas should be imitated. Of this *yin* there exist more than ten varieties. There is the *chhang-yin*,[2] a drawn-out vibrato, which should recall 'the cry of a dove announcing rain'; the *hsi-yin*,[3] a thin vibrato, which should make one think of 'confidential whispering'; the *yu-yin*[4] or swinging vibrato, which should evoke the image of 'fallen blossoms floating down with the stream', etc. Remarkable is the *ting-yin*;[5] where the vacillating movement of the finger should be so subtle as to be hardly noticeable. Some handbooks say that one should not move the finger at all, but let the timbre be influenced by the pulsation of the blood in the fingertip, pressing the string down on the board a little more fully and heavily than usual.

Such a description suggests the infinite subtlety with which any given note could be played. Indeed, even today an expert *chhin* player will himself remain intently listening long after a note has become inaudible to other listeners. As Taoist thought put it:[a] 'The greatest music has the most tenuous notes (*ta yin hsi shêng*[6]).'

This was by no means an aestheticism without basis in physical fact. The ancient zither (*ku chhin*[7]) is the only musical instrument in any culture which has no frets and actually marks the nodes of vibration on the board. Recognition of individual harmonics, 'floating sounds' (*fan yin*[8]), using the same string, was already well advanced in the time of Hsi Khang[9] (+223 to +262). In Europe on the contrary this came very late, not before the +18th century. Indeed, the technique of playing the *chhin* mainly depends on exploiting the production of different timbres at the same pitch, and this was already developed to perfection by the later Sung (+12th century).

Nevertheless, the question remains, what did early Chinese thinkers believe sound to be? Their contemporaries in ancient Greece set themselves this question and tried to answer it. The Pythagoreans, for example, believed sound to be what Laloy describes[b] as 'la chose numérale par excellence'.[c] Theon of Smyrna,[d] about +150, attributes to Hippasus and Lasus (−5th century) the establishment of a relation between sound and speed, sound being something which is thrown so quickly that like a rapid discus it cannot be perceived in flight, but only on the instant of 'landing'. Archytas (*fl.* −370) went further and defined sound as speed itself.[e]

In ancient China, on the contrary, no parallel analysis and abstraction was made. Sound was regarded as but one form of an activity of which flavour and colour were others. The background for Chinese acoustic thinking was largely determined by a

---

  a *Tao Tê Ching*, ch. 41.
  b (1), p. 52.
   c Cf. the remark of Leibniz, quoted by Archibald (2): 'Music is a hidden exercise in mathematics by minds unconscious of dealing with numbers.'
   d *On the Uses of Mathematics for the understanding of Plato*, ch. 12 (ed. Bouilland, Paris, 1644). Cf. Freeman (1), pp. 86 ff.
   e Theon of Smyrna, ch. 13; Laloy (1), p. 64; Freeman (1), pp. 237 ff.

¹ 吟   ² 長吟   ³ 細吟   ⁴ 遊吟   ⁵ 定吟   ⁶ 大音希聲
⁷ 古琴   ⁸ 泛音   ⁹ 嵇康

concept which stemmed from the vapours of the cooking-pot, with its fragrant steam for which the word was *chhi*.[1] We have already had occasion to enlarge on the significance of this basic concept of Chinese pneumatism.[a] Karlgren gives for the word in Chou times the meanings 'vapour, air, breath, vital principle, temperament, to present food, to pray, beg or ask'.[b] It was clearly of wide connotation, and will be used in this Section (as in others) as a technical term for which there is no English equivalent. It moulded Chinese thinking from the earliest times, just as form and matter dominated European thought from the age of Aristotle onwards. For this reason one must have as good an understanding of its connotations as possible. Without this a European reader might consider the commentaries of many acute Han scholars writing on musical subjects as loaded with acoustic observations of a superstitious or nonsensical kind.

The common context, then, of the meanings of the word *chhi* given above is that of sacrifice to the ancestors. They are prayed in the *Shih Ching* to return and reinvigorate their descendants and their crops:

> Sonorous are the bells and drums. Brightly sound the stone-chimes and flutes.
> They bring down with them blessings—rich, rich the growth of grain!
> They bring down with them blessings—abundance, the abundance![c]

The ancestors are tempted to return to earth not only by the prayers of their descendants chanting liturgical phrases, but by the sounding of musical instruments and the delicious emanations which rise up from magnificent bronze cooking-vessels. When they arrive their eyes are also feasted with the sight of an assembly dressed in ceremonial clothing, furs and emblems all conforming to traditional themes of colour. From the earliest historical periods the Chinese were concerned with a synthesis of sound, colour and flavour, responding to the synthesis[d] of Nature manifested in thunder, rainbows and spicy herbs. One *chhi* rises up from the earth to heaven like steam from cooking-pots; another descends from heaven to earth, like ancestors spreading their reinvigorating influence. Their intermingling produces wind,[e] wherewith heaven makes music,[f] and brings into being not only rainbows which are heaven's colours, but the flowers of the changing year and with them the flavouring herbs in due season. All were signs and symbols of those great climatic processes on which the life of the ancient Chinese people depended, balancing ever between flood and drought.[g] Such was the environment which brought forth their organic philosophy.[h] A purely analytic treatment of sound would hardly have been consistent with it.

---

[a] E.g. Vol. 2, pp. 22, 41, 76, 150, 238, 275, 369; Vol. 3, pp. 217, 222, 411, 467, 480, 636.   [b] K 517.
[c] *Shih Ching*, cf. Legge (8), pt. IV, i (1), no. 9; Mao, no. 274; tr. auct. adjuv. Karlgren (14), p. 243; Waley (1), p. 230. Perhaps −7th century.
[d] Almost an orchestration. Cf. p. 164 below.
[e] *Chhien Han Shu*, ch. 21A, p. 4a: 'The *chhi* of heaven and earth unite and thereby produce wind.'
[f] *Chuang Tzu*, ch. 2: 'If Earth pipes, it is with all its apertures. If Man pipes, it is with the collected bamboos.' Cf. Vol. 2, p. 51.
[g] Cf. Vol. 1, pp. 87, 96, 114, 131, etc.; Vol. 3, p. 462 ff., 472 ff. Cf. Sect. 28 below.
[h] Cf. Vol. 2, pp. 51 ff.; 281 ff., 472 ff.

[1] 氣

(3) The Concept of *CHHI* in relation to Acoustics

Chhi, then, had two main sources. It could go up from earth to the ancestors, and it could come down from heaven with the ancestors to earth. A third but very important source was in man himself, in his breath. With increasing sophistication *chhi* is thought of as something more rarefied than steam or breath. It becomes an emanation, a spirit, a *pneuma* (πνεῦμα). Naturally any attempt to trace the development of this idea must be rather hypothetical, but some form of hypothesis is necessary if Chinese acoustic thinking is to be understood.

In one of the early passages of the *Shu Ching*, which Karlgren[a] places not later than −600, there occurs the statement: 'Eighth: the several manifestations (*shu chêng*[1]). They are called rain, sunshine, heat, cold, wind, and their seasonableness.' With this may be compared a passage from the *Tso Chuan*[b] of perhaps some two centuries later: 'The six *chhi* are called Yin, Yang, wind, rain, darkness and brightness.' It is reasonable to suggest that the latter is a more sophisticated version of the idea contained in the former. The *Shu Ching* text would have a straightforward appeal to anyone engaged in farming, whereas the latter has the neat antithesis of the scholar. Its statement of the six *chhi* follows, by way of explanation, another: 'There are the six *chhi* of heaven. Their incorporation (*chiang*[2]) produces the five flavours; their blossoming (*fa*[3]) makes the five colours; they proclaim themselves (*chêng*[4]) in the five notes.'[b] The word *chhi*, then, is used sometimes in a general way for that form of emanation which goes up to and comes down from heaven, and sometimes for a particular form of its descent. Elsewhere in the *Tso Chuan*[c] it is stated that the *chhi* themselves make the five flavours. It is hard to know quite what is implied in 'descent', but the term 'six channels' (*liu thung*[5]) is sometimes used as a synonym for the six *chhi*.[d] Now this suggests a connection which is important for early Chinese acoustic theories, for if *chhi* is something which can be canalised or piped off,[e] the obvious instrument for the purpose would be a bamboo tube, such as is used in China for irrigation. Consequently, it is not surprising to find early references to the shaman-musician piping off his own *chhi* through bamboo tubes in an attempt to alter the

---

[a] (12), p. 33 (ch. 24, Hung Fan).

[b] Referring to −540; Duke Chao, 1st year (tr. auct. adjuv. Couvreur (1), vol. 3, p. 37).

[c] Duke Chao, 25th year (−516), tr. Couvreur (1), vol. 3, p. 380. 'The (six) *chhi* make the five flavours. (Their) manifestation makes the five colours. (Their sound) patterns (*chang*[6]) make the five notes.' The word *chang* means a pattern, signal, rule, to manifest, abundant. See K723, where the explanation of the graph is said to be uncertain. We suggest that the Shang bone character represents a flute or pipe with a drum. Compare this graph with K251 'a big flute', and K147c almost certainly a pellet-drum.

K723   K251b   K147c

[d] *Chuang Tzu*, ch. 33 (Thien Hsia); cf. Legge (5), vol. 2, p. 216.

[e] Ssuma Chêng in the +8th century, commenting on the *Shih Chi*, says indeed that a pitch-pipe is that by which one canalises *chhi* (*lü chê so i thung chhi*[7]).

---

¹ 庶徵   ² 降   ³ 發   ⁴ 徵   ⁵ 六通   ⁶ 章
⁷ 律者所以通氣

processes of Nature—of heaven's *chhi*—by sympathetic magic. Should we not see a rather late echo of this practice in the story of Tsou Yen[1] (−4th century) blowing on his pitch-pipes for the benefit of the crops?[a]

To call these magical tubes pitch-pipes, however, is probably an anticipation of subsequent developments, as will be made clear later. Probably 'humming tubes' would be more appropriate. They are referred to by various names, *lü*,[2] *thung*,[3,4] *kuan*,[5,6] and the other *lü*,[7] a term that is generally translated 'pitch-pipe', but the essential meaning of which is regular steps or regularity.

Here we should pause for a moment to realise the wider significance of the fact that Chinese acoustics (like other branches of physics) was from the first, if not analytical, highly pneumatic. Parallel lines of thought have already been described in meteorology[b] and in geology,[c] while later on we shall see how important the concept of *chhi* was in medicine.[d] Filliozat (1) has shown convincingly that Greek pneumatic medicine (e.g. in the Hippocratic *De Ventis*), of about the −5th century, derives from the same sources as that of the Indian *Suśruta Saṃhitā* and *Caraka Saṃhitā* (+ and −1st century).[e] Though the earliest extant expressions of these ideas in their simplest form occur in the Vedic literature, now regarded as contemporaneous with the late Shang period (−13th century), it seems overwhelmingly probable that their origin was Mesopotamian. From Babylonia they would have radiated to the south-east and north-east as well as to the west. Later in this Section[f] we shall find cause for thinking that China received from the Fertile Crescent certain information about sound much more precise than the stimulus to think about *chhi*[8] and *fêng*.[9] Moreover, as we saw above,[g] acoustic examples were frequently adduced by naturalist thinkers in ancient China to support their characteristic conception of a universal continuum and the reality of action at a distance by wave transmission therein. When we say, therefore, that the acoustics of the old Chinese philosophers was highly pneumatic, we must not forget that they thought of *chhi* as something between what we should call matter in a rarefied gaseous state on the one hand, and radiant energy on the other. Though all our assured knowledge gained by experiment makes us infinitely richer than they, is the concept of 'wavicles' in modern physical theory so much more penetrating? At any rate, the interconversion of matter and energy would hardly have been a surprise to them.

(i) *Conduits for* chhi; *the military diviner and his humming-tubes*

The fact that in Chou texts the number of the *chhi* should be six, and the number of the *lü* should also be six, is probably not a coincidence, and there is early authority for the belief that some tubes respond to a Yang *chhi* and others to a Yin, a different term

---

[a] We have already referred to this on p. 29 above. For the character of Tsou Yen see Vol. 2, pp. 232 ff. Cf. Dubs (5), p. 65.

[b] Vol. 3, pp. 467, 471, 479, 481, 491, etc.

[c] Vol. 3, pp. 637 ff.

[d] Section 44 below.

[e] Renou & Filliozat (1), vol. 2, pp. 147, 150.

[f] Pp. 177 ff. below.

[g] Pp. 29, 32.

[1] 騶衍　　[2] 呂　　[3] 同　　[4] 筒　　[5] 管　　[6] 筦　　[7] 律

[8] 氣　　[9] 風

being used to distinguish them. For example, in the *Chou Li* it is said:[a] 'The Grand Instructor (Ta Shih[1]) takes the Yin-tubes (*thung*[2]), and the Yang-tubes (*lü*[3]), listens to the army's note, and predicts good fortune or bad (*i thing chün shêng chao chi hsiung*[4]).'[b] In this passage the number is not specified, but in many instances they are referred to as the 'twelve pipes (*kuan*)', 'the six (Yang) *lü* and the six (Yin) *lü*', or sometimes quite simply as 'the six *lü*'.[c] Any enquiry into this subject is complicated by the fact that the literature spans many centuries during which musical evolution was rapid, and musical terms of necessity changed their meaning, as also by the fact that the number of tubes used was not necessarily the same at all periods. If a sketch of their evolution were to be attempted, one would postulate first an instrument of two tubes tied together, possibly one open and the other stopped, such as is suggested by the graph of the word *yung*[5] (K1185*b*, *c*) (mean-

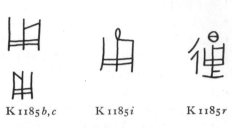

K1185*b,c*   K1185*i*   K1185*r*

ing—an instrument, to use); and its development in the word *yung*[6] (K1185*h*, *i*) which means the loop of a bell, or (*tung*) a flute. This graph was further developed in the word *thung*[7] (K1185*r*, *s*) meaning a channel or communication, previously referred to as a synonym for *chhi*; and in a later work (the −3rd-century *Han Fei Tzu*) by the addition of the bamboo radical to mean a tube (*thung*[8]).[d] All of these words belong to one common phonetic group.

For the next stage one would expect an increase in the number of tubes as the shaman himself develops nicer powers of discriminating between different sorts of *chhi*. In the *Tso Chuan* there is an instance where apparently four were used. The passage[e] describes how the officials of the State of Chin ask the Music-Master Khuang[9] about the outcome of a campaign if the troops of the southern State of Chhu besieging Chêng should march north. The Music-Master replies:

> There is no harm. I repeatedly hummed the northern 'wind'; I also hummed the southern 'wind'. The southern 'wind' was not vigorous. The sound signified great slaughter. (The State of) Chhu will inevitably fail to gain a victory.

Chêng Chung[10] (*fl.* +70), commenting on this passage, says that the northern 'wind' is Chia-chung[11] and Wu-yi,[12] the southern 'wind' Ku-hsien[13] and Nan-lü.[14] These are

---

[a] *Chou Li*, ch. 6, p. 14*a* (ch. 23); tr. Biot (1), vol. 2, p. 51.
[b] Cf. Vol. 2, p. 552, above.
[c] E.g. *Li Chi*, ch. 9 (Li Yün): 'The 12 tubes in turn act as fundamental (*Shih-erh kuan huan hsiang wei kung*[15])'; *Chhien Han Shu*, ch. 21A, p. 3*b*: 'The tubes are 12 in number; the Yang six compose the *lü* and the Yin six compose the (other) *lü* (*Lü yu shih-erh, Yang liu wei lü, Yin liu wei lü*[16])'; *Tso Chuan*, Duke Chao, 20th year (−521): 'The 5 notes, the 6 tubes..., etc. (*Wu shêng, liu lü...*[17]).'
[d] To this group of characters concerning tubes one could also add *sung*,[18] which means to croon or to sing to oneself.
[e] *Tso Chuan*, Duke Hsiang, 18th year (−554); Couvreur (1), vol. 2, p. 342.

[1] 大師   [2] 同   [3] 律   [4] 以聽軍聲詔吉凶   [5] 用   [6] 甬
[7] 通   [8] 箭   [9] 曠   [10] 鄭衆   [11] 夾鐘   [12] 無射
[13] 姑洗   [14] 南呂   [15] 十二管還相爲宮
[16] 律有十二陽六爲律陰六爲呂   [17] 五聲六律   [18] 誦

the names of four notes of the regular gamut of his day, which contained twelve notes in all.[a] In the first century of our era these twelve notes were divided into two groups of six, one group of which was regarded as Yang and the other as Yin. But by no 1st-century classification can the notes mentioned above be made to agree with this Yin and Yang division. According to the orthodox system described in the *Lü Shih Chhun Chhiu*, for example, Chia-chung is Yang and Wu-yi is Yin. In the *Chou Li* the opposite is the case. This fact, and also that four notes are named out of a possible twelve, suggests that Chêng Chung was drawing on some genuine tradition concerning the ancient art of divination by means of humming-tubes.

Before pursuing the idea of the evolution of simple pairs of humming-tubes into complex sets of detachable pan-pipes used for giving the pitch for a gamut of twelve notes, it will be worth while to examine more closely this remarkable passage from the *Tso Chuan* for its bearing on the concept of *chhi*. Since many of the terms are far from clear let us look at the passage in the original.[b] The two terms which cause most difficulty are *ko*[1] and *fêng*,[2] translated above as 'hum' and 'wind' respectively. In support of this we may quote the commentary of Fu Chhien[3] (+2nd century):[c]

> The southern pitch-pipe emanation (*Nan-lü chhi*[4]) did not come up (to its full strength). Therefore the note signified great slaughter.
>
> In speaking of the blowing of pitch-pipes, why do we say 'hum' and 'wind'? The note produced is the 'hum' (*Chhui lü erh yen ko yü fêng chê? Chhu shêng yüeh ko*[5]).
>
> Since the pitch-pipes are also the tubes (used for the practice of) 'observing the *chhi*' (see pp. 186 ff.), the emanation is called 'wind'. This is why we talk of the 'hum' and the 'wind'. (*I lü shih hou chhi chih kuan, chhi tsê fêng yeh. Ku yen ko fêng.*[6])

There can be little doubt that the Chinese of these early centuries believed they knew a way of divining the outcome of a battle by some peculiar process of blowing or humming through tubes.[d] There are other references to it besides those given above. For example, Ssuma Chhien quotes[e] a saying that

> on seeing the enemy from afar it is possible to know in advance what the outcome of a battle will be, for better or for worse. On hearing the sound it is possible to know whether there will be victory or defeat. Such is the method which has not varied under a hundred kings.

The use of hollow tubes, bones or branches as speaking trumpets for disguising or amplifying the voice of the shaman is widespread among primitive peoples.[f] That it

---

[a] Cf. below, p. 171.
[b] *Pu hai. Wu tsou ko pei fêng, yu ko nan fêng. Nan fêng pu ching. To ssu shêng. Chhu pi wu kung.*[7]
[c] *Chhun Chhiu Tso Chuan Chieh I*, in *YHSF*, ch. 34, p. 23 b; tr. auct.
[d] See again Vol. 2, pp. 551 ff. The close connection between warfare and music is attested by the fact that the same character means both an army and a music-master (*shih*[8]).
[e] *Shih Chi*, ch. 25, p. 1 b; tr. Chavannes (1), vol. 3, p. 294, eng. auct.; cf. Sachs (1), p. 25.
[f] Cf. Vol. 2, pp. 132 ff.

[1] 歌　　[2] 風　　[3] 服虔　　[4] 南律氣
[5] 吹律而言歌與風者出聲曰歌　　　[6] 以律是候氣之管氣則風也故言歌風
[7] 不害吾驟歌北風又歌南風南風不競多死聲楚必無功　　　[8] 師

should occur in China is not remarkable. What is remarkable is that the Chinese should have attempted in this, as in so many other of their activities, to reduce the practice to a clearly regulated and classified system.

In a lost 'Book of War' (*Ping Shu*[1]) quoted by Chang Shou-Chieh[2] in his Thang commentary on this passage of the *Shih Chi*, five different states of morale are listed, all of which can be known by the skilful diviner. Every man has within his body his own *chhi*. The diviner uses his to set up a disturbance in the outside world when he blows through his humming-tube. One *chhi* will then 'by a kind of mysterious resonance'[a] react on another *chhi*, just as one musical instrument will touch off another which is in tune with it. In an army where many men are massed together there is a 'collective *chhi*' which floats above it, and which can be seen as a coloured cloud,[b] and heard as a note or sound. As the Thang scholar Ssuma Chên says in his commentary on the passage already cited:[c]

Above every enemy in battle array there exists a vapour-colour (*chhi-sê*[3]). If the *chhi* is strong, the sound (note) is strong. If the note is strong, his host is unyielding. The pitch-pipe (or humming-tube) is (the instrument) by which one canalises (or communicates with) *chhi*, and thus may foreknow good or evil fortune.

There is a certain reasonable basis for this strange belief. If the divination were merely to discover the outcome of a battle it might not deserve much consideration. But, as can be seen from the passages quoted, it was primarily to discover the enemy's morale, and thus to deduce the chances of victory. This is a very different matter from the Roman practice of auguring victory by the observation of the flight of birds or the entrails of animals. In the days of close combat every commander was anxious to study his enemy for signs of morale. There is, for example, Thucydides' famous account of the defeat and death of Cleon, when his opponent Brasidas exclaimed: 'Those fellows will never stand before us, one can see that by the way their spears and heads are going.'[d] In the nervous tension which precedes a battle it would have been easy for the shaman to imagine the *chhi* which he believed to emanate from every individual in the opposing host to be gathering like a cloud above them; moreover, we ourselves are all in some measure able to judge the mood of persons we know well by the timbre of the human voice, a slight shrillness betraying anxiety, harshness, anger, and so on. If, then, as the ancient Chinese believed, sound is produced by *chhi*, and the *chhi* rising up from an army would have been considerable, how could the *chhi* from an army have failed to produce a sound? 'There is', as Ssuma Chhien concludes, 'nothing marvellous in this. It is quite natural.'[e]

It would be interesting to know what exactly was the method by which the different types of sound were distinguished. From the *Tso Chuan* passage it would seem to

---

[a] Cf. Vol. 2, p. 304.
[b] Cf. *Chin Shu*, ch. 12, pp. 9*b*ff., tr. Ho Ping-Yü (1).
[c] Tr. auct. Not in Po-na Pên ed.; *KHCP* ed., vol. 3 (p. 76).
[d] *History of the Peloponnesian War*, tr. Crawley, v, ch. 15.
[e] *Shih Chi*, ch. 25, p. 1*b*; cf. Chavannes (1), vol. 3, p. 294.

[1] 兵書   [2] 張守節   [3] 氣色

have been mainly a matter of whether the sound was vigorous or not. If it did not come up to full strength the notes indicated 'death'. 'Death' sounds implied that the army concerned had poor morale and would be defeated; vigorous sounds on the other hand implied success. If we knew exactly what these tubes were like it might be possible to have a clearer idea of what was meant by the term 'death' sounds.

In the *Li Chi* occur the words:[a]

The five notes, the six fixed pitches and the twelve pipes take it in turn to act as fundamental (*wu shêng liu lü shih-erh kuan huan hsiang wei kung*[1]).

The word for pipe here is *kuan*.[2] So many different descriptions of it are given by various commentators that one is forced to the conclusion that it was frequently used merely as a generic term. But one scholar's opinion deserves particular attention, for though late—he was born in +1536—his understanding of ancient music was exceptional. This is Chu Tsai-Yü,[3] about whom more will be said in due course.[b] He gives it as his opinion[c] that *kuan* was the name for the *lü* when several were tied together, and that they were originally simple open notched pipes. Now notched pipes such as are pictured by Chu in his book, having a small semi-circle cut in the upper edge of one end of the tube, across which the player blows at right angles, are the earliest and most primitive of all such pipes.[d] If Chu is correct in his assumption that the early pitch-pipes were of this sort, one can begin to understand how a diviner could sometimes evoke 'death' notes, and sometimes notes which were 'vigorous'. For of all musical pipes those with a notch are the most difficult to play. As Chu says:[e]

The thoughts must be serious, the mind peaceful, and the will resolute....Open the lips and emit lightly a small (jet of) breath in blowing, causing the air to enter the tube continuously; then its correct note will be sounded....For persons to blow the pitch-pipes, do not employ the old or the very young; their *chhi* is not the same as that of (persons who are) youthful and strong.

It is probable that in the tense moments before a battle, the shaman might from anxiety or excitement fail to emit the small jet of breath at exactly the right angle, or with the right degree of force, or with the constancy which was required, so that fluctuating, feeble or 'dead' notes would result. This then would have provided the basis for the divination, since the variation in the sounds, though arising from the state of morale of the shaman's own side, could well have been attributed, by means of the 'resonance' theory, to the *chhi* of the enemy. Pitch differences were not an essential part of the response.

[a] Ch. 9 (Li Yün), p. 61*b*; cf. Legge (7), vol. 1, p. 382.
[b] Cf. pp. 220ff. below.
[c] *Lü Lü Ching I*[4] (*Nei Phien*), ch. 8.
[d] Kunst (1), p. 57, after discussing this, mentions the Javanese *chalintu* as a modern example.
[e] *Lü Hsüeh Hsin Shuo*,[5] ch. 1, p. 19*a*; tr. auct.

[1] 五聲六律十二管還相爲宮　　[2] 管　　[3] 朱載堉　　[4] 律呂精義
[5] 律學新說

Later a more detailed appraisal of the enemy's morale was attempted by a fivefold division of sound, the terms for which were *kung*,[1] *shang*,[2] *chio*,[3] *chih*[4] and *yü*.[5] These are the names of the five degrees of the pentatonic scale, if we give them the meaning which they have in works of the −4th century onwards. But clearly there was an earlier period when they did not have this meaning only, for if *kung* is translated as 'fundamental', the others being, for example, as in the standard first mode, major second, major third, perfect fifth, and major sixth; it is impossible to make sense of the statement in the *Kuo Yü*:[a]

In the affair of Mu Yeh (the battle in which the old Shang kings were overthrown by King Wu of the Chou) the sounds all exalted the fundamental (*yin chieh shang kung*[6]).

For *kung* was not the name of a fixed note like our middle C, but any note could be *kung*, in the manner of our movable doh, as is clear from the above quotation from the *Li Chi*.

This is even clearer in the *Shih Chi* account, in which it is said:[b]

When King Wu attacked Chou Hsin he blew the tubes and listened to the sounds. From the first month of spring (i.e. from the longest tube) to the last month of winter (i.e. the shortest tube) a *chhi* of bloody slaughter (was formed by their) joint action, and the ensuing sound gave prominence to (the distinctive quality of) the *kung* note.[c]

The fact that from Han times onward *kung*, *shang*, *chio*, *chih* and *yü* became the normal terms for the five notes of the scale, that is to say terms for distinguishing relations of pitch, helped to make these passages musically unintelligible; and the fact that Chavannes wrote off Chinese explanations of their own theories of *chhi* as 'pur pathos' denied subsequent writers the key to the problem. Yet it is sure that in divination these five terms are not concerned exclusively or even primarily with pitch, but rather with a certain quality of sound, or timbre. What that quality may have been is suggested by the words of Tung Chung-Shu in the work quoted at the beginning of this Section:[d]

Violent winds in summer correspond with the note *chio*...crashing peals of thunder in autumn correspond with the note *shang*...autumn lightning flashes correspond with the note *chih*...cloudbursts of rain in spring and summer correspond with the note *yü*... rumbling thunder in autumn corresponds with the note *kung*.

The 'Book of War', previously mentioned, states[e] how these five qualities of sound may be interpreted.

---

a Cf. *Chou Yü*, ch. 3, p. 36*b*.
b Ch. 25, p. 1*b*; tr. Chavannes (1), vol. 3, p. 294, eng. auct.
c *Thui mêng-chhun i chih yü chi-tung, sha chhi hsiang ping erh yin shang kung.*[7]
d *Chhun Chhiu Fan Lu*, ch. 64; tr. auct. adjuv. Hughes (1), p. 308.
e It is in the commentary of *Shih Chi*, ch. 25, p. 1*b*; tr. auct.

[1] 宮   [2] 商   [3] 角   [4] 徵   [5] 羽   [6] 音皆尚宮
[7] 推孟春以至于季冬殺氣相并而音尚宮

The Great Instructor blows the tubes, uniting the sounds. If it is *shang* there will be victory in the fight; the soldiers of the army are strong. If it is *chio* the army is troubled; many vacillate, and lose their martial courage. If it is *kung* the army is in good accord; officers and men are of one mind. If it is *chih* there is restlessness and much irritation; the soldiers are tired. If it is *yü* the soldiers are soft, and little glory will be gained.

The diviner was apparently able to learn the morale of his own troops by blowing the pipes on the first day of the campaign, and of the enemy by blowing them before battle was joined. If then the emanation over Chou Hsin's army was held to emit the qualities of a *kung* note, showing that both armies were in good heart, that version of Chinese history which, in opposition to Mencius,[a] maintained that the Shang dynasty was overthrown only after bloody slaughter would seem more likely to be correct.

We are forced then to conclude that, as a development of the pseudo-science of divination, *kung, shang, chio, chih,* and *yü* were at one time names connoting qualities describing the volume or timbre of certain sounds. This enquiry conveniently introduces the subject of timbre in early Chinese thought and practice.

(4) CLASSIFICATIONS OF SOUND BY TIMBRE

The preceding paragraphs have attempted to show how early Chinese ideas of the nature of sound were based upon the concept of *chhi*. In fact it persisted until recent times, and was abandoned in scholarly circles only under the influence of physical theories of wavelength during the modern period. We shall now see how the Chinese advanced from an early stage in which they were concerned with a general quality of sound, boding good or ill, to an exact appreciation of how one sound may differ from another in timbre, volume and pitch. In doing so we shall see how the classification of sounds gradually became standardised, while sounds themselves were correlated with other phenomena. Today we regard timbre as that which distinguishes one note from another not by volume or pitch but by complex blends of overtones. The ancient Chinese music-masters would not have expressed themselves in this way, for the different elements which make up a sound were probably not thought of in isolation, but timbre was very important for them. Indeed the Chou classification of musical sound heralded that sensitivity to tone which was mentioned in the introduction to this Section. Since many European writers on Chinese music have been inclined to an opposing belief, it is interesting to find Juan Gonzales de Mendoza quoting the testimony of the Austin friars[b] that the Chinese 'do tune their voyces unto their instruments with great admiration'.[c] It may be that +16th-century Europeans, whose ears had not yet become accustomed to the rigid tuning of modern equal temperament, were more tolerant of other systems.

[a] *Mêng Tzu*, III, 2, v (5); tr. Legge (3), p. 149.
[b] See p. 225 below.
[c] Parke tr. p. 140.

(i) *Material sources of sound*

The phrase by which the Chinese designate the orthodox grouping of instruments is *pa yin*[1] or 'the eight (sources of) sound'. This is a convenient shorthand term for the eight different kinds of material which featured most prominently in the construction of the different types of instrument. One of the earliest texts in which they are catalogued is the *Chou Li*, where they are listed[a] as metal, stone, earth (or clay), skin, silk (threads), wood, gourd, and bamboo.[b] Instruments in which the sound-producing agents are so varied produce a variety of timbres. The Graeco-Roman classification into three genera, wind-instruments (*pneumatikon*, πνευματικόν), stringed instruments (*enchordon*, ἔγχορδον), and percussion instruments (*kroustikon*, κρουστικόν), was perhaps more scientific, and gave place only in modern times to the fivefold classification[c] of idiophones, membranophones, chordophones, aerophones, and electrophones.

The great variety of timbre in Chinese music in general, and in the instruments of the Confucian temple orchestras in particular, has often been emphasised.[d] Balinese *gamelan*[e] and Dayak long-house gong ensembles in Sarawak[f] preserve something of the spirit of the early Confucian orchestras,[g] while the court music of the Thang still lives in Japan.[h] This unaccustomed richness in variety of timbre was somewhat baffling to the first Europeans of modern times who experienced it, as may be seen, for instance, in the report of Matteo Ricci after attending a rehearsal at the Nanking *wên-miao*[2] (Confucian temple) in +1599.[i] Yet the most characteristic acoustic features of the music of the Chinese culture-area (extending as it does from Korea to Indonesia)[j] may be defined as twofold: the high proportion and multiplicity of chime-idiophones on the one hand, and the prominence of the bamboo plant (and the pitch-pipes derived from it) on the other.

It is clear that the eightfold Chinese classification of sounds was only arrived at gradually. In the *Yo Chi* (Record of Ritual Music and Dance), a book certainly com-

---

[a] Ch. 6, p. 12*a* (ch. 23), tr. Biot (1), vol. 2, p. 50.

[b] Probably the oldest extant classification of musical instruments in any civilisation, says Schaeffner (1), p. 124.

[c] Mahillon (1); Galpin (1), p. 25; Montandon (1), pp. 695ff.; Schaeffner (1), pp. 143ff., 179ff.

[d] E.g. by Picken (3).

[e] McPhee (1, 2); Picken (2), pp. 170ff.

[f] Private communication from K.R. (1957).

[g] Representations of orchestras from the Warring States period have come down to us, notably on bronze vessels such as the bowl from Huihsien (Fig. 299) preserved in the Archaeological Institute of Academia Sinica (cf. Yang Tsung-Jung (*1*), pl. 19), and the magnificent vase of the −4th or −3rd century known as the Yen-Yo Yü-Lieh Thu Hu[3] (Fig. 300) which may be seen in the Imperial Palace Museum at Peking (cf. Yang Tsung-Jung (*1*), pl. 20). Of Han representations one of the best is that in the I-nan tomb reliefs (Fig. 301); cf. Anon. (7), figs. 27*b*, *c*; Tsêng Chao-Yü *et al.* (*1*), pl. 48.

[h] Harich-Schneider (1); Picken (2), pp. 144ff.

[i] Trigault, tr. Gallagher, pp. 335ff.; d'Elia (2), vol. 2, pp. 70ff. Cf. Vol. 2, pp. 31ff.

[j] This is not only, or mainly, because of influences radiating from China in relatively late times, but rather because common cultural elements helped to shape Chinese music in the Chou period. Cf. Picken (2), pp. 180ff.; and Vol. 1, p. 89, above.

[1] 八音　　　[2] 文廟　　　[3] 燕樂漁獵圖壺

Fig. 299. An orchestra of the Warring States period (*c.* −4th century): the Huihsien bronze bowl (Anon. (*4*); Yang Tsung-Jung (*1*), etc.). Musicians beating upon suspended L-shaped chime-stones can be seen at the top; on the left, beyond the picture of a building, others are striking a row of suspended bells.

Fig. 300. An orchestra of the Warring States period (*c.* −4th century): the Yen-Yo Yü-Lieh Thu Hu bronze vase preserved in the Imperial Palace Museum, Peking. In the fourth row from the top, on the right, there are three ringers of bells, one musician in charge of the row of suspended chime-stones, one drummer with stand-drums and one person playing on a wind-instrument. The frame for the instruments is upheld by two large carved birds, and immediately to the right of it a kneeling figure appears to be playing the 'tiger-box' (see p. 150). The two turtles and the small bird seem to have strayed in from the scene of fishing and fowling on the extreme left of the same row, but the deer-like animal appears to be responding to the orchestra, like the dancing-girl with long sleeves caught in airy leap above the 'tiger-box' player. From Yang Tsung-Jung (*1*).

PLATE CV

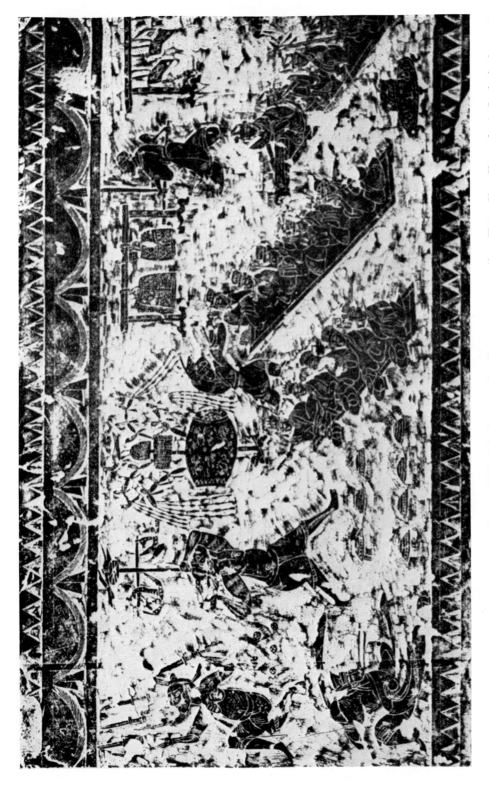

Fig. 301. An orchestra of the Han period depicted in the reliefs of the I-nan tombs, Shantung, c. +193 (from Tsêng Chao-Yü et al., 1). In the background, stand-drums, bells and ringing-stones; in the front row, hand-drums; in the second row, pan-pipes; and in the third row, zither, ocarina and hand-organ with reeds. On the left, jugglers and acrobats.

piled from Chou sources, the following eight instruments of 'music' are listed:[a] bells, drums, pipes, flutes, ringing-stones,[b] feather (wands or dresses), shields, and axes. Only four 'sources of sound' occur here, namely metal, skin, bamboo, and stone (Fig. 302). In another passage in the same text it is said:[c]

As the *Shih Ching* says: 'Guiding the people is very easy.' That is why the sages established the pellet-drum (*thao*[1,2]) and the stand-drum (*ku*[3]), the instrument which starts the miming (*chhiang*[4]) and the instrument which stops it (*chieh*[5]), the globular flute (*hsüan*[6]) and the flute (*chhih*[7]).[d] These six (instruments) gave the notes of numinous music charged with morality (*tê yin chih yin*[8]). After that (they established) bells (*chung*[9]), ringing-stones (*chhing*[10]), the blown pipe (*yü*[11])[e] and the (silk-stringed) zither (*sê*[12]) to go with them.

Elsewhere in the *Yo Chi* it is more specifically stated[f] that the instruments of music are the four sources of sound—metal, stone, silk and bamboo (*chin, shih, ssu, chu, yo chih chhi yeh*[13]).[g] From this and also other references it seems clear that there was a period from which many early texts derive, in which the 'eight sources of sound' were not yet classified and settled. This is reinforced by the fact that the *Tso Chuan* has only a single reference (−717) to the 'eight sources of sound'.[h] The passage states that 'dancing is that by which one regulates the eight sources of sound, and thereby conducts the eight winds (*wu so i chieh pa yin, erh hsing pa fêng*[14])'. References to the

[a] Para. 2, in *Shih Chi*, ch. 24, p. 11 *b*; cf. Chavannes (1), vol. 3, p. 248.
[b] The sounds produced by striking flat L-shaped chime-stones of various sizes were among the most characteristic features of the orchestras of ancient China (Figs. 304, 305). We had occasion to mention them before in connection with centres of gravity (p. 34 above). For the conclusions of modern scholars on the geometry of their shaping, see Wu Nan-Hsün (1), pp. 127ff., Chhen Wên-Thao (1), pp. 67ff.
Kuttner (1) believes that they originated from the flat annular stone symbols called *pi*[15] of which we had to say so much in Vol. 3, pp. 334ff., supposing that the *pi* itself was first struck to make music and that the 'dissection' of its annulus into fragments came about by removal of pieces in a tuning process.
[c] Para. 8, in *Shih Chi*, ch. 24, p. 32 *a, b*; tr. Chavannes (1), vol. 3, p. 276, eng. et mod. auct. See Figs. 302 and 303.
[d] The *chhih* was a transverse pipe blown by a centrally placed hole, with finger holes on each side.
[e] The *yü* is generally considered a large 36-pipe form of the mouth reed-organ or *shêng*. But there was also the vertical pipe (*ti*[16]) with six holes and a back-stop. In the collection of my friend Mr R. Alley I have seen a bronze *ti* of the Warring States period, originally gilded, in which the holes are all situated at the lowest parts of as many regular annular constrictions, the diameter of the pipe swelling between them. The blown end or mouthpiece is shaped in dragon-head form, and the lower orifice is surrounded by ridges and indentations as if to secure a leather or wooden trumpet-like termination. Presumably this wavy form of tube had some acoustic significance.
[f] Para. 6, in *Shih Chi*, ch. 24, p. 25 *b*; cf. Chavannes (1), vol. 3, p. 266.
[g] In these ancient times the metal category refers in general to bells. The different varieties of gongs (*lo*,[17] *tien tzu*[18]) did not originate in China, as has been shown by Kunst (3), but rather in Central Asia. Not until the Thang did they become common. Afterwards special types were evolved, including that which has become known in the West as the 'Chinese crash-cymbal' (*po*[19]), so named because of the brilliant crash it makes when struck with a drumstick.
[h] Duke Yin, 5th year; tr. Couvreur (1), vol. 1, p. 34, eng. auct.

1 鼗	2 鼖	3 鼓	4 椌	5 楬	6 壎	7 篪
8 德音之音		9 鐘	10 磬	11 竽	12 瑟	
13 金石絲竹樂之器也			14 舞所以節八音而行八風			
15 璧	16 笛	17 鑼	18 點子	19 鈸		

10

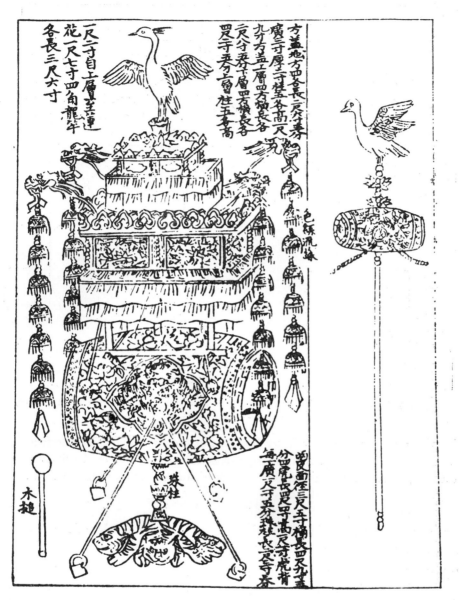

Fig. 303. To the right the pellet-drum (*thao*); to the left the great stand-drum
(*chien ku*). *Akhak Kwebŏm*, ch. 6, pp. 5*b*, 9*b* (+1493).

Fig. 302 (opp.). A late Chhing representation of the instruction of musicians by the legendary music-master Hou Khuei. Before him, on the table, are the zithers *chhin* and *sê*; to his right the reed-organ (*shêng*), the globular flute (*hsüan*) and the transversely blown straight flute (*chhih*); to his left the pan-pipes (*hsiao*) and the vertically blown bamboo flute open at both ends (*yo*). In the background the stand of bells and the great chime-stone; in the foreground the stand of chime-stones and the standing-drum. In front of Khuei's table, on the ground, are the percussion tub (*chu*) and the tiger-box (*yü*). Tea is being served. From *SCTS*, ch. 2, Shun Tien (Karlgren (12), p. 7).

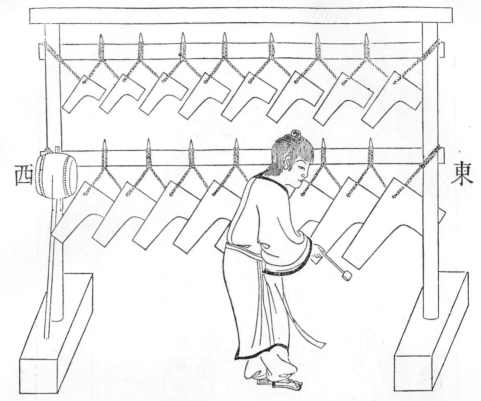

西   東

Fig. 304. The stand of chime-stones (*chhing*). *Hsiang Yin Shih Yo Phu*, ch. 1, p. 13*a*,
in Chu Tsai-Yü's *Yo Lü Chhüan Shu* (+1620).

eight winds, on the other hand, are frequent in the *Tso Chuan*. In the fact that the
ancient Chinese correlated their sources of sound with winds evidently lies the clue
to the early development of the four sources of sound into an ultimate eight.

(ii) *Winds and dances*

Mention was made earlier in this Section[a] of the great annual climatic cycle on which
the life of the early Chinese people depended. The *Yo Chi* specifically relates it to music:[b]

It is the Tao of heaven and earth that if cold and heat do not come at the right time there
will be epidemics; if wind and rain do not come in due proportion there will be famine.
(When the ruler) teaches (what is required by means of ritual mimes), that is the people's
cold and heat. If his teaching does not come at the right time he may blast a whole generation.
(When the ruler) acts that is the people's wind and rain. If his actions do not observe due
proportion they will be without effect. That is why the former kings organised (the ritual
mimes accompanied by) music, and so governed by force of example (i.e. by sympathetic
magic). If these were good, the activity (of the people) mirrored his moral power.

There were many different dances in these ritual mimes, but all fell under two heads,

[a] P. 133 above.
[b] Para. 4, in *Shih Chi*, ch. 24, pp. 16*b*, 17*a*; tr. Chavannes (1), vol. 3, p. 256; eng. et mod. auct.

PLATE CVI

Fig. 305. Great chime-stone and stand of chime-stones in the main hall of the Temple of Confucius at Chhü-fou, Shantung (orig. photo., 1958). In the background a great stand-drum, behind the stones a reconstructed *sê*, in front of them a pellet-drum and a hand-drum. Lying flat on the table in the foreground, the pan-pipes.

pyrrhic (warlike) and peaceful. The pacific dances included beast dances and rain-making dances. Evidence of the latter is found in the *Yo Chi* where it is stated that 'the evolutions (of the dancers) symbolise wind and rain (*chou huan hsiang fêng yü*[1])'. There are many accounts in early texts of grand performances of music, with bells, stone-chimes, etc., followed by a great wind or a storm with thunder and rain.[a] But precise details of the dances themselves are scarce.

The winds, however, were certainly eight in number, one from each cardinal and one from each intermediate point. Chêng Hsüan says that in the 'cap dance' (*huang-wu*[2]) for the four cardinal points feathers were worn covering the top of the head, and the clothing of the dancers was adorned with kingfisher feathers. This dance was performed in time of drought.[b] It is not hard to see a connection between times of drought and kingfishers, birds which frequent rivers and watery places. He adds that these costumes had the brilliant colours of the plumage of the so-called 'phoenix' (*fêng*[3]). Feathers were used in all three of the pacific dances. Cranes were also imitated. The names of these ancient songs and dances are almost all that survive to tell the tale of rain-making music—'The South Wind' (according to Ssuma Chhien a song of birth and growth),[c] 'Receiving the Clouds', 'The White Clouds', and others.

In order to control the dances various instruments were used. Two have already been mentioned,[d] the *chhiang*, which was a hollow wooden beaten instrument used for starting the dances, and the *chieh*, for stopping them. Two others with the same function, and which perhaps they resembled, are the *chu*[4] and the *yü*.[5] The former derives from an agricultural pestle-and-mortar or wooden tub for crushing grain. The latter is a hollowed block of wood shaped so as to resemble a tiger with a serrated back (Fig. 306) and gives a rasping noise when brushed or struck smartly with a stick (*chen*[6]) split at the end into twelve leaves.[e] The way in which the dances were controlled is suggested by the following description in the *Yo Chi*:[f]

In the ancient mimes the dancers advanced in ranks and retired in ranks (*chin lü thui lü*[7]),

[a] See, for example, the story of Duke Phing of Chin (−557 to −532) and the music-masters, *Shih Chi*, ch. 24, pp. 38*b*ff. (tr. Chavannes (1), vol. 3, pp. 288ff.); and also Legge (1) in his preface to the *Shu Ching*, p. 115, concerning the *Annals of the Bamboo Books*.

[b] See his commentary on the *Chou Li*, ch. 6, p. 7*b* (ch. 22), on the six types of dance; cf. Biot (1), vol. 2, p. 41.

[c] *Shih Chi*, ch. 24, p. 38*a*, cf. Chavannes (1), vol. 3, p. 287.   [d] P. 145 above.

[e] Later on, the Buddhists developed other hollow wooden instruments, such as the 'fish' (*yü pang*[8]) struck to assemble the monks; and the slotted box, flat below and convex above (*mu yü*,[9] *ao yü*[10]) used for accompanying the chanting of the *sūtras*, which found its way into Western bands in modern times. As late as World War II its sound was one of the most characteristic features of Chinese cities at night.

Another very curious, and hardly musical, instrument was the *ku tang*[11] which 'consists of a glass bulb, somewhat like a wine-decanter. As the bottom is extremely thin, when the mouth is applied to the open end and the breath drawn quickly in and out, it vibrates with a loud crackling sound' (Moule (10); cf. Bodde (12), p. 79). In view of the evidence given above concerning glass-blowing, this may have been a very late development.

[f] Para. 8, in *Shih Chi*, ch. 24, p. 30*b*; tr. Chavannes (1), vol. 3, p. 273, eng. et mod. auct. The technical terms in the second phrase are not certain, and have generally been taken to mean '(the music is) harmonious and correct with fulness', as in Chavannes' translation.

[1] 周還象風雨	[2] 皇舞	[3] 鳳	[4] 柷	[5] 敔	[6] 籈
[7] 進旅退旅	[8] 魚梆	[9] 木魚	[10] 鰲魚	[11] 鼓噹	

keeping together with perfect precision, like a military unit. The strings, gourds, and reed-tongued organs all waited together for (the sound of) the tambourines and drums. The music was started by a (note from a) pacific instrument (drum). Conclusions were marked by a (note from a) martial instrument (bell). (These) interruptions were controlled by means of the *hsiang*[1] drum. The pace was regulated by means of the *ya*[2] drum.

From this brief extract the impression emerges that what the ancient Chinese were seeking above all in their dance rituals was control. Whereas the Greeks seem from early times to have paid great attention to melody and the tuning of lyres, the Chinese were primarily concerned with rhythm and the control of movement by which the elements were to be influenced. Later the movements of men's minds, their passions, were to be controlled in the same way but symbolically and by suggestion.

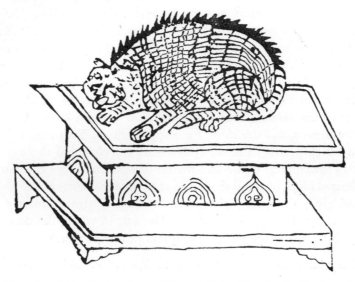

Fig. 306. Tiger-box (*yü*). From a Sung edition of Chhen Yang's *Yo Shu* (+11th century), cf. *Akhak Kwebŏm*, ch. 6, p. 11 *b*.

The belief in the ability of the ritual mimes to control the weather[a] is specifically expressed in the *Yo Chi* where it says:[b] 'The eight winds follow (obey) the *lü* and are not turbulent (traitorous) (*pa fêng tshung lü erh pu chien*[3]).' The exact meaning of this statement depends on how the word *lü* is translated. The first and most probable alternative is that *lü* here has its earliest meaning of regulated dance steps.[c] The second interpretation would take *lü* to mean humming-tubes through which the shaman

---

[a] Widely distributed, of course, among primitive peoples. For instance, Frazer (1) says: 'The Motumotu in New Guinea think that storms are sent by an Oiabu sorcerer; for each wind he has a bamboo (tube) which he opens at pleasure' (vol. 1, p. 327). Frazer gives much other material on the magical control of the wind. Mem. *Macbeth*, Act 1, sc. iii, as Miss Su Lin reminds us.

[b] Para. 6, in *Shih Chi*, ch. 24, p. 24 *b*; tr. Chavannes (1), vol. 3, p. 265, eng. et mod. auct.

[c] Cf. Vol. 2, pp. 551 ff. above, regarding the word *lü* (regulations, and standard pitch-pipes).

[1] 相   [2] 雅   [3] 八風從律而不姦

'whistles for a wind'. The third possibility is that *lü* has its normal later meaning of blown pipes by which the pitch was given to other instruments. The statement that 'the eight winds obey the pitch-pipes' would make sense if it could be shown not merely that certain pitch-pipes were associated with certain winds in a vague and undefined way, as was done in Han times,[a] but that a given wind was induced by its appropriate dance, that a given dance was initiated or regulated by its appropriate instrument, and that a given instrument, or the position occupied by a given source of sound, was associated with a certain pitch or note in a scale. In this way the regulated movements of the dance, the regulated positions of the instruments, and the regulated intervals of the scale, would all be connotations of the very broad concept of *lü*. Support for the idea that different instruments may have been associated with different notes is found in a statement in the *Chou Li*[b] that

the Great Music-Master...weaves the gamut into patterns using the five notes (of the scale) ...and distributes the notes using the eight sources of sound (*Ta Shih...chieh wên chih i wu shêng...chieh po chih i pa yin*[1]).

An exact description of how some of the eight sources of sound were allotted to the five notes is given in the *Kuo Yü*.[c]

The zithers *chhin* and *sê* exalt (or follow) the *kung* note. Bells exalt the *yü* note (*Chhin sê shang kung. Chung shang yü*[2]). Ringing-stones exalt the *chio* note. (Instruments of) gourd and bamboo exalt the note which is appropriate to them (*Shih shang chio. Phao chu li chih*[3]).

This difficult passage apparently means that instruments which do not produce a great volume of sound are principally used for music in which the mode uses the deeper notes of the gamut, these being loud (*ta*[4]) and the higher notes soft (*hsi*[5]). In this way loud instruments such as bells do not drown soft instruments such as zithers, and the music is level (*phing*[6]).[d]

The argument at this stage may be summarised as follows. The earliest known system by which the Chinese classified sounds was according to the materials from which their instruments were made. These originally numbered four—stone, metal, bamboo, and skin or leather. Their number was later increased to eight. This accorded with the fact that eight different winds were recognised. Each wind, it was thought, could be induced by a particular type of dance, and each dance was controlled by a particular musical instrument. It will now be shown that an attempt was made to form an eightfold classification of instruments or sources of sound correlated with the eight directions or sources of wind. From this division of the instruments of the orchestra it will later be suggested that there arose a further classification of sounds

---

[a] *Shih Chi*, ch. 25, pp. 4*a*ff.; cf. Chavannes (1), vol. 3, pp. 301ff.

[b] Ch. 6, p. 3*b* (ch. 22), tr. auct. adjuv. Biot (1), vol. 2, p. 32.

[c] *Chou Yü*, ch. 3, pp. 23*b*, 24*a*; tr. de Harlez (5), eng. auct.

[d] *Chou Yü*, ch. 3, p. 24*b*.

[1] 大師 … 皆文之以五聲 … 皆播之以八音   [2] 琴瑟尚宮鍾尚羽
[3] 石尚角匏竹利制   [4] 大   [5] 細   [6] 平

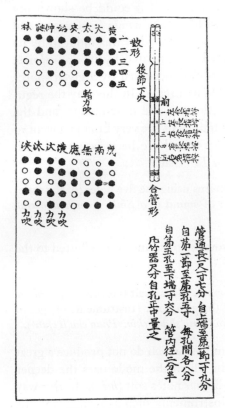

Fig. 307. The double pipe (*kuan*), from *Akhak Kwebŏm*, ch. 6, p. 12*b* (+1493).

Fig. 309. Musicians playing on reed mouth-organs (*shêng*). *Hsiang Yin Shih Yo Phu*, ch. 1, p. 13*b*, in Chu Tsai-Yü's *Yo Lü Chhüan Shu* (+1620). The nearest figure is marking the time with a sort of clapper (*chhung-tu*) consisting of twelve slips of bamboo strung together on a leather thong (like an ancient book); this is held in the right hand and struck with the left (cf. Moule (10), p. 12).

PLATE CVII

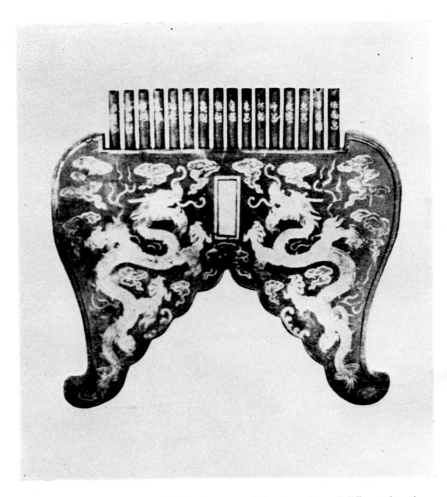

Fig. 308. The pan-pipes (*hsiao*), sixteen small bamboo pipes of different lengths fixed in a row (from Chhien Chün-Thao, *1*).

according to pitch. For an instrument which had originally served only as the starting signal of the dance, or to keep the beat going, was capable eventually of convenient use for giving the pitch-note as well.

(iii) *Correlation of timbre with directions and seasons*

The association of the qualities of musical sounds with various moods is highly subjective, and varies from individual to individual. But in any one culture there may be stereotyped reactions to particular noises. Thus the *Yo Chi* describes five of the sources of sound as having such standard associations:[a]

The sound of bells is clanging. Clangour produces a call as if to arms. Such a call gives rise to wild excitement. Wild excitement produces warlike emotion. When the *chün tzu* (man of breeding)[b] listens to the sound of bells he thinks of heroic military officers.

Chêng Hsüan observes that the effect of a bell is that of a warning to rouse the people, and explains that it causes a person's *chhi* to become abundant. The text proceeds:

The sound of ringing-stones is a tinkling. Tinkling sets up a power of discrimination. Discrimination enables men to press on to their deaths. When the man of breeding listens to the sound of ringing-stones he thinks of loyal officials who have died on the frontiers.
The sound of silken strings is a wailing. Wailing stimulates integrity. Integrity establishes resolution. When the man of breeding listens to the sound of the (silk-stringed) zithers *chhin*[1] and *sê*,[2] he thinks of resolute and righteous ministers.
The sound of bamboo flutes is a gurgling like flood waters. Flood waters entail levies. Levies involve gathering the people (for their tasks). When the man of breeding listens to the sound of the bamboo pipes *yü*,[3] *shêng*,[4] *hsiao*,[5] and *kuan*,[6] he thinks of officials who have been the shepherds of the people.[c]
The sound of the stand-drums and the tambourines is rowdy. Boldness (of spirit) sets up (physical) activity. Physical activity sets the people marching. When the man of breeding listens to the sound of drums and tambourines he thinks of great generals leading out armies.
So when the man of breeding listens to the timbre (of different sorts of instruments) he does not listen merely to their clanging and tinkling, but he is also sensitive to their associations.

In the above translation it has been difficult to keep the effect of the Chinese definitions, the words of which are highly onomatopoeic as well as fully charged with meaning. For the Chinese of this archaic period the timbres of these five sources of sound— metal bells, chiming slabs of stone (Fig. 310), silken-stringed zithers, bamboo flutes, and drums of stretched leather—were respectively summed up by the words *k'εng*, *k'ieng*, **·ər*, **glâm*, *χwân* (i.e. *khêng*,[7] *chhing*,[8] *ai*,[9] *lan*,[10] *huan*[11]).[d]

---

[a] Para. 8, in *Shih Chi*, ch. 24, pp. 32*b*ff.; tr. Chavannes (1), vol. 3, pp. 277ff., eng. et mod. auct.
[b] On this conception, see Vol. 2, p. 6.
[c] The *yü* was a single pipe and the *kuan* a double one, the *hsiao* was the pan-pipes and the *shêng* a mouth-organ with reeds; see Figs. 307, 308, 309.
[d] Respectively K 1252, 832, 550, 609, 158. The phonetic transcription of the ancient Chinese sounds used here and hereafter is that of Karlgren (1).

1 琴	2 瑟	3 竽	4 笙	5 簫	6 管	7 鏗
8 磬	9 哀	10 濫	11 讙			

泗濱浮磬圖

Fig. 310. A late Chhing representation of the making of chime-stones from the rocks of the Ssu River, described in the *Shu Ching*. From *SCTS*, ch. 6, Yü Kung (Karlgren (12), p. 14; cf. Vol. 3, p. 500).

The interesting point about this list of 'sources of sound' is that they number five, in accord with the symbolic correlations of Five Element theory, and not, as was usual later, eight. The latter are tabulated in most works of reference as shown in Table 44.

Table 44. *Traditional list of the eight sources of sound*

Source of sound	Compass-point	Season	Instrument
1. Stone	North-west	Autumn–Winter	Ringing-stone
2. Metal	West	Autumn	Bell
3. Silk	South	Summer	Lute or zither
4. Bamboo	East	Spring	Flute and pipe
5. Wood	South-east	Spring–Summer	Tiger-box (*yü* [1])
6. Skin	North	Winter	Drum
7. Gourd	North-east	Winter–Spring	Reed-organ (*shêng* [2])
8. Earth	South-west	Summer–Autumn	Globular flute

Comparing this orthodox list of eight with the five quoted from the *Yo Chi*, one is first struck by the fact that the reed-organ (*shêng* [2]) is here placed under the very unusual material 'gourd', whereas in the early text it is where one would expect to find an instrument which consists of a series of blown bamboo tubes (even if they happen to have a gourd serving as a wind-chest), namely under the material bamboo. Then again the order in which the points of the compass are listed is peculiar, for one would expect to find 'south-west' placed between south and west, and not attached at the end almost as an afterthought. The seasons also are strangely erratic. But the reason for this is probably quite simple. An earlier classification in fours was stretched[a] to harmonise with a later system of eights. The earlier was thus:

Compass-point	Season	Instrument
West	Autumn	Chimes (bells and ringing-stones)
South	Summer	Zithers
East	Spring	Pipes
North	Winter	Drums

Now in each of these cases there is a clear connection between the instrument and its corresponding quarter. First, autumn is the season when the Yang forces of nature

[a] The tiger-box and similar instruments of punctuation were scarcely of the same rank as the four great types of instrument.

[1] 敔   [2] 笙

are in retreat, and bells or metal slabs were the instruments sounded when a commander ordered his troops to retire.[a] In winter there occurred one of the most solemn
ceremonies of the year, when the sun was assisted over the crisis of the solstice by the
help of sympathetic magic. The primeval instrument, the drum, was essential to this
ceremony, and there could be none more fitting to announce the sun's renewed
advance than the drum which also sounded the advance in human conflict and battle.
In spring when men desire trees to bud and crops to grow, the most potent instrument
would naturally be one made of bamboo, a plant of such vitality that it remains green
even in winter. The various pipes of bamboo, then, through which men's *chhi* causes
a similar *chhi* in Nature to respond, were the instruments of spring, and even in the
orthodox eightfold classification the other vegetable substances, wood and gourd,
were associated with this season. Finally, in summer when the silkworms are fattening
on mulberry leaves, or spinning their cocoons, it was appropriate to play an instrument whose strings were of silk. Moreover, summer was the time when drought was
to be feared, and the zithers which accompanied rain-making songs were believed
to be excellent implements of magic. The association of the instruments with the
points of the compass was no less straightforward. If autumn is the season of
decline, the west is its direction, whereas spring and the east are contrary.[b] Similarly
the north and winter must be associated with cold, and the south and summer with
heat.

Since music was in ancient times a part of government, and intimately associated
with agriculture, changes in the calendrical system necessitated changes in music.
Hence an increase in the 'sources of sound' became inevitable. But the introduction
of a fifth factor to be worked into the system during the rise of the Five Element
theory gave rise to many complications.[c] It is to the credit of the Chinese as system-
atisers that they achieved a synthesis, where the Greeks, apart from occasionally
including æther as a fifth element, left behind them no such pattern as we find in
the *Yo Chi*.[d]

The ethical characteristics which the *chün-tzu* (man of breeding) associated with
the five sources of sound deserve comparison with similar qualities attributed to the
five notes. Our next study must therefore concern itself with pitch.

[a] Cf. Vol. 2, p. 552. Abundant evidence shows that in ritual music there was the closest association
between bells and ringing-stones. Fig. 311 shows the stand of bells (cf. Hett's description (1) of Confucian ceremonies at Seoul (Korea) in our own time). The chime-stones were among the oldest of Chinese
instruments, as is attested by the find (in 1950), in a great royal tomb at Anyang dating from about the
−14th century, of a musical slab of grey limestone (Hsia Nai (1); Li Shun-I (1), p. 38). This is perfectly preserved and gives out a clear ringing note when struck. A conventional tiger, beautifully engraved, decorates its face (Fig. 312). On the Confucian temple ceremonies themselves see G. E. Moule
(2); Shryock (1); Johnston (1); and Vol. 2, pp. 31 ff. Sets of seven and of nine chime-stones from the
Warring States period are figured in Thang Lan (1), pl. 65. A set in use is depicted in the Wu Liang
tomb-shrine reliefs of +147 (Jung Kêng (1), indiv. rubbing Hsin, 1); cf. Figs. 299, 300.
    [b] See Vol. 2, p. 262.
    [c] Cf. Vol. 2, pp. 242 ff. The note *kung* appropriate to zithers was given the central place of Earth,
while bells and ringing-stones were separated, the note *yü* of the former going to Water and the note
chio of the latter going to Wood. At the same time *chih* was assigned to Fire and *shang* to Metal.
See Vol. 2, p. 262.
    [d] Cf. Vol. 2, p. 246.

PLATE CVIII

Fig. 311. Great hanging bell and stand of bells in the main hall of the Temple of Confucius at Chhü-fou, Shantung (orig. photo., 1958). In the background a great stand-drum, behind the bells a *ku chhin*, in front of them a pellet-drum and a hand-drum, with a small *shêng* on the nearer table.

PLATE CIX

Fig. 312. One of the oldest and most magnificent specimens of a chime-stone, from a royal tomb of the Shang period at Anyang (*c.* −14th century). A stylised tiger is engraved on its face. Imperial Palace Museum, Peking (Anon. *26*).

(5) CLASSIFICATIONS OF SOUND BY PITCH

By 'pitch' is meant that quality of sound which is determined by the frequency of vibration of an elastic body and of the air surrounding it, fast vibrations producing one type of auditory sensation and slow vibrations another. Language deals with these sensations by curious metaphors. Very rapid vibrations produce a less pleasant sensation than slow ones. The Romans described such notes as 'cutting' or 'sharpened' (*acutus*). Notes at the opposite end of the scale, however, were not 'blunt' but 'heavy' (*gravis*). English uses a consistent metaphor based on the scale or 'ladder' in which sounds at one end are said to be 'high' and at the other 'low'. Chinese uses a metaphor which is not surprising for a people whose economy was so bound up with hydraulic engineering, namely—clear (*chhing*[1]) and muddy (*cho*[2]).

In Greece names were invented which bore a clear relation to the strings of the lyre; for example, the lowest note, which was sounded by the top string as the lyre was held for playing, was called *hypatē* or 'uppermost'. Its octave was *neatē* or 'lowest', while between the two on the primitive lyre was a *mesē* or 'middle' (string). Later other names were invented, such as *lichanos* or 'first finger' (note), and *tritē* or 'number three'.

There is no such simplicity in the etymology of the original five notes of the Chinese scale. Their names, as stated on p. 140 above, are *kung, shang, chio, chih* and *yü*. Previous writers on Chinese music[a] have been content to say that these names seem to hold 'traces of an ancient symbolism'. But it is quite certain that these terms had symbolical associations. In the *Yo Chi*, for example, it is said[b] that

kung acts as the prince, *shang* as the minister, *chio* as the people, *chih* as affairs, and *yü* as beings (animate and inanimate) (*kung wei chün, shang wei chhen, chio wei min, chih wei shih, yü wei wu*[3]).

To this Chêng Hsüan adds by way of commentary that

in general, notes which are deep in pitch are noble, while those which are high in pitch are humble (mean) (*fan shêng, cho chê tsun, chhing chê pei*[4]).

These statements shed some light on the evolution of Chinese ideas about pitch.

The name for the note *shang*[5] (K734) was pronounced **śiang* in archaic Chinese, and also had the meanings of 'discuss, debate, trade'. This character was anciently interchangeable with *hsiang*[6] (K715), archaically pronounced **χiang*, which meant 'a window facing north, turn towards, formerly'.[c] With this may be compared *hsiang*[7] (K714*c, d*) having the same archaic pronunciation as the preceding word, and a set of

---

[a] E.g. Laloy (2), p. 54.
[b] Para. 1, in *Shih Chi*, ch. 24, pp. 5*b*, 6*a*; tr. Chavannes (1), vol. 3, p. 240, eng. auct.
[c] Cf. Chavannes (1), vol. 3, pp. 278, 294.

[1] 清   [2] 濁   [3] 宮爲君商爲臣角爲民徵爲事羽爲物
[4] 凡聲濁者尊清者卑   [5] 商   [6] 向   [7] 鄉

similar meanings, 'facing towards, turn towards, a little while ago'.[a] The last and most interesting link is that this character *chhing*[1] (K7140, *p*), archaically pronounced **k'iǎng*, which meant a minister. In spite of the roundabout etymology the general connection is clear. In State discussion the minister turned towards his prince, like his echo *hsiang*[2] (**χiǎng*, K714*n*). In this context the word *hsiang*[3] (K731), meaning 'to look at', 'mutually', may be noted, for (as we saw, p. 150) it was also the name of a drum which marked

K714*d*

the intervals in music, thereby playing the role of 'a firm and just official' by seeing that the music is 'correct and just'. In addition it may be noted that *chhing-hsiang*[4] was a familiar compound term meaning minister.

From the above analysis it will be clear that originally it was not so much that the note *shang* symbolised the minister as that 'minister' was the name of one of the notes. How it got that name may become clear from further examination of the other four note names. The first of these is *kung*,[5] which 'acts as prince'. *Kung* in Chou times meant a house. By a specialisation natural to an age in which princes lived in houses and the common people in huts and hovels, *kung* later came to mean a palace. Now it is stated in the *Chou Li*[b] that the Junior Aides (Hsiao Hsü[6])

regulate the position of the musical instruments which are hung on frames. A prince has frames in the form of a house (*kung*[5]). Feudal lords have frames in the form of a chariot. Ministers and great officials have frames in the form of divided (walls). Ordinary gentlemen have single frames.

By this is meant that at musical court ceremonies stands or frames for sets of bells, ringing-stones or drums were set up, four frames enclosing a hollow square like the walls of a house. At the inferior courts named above, the southern wall, the northern and southern, and the northern, southern, and western walls respectively, were absent. The *kung* or house was, therefore, the name of a musical instrument placed in a certain position for the making of music at a princely court.

Two further note names seem to have derived from stands for suspended instruments, namely *chio* (horns) and *yü* (feathers). One of the odes in the *Shih Ching* has these lines:[c]

There are blind musicians, there are blind musicians, present and ready in the Chou palace yard.
(Their assistants) set up the serrated boards (to support the instruments), set up the drum posts, raising the tusks, planting the feathers (*shê yeh shê chü chhung ya shu yü*[7]).

---

[a] The graph shows two men sitting turned towards one another, with a food vessel between them, hence the better known later meaning of village and by extension country.

[b] *Chou Li*, ch. 6, p. 11*a* (ch. 22); tr. Biot (1), vol. 2, p. 47, eng. auct.

[c] *Shih Ching*, pt. IV, i (2), no. 5; Mao, no. 280; tr. auct. adjuv. Legge (8); Karlgren (14), p. 245; Waley (1), p. 218.

[1] 卿   [2] 響   [3] 相   [4] 卿相   [5] 宮   [6] 小胥
[7] 設業設虡崇牙樹羽

The exact meaning of some of the terms in this poem has been the subject of much comment, but the point is that certain posts from which drums and other instruments hung were ornamented with feathers; furthermore, either part of the wooden framework was cut so as to make sharp points (teeth) or angles (horns), or tusks or horns were fastened on to it.[a] Thus three of the five note names were terms used for describing stands supporting drums, bells or ringing-stones.

The hypothesis may now be suggested that the terms *kung*, *shang*, *chio*, *chih*, and *yü* originally referred to the positions occupied by certain instruments used in controlling the music and dancing. A number of references suggest that the earliest Chinese conception of a scale was not, as in the West, that of a ladder ascending from low to high or descending from high to low pitch, but of a court in which the notes are ranged on either side of the chief or *kung* note. As the commentator of the *Huai Nan Tzu* book succinctly puts it: 'The *kung* note is in the middle; therefore it acts as lord (*kung tsai chung yang, ku wei chu yeh*[1]).'[b] This refers to a statement in the text which says of the five notes that *kung* is their lord. It will also be recalled that in the quotation from the lost 'Book of War' (p. 141 above) the five notes or qualities of sound are listed in the order *shang*, *chio*, *kung*, *chih*, *yü*. This tradition of the notes being ranged not by pitch but by some sort of ceremonial array lends support to the view that at some time there were five stations round the dancing floor for posts and frames on which were hung the instruments for controlling the ritual mimes, the *kung* position for the house-frame, which was lord; the *shang* position for the *hsiang* drum regulating the proceedings like a just minister; the *chio* position for a stand dressed with horns; the *chih*[2] or summoning position, possibly associated with the *ying*[3] bell or drum[c] (both are known by name); and the *yü* position where the post or stand was adorned with feathers.

That these terms should later refer to the pitches of notes appears almost inevitable, for the instruments hung on the posts and frames were in fact the pitch-giving instruments, i.e. the ringing-stones, the bells, and, as is clear from early paragraphs of this Section, the drums. There is abundant evidence that from the earliest times it was these instruments which regulated the music. Two quotations will suffice. The first from the *Shu Ching*, where Khuei[4] (the great legendary musician), describing the arrangements for one of the ritual 'beast dances', says:[d] 'I strike the sounding-stone, I gently strike it, and the various animals lead one another on to dance.' The second is from the *Shih Ching* where a vivid description of ritual music contains the lines:[e]

[a] A comment of Chêng Hsüan's on drum-posts in the *Chou Li* relevant to this passage is here quoted in some editions of Mao's version of the *Shih Ching*; namely that 'they put (feathers) in the horns on the top of the uprights of the bell-frames'.

[b] *Huai Nan Tzu*, ch. 4, p. 8a. The comment is by Kao Yu[5] (*fl.* +210).

[c] *Ying* is an essential term in Chinese acoustics meaning resonance, especially that mysterious resonance referred to in connection with *chhi*, cf. Vol. 2, pp. 282, 304, 500, etc. The relation between the note *chih*, also read *chêng*, to summon (K891), and *ying* to respond (K890) is also etymologically close, *tjəng* and *tjəng*.

[d] *Shu Ching*, ch. 2 (Shun Tien), tr. auct. adjuv. Karlgren (12), p. 7.

[e] *Shih Ching*, pt. IV, iii, no. 1, Mao, no. 301; tr. auct. adjuv. Legge (8); Karlgren (14), p. 262; Waley (1), p. 225.

[1] 宮在中央故爲主也   [2] 徵   [3] 應   [4] 夔   [5] 高誘

'Then we bring (the instruments) together in time and pitch, relying on the notes of our ringing-stones (*chi ho chieh phing i wo chhing shêng*[1]).' Thus one might say that the musical interest of the Chinese in early Chou times was mainly focused on timbre and association. Exact pitch probably did not become a dominating factor among them till Babylonian influence made itself felt at the beginning of the −4th century.[a]

(6) THE DEVELOPMENT OF PHYSICAL ACOUSTICS

(i) *The pentatonic scale*

With the recognition of pitch intervals and the naming of notes, accurate measurement, observation and test become possible, and the science of acoustics has been born. One cannot say precisely when the Chinese first gave names to their notes, but the *Tso Chuan*, in passages for which the −4th century is a probable date, contains five references to the fact that the notes of the scale were five in number. Nowhere, on the other hand, does it refer to the notes by name. One might accept the previously quoted passage from the lost 'Book of War' as possibly the earliest instance of the notes being named in a surviving text,[b] but our argument has suggested that these names did not at that time necessarily refer to pitch. The same reservation holds true for a passage in Mencius (*fl.* − 350), in which *chih* and *chio* are mentioned as follows:[c]

(Duke Ching) called the Grand Music-Master and said: 'Make for me music to suit a prince and his minister pleased with each other.' And it was then that the Chih-shao[2] and Chio-shao[3] were made.

Legge says of this passage, 'The Chih-shao and Chio-shao were, I suppose, two tunes or pieces of music, starting with the notes *chih* and *chio* respectively'. If Legge's supposition can be accepted, this passage would provide a date to work from, but the evidence remains scanty at this period. Some fifty years later, however, there is no longer any doubt that *kung*, *shang*, *chio*, *chih* and *yü* were being used to distinguish different notes on stringed instruments.[d] This is attested by the definitions which open the chapter on music in the *Erh Ya* encyclopaedia.[e]

---

[a] A possible explanation for the origin of the names of the five notes has been sketched in some detail here, because anyone dependent on European works on Chinese music, or even on the orthodox Chinese accounts based on the dynastic histories from Han times onwards, will inevitably form the opinion that the ancient Chinese made a great point of absolute pitch, fixed by special pitch-pipes, if not indeed from the mythical age of the Yellow Emperor, then at least from high antiquity. This view is quite mistaken, as will be made clear in what follows.

[b] P. 141 above.   [c] *Mêng Tzu*, I (2), iv, 10; tr. Legge (3), p. 37.

[d] The *Chuang Tzu* book (ch. 2) cannot give evidence here, as has sometimes been thought. In connection with the legendary skill of the lutanists Chao Wên[4] and Shih Khuang[5] we read: 'Even the most skilful zither player, if he strikes the *shang* (note) he destroys the *chio* (note), if he vibrates the *kung* (note) he neglects the *chih* (note). It is better not to strike them at all; then the five notes are complete in themselves.' This extremely Taoist thought might be interpreted in our own idiom as a preference for 'piping to the spirit ditties of no tone', or for a totality in music which cannot be achieved when it is merely played. But we are in the +8th century, not the −4th, for the passage occurs not in the text, but in the Thang commentary of Chhêng Hsüan-Ying;[6] see *Chuang Tzu Pu Chêng*, ch. 1B, p. 18a.   [e] Ch. 7, p. 1a.

[1] 既和且平依我磬聲   [2] 徵招   [3] 角招   [4] 昭文   [5] 師曠
[6] 成玄英

Still later (*c.* − 150) there comes the work of Tung Chung-Shu to which reference has already more than once been made[a] on account of his statement that instruments are tuned to certain notes such as the *kung* and the *shang*, and that strings similarly tuned will sound in sympathetic resonance. It is remarkable that at this early period the Chinese were tuning even their drums and noting this phenomenon when they were struck.[b] In Europe, on the other hand, as late as the + 16th century such writers on music as Virdung were content to describe drums as 'rumbling tubs'.[c] By − 120 the *Huai Nan Tzu* book gives us an explicit statement not only that the five notes are named *kung, shang, chio, chih,* and *yü,* but that in combination with the twelve absolute pitches of the fixed gamut, sixty 'mode-keys' can be formed:[d]

(Given) a single note of fixed pitch, one can then interpret it as the keynote of five (distinct modes).[e] (Given) twelve notes of fixed pitch, one can then elicit the keynotes of sixty (distinct mode-keys). (*I lü erh sêng wu yin, shih-erh lü erh wei liu-shih yin.*[1])

In spite of the scarcity of early evidence concerning notes in relative pitch, it is not suggested that there was no differentiation of pitches before the − 4th century. There may well have been different terms for use with different instruments, a flute-player teaching a pupil tunes by naming the finger-holes on the flute, and a *chhin* player naming the different strings. If Ssuma Chhien (*fl.* − 100) is to be believed there was even a system of notation for stringed instruments as early as the − 6th century, for in the famous story of Duke Ling and the dancing cranes,[f] it is stated that he made his Music-Master Chüan[2] (*fl.* − 500) write down the tune of the kingdom-destroying music composed by Music-Master Yen[3] in an earlier age. This was mysteriously borne to their ears when they were resting one night on the banks of the river where Yen had drowned himself after the Shang dynasty had fallen.

We may therefore conclude that by the − 4th century a scale of five notes was without doubt used, and that the relations of the notes in this scale were designated by the terms *kung, shang, chio, chih,* and *yü.* We cannot say precisely what were the intervals between these five notes, however, without further information such as the relative lengths of five tuned strings of the same material at the same tension, or of five bamboo pipes of known dimensions identically blown. This precise information is not given in Chou texts. Some assistance may be looked for from archaeology however, for though excavated bells may no longer ring true on account of corrosion, and blown instruments such as globular flutes[g] may be misinterpreted because we cannot be

[a] Pp. 130, 140 above, and Vol. 2, p. 281.

[b] With regard to drums, we are fortunate in possessing (as already mentioned) a Thang work on their history and use, the *Chieh Ku Lu*[4] by Nan Cho[5] (+ 848).

[c] *Musica getutscht* (+ 1511); see Galpin (2), p. 26.

[d] Ch. 3, p. 13*a*; tr. auct. adjuv. Chatley (1), p. 27. Cf. Liu Fu (2).

[e] A mode is a pattern of intervals depending on the distribution of semitones and gaps between tones, in scales used for forming melodies. See further p. 169 below.

[f] *Shih Chi,* ch. 24, pp. 38*b*ff.; see Chavannes (1), vol. 3, pp. 287ff.

[g] These *hsüan*[6,7] were generally made of pottery. Ocarinas from the Shang period are preserved in museums at Chêngchow and Peking; cf. Li Shun-I (1), pp. 33, 47.

[1] 一律而生五音十二律而爲六十音   [2] 涓   [3] 延   [4] 羯鼓錄
[5] 南卓   [6] 壎   [7] 塤

certain of the method of fingering used, the instrument for which ancient China was most remarkable, the sets of ringing-stones made of imperishable jade and other hard minerals, would provide a certain means of knowing ancient scales provided that on excavation a set was found to be intact and complete. There are many references in later Chinese history to the recovery of lost sets of ancient bells and stones, which were used for tuning those of later manufacture, so this knowledge may yet be forthcoming.

Remarkable discoveries of this kind have recently been made. The Archaeological Institute of Academia Sinica at Peking has a set of three *to* of different sizes, and another set of ten *ling*,[a] all in bronze of the Shang period from Anyang.[b] Ampler, though later, is the magnificent set of thirteen *chung* discovered in 1957 in a princely tomb of the Warring States period in the Huai River valley north of Hsinyang. This is preserved in the Honan Archaeological Institute at Chêngchow.[c]

Table 45. *Li Shun-I's frequency tests for ringing-stones and bells*

	Ringing-stones		Bells	
Frequencies	Theoretical vibrations/sec.	Tested vibrations/sec.	Theoretical vibrations/sec.	Tested vibrations/sec.
1	711·45	—	562·2	562·2 (given)
2	762·88	—	632·9	688·4
3	858·24	—	712·07	—
4	948·60	948·6 (given)	801·07	—
5	1017·17	1046·5	843·3	—
6	1144·32	—	949·38	915·7
7	1287·36	1278·7	1068·1	—

Experimental workers are now beginning to test the frequencies of such archaic ringing-stones and bells which evidently formed sets in series. Li Shun-I gives[d] results of this kind for three Shang ringing-stones from Anyang and three Shang bronze bells preserved in the Imperial Palace Museum (Table 45). In this table the observed frequencies are correlated with the theoretical values of the set obtained by the usual method of superior and inferior generation (cf. p. 173). For the stones, no. 5 is about a quarter tone out, which would be perceptible, but no. 7 only 9 vibrations/sec. flat, which at that pitch might be regarded as correct. This suggests that in Shang times stones were not more than approximately tuned to satisfy the ear. Of the bells, no. 2 would seem to be more than a quarter tone out, but

[a] For the definitions of these technical terms for different kinds of bells see p. 194 below.
[b] Chou sets of as many as nine are figured in Thang Lan (1), pls. 34, 54, 55, 56. Cf. Anon. (17), pls. 18, 19.
[c] Where I had the pleasure of examining it during the summer of 1958. The music of a copy of this chime of bells has been recorded and played over the radio.
[d] (1), pp. 24, 26. Cf. (2).

no. 6 only 44 vibrations/sec. flat, which might pass. In any case the tuning of a bronze bell will be affected if it is pitted by corrosion, so that sets of ringing-stones will be the best material for these studies, and definite conclusions will no doubt before long be possible.

(ii) *The heptatonic scale and later elaborations*

A pentatonic scale or a number of different pentatonic scales were thus in use in China by the −4th century. But there is also a tradition of heptatonic music, invented, as Chêng Hsüan and other commentators assert, by the Duke of Chou, the great minister at the founding of the Chou dynasty. It is strange that 'the seven notes' (*chhi yin*[1]) should only be referred to twice in the *Tso Chuan*,[a] and each time merely in a numerical catalogue, as, for example, 'the five tones (*shêng*[2]), the six pitches (*lü*[3]), the seven notes (*yin*[4]), the eight winds (*fêng*[5]), the nine songs (*ko*[6])'. In the above passage *shêng* has been translated as 'tones',[b] in order to distinguish it from the word *yin* here interpreted as 'notes'. These two words seem at one time to have meant quite different sorts of sound. The former in its earliest written form suggests the sound produced when one of the ringing-stones is struck, and the latter that produced by blowing through a flute or pipe. By the −2nd century, however, either *wu shêng* or *wu yin* is used for the 'five notes' of the scale. Moreover, the word *lü* (pitch) is also used as a synonym for note. We find frequent references to the seven pitches at this time, meaning the seven notes of a scale. This even occurs in the *Kuo Yü*, in a text possibly as early as the −4th century, where it is stated[c] that when Wu Wang attacked Chou Hsin in order to overthrow the house of Shang 'there were then seven pitches' (*yü shih hu yu chhi lü*[7]). Khung Ying-Ta (*c.* +600), commenting on the *Tso Chuan* passage quoted above, states that the seven notes were introduced at the start of the Chou dynasty. Chêng Hsüan, writing in the +2nd century, also comments on this line, and identifies the seven notes by reference to the fixed pitches of the gamut of his day, from which it is clear that he believed the seven-note scale to have a structure which in our modern notation, if one pitched the *kung* note on middle C, for example, would read C D E F♯ G A B, the 'five notes' being C D E G A, and the two attributed to the Duke of Chou being semitone notes known as *pien chih*[8] (F♯ in this instance) and *pien kung*[9] (B) respectively.[d]

This word *pien* means change,[e] or 'to become, on the way to', and the term itself suggests what a natural musical evolution would lead one to expect, namely, that these two notes were used as a help in passing from one note to another in a 'gapped scale', at any rate at first, though when the ear had become conditioned to this new refine-

---

[a] Duke Chao, 23rd and 25th years (−518 and −516); tr. Couvreur (1), vol. 3, pp. 355 ff.
[b] Cf. Sect. 2 (Vol. 1, p. 36) and the Section on phonetics and linguistics in our concluding volume.
[c] *Kuo Yü* (Chou Yü), ch. 3, pp. 33*b*, 36*a*.
[d] Cf. *Lü Lü Hsin Lun*, ch. 1, pp. 18*b*ff.   [e] Cf. Vol. 2, p. 74.

[1] 七音   [2] 聲   [3] 律   [4] 音   [5] 風   [6] 歌
[7] 於是乎有七律   [8] 變徵   [9] 變宮

ment in sound, and would readily tolerate the presence of semitones, truly heptatonic music would be free to develop.

Whether or not truly heptatonic music was used in Chou times cannot be known, for no examples survive, yet something more than a hint is contained in many references to a 'New Music'. This was considered to be a great scandal, undermining the foundations of the ancient ritual. The *Shih Chi*, for example,[a] records how in the −4th century Prince Wên of Wei[1] observed that when he heard the ancient music his only fear was that he might fall asleep. When he listened to the tunes of the States of Chêng[2] and Wei,[3] on the other hand, this effect did not occur. One objection to the 'New Music' seems to have been that men and women mingled in the dance,[b] and another that the tempo was too quick;[c] but in certain passages it is specifically stated that the tones were wrong. Confucius, for example, in a famous passage in the *Lun Yü*, criticises not the mime-music (*yo*[4]), nor the songs (*ko*[5]), but the notes or tones (*shêng*[6]):[d] 'I hate the way that russet corrupts true red. I hate the way that the tones of Chêng confuse the orthodox music....'[e]

There was thus in ancient China a period of struggle between two different forms of music, an earlier one which used five regular tones, and a later one (stimulated perhaps by the infusion of new ideas from the western borders when the Shang were overthrown by the Chou), in which two auxiliary notes or semitones were used. To this day heptatonic music is stronger in the north of China than in the south. It is even maintained[f] that in the north the heptatonic scale predominates over the pentatonic.

While heptatonic influence was reinforced from the west more than once in later Chinese history, further elaborations were made in the division of the scale in China itself. In the Sui and Thang periods China was very receptive of influences from

---

[a] Ch. 24, p. 30*a*, *b*; cf. Chavannes (1), vol. 3, p. 272.

[b] This betrays a class prejudice, for the participation of the two sexes in dancing had been universal among the mass of the people since high antiquity (cf. Granet, 1, 2).

[c] Hawkes (1), p. 6, suggests that the essence of the innovation was the preponderance of pipes and flutes. Perhaps 'the mainly percussive music associated with the *Shih Ching* now gave way to a type of music dominated by various kinds of wood-wind'. This would have been mournful, erotic or languorous in slow movements, and rather excited or hysterical when fast—just the qualities which were criticised. Hawkes also associates the 'New Music' with the prosodic inventions of the Chhu Tzhu song style.

[d] *Lun Yü*, XVII, xviii; tr. auct. adjuv. Legge (2), p. 190.

[e] The parallelism between musical notes and ritual colours is evident, and just as some colours such as russet and purple are intermediate (*chien*[7]) between the five 'correct' spectral colours, so, it would appear, some notes are intermediate between the five 'correct' notes. Colours and sounds were described in Han times as 'tallies' (*fu*[8]). Lacking the prism and the optical ideas of the Renaissance the Chinese could hardly have arrived at Newton's analogy between the colours of the spectrum and the notes of the diatonic scale by measurement. What is remarkable is that in formulating a similar analogy intuitively, they chose correctly the three classical primary colours red, yellow, and blue. Where other peoples reached a total of five by including silver and gold or other 'false' colours, the Chinese included the two 'hueless colours' black and white to make their scale of five. Moreover yellow, which in fact occupies the middle of the spectrum, was regarded by the Chinese as the colour of the centre, the royal colour, which underlies all the others. That rainbows and the yellow earth of loess China may have assisted their thinking on these lines hardly detracts from the achievement. On the theory of synaesthetics in general, with a Chinese reference, see Ogden & Wood (1). Cf. p. 133 above.

[f] Hartner (7).

---

[1] 魏   [2] 鄭   [3] 衞   [4] 樂   [5] 歌   [6] 聲   [7] 間   [8] 符

PLATE CX

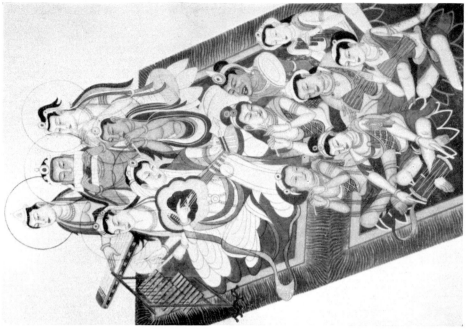

Fig. 314

Fig. 313

Figs. 313, 314. An orchestra of heavenly musicians as imagined in the Thang period: frescoes from cave no. 220 at Chhien-fo-tung, near Tunhuang, painted *c.* +642. Besides the instruments which have appeared in previous illustrations and will easily be recognisable, we find the cymbals (*po*, cf. Moule (10), p. 24), the harp (*khung-hou*), the stand of metal plates (*fang-hsiang*), the true short lute (*phi-pha*) from Persia, and various hand-drums of Indian type. It is interesting that while the Chinese and Indian instruments are all played by spirits very lightly clothed in the manner of Buddhist apsaras, and some of dark complexions, the *phi-pha* and *fang-hsiang* players wear robes more Turkic or Persian in style. Anon. (*10*).

abroad, and not least in music.[a] We have already mentioned the playing of Japanese music at the court of Sui Wên Ti,[b] and the success of exponents of Indian music such as Tshao Miao-Ta[1].[c] This fascinating period has been the subject of a notable book by Hayashi Kenzō (1), translated into Chinese by Kuo Mo-Jo. The twelve ritual melodies of the Thang period which have survived have been carefully studied by Picken (4); all are in two heptatonic modes. Other scholars[d] have investigated the orchestras which played in the Thang[e] and Sung.

Levis describes[f] how the musical notations of Chiang Khuei[2] (+1155 to +1229), of the Sung dynasty, reveal that the scale was enlarged from its five- and seven-tone basis by cadential sharpening of one or two notes so as to comprise nine tones, in which the additional two were of an auxiliary nature. Chiang Khuei was certainly not alone in the use of these more complicated scales, and several other innovators departed from the traditional musical uses of their time. All the tunes of Chiang Khuei's *Yüeh Chiu Ko*[3] (Nine Songs for Yüeh)[g] of +1202 have now been fully transcribed and studied by Picken (5). As an example of the nine-tone scales actually used in Chiang Khuei's songs, we may quote the scale in C, where a microtone— *chê tzu*,[4] or turning-note, to give it Chiang Khuei's term—occurs between E and F. The scale then runs: C D E E[2] F G A B♭. It would be a mistake, however, to believe, as Europeans frequently do, that Chinese music is characterised by 'quarter-tones'. The opposite is true; five-toned music is the rule, the use of semitones is met with, especially in the north, and microtones are quite exceptional.

(iii) *The twelve-note gamut and the set of standard bells*

The evolution of Chinese acoustic theory leads from the formation of scales in relative pitch to that of a gamut of notes of fixed or absolute pitch. The five-note scale *kung, shang, chio, chih, yü* may be compared to a movable doh scale in Western music, in

[a] On the general question of the musical relations between China and the West throughout the centuries, the valuable summary of Wang Kuang-Chi (1) may be consulted. Though foreign influences in China were many and great, Chinese music always retained its own very characteristic ethos, firmly fixed in appreciation and aesthetic.

[b] Vol. 1, p. 125, following Goodrich & Chhü Thung-Tsu (1). Twitchett & Christie (1) have translated the detailed and interesting account in *Hsin Thang Shu*, ch. 222c, pp. 9aff. of the Burmese orchestra presented to the court in +802.

[c] Vol. 1, p. 214.   [d] Yin Fa-Lu (2); Trefzger (1).

[e] We must not forget the remarkable representations of orchestras which are depicted on the frescoes at the Tunhuang cave-temples (cf. Figs. 313, 314). Reproductions and discussions will be found in Anon. (10), pls. 38, 39, 47, 48, 49 (for caves no. 172, 220 and 112); Phan Chieh-Tzu (1), pp. 67, 104 (for caves no. 113 and 144); and Chhang Shu-Hung (1), fig. 12 (also no. 220). The composition is interesting. While instruments of the *shêng* (mouth-organ) and *chhin* (zither) types hold their own alongside the *phi-pha* lutes and a great variety of drums and cymbals, the bells and chime-stones of Confucian antiquity are conspicuous by their absence. A stand of metal plates (*fang-hsiang*[5]) deputises, however (Fig. 315). A prominent newcomer is the large yet portable harp (*khung-hou*[6]). At Mai-Chi Shan (e.g. in cave no. 51) the sistrum (*yün-lo*[7]), a hand-held framework of from three to ten tinkling cymbals (cf. Wang Kuang-Hsi (1), vol. 2, p. 52) is prominent, besides vertical flutes or pipes (*ti*[8]), in frescoes of the Northern Wei period.   [f] (1), p. 75.

[g] Contained in his *Pai Shih Tao-Jen Shih Chi Ko Chhü*[9] (Collected Poems and Songs of the White-Stone Taoist). Cf. Picken (1), p. 109; Yang Yin-Liu (1); Yang Yin-Liu & Yin Fa-Lu (1).

[1] 曹妙達   [2] 姜夔   [3] 越九歌   [4] 折字   [5] 方響   [6] 箜篌

[7] 雲鑼   [8] 笛   [9] 白石道人詩集歌曲

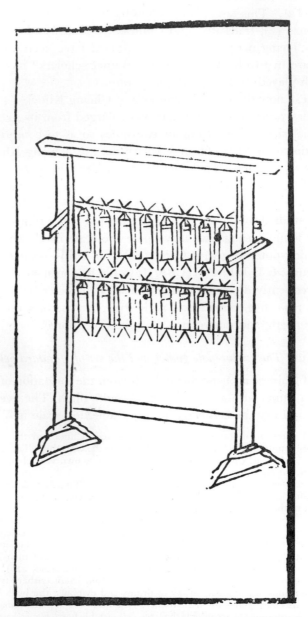

Fig. 315. The stand of metal plates (*fang-hsiang*), sixteen rectangular pieces of steel. From a Sung edition of Chhen Yang's *Yo Shu* (+11th century), cf. *Akhak Kwebŏm*, ch. 7, p. 1*a*, *b*. Ma Tuan-Lin (*c.* +1300), classified the *fang-hsiang* as of foreign origin, like the *phi-pha*, quoting a Thang book, the *Ta Chou Chêng Yo* (see p. 193), as saying that it came in under the Hsi Liang (*Wên Hsien Thung Khao*, ch. 134 (p. 1195.1), cf. Moule (10), p. 146). The Western Liang State ruled in north-western Kansu during the first quarter of the +5th century.

which approximate equivalents would be doh, ray, me, soh, lah, in the standard or *kung* mode. But it would be misleading to press this analogy, for the movable doh system applies to a gamut in which the semitones are almost equal, and this is a fairly recent invention. In all major and minor scales the pattern of tones and semitones is identical, but the modes or patterns of our scales have been reduced, generally speaking, to two. Furthermore, a given note in one scale has exactly the same pitch or frequency in all other scales, no matter what the interval may be which it forms with other notes. For example, the note E may have a frequency of 644. It will have this frequency no matter whether it is regarded as a major third from middle C, or a minor third from C♯, or a major second from D, or any other interval. The pitches according to this system have been made identical for each key, the advantage being that a musician can now modulate freely from one key to another, without having to retune his instrument or adjust his playing to the altered pitch requirements of a different key.

This seems so obvious and desirable an arrangement today that we are likely to overlook its revolutionary nature and to forget the price which has been paid for the convenience, i.e. the sacrifice of some of the distinctive character between keys (cf. p. 215). In all early music these differed qualitatively, and still more so did the modes. Akin to this difference is that which we still appreciate in our major and minor.

The acoustic basis for the qualitative differences between keys in early music is as follows. There are certain musical intervals which are universally acceptable to the human ear; for example, the octave, the fifth and the fourth.[a] They are said to be 'just', when their frequencies form part of a series in arithmetic progression. If the frequencies of two sounds are in the relation of 1:2 they will form an octave; if of 2:3 a perfect fifth; if of 3:4 a fourth; if of 4:5 they will form a major third, and so on. Knowledge of this enabled the Greeks to calculate the pitch of strings of the same tension and thickness from their length, and had a decisive influence on the development of their acoustic theory. The 'just' intervals used in European music until the +17th century derived from Greek theory and were based on these proportions. In any scale, therefore, the frequencies of the octave, the fifth, the fourth and the major third were as just stated, while those of the minor third were as 5:6, of the major second as 8:9, of the major sixth as 3:5, and of the major seventh as 8:15. These are the proportions required for calculating diatonic scales[b] in just intonation. Given, for example, a fundamental note with a frequency of 200, its octave will be as 2:1, i.e.

---

[a] Everyone who has become personally familiar with Chinese music will have recognised the presence of a distinct 'melodic fund' quite different from that of occidental music, but just as pleasing aesthetically. As Picken (3) has pointed out, this is because in Chinese music the characteristic interval is always the fourth, and tunes are built up of chains of fourths, unlike European tunes, which are generally chains of thirds, though many Western folksongs were pentatonic. The watchword of Chinese music is 'order without mechanical symmetry', and while the West developed the harmony of simple melody, Asia developed complex melody. On the structure of Chinese instrumental music see von Hornbostel (2) and Picken (1), p. 125. In China music was never divorced from other activities, whether landscape painting, lyric poetry, alchemy or even mineralogy—thus the alchemical prince Ning Hsien Wang published in +1425 a famous collection of musical pieces using the entablature notation system (Picken (1), p. 118). Cf. on him, Vol. 3, pp. 513, 705 above. As for the connection between music and alchemy in the West, compare Read (2) and Tenney Davis (1) on Michael Maier (+1568 to +1622) and his *Atalanta Fugiens*.

[b] I.e. scales containing series of tones and semitones.

400; its perfect fifth will be as 3:2, i.e. 300. We may now compare the frequencies of two scales in just intonation, note for note, and see why it is that in this system the 'same' note, e.g. the note A, does not have the same frequency in the scale of D as it has in the scale of C. For this purpose the frequency of C will be taken to be 512. This represents an octave above middle C at 'philosophic pitch', by which reckoning an imaginary sound with a frequency of 1 is regarded as the lowest possible note, and middle C with a frequency of 256 as its eighth octave.

Table 46. *Frequencies of 'just' intervals in the scale of C major*

| Notes | INTERVALS IN JUST INTONATION CONSTRUCTED | | | |
| | from C as fundamental | | from D as fundamental | |
	Frequency (vibrations/sec.)	Intervals (C to x)	Frequency (vibrations/sec.)	Intervals (D to x)
C	512			
D	576	Major second	576	
E	640	Major third	648	Major second
F	682·3	Fourth	691	Minor third
G	768	Fifth	768	Fourth
A	853·3	Major sixth	864	Fifth
B	960	Major seventh	960	Major sixth
C	1024	Octave	1080	Minor seventh

From Table 46 it is possible to see what intervals are formed by the different notes in the scale of C major with their fundamental note C, and their frequencies obtained by multiplying the frequency of the fundamental by the appropriate proportion; then by looking along to the same note in the right-hand column, to see the difference in frequency. The note E, for example, is a major third from C and a major second from D. Today we consider it to be the same note occurring in different keys, but in ancient times it was a different note, for the frequency of C multiplied by 5/4 is not the same as the frequency of D multiplied by 9/8. In just intonation, then, it was not possible to transpose a melody from one key to another, e.g. from C to D, without considerably altering its character, for the pitch relations within different keys are not identical.[a]

The musicians of ancient China were particularly sensitive not only to the obvious changes in character of music caused by the displacing of the semitones in heptatonic

[a] See Geiringer (1), p. 26. Cf. p. 216 below.

modes, and of the 'gaps' in pentatonic modes, but also to the subtler changes in character caused by transposition of a melody from one key to another within the same mode. Whether or not they were all used is uncertain, but there are references to sixty (pentatonic) and to eighty-four (heptatonic) mode-keys (tiao[1]). For example, in the *Huai Nan Tzu* book we have the passage already quoted[a] to the effect that if one has twelve notes of fixed pitch (lü), one can build on them the keynotes of sixty distinct mode-keys.

The heptatonic modes are familiar to the West under the names Dorian, Phrygian, Lydian, Mixolydian, Aeolian, Ionian and Locrian. Although there is no equivalent in classical Chinese for our modern debased and restricted conception of 'mode', the mode-keys were named in a simple and unambiguous way.[b] As Hartner says,[c] 'A method of indicating the shifting intervals of the five modes of the pentatonic scale...and of their eleven transpositions into all possible keys, was very conveniently obtained by combining the syllables of the ancient five-tone notation: *kung, shang, chio, chih, yü*, with the first syllables of the twelve lü, i.e. *huang, ta, thai, chia, ku, chung, jui, lin, i, nan, wu, ying*.' The names of the twelve lü[d] may be regarded as equivalents for our letters of the alphabet from A to G, including black notes. Thus where we are obliged rather awkwardly to say, for example, that a melody was in the key of C in the Lydian mode, the Chinese simply say 'the melody used *kung-huang*'.

So far we have followed the establishment of scales in relative pitch, which give a particular form to a melody regardless of the actual note on which it is pitched. But the existence of sixty or even eighty-four different scales, deriving from the five or seven different modes respectively, implies a fixed gamut of twelve semitones, such as is familiar to us on our keyboard instruments. Western manufacturers of these were faced with a dilemma. They wished to provide musicians with a keyboard capable of sounding all notes correctly for just intonation (which in theory means some eighty-four notes per octave) but they were physically unable to compress these into the natural span of a musician's hand. Chinese technicians were also dogged by the same dilemma, though it was not keyboard instruments that concerned them, but bells suspended on frames, a far more expensive and cumbersome proposition. Over 150 bells for a compass of three octaves would require the prowess of an athlete in the striker and the wealth of a prince in the purchaser. Of course, such elaboration was contemplated only in theory.

How the Chinese gamut of twelve notes came to be formed is intimately associated with the history of Chinese bells. In early orchestras they had a double use—for giving the pitch and for starting the music. As is said in the *Kuo Yü*:[e] 'Furthermore the bell does not (ring) false so we use it to lead off the notes (*chieh fu chung pu kuo, i tung shêng*[2]).'[f]

---

[a] P. 161 above. [b] Cf. Chao Yuan-Jen (3). [c] (7), p. 82.
[d] See p. 171 below. [e] *Chou Yü*, ch. 3, p. 21 b.
[f] A modern parallel for this practice is to be found in Indonesia. Sachs (1) has noted that 'when one asks for Javanese or Balinese tuning methods one is told that some old gong-founder owns a few highly respected metal bars inherited from a remote ancestor which he uses with more or less accuracy'.

[1] 調   [2] 且夫鐘不過以動聲

By the time of the Chou period at latest, the Chinese had advanced beyond the stage of striking lumps of ringing-stone or slabs of bronze which gave out the desired note by chance, and they were producing bells which they were able to tune with accuracy.[a] The earliest works frequently mention bells, their consecration with blood,[b] and their importance in musical performances. Often they have names, and the names are many and various. We can only guess at their meanings. Han commentators two thousand years ago also had to make such guesses. It is safe to say that the names were connected with the ceremonies in which they were used, the sympathetic magic of the name perhaps adding to the efficacy of the rite. Synonyms abounded; thus Lin-chung,[1] literally 'forest bell', is referred to in the *Chou Yü* passage just quoted as Ta-lin,[2] literally 'great forest'. Chêng Hsüan takes a bell named Han-chung[3] to be the same as Lin-chung, because in a list of pitch bells named in the *Chou Li*[c] its position in the sequence is that which would have been taken by Lin-chung some centuries later when the names of the instruments giving the fixed pitch scale had been standardised. The nomenclature of the gamut is complicated by such anachronistic interpretations of Han commentators, as well as by the general fluidity of names forming the system at the time when the gamut was being evolved. But the general process of evolution is fairly clear.

To accompany a singer in just intonation, or in any temperament other than Equal Temperament, on bells, would have required very large sets, but as we have seen the main purpose was to use bells for the initial note to set the pitch, or sound the keynote. Thus the most useful set of bells would in time be found to be one giving the gamut of twelve consecutive semitones which served as keynotes. It should not be imagined that this gamut ever functioned as a scale, and it is erroneous to refer to the 'Chinese chromatic scale',[d] as some Western writers have done. The series of twelve notes known as the twelve *lü* were simply a series of fundamental notes from which scales could be constructed.

It is not possible to say when the process of standardisation of bells and pitches was first completed, but the earliest reference in literature to the full set of twelve bells may be that in the *Kuo Yü*,[e] where they are mentioned in a discussion said to have taken place in the year −521.[f] Alternatively, if the *Yüeh Ling*[4] (Monthly Ordinances of the Chou Dynasty) really dates from as early as −600, the list in this text may take

[a] More will be said about actual tuning methods later (pp. 184ff.).
[b] Cf. *Mêng Tzu*, I (1), vii, 4; cf. Legge (3), p. 15.
[c] *Chou Li*, ch. 6, p. 12*a* (ch. 23); cf. Biot (1), vol. 2, p. 49.
[d] A chromatic scale is one composed of a continuous series of semitones.
[e] *Chou Yü*, ch. 3, pp. 21*b*, 26*a*ff.
[f] The High King of Chou whose name was Ching[5] wished to have a bell melted down and converted into another bell of lower pitch. His minister Shan Mu Kung[6] remonstrated with him, adducing many good reasons why this should not be done, one of the most compelling of which was that the smaller bell would not produce enough metal to make it. The sovereign, nevertheless, had his way and the bell was cast, but after his death it was found that in fact the bell was out of tune. The full list of bells with their names, qualities and definitions occurs in a conversation of this potentate with another of his acoustic advisers, Lingchou Chiu.[7] Cf. p. 204 below.

[1] 林鐘   [2] 大林   [3] 函鐘   [4] 月令   [5] 景   [6] 單穆公
[7] 伶州鳩

precedence.[a] Like bronze mirrors, bells were regarded in Chou times as instruments of high magical potency, their special merit being to attract or collect the emanations and essences generically known as *chhi*.[b] This *chhi*, it will be remembered, had six forms, Yin and Yang, wind and rain, darkness and brightness; the Yin and Yang being two antitheses into which all the others were in the last resort subsumed.

The types of bells naturally divided into two analogous groups, Yin and Yang. The *Kuo Yü* lists them as in Table 47.[c]

Table 47. *Classification of bells in the* Kuo Yü

Yang bells		Yin bells	
Huang-chung[1]	'yellow bell'	*Ta-lü*[7]	'great regulator'
Ta-tshou[2, d]	'great budding'	*Chia-chung*[8]	'compressed bell'
Ku-hsien[3]	'old and purified'	*Chung-lü*[9]	'mean regulator'
Jui-pin[4]	'luxuriant'	*Lin-chung*[10]	'forest bell'
I-tsê[5]	'equalising rule'	*Nan-lü*[11]	'southern regulator'
Wu-yi[6]	'tireless'	*Ying-chung*[12]	'resonating bell'

The names of these twelve bells thus stabilised became the names of the twelve notes which formed the classical Chinese gamut. The fact that the Yang scale is referred to in the *Kuo Yü* as pitches (*lü*), whereas the Yin notes are called *hsien*,[13] interstitials, i.e. notes which come between the regular pitches, strongly suggests that the standardised gamut of twelve semitones already existed at that date. No details are given in the text about the exact method by which the intervals were calculated, but the order in which the names appear represents an intermediate stage before their final form first recorded in the *Lü Shih Chhun Chhiu*[e] and a more primitive version of twelve pitches grouped in sixes which is preserved in the *Chou Li*.[f]

(iv) *The introduction of the arithmetical cycle*

In tracing the evolution of the gamut, three stages have been mentioned so far. First, there was the primitive stage preserved in the *Chou Li*, in which the notes had names,

[a] See Vol. 3, p. 195.

[b] In connection with the fancied relation between square earthly Yin bells and hollows, round heavenly Yang bells and mounds, etc., one may quote a Yüeh Ling commentary: 'Bells are hollows. The inside of the hollow receives *chhi* abundantly.' *Chhi* in subsequent acoustic theory is discussed below, pp. 202 ff.

[c] *Chou Yü*, ch. 3, pp. 26 *b* ff.   [d] Also *Thai-tshou*.[14]

[e] Ch. 27 (vol. 1, p. 54); tr. Wilhelm (3), pp. 69 ff.

[f] Ch. 6, pp. 11 *b*, 12 *a* (ch. 23); cf. Biot (1), vol. 2, p. 49. In concluding this subsection one can hardly avoid referring to the experiments in dodecaphonic music made in the present century by Schönberg, Webern, Alban Berg and others. It would be interesting to know whether the origins of this had anything to do with ancient twelve-note series such as that of China.

[1] 黃鐘   [2] 大蔟   [3] 姑洗   [4] 蕤賓   [5] 夷則   [6] 無射

[7] 大呂   [8] 夾鐘   [9] 仲呂   [10] 林鐘   [11] 南呂   [12] 應鐘

[13] 間   [14] 太蔟

though some of them differed from those ultimately adopted.[a] Secondly, we have the twelve bells listed in the *Kuo Yü*, also divided into sixes, and here all the names agree with those of the ultimate orthodox gamut. We cannot say anything positive about the intervals of this gamut or the way they were arrived at, still less about the frequencies of any of the notes. But with the series of twelve notes described in the *Lü Shih Chhun Chhiu* about −240 a new stage is reached, for though the frequencies remain unknown, it at last becomes possible to see how the series of notes was obtained.

As this gamut of twelve notes bears certain resemblances to the so-called Pythagorean scale, the difference in construction between that scale and the Chinese is of great interest. The instruments on which the Greeks evolved their scales were the lyre and the cithara,[b] not, as in China, bells and ringing-stones. The framework of Greek scales was the octave made by tuning the two outside lyre strings, after which two inner strings were tuned to the intervals of the fifth and the fourth. In Homeric times all strings were tuned by ear. It was not until the −6th century that the quantitative discovery attributed to Pythagoras was made, concerning the half, two-third and three-quarter length strings needed if the octave, fifth and fourth were to be calculated. And the discovery that the interval of a major tone is that which lies between the fourth and fifth was not made until a century later by Philolaus.[c] There was then a parting of the ways in Greek music, one school, of which Aristoxenus of Tarentum was the leading exponent, maintaining that musical intervals should be judged by ear, the other, the Pythagorean, asserting that musical intervals were essentially mathematical. The mathematics of the Pythagorean scale, which are considerably more complicated than those of the Chinese gamut, and require a knowledge of means, were set forth by Plato,[d] though for metaphysical rather than musical reasons; and a full description of its most extended form was given by Euclid.[e]

The Chinese gamut of pitches, on the other hand, requires only the simplest mathematics and does not use the octave as a starting-point. Indeed, it does not even include a true octave at all. The only mathematical operation needed is the multiplication of certain figures by 2/3 and 4/3 alternately.[f] The frequency of a fundamental note multiplied by 3/2 produces a perfect fifth higher. Before the idea of frequency existed, however, the same relation was expressed simply in terms of length, the length of a resonating agent multiplied by 2/3 being equivalent to the frequency multiplied by 3/2. The length of a zither string, then, multiplied by 2/3 gives a note which when struck is a perfect fifth higher than its fundamental. This is the first step (or *lü*) in a process which evolves an unending spiral of notes. The length of the resonating agent which sounds the perfect fifth is then multiplied by 4/3, the resulting

---

[a] If indeed they were the same notes, though the possibility must be borne in mind that the passage describes two distinct scales in different registers.

[b] A lyre in which the sound-box cavity was continued into the arms, these also being hollow.

[c] See the (prob. +2nd-century) *Encheiridion Harmonices* of Nicomachus of Gerasa (Meibom's ed.), Bk. I, pp. 13, 17, 27.

[d] *Timaeus*, 35 B; Archer-Hind tr., p. 107.

[e] *Euclidis Introductio Harmonica* (Sectio Canonis), 'Canonem designare secundum systema, quod vocatur immutabile' (Meibom's ed.), p. 37.

[f] Cf. *Lü Lü Hsin Lun*, ch. I, pp. 13 a ff.

note being a fourth below the perfect fifth, and therefore a major tone above the fundamental, since $1 \times 2/3 \times 4/3 = 8/9$, i.e. the same interval that Philolaus found by a different method to exist between the two tetrachords, e.g.

$$C \text{————} F \text{ (tone) } G \text{————} C.$$

The Greeks used the interval of the tone as the basis of their scale structure, an octave being subdivided into a tone and two tetrachords, and a tetrachord being sub-divided into two tones and a Pythagorean semitone or diesis. The Chinese did not become involved in the complications of the Greek semitones *apotomē* and *leimma*, but having advanced two steps from their fundamental note (the Huang-chung already referred to), went on to calculate a fourth note in the series by multiplying the length of the resonating agent of their third note by $2/3$, which gave a length $16/27$ that of the fundamental. This was their sixth. From the sixth a major third was produced by multiplying by $4/3$, the product being $64/81$. This note, it will be observed, is not justly tuned, for in just intonation the fraction would be $4/5$, but it agrees with the Pythagorean major third. This process of multiplying by $2/3$ or $4/3$, whichever was required to keep the gamut within the compass of a single octave, was continued up to the twelfth note, these being the twelve *lü* (Fig. 316). The Chinese describe it as 'generation' (*shêng* [1]), the notes being like 'mothers' giving birth to 'sons'.[a] Notes produced by $4/3$ multiplication were said to be of 'superior generation', while multiplication by $2/3$ yielded 'inferior generation'. The *Lü Shih Chhun Chhiu* contains our earliest description (-239) of the system by which the notes of the Chinese gamut were generated.[b]

Our oldest source for any actual lengths calculated according to this 'up-and-down' principle is the *Shih Chi* (c. -90) of Ssuma Chhien. He is speaking of blown pipes, and gives the length of the Huang-chung pipe as 81 (tenths of an inch).[c] This is obviously a good figure to start from when one is calculating with $2/3$ and $4/3$ fractions. Correcting certain obvious errors,[d] the lengths of the pipes are given in Table 48. The actual lengths of these pitch-pipes are of no great value in themselves, for without further data, such as their diameters, we cannot calculate the frequencies obtained. But the manner in which these lengths are expressed is of great interest, for the use of a decimal system in conjunction with a system based on thirds has a strikingly Babylonian flavour.[e] To this point we shall return.

Before he lists the actual pitch-pipe lengths Ssuma Chhien gives the formula on which his calculations were based. It will be useful now to compare his proportions

[a] Cf. the Chinese terminology for arithmetical fractions, Vol. 3, p. 81.
[b] Ch. 27 (vol. 1, pp. 54ff.); tr. R. Wilhelm (3), pp. 69ff.
[c] Ch. 25, pp. 8bff. Chavannes (1), vol. 3, pp. 313ff. Cf. *Chhien Han Shu*, ch. 21A, pp. 3bff.
[d] See Robinson (1), pp. 44ff. and Chavannes (1), vol. 3, pp. 631ff. (Appendix II), where the errors are examined. They were first pointed out by the great Sung scholar Tshai Yuan-Ting [2] ($+1135$ to $+1198$) in his *Lü Lu Hsin Shu* [3] (New Treatise on Acoustics and Music), which is preserved in the *Hsing Li Ta Chhüan* (cf. Vol. 2, p. 459). On him see Forke (9), pp. 203ff. On the whole subject see further the paper of Yabuuchi (18), and Wu Nan-Hsün (1), pp. 73ff., 115ff., 204.
[e] Cf. Vol. 3, p. 82 for the use of the words *lê* [4] or *lê* [5] as $1/3$. The *Shih Chi* in the present passage uses the more normal expression *san fên i*.[6]

[1] 生   [2] 蔡元定   [3] 律呂新書   [4] 阞   [5] 仂   [6] 三分一

for the notes of the gamut, twelve in all (to which a thirteenth, the octave, may be added by simply continuing the calculation an additional step), with the proportions of the eight notes of Timaeus' Pythagorean scale, so that their similarities and differences may be observed (Table 49).

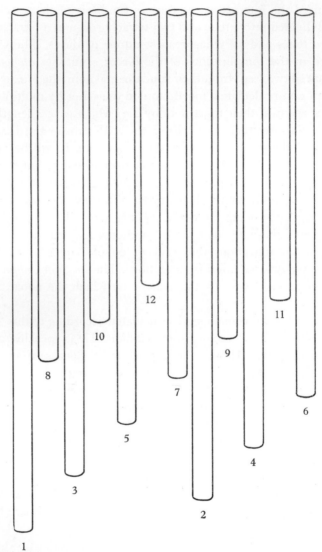

Fig. 316. The orthodox standard pitch-pipes drawn to scale. Reconstruction by K.R. to show the principle of superior and inferior generation. 1, *Huang-chung*; 2, *Ta-lü*; 3, *Ta-tshou*; 4, *Chia-chung*; 5, *Ku-hsien*; 6, *Chung-lü*; 7, *Jui-pin*; 8, *Lin-chung*; 9, *I-tsê*; 10, *Nan-lü*; 11, *Wu-yi*; 12, *Ying-chung*.

Like the twelve standard bells, the twelve pitch-pipes were also divided into two companies, Yang and Yin. Chêng Chung, commenting on the *Chou Li*[a] in the +1st

---

[a] Ch. 6, p. 16a, b (ch. 23).

Table 48. *Ssuma Chhien's calculations of the lengths of pitch-pipes*

Name	Inches	Tenths	Thirds of hundredths	Total (uncorrected)	Total (corrected)
Huang-chung	8	1	—	8·1	8·1
Ta-lü	7	5	1	7·53	7·585
Ta-tshou	7	2	—	7·2	7·28
Chia-chung	6	1	1	6·13	6·742
Ku-hsien	6	4	—	6·4	6·4
Chung-lü	5	9	2	5·96	5·993
Jui-pin	5	6	1	5·63	5·689
Lin-chung	5	4	—	5·4	5·4
I-tsê	5	4	2	5·46	5·057
Nan-lü	4	8	—	4·8	4·8
Wu-yi	4	4	2	4·46	4·495
Ying-chung	4	2	2	4·26	4·266

Table 49. *Comparison of the proportions of the Chinese and Greek (Pythagorean) scales*

	Chinese	Greek (Pythagorean)		Chinese	Greek (Pythagorean)
C	1	1	G	$\frac{2}{3}$	$\frac{2}{3}$
C♯	$\frac{2048}{2187}$	—	G♯	$\frac{4096}{6561}$	—
D	$\frac{8}{9}$	$\frac{8}{9}$	A	$\frac{16}{27}$	$\frac{16}{27}$
D♯	$\frac{16384}{19683}$	—	A♯	$\frac{32768}{59049}$	—
E	$\frac{64}{81}$	$\frac{64}{81}$	B	$\frac{128}{243}$	$\frac{128}{243}$
F	$\frac{131072}{177147}$	—	C	$\frac{262144}{531441}$	—
	—	$\frac{3}{4}$		—	$\frac{1}{2}$
F♯	$\frac{512}{729}$	—			

The notes in the left-hand columns are only arbitrarily selected by way of illustration.

century, says that the Yang pitch-pipes (*lü*[1]) were made of bamboo, but the Yin ones (*thung*[2]) of copper or bronze, the former material corresponding in the system of symbolic correlations to heaven, the latter to earth.[a]

It will be seen that the Pythagorean scale and the Chinese gamut (the spiral of fifths) are not identical, either in the general manner of their construction or in the particular proportions of certain notes, i.e. the octave and the fourth. Nevertheless, their resemblances were sufficiently striking to cause a misapprehension which has persisted for almost two hundred years.

(v) *Pythagoras or Ling Lun?*

The earliest account of the theoretical basis of Chinese music available in a European language is that written by the Jesuit Joseph Amiot in Peking in +1776, and published in Paris in +1780. Amiot accepted the traditional datings of Chinese history, and therefore believed that music in China originated in the year −2698. By this reckoning the Chinese would have had a gamut closely resembling the Pythagorean scale in many of its intervals more than eleven centuries before the birth of Pythagoras. He concluded that the Pythagorean claim for the invention of this scale was nothing less than an 'act of robbery'.[b] How precisely the plagiarism was carried out he did not explain, but assumed that Pythagoras, who was noted for his travels, must either have gone to China, or met someone from that country who transmitted the secrets of the scale. Noting that the Greek scale differed somewhat from the Chinese gamut, he concluded that the Greek version was a degenerate one.[c]

With the decline of China's prestige abroad during the 19th century, and the great revival of Hellenism, it was only to be expected that this judgment would be reversed. Chavannes, considering that there was no textual reference to the Chinese gamut earlier than the −3rd or −4th century, wrote:[d] 'Ce même système musical avait été exposé par les Grecs plus de deux siècles avant l'époque où les Chinois le connurent. N'est-ce pas aux Grecs que les Chinois l'ont emprunté?' And Chavannes attempted to explain how it was that the Chinese came to 'borrow' this acoustic system:[e] 'Sur la lourde vague de civilisation que l'expédition d'Alexandre avait fait déferler aux pieds des Pamirs surnagèrent les douze roseaux en qui chantait la gamme de Grèce.' Guesses of this sort take us no further forward than Amiot's mythology. Yet

[a] Biot (1), vol. 2, p. 56 has these materials interverted and should be corrected. The usual later name for the Yin pipes was *lü*.[3]

[b] Amiot (1), p. 8: 'l'Heptacorde des Grecs anciens, la lyre de Pythagore, son inversion des tétracordes diatoniques, et la formation de son grand système, sont autant de larcins faits aux Chinois du premier âge.' We find echoes of this point of view even today, e.g. in Hogben (1), p. 113, where he says: 'The Tyrian parentage of Pythagoras gives us a clue to the clear signs of Chinese influence in his teachings. He travelled in Asia.' And perhaps the work of Kuttner (3) will reopen the question.

[c] Cf. Robinson (1), pp. 48ff., who elucidates the part played by the Abbé Roussier in initiating the controversy.

[d] (1), vol. 3, p. 638.

[e] (1), vol. 3, p. 644.

[1] 律   [2] 同   [3] 呂

they have been accepted for the last fifty years. For Amiot it may at least be said that in his day the scales were not regarded as identical, though since Chavannes loosely described them as 'ce même système' the mistake has spread.[a] Chavannes himself was aware that differences existed, but attributed them to lack of understanding by the Chinese, a people of whom, he adds in a conclusion unworthy of a great scholar, 'le caractère tapageur et monotone de leur musique est d'ailleurs bien connu'.[b]

Chavannes' hypothesis must be dismissed not merely because the Chinese were tuning sets of twelve bells in the same century as that which is said to have seen the lifetime of Pythagoras,[c] and in any case long before any possible influences of Alexander's expeditions could have brought the Greek formula into Chinese literature; but also because the Chinese gamut is in its structure essentially unlike the Pythagorean scale. Yet Amiot's notion that a transmission took place in the other direction at such an early date can no longer be taken seriously either. The simplest alternative hypothesis for which good reason can be found is that there radiated east and west from Babylonia the germ of an acoustic discovery which was developed in one way by the Greeks and in another by the Chinese; namely, that the pitch of notes emitted by strings when plucked is in part determined by their length. More particularly, the Babylonians, who had many highly developed stringed instruments, would have made the observation that one string half the length of another at the same tension will sound its octave, that a string two-thirds the length will sound its fifth, and that a string three-quarters the length will sound its fourth. Knowledge of these proportions is all that was needed to develop the Chinese 'spiral of fifths', and it is also the sum total of the acoustic discoveries which the ancient Greeks attributed to Pythagoras either as inventor or transmitter. The intricate developments of the Pythagorean scale in later centuries, which include the subdivision of the octave into tetrachords, the definition of the tone, and at some time not earlier than the -4th century, the subdivision of the tetrachords, are all specifically Greek discoveries; and for 'Timaeus' construction of a scale not by a series of perfect fifths, but by finding the arithmetic and harmonic means between the numbers of the Pythagorean *tetractys* ($1 - 2 - 3 - 4 - 8 - 9 - 27$), there is no Chinese equivalent.

It must be stressed that a Babylonian origin for these discoveries is hypothetical, for of Babylonian music we know very little. Yet such evidence as survives seems to indicate that this is the answer to the problem.

First of all, it is interesting and may be significant that both Greek and Chinese traditions gave credit for the origin of the acoustic systems to a foreign country. Greek authors writing before the capture of Babylon by Alexander the Great aver that Pythagoras visited Egypt,[d] while later authors say that in his travels he went to

---

[a] For example, Apel (1), p. 618, states categorically that the 'spiral of fifths' was 'invented by Pythagoras'. And even Chinese scholars such as Chao Yuan-Jen (2), p. 85, have been misled into thinking that 'the circle of fifths gives a Pythagorean scale'.

[b] (1), vol. 3, p. 642.

[c] Cf. the quotations from the *Kuo Yü*, above, pp. 151, 170.

[d] This is implied by Aristotle in *Metaph.* I, 1, and is stated by Isocrates (*fl.* -380), in *Laud. Busir.* XI, 28.

Babylon.[a] Iamblichus goes so far as to say[b] that knowledge of the 'musical proportion'[c] was brought to Greece from Babylonia by Pythagoras. Certainly both Egyptians and Babylonians knew and used the 2/3 and 1/3 fractions.[d] Knowledge of the harmonic progression was enshrined by the Egyptians in the dimensions of a box described by the priest Ahmes in a papyrus in the Rhind Collection of the British Museum, dating from a time between −1700 and −1100.[e] But whatever the route by which the musical discoveries attributed to Pythagoras came to Greece, it is certain that they were based on facts long known to the world of the Fertile Crescent. As Burnet says:[f] 'The use of Babylonian as an international language will account for the fact that the Egyptians knew something of Babylonian astronomy.' Before Alexander's invasion such knowledge as the Greeks had of Babylonian science came to them by way of Lydia and Egypt. After the fall of Babylon it was realised that the fountain-head was in that city, and the legends were naturally adapted.

The story of Pythagoras' journey eastward to Babylon has a striking parallel in the legend of the westward journey of a certain Ling Lun, minister to Huang Ti, the mythical Yellow Emperor, who was supposed to have reigned for a hundred years in the −27th century. According to the legends various duties were assigned to the ministers of this ruler, and Ling Lun[g] was commissioned to establish the correct pitch for music.[h]

Anciently [says the *Lü Shih Chhun Chhiu*], Huang Ti ordered Ling Lun[1] to make pitch-pipes. So Ling Lun, passing through Ta-Hsia[2] towards the west,[i] travelled to the northern slopes of the Juan-yü[3] mountains,[j] and there in the valley of Hsieh-chhi[4] found bamboos with stems of which the hollow (part) and the thickness (of the walls) were uniform. Cutting one between the nodes to a length of 3·9 in.,[k] he blew it, and took its fundamental note (*kung*[5]) to be that of the Huang-chung tube. Blowing again, he said 'This is good enough',[l] and proceeded to make all the twelve pipes (*thung*[6]). Then at the foot of the Juan-yü

　[a] This is first found in Strabo (*fl.* −25), XIV, 1, 16.
　[b] *Fl.* +300 and later, *Introductio Nicomachi Arithmet.* pp. 141–2, 168 (Tennulius' ed.).
　[c] In the musical proportion the second term is the arithmetical and the third the harmonic mean, e.g. 6:9::8:12, i.e.

$$a : \frac{a+b}{2} :: \frac{2ab}{a+b} : b.$$

　[d] See Heath (6), vol. 1, pp. 27 ff.
　[e] See Warren (1), p. 48.　　　　　　　　　　　　　　　　　　　　　　　[f] (1), p. 20.
　[g] The name Ling Lun seems to be artificial, *ling* meaning music and *lun* a rule (Haloun, 6, 7). But this does not exclude the existence of a real person behind the legend.
　[h] The fullest version of the story is given in the *Lü Shih Chhun Chhiu*, ch. 25 (tr. R. Wilhelm (3), pp. 63 ff.), vol. 1, p. 49; but other references are quite frequent, as in *Chhien Han Shu*, ch. 21A, p. 4a; *Lü Lü Ching I*, ch. 8, p. 9b.
　[i] It will be remembered that Ta-Hsia[2] was the ancient name for Bactria (many references in Vol. 1).
　[j] The *Chhien Han Shu* text reads Khun-lun[7] here, i.e. the northern ranges of the Tibetan massif.
　[k] This figure for the length of the fundamental pipe is very curious and has much exercised the commentators. The obvious emendation to 8·1 in. in accord with the *Shih Chi* has no authority and seems unlikely as a copyist's error. But the difference between the longest and shortest pitch-pipe is in fact 3·9 in. (8·1–4·2), and so is the length of the octave Huang-chung or thirteenth note, i.e. Chung-lü (5·9 × 2/3). The text perhaps became too compressed here.
　[l] Here the text early became corrupt and commentators have never agreed on an assured version.

[1] 伶倫　　　[2] 大夏　　　[3] 阮隃　　　[4] 嶰谿　　　[5] 宮　　　[6] 筒
[7] 昆侖

mountains, he listened to the singing of the male and female phoenix[a] and divided the pitch-pipes accordingly (into two groups), the male notes making six and the female also six. In order to bring them together, the Huang-chung fundamental harmonised them. Indeed the Huang-chung fundamental (kung[1]) is capable of generating the entire (series). Therefore it is said that the Huang-chung fundamental is the source and root of the male and female pitch-pipes (lü lü[2]).[b]

Thus Ling Lun cut one of these non-tapering bamboo stems between the nodes to make his Huang-chung pipe, after which all the rest took their places in the series of twelve standard pitch-pipes. The *Lü Shih Chhun Chhiu* continues:

(Upon his return) Ling Lun, together with Jung Chiang,[3] was ordered by Huang Ti to cast twelve bells in order to harmonise the five notes (i ho wu yin[4]), so that splendid music might be made. It was on an *i-mao* day in the middle month of spring, with the sun standing in Khuei *hsiu*, that these were finished and presented. Order was given that this (set of bells) should be called Hsien chhih[5].[c]

This is of great interest as showing that all the other musical instruments were to be tuned in accordance with the pitch of the five notes emitted by the unaltering standard bells.

The truth enshrined in this strange story may be not only that in early times bells were used for giving the pitch to instruments in need of tuning,[d] but that the bells themselves were tuned[e] by strings, the lengths of which were determined by certain standard lengths of bamboo, just as the ratios of the octave and the fifth (which are in harmonic progression, e.g. 6:4:3) were preserved in the pyramid box or coffer described by the priest Ahmes. To keep certain bamboos of precise length as standard measures was a rational act for an early people, and foreshadowed our own practice of keeping standard measures in metal.[f]

No doubt the acoustic implications of the harmonic progression were not at first properly understood, for both in China and in Greece we find the formula for tuning strings applied in cases where it is quite inappropriate. Amiot, for example, says that he examined and measured some ringing-stones which he saw at the imperial court.[g] They had been made in the Sung period, and their four straight sides formed certain proportions of the *lü*, namely 27 in., 18 in., 9 in. and 6 in., which between them form octaves and fifths. Amiot observes that stone-chimes of more recent make no longer used these proportions. To shape a slab of stone so that its linear dimensions form octaves and fifths may have magical or possibly mnemonic uses, but represents a

---

[a] Cf. *Tso Chuan*, Duke Chuang, 22nd year (−671) where the male and female phoenix are said to sing together with gem-like sounds (chhiang-chhiang[6]); Couvreur (1), vol. 1, p. 179. Here the comparison is with chime-stones rather than pitch-pipes.

[b] Tr. auct. adjuv. Wilhelm (3).   [c] Tr. auct. adjuv. Wilhelm (3).

[d] Cf. p. 170 above.   [e] As will be shown later, p. 185 below.

[f] Cf. what was said above in Sect. 20g on the gnomon shadow template; Vol. 3, pp. 286 ff.

[g] 'Essay on the Sonorous Stones of China' (an appendix to his *Mémoire*), p. 264.

¹ 宮   ² 律呂   ³ 榮將   ⁴ 以和五音   ⁵ 咸池   ⁶ 鏘鏘

complete misapplication of acoustic laws, since the pitch of plates and of discs such as gongs is not determined in the same way as the pitch of elastic resonating agents such as strings and columns of air (cf. pp. 195, 213).

An even more curious application of the knowledge of the proportions required for producing musical intervals occurs in connection with details of foundry technique given in one of the later parts of the *Chou Li*, the Khao Kung Chi (Artificers' Record).[a] The passage, which will be studied in Section 36 on metallurgy, is one of the most venerable relics of the bronze-founder's art in any civilised literature, for it cannot be later than the − 3rd century and may be a great deal earlier. It describes systematically the properties and uses of a whole series of alloys and defines the proportions of the metals composing them. Modern archaeological research has shown that such knowledge must have been possessed in considerable measure by the bronze-founders of the Shang period.[b] In any case it is curious to find that the proportions of a string required to sound the minor third, major third, fourth, fifth, major sixth and octave, namely 5/6, 4/5, 3/4, 2/3, 3/5 and 1/2, in just tuning, here appear in terms of copper content. How far the proportions of tin and copper in the making of the various vessels and implements is in accord with modern metallurgical knowledge on the one hand, and with what we can tell of ancient practice by analyses of alloys in existing specimens on the other, will be discussed in the appropriate place. The point here is the appearance of an acoustic series (if this set of simple fractions is not merely coincidental) in a metallurgical text.

That misapplications of harmonic laws were not exclusively Chinese, however, can be seen from an anecdote concerning Pythagoras. It was first recorded by Nicomachus of Gerasa[c] (*fl.* +100), and repeated by Iamblichus,[d] Boethius,[e] and others, to the following effect. Pythagoras, passing by a forge, heard the hammers ringing out to form the intervals of the octave, fifth, and fourth. After inspecting them he realised that this was due to the different weight of the heads of the hammers, which produced different notes according to their mass. He therefore made four similar weights the basis of his experiments, but no matter what he tested, strings by tension, vases by striking, flutes or monochords by measurement for length, he always found that the numbers 6, 8, 9, 12 formed the proportions of the consonances, 6:12 the octave, 8:12 the fifth, 9:12 the fourth. The statement that the consonances in the forge were produced by the proportionate weights of the heads of the hammers can no more be true than that the pitch of ringing-stones depended on the proportionate lengths of their sides. In Nicomachus' day this must have been fully recognised, for the acoustic properties of objects had long since been subjected to exhaustive tests. But that he and other experts should have repeated the story suggests that it was of respectable tradition, and inclines one to the belief that just as Thales used his partial knowledge of Babylonian astronomy to make some lucky predictions, so Pythagoras also may have

[a] On this see Vol. 1, p. 111. The passage referred to is in ch. 11, p. 20*b* (ch. 41), tr. Biot (1), vol. 2, pp. 490ff.
[b] Li Chi (3), p. 48.
[c] *Encheiridion Harmonices* (Meibom's ed.), Bk. 1, p. 10.
[d] *Vit. Pythag.* 1, 26.   [e] D. +524; *De Mus.* x.

introduced a limited amount of Babylonian acoustic information which at first was
not properly understood. But armed with the monochord for measuring intervals, the
Greeks soon made progress far beyond a knowledge of those three consonances which
we believe to have been their inheritance from Babylon.

The use of a sexagesimal cycle in calendar-making is very probably an example of
Babylonian influence on China.[a] It is interesting to find that according to the legend,
when Huang Ti sent Ling Lun to the west to fix the musical pitches, he entrusted
Ta Nao[1] with the elaboration of the sixty-year cycle, and Jung Chhêng[2][b] with the
redaction of a 'harmonious calendar',[c] as well as the division of the officials into five
classes.[d] The association of the calendar with music is particularly significant, for we
learn from a Western source that this also was Babylonian. Plutarch wrote:[e]

> The Chaldeans say that Spring stands to Autumn in the relation of a Fourth, to Winter in
> the relation of a Fifth, and to Summer in the relation of an Octave. But if Euripides makes
> a correct division of the year into four months of Summer, and of Winter a like number, of
> 'well-loved Autumn a pair, and of Spring a like number', the seasons change in the octave
> proportion.

The numbers which give these proportions are in fact spring 6, autumn 8, winter 9,
and summer 12, the numbers used by Pythagoras for the musical consonances. From
these proportions the seasons in Babylonia may be calculated as spring 2·1 months,
autumn 2·7, winter 3·1, and summer 4·1. The fact that a brief spring and prolonged
summer are more typical of Babylonia than of Greece enhances the value of this text.

It will now be convenient to summarise the present argument. The Chinese gamut
is essentially different from the Pythagorean scale, though similarities led +18th-
century writers to regard one merely as a degenerate form of the other. A more
satisfactory hypothesis is that the Babylonians discovered the mathematical laws
governing the necessary length of strings forming the octave, fifth, and fourth intervals.
This knowledge spreading both west and east was used by the Greeks and the Chinese
independently, the former for constructing their acoustic theory by subdivision first
of the octave and later of the tetrachord, the latter for developing a spiral of notes by
an alternating series of fifths and fourths from a given fundamental.

If this hypothesis is correct, it helps to explain why certain ideas are common to
both Greeks and Chinese, and others not. The Chinese held, as did the Pythagoreans,
that number is the basis of musical notes. Apart from the numerological cosmogonic
passages in the *Tao Tê Ching*[f] and the *Huai Nan Tzu*,[g] the *Shih Chi* plainly declares[h]

---

[a] Cf. Vol. 3, pp. 82, 256, 397 above.   [b] Cf. Vol. 2, pp. 148, 150.

[c] References to these labours were collected by Chavannes (1), vol. 3, p. 323, but the most important
study is that of Chhi Ssu-Ho (1); cf. Vol. 1, pp. 51 ff.

[d] These were designated by names of clouds coloured blue-green, red, white, black, and yellow,
symbolising the four seasons and the 'mid-season'. Cf. Vol. 2, p. 238.

[e] *Moralia*, 'Creation of the Soul', 1028F. See also the translation of John Phillips (+1694), p. 217,
which, however, has many inaccuracies.

[f] Ch. 42 (Waley (4), p. 195).   [g] Ch. 3, p. 11 *b* (Chatley (1), p. 23).

[h] Ch. 25, p. 11 *b*; cf. Chavannes (1), vol. 3, p. 317.

[1] 大橈   [2] 容成

that 'when numbers assume form, they realise themselves in musical sounds (*shu hsing erh chhêng shêng*[1])'. Again, Sumerian harps occur with the bull, sheep or goat carved in their sounding boards,[a] while in China there was an association of the five notes with the five sorts of (domesticated) animals.[b] On the other hand, we do not find any theory of the harmony of the spheres in Chinese literature, and understandably so, for it was the child of Greek reasoning proceeding from the assumption that motion necessarily produces sound.[c] The Chinese, like the Babylonians, merely associated certain numbers with planets, and certain musical notes with numbers.

But the real reason why, starting from a common origin, Chinese acoustic theory took so different a road from that of the Greeks, must surely be that the Babylonian theory of proportions was applied to the music and the scales which actually existed in Greece and in China at the time, and naturally enough these were different, as also were the instruments on which the music was performed. The importance of the lyre and cithara in the history of Greek tuning is matched rather by the bells and stone-chimes of the Chinese than by any of their stringed instruments. In the need for tuning too, there was a world of difference, the former requiring constant adjustment and intimately associated with the pitch of the human voice, the latter immutable once out of the maker's hands.

There seems to have been a remarkable exchange of blown instruments between East and West in the centuries immediately preceding our era. The double-reed pipe or *aulos*[d] was used in Greece in classical, and the pan-pipes in post-classical times;[e] whereas in China the *kuan*[2] was known only in Han,[f] the pan-pipes (*hsiao*[3]) long before Han times. Pan-pipes are found today in a great arc stretching from northwest Brazil and Peru through Oceania across to Equatorial Africa,[g] a diffusion which indicates a very early origin. Von Hornbostel has suggested[h] that there was at one time a gamut of twenty-three *lü* or steps produced by over-blowing twelfths on a pipe

---

[a] Woolley (3), vol. 2, pls. 109, 111, 112.

[b] *Kuan Tzu*, ch. 58, p. 2a; cf. Yin Fa-Lu (1). Indian parallels in the *Bṛhaddeśi* of Mataṅga-muni quoting Kohala (+1st century); Trivandrum Sanskrit series, no. 94, p. 13. Galpin (1), p. 59 gives a Sumerian one. Cf. Vol. 2, pp. 262, 263.

[c] Theon of Smyrna attributed the establishment of a relation between sound and speed to Lasus (*fl. c.* −500). The spheres were first suggested as a hypothesis by Eudoxus of Cnidus (−406 to −355), Plato's contemporary and associate (Berry (1), p. 28). We have already discussed them in relation to Chinese astronomical learning in Vol. 3, pp. 198, 220, etc.

[d] See Schlesinger (1).

[e] According to Galpin (1), p. 14, the pan-pipes were unknown in ancient Mesopotamia and did not appear in Egypt until the −4th century. In China the oldest mention of the *hsiao*[3] is no doubt in one of the *Shih Ching* odes which may be dated as of the −8th century (Mao, no. 280; Karlgren (14), p. 245 has 'flutes'; Legge (8) 'organ'; Waley (1), p. 218 correctly 'pan-pipes'). The mention in the *Shu Ching* (Historical Classic), ch. 5 (I Chi), tr. Karlgren (12), p. 12, with 'pan-flutes', will not be quite so old.

[f] The earliest reference seems to be that of Chêng Hsüan (+2nd century) who says: 'Two (pipes) are tied together and so blown; the present-day Office of the Grand Revealed Music uses it', *Chou Li Chêng I*, ch. 45, p. 13a. Cf. Robinson (1), pp. 116ff.

[g] For details of the distribution of pan-pipes see Schaeffner (1), pp. 279ff. on 'instruments polycalames'.

[h] In a highly controversial thesis. See the criticisms of Bukofzer (1) and the reply by Kunst (2).

[1] 數形而成聲   [2] 管   [3] 簫

and reducing by an octave. Since an over-blown fifth is slightly smaller (twenty-five cents) than a fifth measured mathematically on a string, twenty-three steps were necessary to form a more or less complete cycle comparable to the arithmetically calculated twelve *lü* of the Chinese. Though it is not very likely that such a cycle ever existed, it is conceivable that early pan-pipe tunings were made on the 'up-and-down' principle by which the Chinese generated their twelve *lü*.

The Babylonian discovery of the proportions of the consonances then became known in China. To a people striving for constant pitch in order that the music and its magical virtue might be retained for the reigning dynasty, the acquisition of this piece of mathematical knowledge must have been electrifying, for as Mencius says:[a]

> When the (sages) had used their power of hearing to the utmost they extended it by means of the six *lü* (mathematical proportions?) to determine the five notes; one cannot exhaust their use. (*Chi chieh erh li yen, chi chih i liu lü, chêng wu yin, pu kho shêng yung yeh.*[1])

And four and a half centuries later his words were echoed by one of the greatest Han experts on acoustics and music, Tshai Yung[2] (+133 to +192). In his commentary on the *Yüeh Ling* he wrote:[b]

> In determining the pitch of bells in antiquity they levelled off their notes by ear. After that when they could go no further they availed themselves of numbers and thereby made their measurements correct. If the figures for the measurements are correct the notes will also be correct.

This empirical and experimental use of number was a refreshing contrast to the numerological games and number-mysticism which fascinated so many scholars of the Chhin and Han.[c] No exact date can, of course, be given for the introduction of the Babylonian formula, but the above reference in Mencius coincides significantly with the development of the New Music to make the −4th century the later end of the bracket.

The conclusions here reached bear close similarity to those of Section 20*e* on astronomy. There it appeared[d] that in all probability the original body of Babylonian ideas and observations, spreading west and east, was developed in one way by the Greeks to form their ecliptic and heliacal system, while the Chinese developed it in quite a different way so as to evolve the polar and equatorial system with its lunar mansions and circumpolar key-constellations. Common origin of a few basic ideas followed by divergent development seems to have occurred in acoustics also.

---

a *Mêng Tzu*, IV, I, (i), 5; tr. auct. adjuv. Legge (3), pp. 165, 166.
b *Yüeh Ling Chang Chü*[3] in *Li Chi Chi Chieh*, ch. 13, p. 64; tr. auct.; also quoted in *Hou Han Shu*, ch. 11, p. 18*a*, commentary.
c Cf. Vol. 2, pp. 287ff.
d Vol. 3, p. 256 e.g.

1 既竭耳力焉繼之以六律正五音不可勝用也   2 蔡邕   3 月令章句

(7) THE SEARCH FOR ACCURACY IN TUNING

The discovery that musical intervals are determined by mathematical ratios put the art of tuning on an entirely new basis. In Plato's *Republic* we can detect a certain contempt for the empirical experimentalist in sound:[a]

'As you will know, the students of harmony make the same sort of mistake as the astronomers; they waste their time in measuring audible concords and sounds one against another.' 'Yes', said Glaucon, 'they are absurd enough, with their talk of "sound-clusters" and all the rest of it. They lay their ears to the instrument as if they were trying to overhear the conversation from next door. One says he can still detect a note in between, giving the smallest possible interval, which ought to be taken as the unit of measurement, while another insists that there is now no difference between the two notes. Both prefer their ears to their intelligence.'

In the history of Chinese acoustics this air of condescending banter is fortunately absent. The musician or scholar with almost miraculous ability to detect small differences of tone was revered. Though un-Hellenic, this attitude bore good fruit in the world of practice.[b]

The Chinese, nevertheless, recognised the physical limitations of the ear, and used the measured steps of the *lü* as a check, as we have just heard Mencius say. Of course, tuning one bell against another by ear would simply lead to the sort of situation which moved Socrates to mirth, one 'expert' saying the two notes were exact, another declaring he could still detect a slight difference. Even if the word *lü* in Mencius meant 'pitch-pipe', that is to say a bamboo tube of such dimensions as to emit a desired note when blown, Socrates' objection would still apply, for only the ear could judge whether the note of a bell and that of a pipe were identical, and this would necessarily be a subjective judgment. The secret of the *lü* at this early stage, we believe, was the physical phenomenon about which so much has already been said in connection with the concept of *chhi*,[c] the 'humming-tubes' for canalising *chhi*,[d] and the interpretation of physical phenomena and human affairs[e]—resonance. If an instrument comparable to the monochord of Pythagoras was used, on which the measured steps, or *lü*, could be calculated mathematically, it would have been possible to tune a bell with absolute precision by means of this phenomenon. A string of measured length and tension being struck, if the proportions of the bell were correct a sympathetic note would be elicited by the string. If no note responded further rubbing and filing would be necessary till the bell's fundamental tone was perfectly tuned.

[a] 531 A. Cornford tr., p. 244.
[b] It is interesting that a form of the Chinese spiral or cycle of fifths has been used by piano-tuners in modern times, the interval of the fifth being, like that of the octave, one which can be fixed with tolerable accuracy by the ear alone. Cf. Closson (1), p. 117.
[c] Pp. 8, 32, 133 ff. above, and, e.g. Vol. 2, p. 369.
[d] Pp. 135 ff. above, and Vol. 2, p. 552.
[e] P. 129 above, and Vol. 2, e.g. pp. 282, 304, 500.

(i) *Resonance phenomena and the use of measured strings*

We have in fact evidence for the existence in Chou times of an instrument capable of serving this purpose. Commenting on the method of tuning bells mentioned in the *Kuo Yü*, Wei Chao[1] (+3rd century) states[a] that 'a board seven feet long (was used) having a string (or strings). They fixed them and so tuned.'[b] It is possible that by Wei Chao's time the principle on which this instrument functioned had been forgotten. His description is certainly far from clear, though he adds that the Office of the Grand Revealed Music of the Han possessed a *chün* or 'tuner'. Unfortunately, the text does not tell us whether one or more strings were used. But the great length of the instrument is interesting, for a long string would emit a good loud tone suitable for producing a sympathetic tone in a bell. Its great length would also make possible more accurate division of the string.

The *Kuo Yü* itself says:[c]

We measure the pitches and (so) tune the bells (*tu lü chün chung*[2]). Every official can describe the principle (*pai kuan kuei i*[3]). We form the series using 3. We tune (the bells) using 6. We complete (the operation) at 12. (*Chi chih i san, phing chih i liu, chhêng yü shih-erh.*[4])

From this one may conclude that when the passage was written twelve bells made a complete set, divided into two groups of six, one Yang, the other Yin. That the series was formed 'using 3' refers to the denominator of the fractions used for calculating by 'superior generation' (4/3), and 'inferior generation' (2/3).

An interesting light on the art of tuning by resonance is given in the *Chin Hou Lüeh Chi*[5] of Hsün Chho[6] (*fl.* +312).[d] This says:

The instrument for tuning the pitch of bells was neglected at the end of the Chou period. In the time of the Han emperors Chhêng (−32 to −7) and Ai (−6 to −1) many scholars devoted themselves to it, but it was again neglected by the end of the Later Han period.

The narrative goes on to describe how Tu Khuei[7] made efforts early in the +3rd century to tune the instruments according to ancient rules, not very successfully. But in the time of Hsün Hsü[8] (*d.* +289), some bells of about four centuries earlier were discovered in a provincial treasury, and it was possible to check them against pipes made with the help of jade measures of Chou time which had also been found.[e]

---

[a] *Chou Yü*, ch. 3, p. 26*a*; cf. p. 22*a* where the same statement is made using *chün*[9] and *hsüan*.[10]
[b] *Chün chê chün chung, mu chhang chhi chhih yu hsien, chi chih i chün.*[11]
[c] *Chou Yü*, ch. 3, p. 26*a, b*; tr. auct.
[d] Quoted in the commentary of the +5th-century *Shih Shuo Hsin Yü*, ch. 20, p. 29*b*; tr. auct.
[e] An excellent account of the work of Hsün Hsü is given by Wu Nan-Hsün (*1*), pp. 145 ff.

[1] 韋昭　　　[2] 度律均鐘　　　[3] 百官軌儀　　　[4] 紀之以三平之以六成於十二
[5] 晉後略記　　　[6] 荀綽　　　[7] 杜夔　　　[8] 荀勗　　　[9] 鈞　　　[10] 弦
[11] 均者均鐘木長七尺有絃繫之以均

Using the standard pitches they gave them their summons, and all (the bells) responded though they had not been struck (*i lü ming chih, chieh pu khou erh ying*[1]). The notes and the sympathetic tones (rhymes) agreed and became one (*shêng yin yün ho, yu jo chü chhêng*[2]).[a]

The *Thang Yü Lin* also gives an account[b] of how Tshao Shao-Khuei,[3] a great acoustic expert, once calmed the fears of a superstitious monk by his understanding of the principle of resonance. The monk had in his room a sonorous stone (*chhing-tzu*[4]) which seemed to produce sounds spontaneously. By filing off small portions of a bell in the monk's room which happened to be of the same frequency, and which was the cause of the trouble, Tshao altered the pitch of the bell so that the ringing-stone no longer responded to its note. Narratives of this kind are quite common in Chinese literature, and understandably so in view of the great philosophical importance of the idea of resonance to which attention has already been drawn.[c]

The search for accuracy in tuning may be traced back to the legend concerning Ling Lun's journey to the West for the rare bamboos. It has been suggested above that this legend could embody a good deal of literal truth if the bamboos which had been specially cut to the correct lengths were first used not for producing a sound by blowing, for which they would in fact be inaccurate,[d] but for measuring the correct distances on the strings of the tuner instrument (*chün*[5]), by which the bells were tuned. A feature of the Ling Lun legend is that on his return bells were tuned by means of the bamboos he brought back. But with such a perishable material doubt would naturally soon arise concerning the exact lengths required, since every time a fresh set was made errors would be liable to occur. With this background in mind one can understand the basis for a remarkable theory and technique which might otherwise be dismissed as pure nonsense.

(ii) *The cosmic tide in buried tubes*

How to verify whether tubes were of the exact length constituted a great problem. Bamboo tubes, as we saw earlier, had from ancient times been used for canalising *chhi*. One of the great manifestations of *chhi* was wind, and the winds of the eight directions were summoned each by its appropriate magical dance, led off by a note from an instrument made from one of the eight sources of sound. There was, therefore, a clear correlation between notes, winds, and directions. Probably no one was ever so simple as to hope that if bamboo tubes were pointed in the right direction the appropriate

---

[a] *Shih Shuo Hsin Yü*, ch. 20, p. 29 *b*, comm., tr. auct. Another version of the story is given in *Sui Shu*, ch. 16, p. 11 *a, b*.

[b] Ch. 5, p. 12 *a*. Parallel stories in ch. 6, p. 6 *a*.

[c] In the Section on fundamental ideas, Vol. 2, pp. 282 ff., 304 above.

[d] The reason for the inaccuracy is that the effective length of a blown pipe is greater than the length of the pipe itself. The frequency of the note of a blown pipe with open ends equals the velocity of sound divided by twice the length of the pipe. But the effective length of the pipe, i.e. the length of its resonating air column, is its geometric length $+0.58 D$, where D is the internal diameter. This is termed 'end-effect'. A vibrating string has no end-effect.

[1] 以 律 命 之 皆 不 扣 而 應   [2] 聲 音 韻 合 又 若 俱 成   [3] 曹 紹 夔   [4] 磬 子
[5] 均

wind would blow through them and sound the right note. But some ancient nature-philosophers set out to trap the *chhi* another way, that *chhi* which rose up from the earth combining with the *chhi* which descended from heaven to produce the different types of wind that blew at different seasons of the year.[a] In the words of the *Chhien Han Shu*:[b]

The *chhi* of heaven and earth combine and produce wind. The windy *chhi* of heaven and earth correct the twelve pitch fixations (*chêng shih-erh lü ting*[1]).

Chhen Tsan[2] commenting[c] some time before the end of the +4th century on this passage says:

The *chhi* associated with wind being correct, the *chhi* for each of the twelve months (causes) a sympathetic reaction (*ying*[3]) (in the pitch-pipes); the pitch-pipes (related serially to the months) never go astray in their serial order (*chhi lü pu shih chhi hsü*[4]).

Thus arose the strange practice termed *hou chhi*[5] (observing the *chhi*) or, more colloquially, *chhui hui*[6] (the blowing of the ashes). Perhaps the clearest statement of the principle of the technique was that of the Neo-Confucian philosopher Tshai Yuan-Ting (+1135 to +1198), notably expert in acoustics and music. In his *Lü Lü Hsin Shu* of about +1180 he wrote:[d]

The (pitch-pipes) are blown in order to examine their tones, and set forth (in the ground) in order to observe (the coming of) the *chhi*. Both (these techniques) seek to (determine the correctness of the) Huang-chung tube by testing whether its tone is high or low, and whether its *chhi* (arrives) early or late. Such were the ideas of the ancients concerning the making (of the pitch-pipes)....

If one desires to find the middle (i.e. the correct) tone and *chhi* without having anything available as a standard, the best thing to do is to cut several bamboos for determining the right Huang-chung length, making some shorter and some longer. Tubes are made for every tenth of an inch within their length range, with nine inches being taken as the (approximate) length standard for all, and circumference and diameter being measured (from this basis) according to the rules for making Huang-chung.

If this having been done one blows them one by one, the middle (i.e. the correct) tone will be obtained, and if one sets them more or less deeply (in the ground), the middle (i.e. the correct) *chhi* may be verified. When its tone is harmonious and its *chhi* responds, the Huang-chung is really a Huang-chung indeed. And once it is really so, then (from it) may be obtained the (other) eleven pitch-pipes, as well as the measures of length, capacity and weight. Later generations, not knowing how to go about this, have sought (to construct accurate pitch-pipes) only by measuring with the foot-rule.

[a] A detailed study of the strange subject here to be unfolded has been made by Bodde (17). We are much indebted to Professor Derk Bodde for his kindness in sending us an advance typescript of this paper. Although our own account was already written we were thus enabled to round it out by several interesting additions.

[b] Ch. 21A, p. 4*a*, tr. auct.

[c] The family name of this commentator is not definitely known; see Yen Shih-Ku's preface to the *Chhien Han Shu*, p. 5*b*.

[d] Ch. 2, sect. 1. Contained in *Hsing Li Ta Chhüan*, ch. 24, pp. 2*b*ff. Tr. Bodde (17), mod.

[1] 正十二律定　　[2] 臣瓚　　[3] 應　　[4] 其律不矢其序　　[5] 候氣

[6] 吹灰

Commenting on this, Bodde says: 'Tshai's preference for mechanical trial-and-error rather than a mathematical formula is, one suspects, typical of a good deal of (medieval) Chinese scientific activity.' Experimentalists may feel that this is not unlike praise, though not so intended. In this particular context, it is true, we are dealing with proto-science, or even pseudo-science, rather than with science itself. But let us not forget that this distinction was far from being as obvious to the early Fellows of the Royal Society as it is to us, and that Kepler cast his own horoscope every year.

What now in concrete detail was the strange technique of 'watching for the *chhi*'? According to the classical account of it given by Tshai Yung, the method by which the length of the tubes was checked was as follows:[a]

The standard practice is to make a single-roomed building with three layers (*san chhung*[1]) (i.e. concentric draught-proof walls). The doors can be closed and barred off (from the world outside), and the walls are carefully plastered so as to leave no cracks. In the inner chamber curtains of orange-coloured silk are spread out (forming a tent over the pitch-pipes), and certain stands are made out of wood. Each pitch-pipe has its own particular stand, set slanting so that the inner side is low and the outer side high,[b] all the pipes being arranged round the (circle of compass-)points in their proper (corresponding) positions.[c] The upper ends of the pitch-pipes are stuffed with the ashes of reeds, and a watch kept upon them according to the calendar. When the emanation (*chhi*) for a (given) month arrives, the ashes (of the appropriate pitch-pipe) fly out and the tube is cleared.

The *Hou Han Shu* adds a little more to this account:

They rely on calendrical calculation and so await (the coming of the emanation); when it arrives the ashes are dispelled; that it is the emanation which does this (is shown by the fact that) its ashes are scattered. If blown by human breath or ordinary wind its ashes would remain together.[d]

That the results were not considered entirely satisfactory or convincing is suggested by later modifications described in the *Sui Shu*[e] in which the tubes were not simply held in stands, but buried in levelled earth so that only the ends were visible. It was then thought that the *chhi* emanation rising upwards like a tide from the Yellow Springs far under the earth would blow the ashes out of the longest tubes first, beginning with Huang-chung, each month a different tube being blown. The most interesting part of this strange experiment is the care which seems to have been taken

---

[a] Quoted from his commentary *Yüeh Ling Chang Chü*[2] in the *Li Chi Chi Chieh*, ch. 13, p. 64; and *YHSF*, ch. 24, p. 31*b*, tr. auct. adjuv. Bodde (17). Parallel passage in *Hou Han Shu*, ch. 11, pp. 17*b*, 18*a*. Paraphrased in *Sui Shu*, ch. 16, pp. 10*a*ff. tr. Bodde (17). Cf. *San Tshai Thu Hui* (+1607), Shih ling sect. ch. 1, pp. 14*b*ff. and many other places.

[b] Thus the pipes all pointed towards the centre of the ring. An alternative interpretation of these words would mean that the pipes were partly buried.

[c] According to the system of symbolic correlations; cf. Vol. 2, pp. 261 ff.

[d] The official history also informs us that within the palace twelve jade pitch-pipes were used and observations made only at the solstices, while at the Imperial Observatory there were sixty pitch-pipes of bamboo (cf. p. 169), with correspondingly more frequent observations.

[e] Ch. 16, p. 10*a*. This source reports (p. 11*a*) a noteworthy failure to make the method work, namely the experiments of Tu Khuei,[3] the famous musician (*d. c.* +223).

[1] 三重   [2] 月令章句   [3] 杜夔

to ensure that no ordinary wind could enter the sealed chamber.[a] A third formulation of the theory states[b] that all twelve tubes were blown twice a year, that is to say the *chhi* reached its waiting tube every fifteen days. The expression 'waiting for, or observing, the *chhi*' (*hou chhi*[1]) has an almost punning connection with the fact that the year contained seventy-two *hou*[2] of five days each, two making a *hsün*[3] and three a fortnight or *chhi*[4].[c] There was thus just room in the year for the sixty pitch-pipes of Ching Fang.[d]

The most extraordinary development came about the middle of the +6th century, when Hsintu Fang, the mathematician, astronomer and surveyor,[e] actually invented[f] certain rotary fans (*lun shan*[5]) which were fixed to the buried pipes so that they should rotate when the *chhi* blew out the ashes.[g] Further information as to the various forms of the technique may be obtained from Chin books still extant in fragmentary form,[h] e.g. the *Mei Tzu Hsin Lun*[6] (New Discourse of Master Mei)[i] which gives directions for preparing the ashes from the inner membranes of plants of the reed family. In a more sophisticated age scholars such as Chu Tsai-Yü (*b.* +1536)[j] and Chiang Yung

---

[a] This sealing is perhaps really the most significant technical feature of the whole story. The precautions against chance breaths of wind and other disturbances reached their greatest degree of elaboration by the middle of the +6th century. Besides the tent of orange silk, gauze covers were fitted for each pitch-pipe individually. According to the descriptions, the stands or holders for the pipes were rather like our retort-stands. And the walls were so arranged that the doors of the inner and outer walls were at the south, while the door of the middle wall was at the north. Thus there were imbricated corridors exactly as in modern photographic dark-room practice. These remarkable details of the pursuit of an essentially unreal phenomenon may be found in the commentary on the *Yüeh Ling* written by Hsiung An-Shêng[7] about +570, and in the *Yo Shu Chu Thu Fa*[8] (Commentary and Illustrations for the Book of Acoustics and Music), due to the eminent mathematician and astronomer Hsintu Fang,[9] his older contemporary. Hsiung's words (tr. Bodde, 17) are preserved in *Li Chi Chu Su*,[10] ch. 14, p. 7a, and *Chin Shu*, ch. 16, p. 10b. Hsintu Fang's fuller account has only come down to us because it was preserved in the *Yo Shu Yao Lu*[11] (Record of the Essentials in the Books on Music and Acoustics) compiled about +670 by the celebrated Wu Huang Hou[12] (Wu Tsê Thien[13]), the empress of Kao Tsung, and later herself sole ruler. There it will be found in ch. 6, pp. 17bff. It is quoted in full (more accessibly) by Hu Tao-Ching (1), vol. 1, pp. 325ff. Finally, a closely similar description was given by Chhen Yang in his *Yo Shu* of +1101 (ch. 102, pp. 4bff.).

[b] *San Tshai Thu Hui*, Jen shih sect. ch. 9, p. 17a; see the diagram given by Robinson (1), p. 113.

[c] Cf. Vol. 3, p. 404.

[d] See p. 218 below. We have just seen that sixty pitch-pipes were used at the Han Observatory, hard though this is to believe.

[e] Often met with elsewhere, e.g. Vol. 3, pp. 358, 632. See above, p. 35.

[f] *Sui Shu*, ch. 16, pp. 9bff. Cf. *TPYL*, ch. 871, p. 6b; tr. Pfizmaier (98), p. 43, and *Ku Chin Yo Lu*[14] (in *YHSF*, ch. 31, p. 8a, following *TPYL*, ch. 565, p. 8a). It is unfortunate that neither of his own fragmentary works (*Yo Shu* and *Yo Shu Chu Thu Fa*) make any mention of the rotary fans, nor can further light on them be obtained from any of his three biographies (*Pei Chhi Shu*, ch. 49, p. 3b, *Pei Shih*, ch. 89, p. 14a, *Wei Shu*, ch. 91, p. 13b).

[g] Perhaps they were small horizontally-rotating vane wheels like the zoetrope (see p. 123 above) or the helicopter top (Sect. 27m below). But the text clearly says that they were themselves buried in the ground, so it is difficult to visualise them. There were twenty-four of them so that there was one pipe for each of the fortnightly *chhi*.

[h] There is much in *TPYL*, ch. 16, pp. 2aff., 5a, 7b, etc. This material, which needs special investigation, gives one the suspicion that the real originator of the method was Ching Fang (*fl.* −45, cf. pp. 213, 218 below), and that at first it had something to do with the practice of weighing 'ashes' to measure the humidity of the air (cf. Vol. 3, p. 471). Bodde (17) brings forward further evidence pointing to Ching Fang.　　[i] *YHSF*, ch. 68, p. 30a.　　[j] *Lü Lü Ching I*, ch. 1, pp. 117a, 127b.

[1] 候氣	[2] 候	[3] 旬	[4] 氣	[5] 輪扇	[6] 梅子新論
[7] 熊安生	[8] 樂書註圖法	[9] 信都芳	[10] 禮記注疏
[11] 樂書要錄	[12] 武皇后	[13] 武則天	[14] 古今樂錄

(*b.* +1681)[a] had no hesitation in dismissing the whole matter as unworthy of belief, 'erroneous and moreover unclassical'.

The development of a sceptical attitude towards the procedure of 'watching for the periodic arrival of the *chhi*', and finally, its frank rejection, raises points of considerable interest. When in +589 the emperor Sui Wên Ti commissioned Mao Shuang[1] and his colleagues to carry it out, and when later on they prepared their report on it, the *Lü Phu*,[2] there was no shadow of hesitation about the technique itself.[b] When the empress Wu wrote her book in the +7th century it was still firmly believed in. That Chhen Yang at the end of the +11th, and even the great Neo-Confucian philosopher Chu Hsi in the +12th[c] still accepted it, is not perhaps surprising. What is more curious, however, is that Shen Kua with all his scientific acumen also had no doubts, and gave instructions for making the process work.[d] By the Ming period, however, scepticism was rampant. Besides the scholars just mentioned, Hsing Yün-Lu[3] (*fl.* +1573 to +1620) made a devastating attack on the *hou chhi* practice about +1600 in his *Ku Chin Lü Li Khao*[4] (Investigation of the Calendars, New and Old).[e] After showing the scientific absurdity of the idea, he did not hesitate to accuse the astronomical-acoustic officials of purposeful deception, saying that they must have had some concealed mechanism analogous to the jack-work of clocks whereby the ashes were blown out of the tubes or the fans made to turn at the proper time. Bodde (17) has found two instances of attempts to use the technique not long before Hsing was writing, and these may well have been in his mind. One was connected with a court official named Chang Ê[5] (+1530 and +1539),[f] the other with Yuan Huang,[6] an acoustics expert, in +1581 or the following year.[g] There was scepticism at the time in both cases, but the second experiment was said to have been successful. Chiang Yung could not account for this, though he did not believe in it.[h] The interest of the whole story is that by the +16th century the procedure was being decisively rejected on scientific grounds. This is a rather striking demonstration of the fact that a rise of critical judgment on matters of natural science went on in the Ming paralleling (if not sometimes even preceding) that development of scientific scepticism which in Europe was the work of the scientific Renaissance. We shall find many other examples of this as we go on, notably among the pharmaceutical naturalists such as Li Shih-Chen (*d.* +1593). Such a parallel process cannot be without significance for the problem of why science in its distinctively modern form did not develop in China, and we shall return to it when in the end we come face to face with that grand enigma.

As for the pitch-pipes buried in the ground, and all their accompanying para-

---

[a] *Lü Lü Hsin Lun*, ch. 2, pp. 23 *b* ff. Cf. Chang Chieh-Pin's *Lei Ching* (*Fu I*), ch. 2, p. 14 *b* ff.
[b] *Sui Shu*, ch. 16, pp. 10 *a* ff. Opinions differed on the phenomenalistic interpretation of the results.
[c] *Chu Tzu Chhüan Shu*, ch. 41, pp. 20 *b*, 26 *a*.
[d] His remarks on the subject in *Mêng Chhi Pi Than*, ch. 7, para. 25 (cf. Hu Tao-Ching (*1*), vol. 1, pp. 325 ff.) have been translated in full by Bodde (17).   [e] Ch. 33 (pp. 525 ff., 528).
[f] See *Hsü Wên Hsien Thung Khao*, ch. 107, pp. 3747.1 and 3748.2.
[g] The episode is related in the preface of one of Yuan Huang's books, the *Li Fa Hsin Shu*, reproduced in Chiang Yung's *Lü Lü Hsin Lun*, ch. 2, p. 24 *a*.
[h] *Lü Lü Hsin Lun*, ch. 2, p. 23 *b*.

[1] 毛爽   [2] 律譜   [3] 邢雲路   [4] 古今律曆考   [5] 張鶚   [6] 袁黃

phernalia, was it not an archaic survival from the time when no one could distinguish cosmic magic from true science? And yet we are tempted to feel that there must have been some genuine natural phenomenon, even if only once observed, which sufficed to keep this strange technique living for a dozen centuries.[a] However that may be, no rational basis for the system can be suggested, but the following paragraphs will attempt to show what the need was which gave rise to it.

Let us summarise the semantic development of the word *lü* from its earliest beginnings. Probably it first meant the rule, regularity, or regulated step, in dancing. Secondly, when bells were tuned by resonance from the string of a 'tuner', the regularity of the *lü* would have been the measured steps or divisions of the string, i.e. the exact length of the string required for a given note, determined by the use of certain standard lengths of bamboo traditionally believed to enshrine the necessary proportions.[b] The discovery of these proportions would seem to have been Babylonian. Later on, a better understanding of the mathematics involved enabled the actual proportions of the *lü* to be preserved, and the scale to be reduced, so that the unwieldy 7-ft. tuner (*chün*) became obsolete.

Though knowledge of the mathematical formula guaranteed the relative proportions of the *lü*, and hence the relative pitch of the twelve notes in a cycle of fifths, it did not guarantee their absolute pitch, if the absolute lengths of the *lü* (measuring-rods) were still in doubt.[c] In an attempt to discover their absolute lengths as well as to check their proportions, a set of tubes was cut, it would seem, resembling the humming-tubes used by the *wu* shaman for canalising *chhi*. It was supposed that if the tubes were of the right length, each one with its opening so many inches above or below the ground according to the method used, the ashes would be blown out of the tube at the exact instant that the *chhi* reached that point. The *chhi* was thought of as ebbing and flowing like an annual tide, and therefore it was supposed that its exact distance from the earth's surface could be calendrically calculated.

In spite of its long persistence, this practice naturally never gave the results sought, and in the third stage we find *lü* used in a new sense. The generic word for a flute or pipe was *kuan*.[1] The standard lengths of bamboo having reassumed their canalisation function first as '*chhi*-detectors' planted in the earth, and then, by an easy transition, as blown pipes comparable to the '*chhi*-detectors' of the *wu* shamans, became in the third stage simply *lü-kuan*[2] or pitch-pipes.[d] Since the formula by which the pro-

---

[a] Early experimentation with vents of natural gas has been suggested, but it seems very improbable.

[b] It is to be noted that though the *Shih Ching* is one of our earliest reliable texts, and abounds in references to musical instruments and the need for their being properly tuned, the character *lü* is never used in it in the sense of 'pitch-pipe'. Evidence was marshalled by Chavannes (1), vol. 3, pp. 638 ff. to show that before the − 4th century *lü* were always bells, not pitch-pipes. Cf., however, Yabuuchi (*18*).

[c] Of course, the medieval Chinese did not think in terms of a continuous band of wave-frequencies, but they were aware that the pitch of the Huang-chung note had varied from dynasty to dynasty, just as we are aware that middle C today is considerably higher than it was in Elizabethan times.

[d] On the significance of the term *kuan* see *Lü Lü Ching I*, ch. 8, pp. 4 a ff.; *Lü Hsüeh Hsin Shuo*, ch. 1, p. 17 a; *San Tshai Thu Hui*, Chhi yung sect. ch. 3, p. 15 b, and many other authorities discussed in Robinson (1), pp. 116 ff.

[1] 管   [2] 律管

portionate lengths of string of the 'tuner' were calculated worked reasonably well for pipes, the *lü*, now meaning pitch-pipes, were, at least from Han times, the orthodox devices for giving the pitch to other instruments. Nevertheless, their adoption for this purpose, and the respect in which they were held on account of the undoubted antiquity of the *chhi*-detecting tubes, and of the whole concept of regularisation summed up in the word *lü*, focused men's attention on the acoustic properties of pipes, which will be considered below.[a] In the meantime it is worth noting that the only really general translation of the word *lü* in its acoustic sense is 'pitch-giver'.[b]

(iii) *Tuning by means of hydrostatic vessels*

There were, however, other methods of tuning instruments which first deserve consideration. We have already noticed[c] the interest which the Chinese (like the Alexandrians) took in hydrostatics, and among their oldest observations must have been the variation in acoustic properties caused by the filling of vessels to varying extents. One of our oldest accounts is that given by Kan Pao[1] (*fl.* +320) in his commentary[d] on a sentence in the *Chou Li*.[e] This sentence says: 'With the metallic instrument *chhun*, the note is given to the drums (*i chin chhun ho ku*[2]).' Chêng Hsüan's commentary on this is merely that this metal instrument is shaped like the end of a pestle, being wider at the top than the bottom, and that music causes it to emit a ringing sound. It associates with the drums and they sound together. But Kan Pao enlarges as follows:

Water is filled in (to the *chhun*) to a height of one foot above the ground, and a container is filled with water and put underneath. The *mang*[3] (an apparatus on which strings were set) is placed between them. If the *mang* is shaken by hand, a tremendous ñoise like thunder is produced.[f]

The Chinese fully exploited the possibilities of water in relation to tuning, with its great advantage, precise control over microtonic adjustments by the addition or removal of small amounts of water.[g] 'In Thang times', we are told, 'bowls (were used) containing water; they added to it or diminished it, and thereby tuned the notes of the scale.'[h] The use of pottery vessels without water in them as musical instruments

---

[a] Pp. 199, 212 ff.

[b] It will be seen that this is closely related to the normative significance of the word in its juridical sense; cf. Vol. 2, pp. 550 ff.

[c] P. 34 above. See also our account of clepsydra physics, Vol. 3, pp. 313 ff.

[d] Quoted in the *Kuang Chhuan Shu Po*,[4] a Sung book by Tung Yu,[5] ch. 2.

[e] Entry for 'Drumming Personnel' (Ku Jen), ch. 3, p. 36*a* (ch. 12); tr. Biot (1), vol. 1, p. 266; Kan Pao's commentary tr. auct.

[f] For one of the rare statements on the *chhun*, see Hsintu Fang's *Yo Shu*, in *YHSF*, ch. 31, p. 20*a*. As may be seen from Fig. 317, it normally faced upwards and had a tongue suspended from a crossbar. Archaeological evidence collected by Umehara Sueji (1) indicates that the *chhun* has affinities with the bronze drums of the Dôngsón culture, and was probably introduced from Indo-China during the Han.

[g] As in the counterpoised cylinders rising and sinking in water, familiar as 'resonance tubes' in elementary textbooks of physics.

[h] Wu Jen-Ching & Hsin An-Chhao (*1*), p. 32. It will be remembered that the characters of the terms used for 'high' and 'low' sounds, *chhing*[6] and *cho*,[7] use the water radical.

[1] 干寶   [2] 以金錞和鼓   [3] 芒   [4] 廣川書跋   [5] 董由
[6] 清   [7] 濁

is undoubtedly of great antiquity in China.[a] There is, for example, the famous reference to the behaviour of Chuang Chou at the death of his wife, making music by drumming on a bowl.[b] Pots were also used as primitive drums, first by themselves and later with a skin stretched over the top. Li Ssu of Chhu, a minister of Chhin State (−3rd century), referred disparagingly to the indigenous music of Chhin as 'beating on earthen jugs and knocking on jars'.[c] This was not, however, peculiar to Chhin, but merely represented a more primitive phase of music which had at one time been known to the Chinese of the central States, for in the *Shih Ching* the first ode of Chhen[d] speaks of the drummer on the mound beating his earthenware jars (*fou*[1]).

The *Ta Chou Chêng Yo*[2] mentions a set of eight tuned earthen vessels,[e] invented by Ssuma Thao[3] of the Thang period in +765 and presented to the throne. Other later sources speak of eight vessels (*shui chan*[4]) which were put on a table and struck.[f] If Ssuma Thao filled his vessels with different amounts of water, as seems very probable, he was a pioneer of this method.[g] For soon afterwards we have a clear account of sets of vessels tuned by the addition and removal of water; the details are given in the +10th-century *Yo Fu Tsa Lu*.[5] Tuan An-Chieh there says[h] that in the year +847 Kuo Tao-Yuan,[6] an official of the Bureau of Music,

used (a set of) vessels (*ou*[7]) of Hsing[8] and Yüeh[9] twelve in number, in which the amount of water was increased or diminished (to tune them); when struck with sticks the vessels gave out sounds better than those of the metal plates (*fang-hsiang*[10]).

Later in the same century (*c.* +870) Wu Pin[11] was also known as a master of this method.

Wooden as well as earthen bowls were used for tuning purposes, and accounts survive of the wonderful skill with which these bowls were made, so that the rim

Fig. 317. The *chhun*, a bronze bell of elliptical section, wide at the mouth and narrowing towards the round base, suspended by a loop usually, as here, in the form of a tiger (*Akhak Kwebŏm*, ch. 6, p. 24*b*). Like the *ling*, it had a tongue, but hung from a crossbar as the *chhun* was open upwards. *Chhun* filled with various amounts of water were sometimes used for tuning purposes.

[a] A number of quotations are collected in *TPYL*, ch. 758, p. 1*a, b*.
[b] *Chuang Tzu*, ch. 18; tr. Legge (5), vol. 2, p. 4; Waley (6), p. 21.
[c] *Shih Chi*, ch. 87, p. 5*a*; tr. Bodde (1), p. 19.
[d] Mao, no. 136; tr. Karlgren (14), p. 87; Legge (8), p. 153.
[e] In *TPYL*, ch. 584, p. 4*a*; the book is probably of Thang date.
[f] See *KCCY*, ch. 47, p. 12*a*.
[g] Cf. p. 38 above on the iron vessels with very smooth flanges studied by Li Kao and Li Yuan in the +8th century.
[h] P. 14*a*. This record, written after the end of the Thang period, is quoted in shortened form in *TPYL*, ch. 584, p. 4*a*.

[1] 缶   [2] 大周正樂   [3] 司馬滔   [4] 水盞   [5] 樂府雜錄
[6] 郭道源   [7] 甌   [8] 邢   [9] 越   [10] 方響   [11] 吳繽

was perfectly even and level.[a] The *Thang Yü Lin*[1] describes a wooden bowl (*chhüan*[2]) which fitted so closely to a smooth lacquer plate that even when inverted full of water none flowed out. It was said that this bowl was used for tuning notes, and that strings so tuned could remain in tune for one month.[b] 'But', as the writer sadly remarks, 'wooden bowls today cannot compete with that old one.'

To summarise the evidence concerning this method, empty earthen pots were used as drums in very early times, and then perhaps in the Han period, and certainly in the Thang, were adapted as tuning devices by being filled with varying amounts of water. The use of water for testing the measurements of standard vessels of capacity, which has already been mentioned,[c] suggests a possible connection.

(iv) *The manufacture and tuning of bells*

The tradition of bell-making in China is so old, and the part which bells played there in music and the tuning of instruments is so important, that the art of the Chinese bell-founders deserves careful study.[d] Description of the processes involved requires the use of various technical terms, so it will be as well to begin with a description of bells in general, and to compare their evolution in Europe and China.

Galpin tells us[e] that some early European bells were formed in four-sided shape, with a ring or handle, by folding or riveting iron plates together and then brazing or bronzing them. As an example he cites St Patrick's Clog of about the +6th century, which is 6 in. high, 5 in. broad, and 4 in. deep. Though this construction was primitive, the Clog represents an advanced stage in campanological evolution, and has taken over features worked out in the course of centuries in China, such as the use of a clapper, the downward pointing mouth, and a suitable means of suspension.

Yetts suggests[f] that the evolution of the bell in China was on the following lines. The small hand-bell named *to*[3] (Fig. 318), with a barrel (in the earliest examples) of diameter greater than its height, was probably the ancestor of all Chinese bells. It was normally held with the mouth uppermost.[g] When the mouth of the *to* began to point downwards the hanging clapperless bells called *chung*[4] (Fig. 319) came into being.[h] Bells with clappers (the generic name for which is *ling*[5]) developed later.

An intermediate stage in the development of the clapper is surmised. The *to* described by Yetts (5) has no means by which a clapper could be suspended inside it. In the earliest times such bells were doubtless struck with a stick or hammer on the

---

[a] Cf. Sect. 26c (p. 38 above).

[b] *Thang Yü Lin*, ch. 5, p. 26a. These methods were associated with the names of Li Kuei-Nien,[6] one of three brothers all famous in music and dance in the Khai-Yuan reign-period (c. +720 to +735), and his disciple Jen Shih-Chün[7] (d. not long after +782).

[c] P. 40. Cf. pp. 199ff. below, esp. p. 201; and Vol. 3, pp. 471ff.

[d] Cf. Moule (10), pp. 35ff.   [e] (2), p. 42.   [f] (5), pp. 78ff.

[g] The normal position of a bell can, of course, be inferred from the ornamentation and the direction of the inscriptions. Cf. p. 200 below.

[h] See Koop (1), pl. 23, showing such a bell of the Chou period, and pl. 42. The former is reproduced in Fig. 319. The earliest dated bell of this type is connected by an inscription with the High King Mu of the Chou, and thus of the −10th century.

[1] 唐語林   [2] 椫   [3] 鐸   [4] 鐘   [5] 鈴   [6] 李龜年
[7] 任使君

PLATE CXI

Fig. 318. Clapperless upward-facing hand-bell (*to*). Chou period. Winkworth Collection
(photo. Koop). Height 17 in.

PLATE CXII

Fig. 319. Clapperless downward-facing hand-bell (*chung*). Chou period, perhaps as early as the −6th century. It bears an inscription saying 'The Elder of Hsing in Ting (district) has made this bell, named Mysterious Harmony, with the note Jui-pin, for use'. Victoria and Albert Museum (photo. Koop). Height 22 in.

outside surface. But some *to* have four grooves running down near the two lateral edges at the top of the bell, and Yetts suggests that splints of bamboo may have been secured in these grooves, each so bowed back that its other end could be fixed in the groove opposite, thus forming a crossed splint over the mouth of the bell. From this a clapper could have hung down inside when it was held in the early upward-pointing position. The later clapper arrangement would then be a modification of this suitable for a downward-pointing mouth.[a]

Developments such as these, which had to take place in the evolution of the bell before it could be satisfactory merely as a noise-making instrument, had in fact been accomplished in China during a period of some thousand years, by the -5th century or earlier. But to make the bell an instrument of music it had to be properly tuned and this is regarded even today as a highly intricate matter. The bell, as Helmholtz said,[b] is a variety of curved metal plate. In both plate and bell the vibration frequency increases with the thickness and elasticity of the metal, changing in inverse ratio to the diameter and the specific gravity. For a given pitch and good tone quality a bell-maker must thus consider the nature and proportions of the metals used,[c] the profile, i.e. the inner and outer contours of the bell and the space between them,[d] the amount of metal needed to fill this, its temperature on pouring, and the rate at which it cools. But the required musical quality may still elude him, for the slight tolerances inevitable in mould-making and casting fall short of the necessary precision. The bell must then be brought to specification by removing small amounts of metal so as to thin it slightly in certain places.[e] This corrects the fundamental frequency, and brings several partial tones (often very noticeable, and liable to be dissonant) into consonant relationship with it.[f]

[a] As to the general question of the Asian origin of European bells and church-bells, which has been raised by Feldhaus (1, 16, 17, 20), Lynn White (1), p. 147, and others, it is at least certain that, since bronze bells were being made in the Shang period (-14th century), Chinese campanology is extremely ancient. Adequate comparisons of these bells with those of Babylonian and Hellenistic times have yet to be made, but it seems sure that during the -1st and $+1$st millennia the art of bell-founding in China was much more advanced than it was in Europe or the Middle East. Small round downward-pointing Assyrian harness-bells of the -7th century in cast bronze up to 4 in. high with hanging clappers of soft iron are in the British Museum, and a number of similar Roman examples of the $+1$st and $+2$nd centuries have been excavated, but none exceed 8 in. in height. As late as $+1000$ no bell more than 2 ft. high had been seen or heard of in Europe, though some were by then of fine workmanship, such as the Mozarabic Christian bell of Córdoba ($+961$). But the bell at Phing-ting in Shansi, an iron-casting of $+1079$, was already four or five times this size. Perhaps for once China taught Mesopotamia in high antiquity.   [b] Cf. Geiringer (1), p. 52.

[c] Most bells, both in China and the West, were always of copper–tin alloys (bronzes), but the Chinese early employed cast iron (cf. Needham, 31, 32).

[d] Western bell-founders have divided the side of the bell into four zones, the English terms being the 'lip', close to the rim; the 'sound-bow', the greatly thickened part just above, where the clapper strikes; the 'waist', most of the remaining flank; and the 'shoulder' at the top of it. We shall see that the Chinese bell-founders made similar distinctions, though we cannot take them to be exactly the same because of the different profiles of Chinese bells.

[e] There is evidence of this on both Chinese and Western bells.

[f] Modern European founders who practise this tuning control five frequencies in a bell. The English names of these, exemplified for a bell sounding C^1, are as follows: 'hum-note', C; 'fundamental', C^1; 'tierce', $E\flat$; 'quint', G^1; and 'nominal', C^2. Bells also have the peculiarity of producing another note which is solely aural, and cannot be picked up on any acoustical instrument. This is called the 'strike-note'. On untuned Western bells it lies close to, but just off, the fundamental; on tuned bells it agrees with the fundamental and reinforces it. The harmonics are excited by striking particular zones of the bell's periphery and tuned by adjusting others.

Chinese practice may now be considered as it is described in the *Chou Li*, though it must be borne in mind that this may represent the ideas of Han scholars on the art of bell-making, possibly a very different thing from the practical rule-of-thumb traditions of actual foundrymen. But even if the information is not strictly accurate, it is still of value in showing the awareness of the Chinese in early times of the many factors involved in the tuning of bells.

The entry on the bell-makers (Fu shih[1]) in the *Chou Li*[a] begins by naming the different parts of the bell. From the rim at the mouth to the loop or handle at the top the surface is divided into four zones. That nearest the rim is called the *yü*,[2] a word which is also used in the *Chou Li* meaning 'to chant'. Above this comes the zone called *ku*,[3] i.e. 'drum'; Chêng Chung[4] (*fl.* +50 to +83) says that this part is the strike-point (*chi-chhu*[5]). More accurately, however, the struck zone, termed *sui*,[6] is inside, and this is said to mean 'mirror' because the curve is concave like a burning-mirror.[b] Above this comes the part of the bell with straight cylindrical walls, as the name *chêng*[7] implies.[c] The fourth and highest zone is called *wu*,[8] which means 'dance'. No commentator offers any explanation of this term. In many bells the zones are separated by narrow bands embossed with studs of metal.

The author of this description then attempts to define the necessary proportions of the bell. He takes as his unit one-tenth of the distance formed by the two horns or points on its rim at the extremities of its oval circumference. To construct a bell after his pattern one would need to know the total height and both the long and short diameters of the zones at different specific points, the height allotted to each zone, the thickness of the metal and its weight, and details concerning the clapper if one was to be used.[d] The thickness is given as one unit, and a few of the diametrical proportions are given, but as a whole the information which has come down to us is quite inadequate. It may have been that the writer wished to give a formula covering all types and shapes of bells, and finding that impossible just set down such items as he believed to be of general validity. But even here one finds that his proportions produce a bulge in the third zone resembling an old-fashioned oil-lamp glass. He covers himself in conclusion, however, with the words:[e]

Thinness and thickness, that is what produces vibration (*chen*[9]) and throbbing (*tung*[10]) (respectively); purity or impurity (of the metal), that is what (causes the sound) to proceed outwards (*yu chhu*[11]) (i.e. from the vibrating walls of the bell themselves); the open or closed

---

[a] *Khao Kung Chi*, ch. 11, pp. 23ff. (ch. 41); tr. Biot (1), vol. 2, pp. 498ff. Wu Nan-Hsün (1), pp. 125ff. has a good discussion of it.   [b] Cf. pp. 87ff. above.

[c] This term came later to mean a small gong or cymbal.

[d] All such questions were studied in detail by the Chhing scholars such as Chhêng Yao-Thien (2) in his *Khao Kung (Chi) Chhuang Wu Hsiao Chi* (in *Huang Chhing Ching Chieh*, ch. 538) *c.* 1805; and his great master the archaeologist Tai Chen (+1723 to +1777) whose *Khao Kung Chi Thu* of 1746 (in *Huang Chhing Ching Chieh*, ch. 563) deals with bell-making on pp. 47ff. On the work of these men, especially on bells, see Kondō Mitsuo (1).

[e] Ch. 11, p. 24*b* (ch. 41), tr. auct. adjuv. Biot (1), vol. 2, p. 501.

[1] 鳧氏   [2] 于   [3] 鼓   [4] 鄭眾   [5] 擊處   [6] 隧
[7] 鉦   [8] 舞   [9] 震   [10] 動   [11] 由出

(form of the mouth), that is what (causes the sound) to proceed upwards (*yu hsing* [1]).[a] For all these things there are (special) explanations.[b]

These observations seem reasonable, and are followed by others to the effect that 'if (the walls of) a bell are too thick (it will sound like) a stone (*shih* [2])'. This may mean either a dull heavy sound as from an ordinary stone, or alternatively the timbre of the chime-stones used by the Chinese for music. The former seems more probable. 'If they are too thin (the sound will be) scattered (as by winnowing) (*po* [3]).' The word originally had the highly onomatopoeic sound *pwâr*. 'An open brim produces spreading (sounds); a closed brim produces choked (sounds) (*chhih tsê tsê; yen tsê yü* [4]).' The words *tsê* and *yü*[c] are metaphors, words originally meaning 'clearing trees' and 'densely wooded'. The idea of the sound being free or muffled is quite clear. Other details follow in which it is said that if a bell is large and short in the barrel its sound will be sickly and brief, but if it is small and long it will be healthy and rolling.

From the above one is forced to the conclusion that though the scholar who recorded some fragments of bell-lore in the *Chou Li* scarcely did justice to the foundrymen, it is evident that there existed in his day a wealth of technical terms and empirical knowledge. This is substantiated elsewhere in the *Chou Li* where twelve different types of sound are enumerated.[d] Chêng Hsüan says that they are bell sounds, but other commentators disagree, for the passage follows one describing the duties of the official responsible for the (pitch-)measuring tubes (Tien Thung [5]), and should apply to all instruments. However, it must be remembered that at this time bells gave the pitch to all the other instruments, so that an accurate enumeration of bell sounds here was quite appropriate. Moreover, a proof that they are bell sounds is that some of the terms are repetitions of the descriptions of bell sounds given above. The twelve are as follows:[e]

The sound (produced in) the upper part (of the bell) is rumbling (*kun* [6]).
The sound (produced in) the straight part (of the bell, i.e. the *chêng*) is slow (*huan* [7]).
The sound (produced in) the lower part (of the bell) is spreading (*ssu* [8]).
The sound (produced by) the parts which curve outward is scattered (*san* [9]).
The sound (produced by) the parts which curve inward is hoarded (*lien* [10]).
The sound (produced by) a part which is somewhat too big is exaggerated (*ying* [11]).
The sound (produced by) a part which is somewhat too small is dark (incomplete) (*an* [12]).
The sound (produced by a bell of?) oval (shape) (lit. 'somewhat round', 'returning') is ample and full (*yen* [13]).
The sound (produced from) an open (mouth or brim) is *tsê* [14] (*tsăk*).[f]

---

[a] This would certainly apply for bells held mouth pointing upwards.
[b] Oral traditions among the skilled master-craftsmen, themselves almost certainly illiterate.
[c] Cf. the medical use of this word to mean stasis in the pores or channels of the body; Sect. 7*j* (Vol. 1, p. 219), and Sect. 44.   [d] Ch. 6, p. 16*b* (ch. 23); tr. Biot (1), vol. 2, p. 56.
[e] Many of the terms used are obscure and the interpretation suggested is that which has seemed best after considering the opinions of all the available commentators.
[f] Three variants for this character occur, meaning 'bamboo cable' (*tso* [15]), 'clearing trees' (*tsê* [16]) or a kind of oak (*tso* [16] or *cho* [16]), and 'suddenly' (*cha* [17]). Chêng Chung says: 'The sound is forced and issues hurriedly'.

[1] 由興   [2] 石   [3] 播   [4] 侈則柞弇則鬱   [5] 典同   [6] 硍
[7] 緩   [8] 肆   [9] 散   [10] 斂   [11] 嬴   [12] 舍   [13] 衍
[14] 筰   [15] 筰   [16] 柞   [17] 咋

The sound (produced from) a closed (mouth or brim) is choked (*yü*[1]).[a]
The sound (produced by) thin (walls) is a staccato shaking (*chen*[2]).[b]
The sound (produced by) thick (walls) is (like) stone (*shih*[3]).[c]

Of these twelve definitions the first three clearly apply to the barrel of the bell. The last four, concerning the thickness of the metal and the shape of the mouth, are treated in terms closely similar to those in the passage quoted above. There seems little doubt therefore that this text contains an analysis of the factors which interested the Chinese some two thousand years ago in the production of suitable harmonics from bells. Timbre can only be described by metaphor, as when we say that a sound is 'rich' or 'sharpened' or 'thin'. Phoneticians even speak of a 'dark L'. The Chinese can hardly therefore be criticised for such terms as 'dark' or 'choked'. On the contrary, the refinement of their investigation into the nature of bell-sounds in this early period is quite remarkable. The two factors given special prominence here are the diameters and contours of the bell, and the thickness of the metal. Four other factors yet remain according to modern practice—the elasticity and specific gravity of the metals used, the proportions of each, and the total mass. All of these were taken into consideration by the Chinese, though naturally they were not thought of quite in these terms. It will be better to use two heads, the nature of the metal used, and how much.

Following the description concerning the proportions of the different parts of the bell the *Chou Li* has a section describing the preparation of the metal by workers called Li shih[4] who made vessels (*fu*[5]) as standard measures of volume.[d] The commentators tell us that the *fu* measure was one-tenth that of the *chung*[6] (a word which the *Chou Li* constantly uses as a homonym for bell). The processes of these artisans applied also to the making of bells. In order that the copper might be quite pure it was melted three times before casting. The process was controlled by observation of the colour of the metal. The proportions of copper and tin used for bells (16–17% tin) are stated at the beginning of the chapter on metal-workers.[e] Weights of metal were checked against the weight of the standard vessels of capacity, and these standard vessels could themselves be checked by the pitch of the notes they gave out when they were struck, as it was contrived that they should emit particular notes of the scale.[f] The 'pitch-pipes' or, as we termed them above, standard measuring tubes, were used as standard rulers for measuring distances, and the twelve pitches, the notes, were used for checking the weight of vessels, or, when vessels were identical in weight,

---

[a] The commentary explains this as 'flurried but unable to escape'.
[b] Commentators say that this word must be taken here in the rather unusual sense of 'shaking' (*tiao*[7]).
[c] Chêng Hsüan says that this means like the sound of musical stone-chimes.
[d] *Chou Li*, Khao Kung Chi, ch. 11, p. 25*b* (ch. 41); tr. Biot (1), vol. 2, p. 503. Cf. *Lü Lü Hsin Lun*, ch. 1, pp. 8*b*ff.
[e] Ch. 11, p. 20*b* (ch. 41); cf. Biot (1), vol. 2, p. 491. Cf. p. 180 above.
[f] Ch. 11, p. 26*a* (ch. 41); cf. Biot (1), vol. 2, p. 505.

[1] 鬱   [2] 甄   [3] 石   [4] 奧氏   [5] 鬴   [6] 鍾   [7] 掉

for checking the material of which they were made, since the composition of the alloy is one of the factors determining pitch.[a]

The information possessed by the Chinese in Chou times on bell-tuning, so far as we know it, may now be summarised. First, there was a wealth of empirical knowledge, not fully recorded, but orally handed down, the *Chou Li* author contenting himself with saying that on certain points special instructions are given.[b] Secondly, the Chinese appreciated the importance of accuracy in determining the diameter of the sound-bow and of other parts of the bell. Third, they seem not only to have listened very intently to the harmonics produced by bells, but to have classified the different sorts of timbre they produce, and to have attempted to attribute defects in timbre to faults in the form of the bell. Fourth, they paid great attention to the preparation, purification and weighing of the metal. We are not told, however, what was done to cure a flaw in tuning. We know indeed, from inspection of the whole field of bronzes, that few modern craftsmen could compare with the ancient Chinese for technical skill in bronze-founding, even the most intricate ornamentation being untouched on leaving the mould, and innocent of the file. So it may well be that cases in which bells required filing in order to tune them were so rare as not to deserve a mention. But elsewhere we are told that ringing-stones were tuned if necessary by filing down the sides or ends,[c] so it is reasonable to assume that filing would also have been resorted to for bells.

(8) Pitch-pipes, Millet-Grains and Metrology

While other early civilisations concerned themselves with linear measure, capacity, and weight in formulating their metrological systems, the Chinese were apparently unique in including pitch-measure (*lü*[1]), and that not merely on a par with, but as the basis of, the other three.[d] As the *Shih Chi* emphatically puts it:[e] 'The six *lü* are the root-stock of the myriad things (*liu lü wei wan shih kên pên yen*[2]).' The *Kuo Yü* describes this systematisation as follows:[f]

For this reason the ancient kings made as their standard the *chung*[3] vessel, (and decreed that) the 'size' of its pitch should not exceed that (produced by the string) of the *chün*[4] (seven-foot tuner), and that its weight should not exceed a stone (*tan*[5]) (120 catties). The measures of pitch, length, capacity and weight originate in this (standard vessel) (*lü tu liang hêng yü shih hu sêng*[6]).

[a] On this subject Chu Hsi's comment may be quoted: 'Quant à la régularité du poids, on considère par exemple que la matière des pierres sonores est ferme ou tendre, pure ou impure, et qu'il y a des sons légers ou graves, des sons hauts ou bas. En conséquence, on se sert encore des douze sons pour régulariser le poids' (tr. Biot (1), vol. 2, p. 58).

[b] One would not wish to exclude the possibility of the existence of written treatises on the technology of bell-making, either in the State of Chhi (whence the *Khao Kung Chi* probably derives) or in the Chhin and Han, but nothing whatever has survived and in any case oral tradition was certainly important.

[c] *Chou Li*, Khao Kung Chi, entry on the Chhing shih[7] (makers of stone-chimes), ch. 12, p. 5a, b (ch. 42); tr. Biot (1), vol. 2, p. 530.

[d] Something has already been said of Chinese metrology in Vol. 3, pp. 82ff.

[e] Ch. 25, p. 1a; cf. Chavannes (1), vol. 3, p. 293.

[f] *Chou Yü*, ch. 3, p. 22a; tr. de Harlez (5), eng. et mod. auct.

[1] 律   [2] 六律爲萬事根本焉   [3] 鐘   [4] 鈞   [5] 石
[6] 律度量衡於是乎生   [7] 磬氏

The standard measuring vessel *chung* is mentioned in the *Chou Li* and has been referred to already.[a] The word occurs in *Lieh Tzu* meaning a wine bowl, in the *Tso Chuan* meaning a grain measure, and is invariably used in the *Chou Li* and many other texts to mean a bell, for which today the term *chung*[1] would be a more normal appellation. The connection between grain-scoops and measures of capacity, bells and pitches is not hard to see. Primitive musicians all over the world used whatever was handy as their earliest instruments. In China the rice pestle-and-mortar existed in the classical orchestra till modern times as an instrument of percussion.[b] What more natural than that the primitive farmer when making music should use his grain-scoop or his bushel bowl and strike it for its rhythm, or if of metal, for its pitch? Standard measures and pitch were thus associated from primitive beginnings, and we have in the grain-scoop the first of all bells, which, as we noted above,[c] were in China originally clapperless. This origin also gives us a clue to the moral significance with which the *lü* were invested, for if the standard measures were not exact, cheating and corruption would follow, trade would be disrupted, disorders would break out, and all under heaven be thrown into confusion. The *Kuo Yü* writer develops his theme still further into the field of ethics and psychology, and in so doing takes up a position very like that of Plato in the *Republic* questing for justice in the State, where he lays it down that children should see and hear only that which is good.[d] The *Kuo Yü* says:[e]

> The ears and eyes are the pivots of the heart (because the heart is moved by what is seen and heard). That is why one should hear harmonious sounds only, and see nothing but what is correct and fitting. In this way hearing becomes clear and sight piercing, the meanings of words are comprehended, virtue shines forth, and men can be grave and firm, spreading this virtue among all the people.

The simple grain scoop having evolved into a bell, and the simple bell into a standardised measure or *chung*[2] of fixed dimensions, capacity and weight, as well as musical pitch, it was natural, when pipes became the standard pitch-givers, that they should inherit the measuring functions which had at one time belonged to bells.[f] Consequently we read of the number of grains of millet which the Huang-chung pipe ought properly to contain. It has sometimes been supposed that precise numbers of millet-grains governed the length and capacity of the Huang-chung tube and thus checked its pitch. Though this may have been so in and after the Han period, it is quite contrary to the earlier doctrine that the *lü* are the basis of all other measurements; for acting in their capacity as templates, they gave the lengths to the string-tuner, which in its turn gave the standard pitches. The standard measure, *chung*,[2] had to emit the Huang-chung note.

---

[a] Cf. above, p. 198.   [b] Cf. p. 149 and Fig. 302.
[c] P. 194.
[d] Cf. the detailed comparison of Phelps (1) on this.
[e] *Chou Yü*, ch. 3, p. 22*b*; tr. de Harlez (5), p. 85, eng. auct.
[f] Or perhaps the two kinds of instrument developed on parallel lines.

[1] 鐽   [2] 鍾

Nevertheless, the use of millet-grains in the reverse role of checking measuring instruments focused thought on the relation of length to diameter in pitch-pipes, and is highly relevant to a study of acoustic theory. The tables relating to millet-grains are given in the *Chhien Han Shu*, where measures of length, capacity and weight are treated in turn.[a] The smallest unit for each is given first, and thereafter four successive multiples. Thus if the *fên*[1] is the unit of length, ten *fên* make one inch; ten inches make one foot; ten feet make one *chang*;[2] ten *chang* make one *yin*.[3,b] Similarly with the other measures. The text says:

The basis (of the linear measure) is the length of the Huang-chung (pitch-pipe) (*pên chhi Huang-chung chih chhang*[4]). Using grains of medium(-sized) black millet the length of Huang-chung is ninety *fên*, (one *fên* being equal to) the width of a grain of millet... (*i tzu-ku chü shu chung chê, i shu chih kuang tu chih chiu shih fên Huang-chung chih chhang...*[5]).
Using grains of medium(-sized) black millet twelve hundred (grains) fill its tube....
The contents of one (Huang-chung) tube, i.e. twelve hundred (grains of) millet, weigh twelve *chu*[6] (half an ounce, *liang*[7]).

Whether in fact the width (or thickness) of a millet-grain or its diameter or length was originally intended to serve as a unit, proved a fruitful source for subsequent disagreement; so also the exact number of grains required to make a foot length.[c] But these scrupulosities concern us only in so far as they formed a justification for varying the standard lengths of pitch-pipes in order to introduce a tempered form of scale.[d] Further refinements are mentioned in the *Chhien Han Shu*, such as levelling off the top of the tube for testing capacity by filling it with pure well-water so that an exact measure of the interstitial spaces and hence of the total volume could be gained.[e]

The chief point of interest in the use of cereal grains is that it indicates an increasing awareness of the need for accuracy. The old measures based on the human body, such as the foot, or the inch measured from the pulse in the wrist to the base of the thumb, were obviously not sufficiently accurate for measurements designed to achieve exact pitch. Precise linear measurements became necessary once an absolute pitch was sought for Huang-chung.[f] The old formula for calculating the *lü* was adequate

[a] Ch. 21A, pp. 9b, 10a, 11a; tr. auct. An account of this interesting metrological system was published in a Western language as long ago as 1879 by Wagener & Ninagawa (1), but Sarton & Ware (1) renewed interest in the subject recently. Though Ware knew of no source earlier than the *Chhien Han Shu*, there is a parallel passage in *Huai Nan Tzu*, ch. 3, pp. 12bff. (tr. Chatley (1), p. 26) which reveals an older, partly duodecimal, system of grain-packing metrology, also associated with the 12 *lü*.

[b] Cf. Vol. 3, pp. 85ff. where the significance of this adherence to the decimal system is emphasised.

[c] There are two diagrams from the *Chiu Ku Khao*[8] (A Study of the Nine Grains) by Chhêng Yao-Thien[9] (3) (Chhêng Chêng-Shih[10]), in *Huang Chhing Ching Chieh*, ch. 551, p. 9a, b. We are indebted to Dr Lu Gwei-Djen for this reference.

[d] The evolution of the tempered scale will form the concluding theme of this Section.

[e] P. 10a. The connection between this ancient practice and the marked interest of later Chinese mathematicians in packing problems (cf. Vol. 3, pp. 142ff. above) should not be overlooked.

[f] It is worth noting that the Han work *Chiu Chang Suan Shu* (Nine Chapters of Mathematical Art) contains (e.g. ch. 6, p. 20b) problems on the volumes of bamboo sections, involving arithmetical progression, and means of attaining desired ratios between items. See Vol. 3, pp. 25ff.

[1] 分   [2] 丈   [3] 引   [4] 本起黃鐘之長
[5] 以子穀秬黍中者一黍之廣度之九十分黃鐘之長   [6] 銖   [7] 兩
[8] 九穀考   [9] 程畬田   [10] 程徵士

provided that a particular absolute pitch of the fundamental was not desired, or could be obtained by applying measuring-rods of known length to the string of the tuner. But once these were lost or in doubt, or alternatively once an absolute pitch for the fundamental became a necessity, some new means of assessing length, volume and weight became imperative. Though millet-grains might vary individually, when large numbers of one given species were used a fairly consistent average would be struck.[a] This method of ensuring against the loss of standard measures traditionally enshrined in wood or metal was probably as practical as any that could have been devised. But of course as the centuries passed, perhaps because of the urge for magically 'making all things new' at changes of dynasties, perhaps because of the long-continuing search for the equal-tempered scale, and doubtless for other reasons also, the standard numbers of grains varied somewhat from time to time.[b]

(9) THE RECOGNITION OF SOUND AS VIBRATION

The introduction of millet-grain counting as a measure of volumes initiated a phase in Chinese acoustic development which can properly be regarded as scientific.[c] It is interesting to compare the progress made in the Roman Empire in the same subject at a time contemporaneous with the Han dynasty in China. Vitruvius (c. −27) gives a great deal of acoustic information concerning the construction of theatres, the nature of the human voice, and the architectural arrangements needed to amplify it, such as the use of bronze vases set between the seats and tuned to different pitches, so that the different pitches of the human voice and its harmonics may be caught and amplified by resonance.[d] Of the nature of sound itself he says:[e] '(The voice) is moved in an infinite number of undulating circles similar to those generated in standing water if a stone is cast into it, when we see innumerable rings spread forth from the centre and travel as far out as they possibly can—extending indeed till they meet the confines of the limited space, or some obstruction which prevents the waves from reaching the outlets.' He speaks, too, of sound being somewhat of the nature of a blow on the taut membrane of a drum.

The Chinese also thought of sound in metaphorical terms deriving from observation of waves in liquid media at this period, though distinct statements of the analogy are rare before the +8th century. The following striking passage from the *Chhun Chhiu Fan Lu*[f] of Tung Chung-Shu (−2nd century) shows a conception of radiating wave-

---

[a] Large and small grains aberrant in size being rejected.

[b] In +589 Niu Hung[1] the jurist and three specialists (Hsin Yen-Chih,[2] Chêng I[3] and Ho Tho[4]) were commissioned to study the history of acoustic and other metrological standards; their results are given in *Sui Shu*, ch. 16, pp. 8b ff. Cf. Courant (2), p. 84.

[c] Modern comparative anatomy and anthropology provide an equivalent in the shot poured into the crania of different types of animal or human races in order to compare the size of brain cavities.

[d] *De Architectura*, v, v, 1 ff.

[e] v, iii, 6 ff. Cf. also Diogenes Laertius, VII, 158, and Plutarch, *Plac. Philos.* IV, xix, 4. On the Stoics see again p. 12 above.

[f] Ch. 81; tr. auct. adjuv. Bodde, in Fêng Yu-Lan (1), vol. 2, p. 57.

[1] 牛弘   [2] 辛彥之   [3] 鄭譯   [4] 何妥

fronts which he boldly applied to all media whatever their viscosity, including the aetheric *chhi* in which even psychological events participated.

> Man's (activity) brings about the growth of the ten thousand things below, and unites him with Heaven and Earth above. Thus it is that in accordance with his good government or disorderly, the *chhi* of movement or rest, of compliance or contrariness, act either to diminish or increase the transformations of the Yin and Yang, and to agitate everything within the Four Seas (*erh yao tang ssu hai chih nei* [1]). Even in the case of things difficult to understand, such as the spiritual (*shen* [2]), it cannot be said to be otherwise. Thus then, if (something) is thrown on to (hard) ground, it is (itself) broken or injured, and causes no movement in the latter; if thrown into soft mire, it causes movement within a limited distance (*hsiang tung erh chin* [3]); if thrown into water, it causes movement over a greater distance (*hsiang tung erh yü yuan* [4]). Thus we may see that the softer a thing is, the more readily does it undergo movement and agitation. The transforming *chhi* is much softer even than water, and (through it) the ruler of men ever acts upon all things without surcease. But the *chhi* of social confusion is constantly conflicting with the transforming (influences) of Heaven and Earth, with the result that there is now no (good) government.
>
> When the human world is well governed and the people are at peace, when the will (of the ruler) is equable and his character correct, then the transforming (influences) of Heaven and Earth operate in a perfect manner, and among the ten thousand things only the finest are produced. But when the human world is in disorder and the people become perverse, or when the will (of the ruler) is depraved and his character rebellious, then (their) *chhi* opposes the transforming (influences) of Heaven and Earth, harming the *chhi* (of Yin and Yang) and so generating calamities and disasters.[a]

This passage may be compared with another which we have already quoted from the astronomer Liu Chih,[b] written about +274. It will be remembered that he there compared the radiating light of the sun with the ripples sent out from the centre of a disturbance on a water surface. Vitruvius is just bracketed in time between Tung Chung-Shu and Liu Chih.

Two words which distinctly indicate a mental connection between waves in water and air are *chhing* [5] and *cho*.[6,c] Their ordinary meaning is 'clear' and 'muddy' respectively, but in acoustic contexts they are technically used. Chêng Hsüan says[d] that *chhing* (clear) means the notes of the gamut from Jui-pin to Ying-chung, i.e. the six upper notes, while *cho* (muddy) means the notes from Huang-chung to Chung-lü, i.e. the six lower notes. If a small stone is dropped into water it produces a sound of relatively high pitch, and sends out small ripples in close concentric circles; moreover, being small it does not much disturb the bed of the lake or stream so that the water

---

[a] Tung Chung-Shu certainly had in mind the theory of phenomenalism (see Vol. 2, pp. 378ff., Sect. 14f above), but he could easily have found telling examples in such matters as the neglect of water-works by bad rulers.

[b] P. 8 above. He was thinking of light and heat. The three Greek and Latin statements just mentioned (Vitruvius, −1st century; Plutarch, +1st; and Diogenes Laertius, Liu Chih's elder contemporary in the +3rd) explicitly refer to sound. Tung Chung-Shu, writing about −130, and therefore the most venerable of them, has all forms of radiating influence or energy in mind.

[c] Cf. p. 157 above.

[d] Commenting in the +2nd century on the *Yo Chi*, para. 6; in *Shih Chi*, ch. 24, pp. 24*b*, 25*a*.

[1] 而搖蕩四海之內   [2] 神   [3] 相動而近   [4] 相動而愈遠
[5] 清   [6] 濁

remains clear. If, on the other hand, a large stone is dropped into water, it produces a relatively loud deep sound, sending out large wide ripples over the surface, and it does disturb the river bed so that the water becomes muddy. Whether or not this theory of the origin of the terms be true, it certainly seems to fit contexts in which the words *chhing* and *cho* occur not entirely divorced from the other associations of sound such as timbre and volume.

The *Kuo Yü* refers[a] to pitch-range and the formation of sound by the human voice with considerable acumen for so early a period. The passage concerns the incident (−522) in which Ching,[1] the High King of Chou, desired to melt down a Wu-yi bell,[b] the second highest in the gamut, and thus falling in the *chhing* (clear) or upper pitch range, and to make from its metal a Ta-lin[b] bell generally known as Lin-chung,[b] the fifth highest of the gamut, also in the upper pitch range. But Wu-yi was a Yang (male) bell and therefore not loud but soft (*hsi*[2]), whereas Lin-chung was a Yin (female) bell and therefore loud (*ta*[3]). Even if the new bell were tuned correctly its volume would be false. Shan Mu Kung[4] remonstrated with the king, pointing out that debasing the pitches was as bad as debasing the currency. The people can only be trained to a precise appreciation of pitch intervals if the sounds which they hear are correctly tuned. That which enters through the ear and eye must be in accordance with correct measures or the heart will be corrupted:[c]

For the measures (*tu*[5]) which are discernible to the eye do not exceed the intervals of the *pu*[6] (6 ft.), the *wu*[7] (3 ft.), the foot, and the inch. The colours which are discernible to it do not exceed the intervals of the *mo*[8] (5 ft., lit. 'dark'), the *chang*[9] (10 ft.), the *hsün*[10] (20 ft.), and the *chhang*[11] (40 ft.). The harmonies which are discernible to the ear lie within the intervals of the pitch-range (*chhing-cho*[12]); the pitch-range which is discernible to it does not exceed the range of the human voice.

The sentence referring to colours is not very clear, for one would expect the four measures given as measures of distance to be in fact measures of saturation of colour or some such distinction. But it is a correct observation that outside the middle range of pitch our ability to judge intervals increasingly diminishes. The other observation, that debasing musical sounds is like debasing the currency, a form of injustice which Plato would have regarded as undermining the State, also deserves attention.

The exposition of Shan Mu Kung continues as follows:[d]

The ear takes in harmonious (*ho*[13]) sounds, and the mouth sends out excellent words... [The mouth takes in tastes, just as the ear takes in sounds (*khou nei wei erh erh nei shêng*[14]). Sounds and tastes generate *chhi* (*shêng wei sêng chhi*[15]).] When *chhi* is present in the mouth,

---

[a] *Chou Yü*, ch. 3, pp. 21 *b* ff.
[b] For the characters see p. 170 and Table 47 on p. 171.
[c] P. 22 *a*; tr. auct. adjuv. de Harlez (5), p. 64.
[d] *Kuo Yü*, Chou Yü, ch. 3, p. 23 *a*; tr. auct. adjuv. de Harlez (5), p. 66. The sentences enclosed in square brackets are absent from some editions.

[1] 景	[2] 細	[3] 大	[4] 單穆公	[5] 度	[6] 步	[7] 武
[8] 墨	[9] 丈	[10] 尋	[11] 常	[12] 清濁	[13] 龢	
[14] 口內味而耳內聲			[15] 聲味生氣			

it makes speech, and when in the eye, intelligent perception. Speech enables us to refer to things in accepted terms. Intelligent perception enables us to take action at the right times. Using terms (correctly), we thereby perfect our government. Carrying out actions at the right times, we thereby bring abundance to (all) living things. When government is perfect and living things have abundance, joy reaches its solstice.[a]

From this passage it is possible again to see the correlative tendencies of the Chinese mind at work, for sound and taste are linked with government not by mere fantasy but by a correlative sequence.[b] The concept of *chhi* must be accepted as the point of departure for the argument, not because it was adequate to reality, but because, like the concepts of Aristotelian form and matter, or Newtonian space and time, there were periods in which it served a useful purpose as a tool for thinking.

The *Kuo Yü* does not attempt to explain more clearly than this how exactly a sound is formed by the action of *chhi*. But a hint of how this process was imagined to take place may be gained from a sublime passage in the *Yo Chi*, where the nature of music is described.[c]

The *chhi* of Earth ascends above; the *chhi* of Heaven descends from the height[d] (*ti chhi shang chi; thien chhi hsia chiang*[1]). The Yang and Yin come into contact; Heaven and Earth shake together (*Yang Yin hsiang mo; thien ti hsiang tang*[2]). Their drumming is in the shock and rumble of thunder; their excited beating of wings is in wind and rain; their shifting round is in the four seasons; their warming is in the sun and moon. Thus the hundred species procreate and flourish. Thus it is that music is a bringing together of Heaven and Earth (*jo tzhu tsê yo chê thien ti chih ho yeh*[3]).

One does not have to look far into these words to see a reflection from early animistic times of the sort of belief which found expression in the story of Danae visited by Jupiter in a golden shower. Metaphors from the magical feather dances are strangely blended with words of awe, but also with tentative explanations, for example, 'The Yang and Yin rub together. Heaven and Earth shake together; their drumming is in the thunder....'

Thirteen or fourteen centuries later there is naturally a more sophisticated approach. In the Sung period we find this idea of rubbing taken up again and developed. Chang Tsai[4] writing (*c.* +1060) in his *Chêng Mêng*[5] on sound says:[e]

The formation of sound is due to the friction (lit. mutual grinding) (*hsiang ya*[6]) between (two) material things (*hsing*[7]), or (two) *chhi* (or between material things and *chhi*). The

---

[a] Note the elegant double-tracked sorites with its unified ending. On this and all other aspects of language in Chinese scientific and philosophical discourse, see Sect. 49.

[b] Cf. Vol. 2, pp. 261 ff. The idea of words being formed by some process connected with tasting might be dismissed as nonsense, had not Paget (2) published the results of his researches on the origin of speech, in which we find a connection between tasting gestures and the world-wide occurrence of such words as *sip*, *soup*, *gulp*, etc.; forms in which the lingual gesture is unmistakable.

[c] Para. 3, in *Shih Chi*, ch. 24, p. 14*a*, *b*; tr. auct. adjuv. Chavannes (1), vol. 3, p. 253.

[d] Cf. Sect. 21*d* (Vol. 3, pp. 467 ff.) on the meteorological water-cycle.

[e] *Sung Ssu Tzu Chhao Shih*,[8] Chang Tzu sect. ch. 7 (Tung Wu[9]), p. 12*b*; tr. auct. Cf. Bodde, in Fêng Yu-Lan (1), vol. 2, p. 487.

[1] 地氣上隮天氣下降   [2] 陽陰相摩天地相蕩   [3] 知此則樂者天地之和也
[4] 張載   [5] 正蒙   [6] 相軋   [7] 形   [8] 宋四子抄釋   [9] 動物

grinding between two *chhi* gives rise to noises such as echoes in a valley (*ku hsiang*[1]) or the sounds of thunder. The grinding between two material things gives sounds such as the striking of drumsticks on the drum. The grinding of a material thing on *chhi* gives sounds such as the swishing of feathered fans or flying arrows. The grinding of *chhi* on a material thing gives sounds such as the blowing of the reeds of a mouth-organ (*shêng*[2]). These are the inherent capacities in things for response (*wu kan chih liang nêng*[3]). People are so used to these phenomena that they never investigate them.

This extract[a] shows both the strength and the weakness of the traditional Chinese approach to such problems. One must admire the ability to classify and distinguish, but making distinctions is not the same as analysing a complex into its component factors.[b] Assuming the validity of the theory that sound is caused by 'friction' of things corporeal and incorporeal, its fourfold classification is admirable. Indeed, this type of thinking may still have a role to play in the classification of acoustic forms the complexity of which defies analysis as, for example, the grading of voices for the purpose of telephony, or the classification of musical timbres. But to say that the formation of sound is due to friction is obviously as much a defect of language as of thought. Perhaps it would not be too sweeping a generalisation to say that medieval science was as much handicapped by the failure of the Chinese language to make transitive and intransitive verb functions always explicit, as it was by the inability of some European languages to resolve verbs into specific physical operations.[c] To define friction as rubbing, and rubbing in terms of moving surfaces, or to define it as a force causing loss of motion would have seemed to Chang Tsai's correlative mind jejune, for friction would have seemed to him a perfect example of the way in which sound, colour and flavour combine. A knife placed against a rotating grindstone produces a harsh sound, a distinctive smell of charring, and yellow coloured sparks.[d]

Nevertheless, even before his day attempts had been made to reach a clearer understanding of the nature of sound. Than Chhiao[4] or some other Taoist writing in the *Hua Shu*[5] during the period of the Southern Thang (+938 to +975) makes the following statement:[e]

Chhi follows sound and sound follows *chhi*. When *chhi* is in motion sound comes forth, and when sound comes forth *chhi* is shaken. (*Chhi tshung shêng, shêng tshung chhi; chhi tung shêng fa, shêng fa chhi chen.*[6])

[a] Parallel passage in *Lü Lü Hsin Lun*, ch. 1, pp. 3aff. Chiang Yung goes on to advocate the experimental and quantitative study of all sources of sound.
[b] Cf. here particularly the argument in Vol. 3, pp. 156ff.
[c] Such questions must of course be left for Sect. 49.
[d] Other aspects of Chang Tsai's thinking in cosmology and astronomy have been discussed in Vol. 3, pp. 222ff.
[e] P. 12a, tr. auct. The place of this book in the history of optics (pp. 92, 116 above) will be recalled.

[1] 谷響　　[2] 笙　　　[3] 物感之瓦能　　　[4] 譚峭　　[5] 化書
[6] 氣從聲聲從氣氣動聲發聲發氣振

This is an important contribution. It is far more advanced than the Pythagorean conception of sound as a stuff composed of number which strikes the ear and becomes audible just as a swiftly travelling discus becomes visible at the moment of landing. It will be noticed that he is not speaking of air as such, for *chhi* is not merely air, though air under certain conditions may be described as *chhi*; when 'heat' dances above a flame, or smoking fumes arise from molten metal, and cooking-pots; the blast of furnace bellows,[a] forests shaken by the wind,[b] speech or music issuing from the human mouth.[c] It is the relation between sound and atmospheric agitation (*chhi*) which is important. The advent of sound transforms still air into *chhi* (air in a state of agitation), and air in a state of agitation produces sound. The use of the transitive verb *chen*[1] 'to shake' is particularly interesting, for it embodies so clearly the idea of vibration. This has occurred already in the description of the different sorts of sound produced by bells of different shapes: 'The sound (produced by bells with) thin (walls) is a staccato shaking.'[d] One can well understand that the Chinese should have realised that sound is produced by the 'shaking' of the air if in fact they gave such close attention to the timbre of bells. For large gongs and bells produce an inaudible pulsation of extremely low frequency but important in combination with the harmonics, which can be experienced when all normal sound has ceased as a faint pressure on the ear-drum. Indeed this is known in English as the 'shake'.[e]

Than Chhiao also anticipated Chang Tsai's theory of the grinding of *chhi* and material things, but by-passed the problem of how sound is formed by the use of a linguistic side-step, in this case 'riding' instead of 'grinding' or 'friction'. Still, his observations on the possible amplification of sound are interesting.[f]

The void (*hsü*[2]) is transformed into (magical) power (*shen*[3]). (Magical) power is transformed into *chhi*. *Chhi* is transformed into material things (*hsing*[4]). Material things and *chhi* ride on one another (*hsing chhi hsiang chhêng*[5]), and thus sound is formed. It is not the ear which listens to sound but sound which of itself makes its way into the ear. It is not the valley which of itself gives out echoing sound, but sound of itself fills up the entire valley.

So far he seems only to be pointing out that sound is not the product of hearing but exists independently of sensation. But his word 'sound' covers both the physical and the psychological. However, he was probably concerned less with the auditory neural stimuli than with the physical disturbance of the air and its impact on the ear-drum. He then continues:

An ear is a small hollow (*chhiao*[6]) and a valley is a large hollow. Mountains and marshes are a 'small valley' and Heaven and Earth are a 'large valley'. (Theoretically speaking, then)

---

a On the 'hard wind' of the cosmologists, see Vol. 3, pp. 222 ff.
b 'On Wenlock Edge the wood's in trouble...'; so also *Chuang Tzu*, ch. 2 (cf. Vol. 2, pp. 50, 51).
c On 'jets musically inclined' see G. B. Brown (1).
d Cf. above, p. 198.
e I am grateful to the late Mr C. K. Ogden for the opportunity of experiencing this phenomenon on his large Burmese gong (K.R.). Cf. p. 195.
f P. 7*b*, tr. auct.

1 振   2 虛   3 神   4 形   5 形氣相乘   6 竅

if one hollow gives out sound ten thousand hollows will all give out sound; if sound can be heard in one valley it should be heard in all the ten thousand valleys.

Here the writer's contention seems to be that if sound is produced by a disturbance of the *chhi*, all *chhi* everywhere will be in a state of disturbance, and therefore capable of being heard wherever a hollow or resonant chamber exists to receive it. With this there can be no disagreement, and the modern use of amplifiers to detect very small 'disturbances in the *chhi*' attests its truth.

Than Chhiao's next words are equally interesting:

> Sound leads (back again) to *chhi; chhi* leads (back again) to (magical) power (*shen*[1]); (magical) power leads (back again) to the void. (*Shêng tao chhi; chhi tao shen; shen tao hsü*.[2]) (But) the void has in it (the potentiality for) power. The power has in it (the potentiality for) *chhi. Chhi* has in it (the potentiality for) sound. (*Hsü han shen; shen han chhi; chhi han shêng*.[3]) One leads (back again) to the other, which has (a potentiality for) the former within itself. (If this reversion and production were to be prolonged) even the tiny noises of mosquitoes and flies would be able to reach everywhere.

Here the word *shen*[1] needs a little explanation. The translation '(magical) power' is rather inadequate. The character consists of a primitive graph apparently representing lightning,[a] modified by a symbol suggesting deity. Almost any single English word such as 'power', 'energy', etc. will carry misleading connotations. It is not difficult, however, to grasp the mechanism visualised by Than. Above the world is the empty void (*hsü*). Within this void there nevertheless exists a potentiality for energy,[b] for out of nothing it can produce power as in lightning. This lightning can produce *chhi* or atmospheric agitation, and atmospheric agitation can produce sound. So far the author has given within reasonable limits quite an accurate account of how the noise of thunder is generated. We would go a little further today, and explain that lightning produces a sudden heating of the air, which therefore expands violently and irregularly, causing pressure waves to travel through the air to the ears of the listener. But even as it stands his statement is rather remarkable for the + 10th century.

The first part is intended to explain how it is that sounds die away. Than Chhiao evidently thought not that the waves of pressure in the air were getting feebler and feebler, but that the sound was changing back again into *chhi*. Gradually the agitation of the *chhi* would diminish, and it would revert to power itself, which would in its turn subside again into the void. It is tempting to regard this statement as an anticipation of modern views on the nature of energy, just as it is tempting to see in Democritus one who anticipated the findings of modern atomic physics and chemistry. Of course, such anticipations must not be overstressed. But Than would have had little difficulty in understanding the way in which today the tiny sound of a mosquito or a fly may be 'led back again' into electric power 'having in it the potentiality for' amplification of sound by means of 'agitation of the *chhi*' through a loudspeaker.[c]

[a] K385. See also Vol. 2, pp. 225, 226.
[b] This Aristotelian phrase seems to us to render not badly the sense of the word *han*, which means literally 'to hold in the mouth', hence 'to cherish', with the undertone of being able to spit it forth or radiate it again.   [c] Cf. van Bergeijk *et al.* (1).

[1] 神   [2] 聲導氣氣導神神導虛   [3] 虛含神神含氣氣含聲

As already suggested, Than Chhiao was probably not concerned so much with the psychological aspect of hearing. Disturbances and waves of pressure in the air do not become 'sound' until so interpreted by the brain, stimulated by nerves from the inner ear receiving the external impulses. That the medieval Chinese did not overlook the psychology of the auditory sensation is shown by the words of another Taoist, Thien Thung-Hsiu[1] (c. +742) in the *Kuan Yin Tzu*.[2] He describes the hearing process as follows:[a]

It is like striking a drum with a drumstick. The shape of the drum is possessed in my person (in the form of the ear) (*ku chih hsing chê wo chih yu yeh*[3]). The sound of a drum is a matter of my responding to it (*ku chih shêng chê wo chih kan yeh*[4]).

To expand the analogy slightly, it seems that he believed that sounds strike the inner ear, in fact the ear-drum, just as drumsticks strike an actual drum; that is to say, they exert pressure. Nevertheless, it is the response (*kan*[5]) of a sentient being which enables one to describe this process as sound.

There is some reason for thinking that experiments with echoes were occasionally made. For example, the following story appears in the Ming book *Hsiang Yen Lu*[6] by Min Yuan-Ching[7] who attributes it to the +5th-century *Shui Ching Chu*.[b]

The city of Chiangling[8] lies on a slope inclining to the south-east, along which the dyke called Chin Thi[9] is built, starting from the Ling Chhi[10] pool. This dyke was built by Chhen Tsun[11] upon the orders of Huan Wên[12] (+347 to +373). Chhen Tsun was very skilful as a military architect. Once he sent someone to beat a drum (on the slope), whereupon, listening to the sound at a certain distance, he deduced the height of the slope. The dyke was built relying on such data, and there was no mistake in the calculations.

It would seem that this story preserves, in garbled form, some study of the speed of travel of echo-sounds, or perhaps the time-interval between the visible action initiating the sound and its arrival at the observer's ear.

(i) *The detection of vibrations*

The *Mo Tzu* book, in its discussion of fortification technology, written perhaps by Chhin Ku-Li[13] in the early part of the −4th century, mentions the use of hollow vessels as resonators in order to determine the presence and direction of tunnelling and mining by the enemy besieging a city. Forke (3, 17) has termed them 'geophones'. The text says:[c]

Should any unusual activity of the enemy be noticed, such as the building of walls or the piling up of earth, or perhaps streams becoming muddy which were not so before, then it is likely that he is sapping and mining. One must then at once make excavations within the

---

[a] Ch. 2, p. 5b, tr. auct.
[b] Cit. *KCKW*, ch. 2, p. 33a.   [c] Ch. 52, p. 9a; tr. Forke (17), eng. auct.

[1] 田同秀   [2] 關尹子   [3] 鼓之形者我之有也   [4] 鼓之聲者我之感也
[5] 感   [6] 湘烟錄   [7] 閔元京   [8] 江陵   [9] 金隄   [10] 靈溪
[11] 陳遵   [12] 桓溫   [13] 禽滑釐

wall and the moat in order to frustrate him. Within the city shafts are to be dug five paces distant from one another, to a depth of fifteen feet below the level of the base of the city wall, until one reaches three foot depth of water. Then large pottery jars are to be prepared each of a size sufficient to hold more than 40 *tou*;[a] their orifices are closed by a membrane of fresh skin, and they are sunk in the shafts. If men with good hearing are then set on watch to listen carefully, they will be able to hear clearly in which direction the enemy is digging. Then countermines are driven to meet them.

Since this would have been written about −370, our information on this Chinese practice[b] comes between dates for which there is evidence from Europe. From Herodotus[c] we learn that use was made of hollow shields as listening-posts at the siege of Barca by the Persians in the late −6th century; and Vitruvius[d] gives details of more refined procedures[e] employed by Trypho of Alexandria at the siege of Apollonia in Illyria in −214. It is hard to believe that this technique did not arise out of independent empirical observations made both in west and east. Its later developments included the use of bronze vases in theatres to improve the acoustics of the buildings, and of pottery vessels built into the walls of medieval structures.[f]

One of the most curious later applications of Chinese vibration detectors was the use of an instrument by the fishermen of northern Fukien for obtaining audible warning of the approach of shoals or 'runs' of fish.[g] At San-Tu-Ao bay, for example, great quantities of yellow fish are caught at those times of the year when they come up to brackish water to spawn. When the fishermen suspect that a run is about to occur, they take a section of bamboo about 2 in. in diameter and 5 ft. long, plunge it in the water to the depth of $3\frac{1}{2}$–4 ft., and apply their ears to the upper orifice beside the boat. Western observers present have heard sounds like a confused distant rumble when the shoal was said by the fishermen to be about a mile away, an estimate which was confirmed by the catch in due course. It is to be presumed that prolonged experience would confer considerable skill in making the estimates. Although no references to this practice have been found in the Chinese literature, there is no reason for doubting its traditional character, and it may be assumed to be an indigenous technique, anticipating in its primitive way the modern use of echo-sounding in the detection of shoals of fish.[h]

[a] I.e. more than 200 litres.
[b] Certain small metal pots with holes in their concave lids, dating from the Chhin or Han, have been regarded as geophones, used perhaps to pick up the ground-transmitted vibrations caused by distant galloping horses. I am much indebted to Mr Rewi Alley for sending me photographs of these objects from Peking.
[c] IV, 200.
[d] X, 16, 10.
[e] Including suspended bronze vessels.
[f] They may subsequently have taken on another function, namely the lightening of the load in vaults (Straub (1), p. 19).
[g] The information on which this paragraph is based was kindly provided by Mr Horatio Hawkins, formerly a Commissioner of the Chinese Customs. Similar techniques are known and used in Malaya; cf. Robinson (4).
[h] Cf. Hodgson (1); Hodgson & Richardson (1); Burd & Lee (1).

(ii) *The free reed*

Much has been said above[a] of the efforts of the *wu* shamans to canalise *chhi* through pipes. The metallurgists were, however, also interested in this (perhaps indeed they were sometimes the same people), and hence in due time the process was bound to be mechanised. We must reserve the bellows and the piston-bellows for our discussions of mechanics and metallurgy (Sects. 27*b* and 30*d* below); here we would only point out that there is a close connection between valve clacks in pumps, and reeds in musical instruments. The beating reed is exactly like the valve in that it can completely close the aperture, but the free reed is able to vibrate within the aperture. The 'mouth-organ' (*shêng*[1]) goes back far into the Chou, since it is mentioned in the *Shih Ching*,[b] and the generally accepted view is that the principle of the free reed came to the West from China.[c] The *shêng* is therefore the ancestor of the harmonica or reed-organ group of instruments (harmonium, concertina, accordion, etc.) and there is concrete evidence that it was transmitted through Russia in the 19th century.[d]

One of the chief uses of piston-bellows in Europe was for musical organs.[e] Although afterwards so closely associated with the Christian liturgy, they were an invention of the Alexandrians, and Vitruvius gives a minute description of them towards the end of the −1st century.[f] At that time, and before, piston-bellows of bronze were used.[g] No such application of these was made anciently in China, but from the Chou onwards, as we have seen, the Chinese had had the little instrument known as *shêng*[1] or *yü*,[2] made of bamboo pipes with free reeds (*huang*[3]) and played by suction. Moule & Galpin (1) have described how in the +13th century a reed organ was brought from the West and created so much interest that it was reconstructed to play the Chinese scale.[h] It was called the Hsing Lung Shêng,[4] and ten or twelve were made for the imperial orchestra during the Yuan dynasty. According to the *Yuan Shih*, it had 90 pipes and was blown by one man (presumably using piston-bellows) while another man played it. The first was presented by Muslim kingdoms between +1260 and +1264, and the adaptation to the Chinese scale was made by Chêng Hsiu[5] of the Bureau of Music. There were slider-valves and an air-reservoir of leather. Since the instrument was a reed organ and not a flute-pipe organ, this Arabic invention pre-

[a] Pp. 135 ff. above.

[b] Mao, no. 161, Karlgren (14), p. 104; Waley (1), p. 192. Cf. Eastlake (2).

[c] Helmholtz (1), pp. 95, 554; Moule (10), pp. 88 ff.; Goodrich (12). The question is bound up with the origin and distribution of the simple 'Jew's harp', on which see Li Hui (1, 1); Picken (2), pp. 185 ff.

[d] Scholes (1), pp. 787, 991; Schlesinger (2). By Fr G. J. Vogler (1749 to 1814), who saw and studied a *shêng* while at St Petersburg.

[e] Cf. E. W. Anderson (1).

[f] x, 8 (old editions, x, 13). Cf. Usher (1), pp. 89 ff., 2nd edn. pp. 136 ff.; Neuburger (1), p. 230.

[g] There is a good illustration in the Perrault edition of Vitruvius, p. 325.

[h] The information is derived from *Yuan Shih*, ch. 71, pp. 4 a ff., *Cho Kêng Lu*, ch. 5, pp. 2 a ff., and the collected writings of Wang I[6] (d. +1373) (*Wang Chung Wên Kung Chi*[7]), ch. 15, pp. 23 b ff.; all translated by Moule & Galpin (1).

¹ 笙　　² 竽　　³ 簧　　⁴ 興隆笙　　⁵ 鄭秀　　⁶ 王禕
⁷ 王忠文公集

ceded by two centuries the invention of the reed organ by Traxdorf in Nürnberg about +1460, while as reconstructed by the Chinese with free reeds ('apricot-leaves') it anticipated the European harmonium by no less than five and a half centuries. It seems to have had a swell mechanism, but in any case free reeds will allow sound variability by varying wind pressure without loss of pitch. One may say, therefore, that this application of air compression and conduction, though Hellenistic in origin, was improved by the Arabs and the Chinese during the late Middle Ages more quickly than in Europe where the lead was regained only after the Renaissance.

(10) THE EVOLUTION OF EQUAL TEMPERAMENT

This review of Chinese acoustic speculation in ancient and medieval times has shown how much was understood of the nature of sound produced by vibrating strings and columns of air. In the Chou period, as we have seen, tuning was done on a large stringed instrument, the *chün*. This fell into disuse towards the end of the dynasty, and the pitch which it had formerly given to the bells, and hence to all the other instruments of the orchestra, was henceforth given by pitch-pipes, the exact measurements of which consequently became a matter of great concern.

(i) *Octaves and spirals of fifths*

In predicting the sound which a pipe will produce when it is blown, it is not enough to know its length. The diameter is also important. Obviously a bamboo pipe a foot long and half an inch in diameter will not produce the same frequency note as one a foot long and two inches in diameter. Ignoring refinements, the pitch from two such pipes might be calculated as 537 and 501 vibrations per second approximately, a difference of more than a tempered semitone. The narrower the diameter the higher the pitch. If pitch-pipes had been in use in China from Chou times as guardians of standard pitch, one would expect to find not only details of the lengths of the different pipes in the *Chhien Han Shu*, but also details of their diameters. In fact there were different schools of thought on the matter.[a] One, of which Chêng Hsüan may be taken as representative, maintained that the diameters (or as he put it—the circumferences) of all the pitch-pipes should be the same.[b] 'The hollow of all pitch-pipes (should be) 9/10 of an inch in circumference (*fan lü khung wei chiu fên*[1]).' But Mêng Khang,[2] who lived slightly later than Chêng Hsüan (c. +220) declared that

Huang-chung should be 9 in. long and 9/10 of an inch in circumference; Lin-chung should be 6 in. long and 6/10 of an inch in circumference; Ta-tshou should be 8 in. long and 8/10 of an inch in circumference.[c]

[a] See *Li Chi*, ch. 6 (Yüeh Ling), the first month, the pitch-pipe of which is Ta-tshou, for commentators' descriptions of pitch-pipe dimensions.
[b] *Li Chi Chi Chieh*, ch. 13, p. 64.
[c] *Chhien Han Shu*, ch. 21A, p. 7a, comm. This tradition recurs in *Sui Shu*, ch. 16, p. 8b.

[1] 凡律空圍九分   [2] 孟康

As far as it goes this suggests a tapering off in diameters from Huang-chung, and gives diameters rather narrower than those used by Chu Tsai-Yü when evolving his equal-tempered system in the +16th century.[a] Tshai Yung specifies[b] dimensions for Huang-chung only, but gives length, circumference and diameter, from which we are able to see that he was using only a crude approximation for π. 'The Huang-chung pipe is nine inches long, 3/10 of an inch in diameter, and 9/10 of an inch in circumference.'

If approximations like this were used in acoustic calculations, it would be profitless to look for such refinements as appreciation of surface-tension, air temperature and humidity in calculating the pitch of pipes, though the use of jade as a material does suggest an attempt to overcome some of the discrepancies due to temperature and humidity variation. But the most important factor for those who wish to calculate the pitch of blown pipes is that 'end-effect' of which an account has already been given.[c] There is no doubt that this factor was appreciated in Han times, though of course the mathematics involved in its calculation were not attempted. Ching Fang[1] (*fl.* −45) specifically states[d] that pipes cannot be used (accurately) for tuning (*chu shêng pu kho i tu thiao*[2]). For this reason he made an instrument called a *chun*,[3] 10 ft. long like a *sê*[4] with thirteen strings, and by its help worked out the proportions for the notes of the system of tuning which he advocated.

Another reason why Ching Fang was right to do his experimenting with strings rather than pipes is provided by Helmholtz's discussion of sympathetic resonance. 'The principal mark of distinction', he says,[e] 'between strings and other bodies which vibrate sympathetically, is that different vibrating forms of strings give simple tones corresponding to the harmonic upper partial tones of the prime tone, whereas the secondary simple tones of membranes, bells, rods, etc., are inharmonic with the prime tone, and the masses of air in resonators have generally only very high upper partial tones, also chiefly inharmonic with the prime tone, and not capable of being much reinforced by the resonator.' Thus the use of strings in tuning experiments enabled the experimenter to get a mathematically perfect octave, the octave string being half the length of the fundamental, or stopped at half its length. A pipe half the length of another pipe, on the other hand, does not necessarily give its octave. The octave must be calculated taking the factors of end-effect and diameter into account.

From the figures given above[f] in which certain intervals of the Pythagorean scale were compared with the Chinese spiral of fifths, the discrepancy between a Chinese 'octave' so produced and the true octave is apparent, the ratio of a true octave being 1:2, and of an 'octave' in the spiral of fifths 262,144:531,441. The ratio, therefore, between a true octave and a Chinese 'octave' is as 524,288:531,441. This is known as the comma maxima, or more frequently today on account of the mistaken association of the name of Pythagoras with the spiral of fifths, the Pythagorean comma.[g]

---

[a] See below, pp. 220–4.
[b] In his commentary *Yüeh Ling Chang Chu*, in the *Li Chi Chi Chieh*, ch. 13, p. 64, tr. auct.
[c] See p. 186 above. [d] *Hou Han Shu*, ch. 11, p. 3a.
[e] (1), p. 45. [f] See Table 49, p. 175. [g] Cf. Grove (1), vol. 1, p. 688.

[1] 京房 [2] 竹聲不可以度調 [3] 準 [4] 瑟

In speaking of 'Chinese octaves' it would nevertheless be misleading to suggest that the true octave was unknown or not used. The true octave is in fact used wherever men and women, or men and boys, sing together, owing to the difference in the natural register of their voices. There is in addition textual evidence that scholars were aware that the octave is produced by halving the length of the resonating agent. Thus Tshai Yuan-Ting[1] (+1135 to +1198) commenting on the manner in which the pitch-pipes were associated with the months of the year, Huang-chung being the pitch for mid-winter and the others in succession through the year as described in the *Yüeh Ling*, says that the pitch for mid-summer is the octave, *shao-kung*,[2] of Huang-chung, and that the length of the Huang-chung pipe being 9 in., that of its octave is $4\frac{1}{2}$ in. Other scholars refused to accept this, not because of any misgivings on the score of end-effect, but simply because the true octave is not part of the 'cycle of fifths' system which had been looked on as orthodox from the time of the publication of the *Lü Shih Chhun Chhiu* if not earlier. But Tshai Yuan-Ting was concerned with problems of temperament, and was in fact one of the pioneers of the equal-tempered system. For this a perfect octave is essential, as will be made clear in the following pages.

(ii) *Western music and Chinese mathematics*

European music has made such remarkable advances during the last five centuries that it is easy for Westerners to forget or ignore the very existence of other musical systems no less rich and no less highly developed in other directions. For example, while Europe learnt to cross its melodies and develop harmony in pitch, Africa concentrated on crossing its rhythms and developing harmony in rhythm.[a]

There are two recent musical developments most characteristic of Europe. First, the high level of technical ability and practice attained in the manufacture of instruments, for example, the drawing of wire with a tensile strength of up to 200 lb., or the use of an iron frame for the pianoforte, the casting of which is described by Scholes[b] as 'one of the most delicate operations in foundry practice'. Secondly, with the development of harmony, the disappearance of the old modes, accompanied by the tendency to modulate ever more freely from key to key, till at last in the twentieth century we have 'atonalism' and the 'twelve-note music' of Schönberg, which attempt to discard the seven-note diatonic system. The following pages will attempt to show, or to suggest (since proof at this distance in time seems no longer likely), that our modern facility in modulating from key to key is yet another instance of Europe's debt to the civilisation of China.

As has already been explained,[c] when just intonation is used in the tuning of instruments, the 'same' note in two different scales will not necessarily have the same frequency. For example, F will have a frequency of 682·3 if it is the perfect fourth

a Cf. A. M. Jones (1), p. 78.
b (1), p. 715.   c P. 168 above.

1 蔡元定   2 少宮

from C, but 691 if it is the minor third from D. This meant that many instruments were capable of playing only in the one key to which they had been tuned by their designer, and if used in alien keys would sound out of tune, for some of the notes would then have the wrong frequencies. In practice, this is something of an over-statement, for an instrument designed for a particular key could normally be used for a small number of related keys as well, the pitch differences in their respective notes being too small to give offence to the ear.

In China, however, this was in theory a particularly serious problem, for the fixed-key instruments *par excellence* were the very expensive chimes of bells and ringing-stones. In theory no less than sixty bells or stones would have had to be cast or trimmed if tunes in all the twelve keys were to be played without the ritual pentatonic melodies offending the ear.

A similar problem arose when it was desired to play with consorts of instruments if those instruments had not all been designed for the same key. Stringed instruments could readily be adapted to instruments of any key, since executants had only to adjust their playing to the keys required, as violinists do today; or alternatively the strings could be retuned to another key during an interval. But for instruments which could not be retuned a different system was required, a compromise system in which a given set of notes would do duty for as many different keys as possible. This could be done by 'tempering' the tuning, that is to say, sharpening some notes and flattening others, so as to make them more generally serviceable.

Rough and ready ways of tempering must have been used by practical musicians from very early times. Certainly there are many references to the need for the careful placing of frets (originally loops of string tied round the neck of the viol) in the +15th- and +16th-century works of such writers as de Pareja and Bottrigari.[a] It is likely that in China, where a rigidly accurate cycle-of-fifths system of tuning would have involved musicians in the same sort of difficulties as just intonation in Europe, bells and ringing-stones were skilfully filed or chipped in such a way that their princely owner would not be involved in the unnecessary expense of superfluous instruments. Similarly with flutes and pipes, a slight displacement of the finger-hole could imperceptibly sharpen or flatten a note to extend its range of usefulness. But these tentative steps towards some more generally useful system of tuning are not in themselves equal temperament of the sort which received its greatest publicity from Bach's *Wohltemperirte Clavier*. When an instrument is tuned by equal temperament every semitone is equal to every other semitone. But the mathematics of this simple statement are relatively complicated.

The best short account of the evolution of equal temperament in Europe is probably that of Ellis (1), the translator of Helmholtz's *Sensations of Tone*, and inventor of the

---

[a] It is important to distinguish between the mathematically calculated system of equal temperament and the purely empirical methods of distributing the Pythagorean comma more or less equally over the twelve intervals, arrived at in Europe at this time, and in China five centuries earlier (cf. p. 219 below, on Wang Pho). For example, Jeans (2), p. 174, speaks as if Bartolomé Ramos de Pareja proposed equal temperament in his *De Musica Tractatus* of +1482. But historians of music (e.g. Eitner (1), vol. 8, under Ramis; Scholes (1) under Temperament 5, p. 924*b*; and Grove (1), vol. 4, p. 322) are agreed that it was a matter of adjusting the positions of frets according to purely practical and empirical rules.

'cents' system[a] of scale definition, whose contributions to musical knowledge have been used so effectively by others that their originator is now often forgotten.[b] He describes the four main systems of tuning: (a) Just Intonation, which derives from the astronomer Ptolemy (fl. + 156), a system in which 'all fifths and all thirds are perfect'; then (b) Pythagorean Temperament, the relation of which to perfect fifths and fourths has previously been described.[c] Mean-tone Temperament (c), was a system perfected by Salinas in + 1577, and based on perfect major thirds with other intervals so adapted that it was passably accurate for a total of about nine keys, but intolerable if one attempted to modulate into the others. This system was used for organs until quite recent times. Finally (d) Equal Temperament, in which 'every fifth without exception is one eleventh of a comma, or V (vibrations) 1 in 885 too flat, and every major third, without exception, is seven elevenths of a comma, or V 1 in 126 too sharp'.

As post-Renaissance music developed, great need existed in Europe for a system of tuning by which instruments of fixed key could transpose their music into as many keys as possible, preferably all, and by which even adaptable instruments, such as viols, could modulate from key to key without pausing to retune or readjust the frets. This was a revolution but it took place gradually. Equal-temperament tuning does not seem to have become general for the pianoforte in England until midway through the 19th century. Broadwood did not adopt it till as late as 1846.[d] Its very gradualness seems to have been one reason why its origin in Europe has been something of a mystery. Scholes rightly points out[e] that, although many people have a vague idea that Bach himself invented the system, there is no justification for this. It is more orthodox[f] to father the invention upon Andreas Werkmeister, who is said to have formulated the system of absolute equalisation between the semitones in + 1691. But this can scarcely be justified, for Mersenne mentions[g] it in + 1636 and gives the correct figures, adding elsewhere[h] that the system 'est le plus usité et le plus commode, et que tous les praticiens avouent que la division de l'octave en douze démitons leur est plus facile pour toucher les instruments'. Commenting on this passage Ellis says[i] that of the ease there is no doubt, but that of the customariness corroboration is required. In support of Mersenne, however, it is to be noted that Johann Caspar Kerll (+ 1627 to + 1693), whose age of creative activity began not long after the publication of Mersenne's work, wrote a duet on a ground bass, passing through every key.[j]

The situation in Europe, then, was that from the + 15th century onward, musicians were writing more and more in a style which made the use of an equal-tempered system inevitable, and that in the earlier writers of this period there are instructions

---

[a] A 'cent' is equal to 1/100 of the tempered European semitone.

[b] That there should be no entry for A. J. Ellis in Scholes (1), though much of its article on equal temperament is to be found verbatim in Ellis's paper here mentioned, appears to be a singular omission.

[c] Pp. 167ff., 172ff., 177, 181.

[d] Harding (1), p. 218.   [e] (1), p. 924.

[f] E.g. Closson (1), p. 56; Levis (1), p. 67.

[g] *Harmonie Universelle*, p. 132 (Bk. 2, prop. xi).

[h] *Harmonie Universelle*, Bk. 3, prop. xii, 'Des Genres de la Musique'.

[i] (1), p. 401.

[j] Scholes (1), p. 924.

on the tuning of lutes which suggest an approximation to equal temperament, but without specific reference or calculations. Ellis states categorically that 'in Europe neither Zarlino (+1562) nor Salinas (+1577) mentions equal temperament'.[a] But by +1636 we find Mersenne in possession of the actual figures, and stating that their use is a commonplace. At what moment did the actual mathematical formula appear, and who was the mathematician responsible? This question may be left for the present inside a bracket formed by the years +1577 and +1636, in order that attention may be given to the parallel development of acoustic and musical theory in China.

The need for a certain measure of compromise when different-keyed instruments are required to play in concert, and the nuisance of continual retuning or exchanging of instruments when music is played in many keys, both exerted a powerful influence on the development of Chinese musical practice. As an example of the former may be quoted the evidence of a European witness in the +16th century, Gaspar da Cruz, who left a description of life at Canton as he saw it in +1556.

> They played many instruments together sometimes, consorted in four voices which make a very good consonancy. It happened one night by moonshine, that I and certain Portugals were sitting on a bench at the riverside by the door of our lodging, when a few young men came along the river in a boat passing the time, playing on divers instruments; and we, being glad to hear the music, sent for them to come near where we were, and that we would invite them. They as gallant youths came near with the boat and began to tune their instruments, in such sort that we were glad to see them fit themselves that they might make no discord; and beginning to sound, they began not all together, but the one tarried for to enter with the other, making many divisions in the process of the music, some staying, others playing; and the most times they played all together in four parts. The parts were two small bandoraes (viols) for tenor, a great one for counter-tenor, a harpsichord that followed the rest, and sometimes a rebeck and sometimes a dulcimer for treble.[b]

The playing of many instruments together had been a characteristic of Chinese music from the earliest times, as can be seen from an ode in the *Shih Ching*, where bells and zithers, reed-organs and ringing-stones are described as sounding together.[c] The *Chou Li*, on the other hand, affords evidence of changes of key, for in describing the ritual of the three great sacrifices it lays down[d] that at the winter solstice there shall

[a] (1), p. 401.

[b] In *Tractado de las Cosas de China* (Evora, 1570), eng. *Purchas his Pilgrimes*, III, p. 81; mod. C. R. Boxer (1), p. 145. The instruments of this party would seem to have been three lutes of the *phi-pha*[1] type, one larger and deeper than the others, one *chêng*[2] following, with one *hu-chhin*[3] violin and a *chhin*[4] or *sê*[5] sometimes joining in. Such a party is described and illustrated in van Aalst (1), pp. 36, 64. The *hu-chhin* has not much entered into our argument so far. Although it is today an extremely popular instrument, closely associated with the classical opera, it came into China late (much later than the *phi-pha*), probably from Mongol culture. Commonly known as the Chinese violin, it has a small sound-box, a long neck with prominent pegs, and one or two double strings through which passes the hair-strip of the non-detachable bow. Further details will be found in Moule (10), pp. 121 ff.

[c] Ku Chung;[6] Mao, no. 208, tr. Legge (8), p. 280; Karlgren (14), p. 160; Waley (1), p. 140.

[d] Entry for the Grand Music-Master (Ta Ssu Yo[7]), ch. 6, pp. 4*b*, 5*a* (ch. 22); tr. Biot (1), vol. 2 p. 34.

[1] 琵琶   [2] 箏   [3] 胡琴   [4] 琴   [5] 瑟   [6] 鼓鐘
[7] 大司樂

be six changes of melody using three modes, at the summer solstice eight changes of melody with four modes, and at the sacrifice to the ancestors nine changes of melody with three modes.[a] These three ceremonials thus employed all the five pentatonic modes distributed over eight of the possible sixty 'mode-keys'.[b]

It is therefore not surprising to find evidence of very early attempts to temper the scale, such as that recorded in the *Huai Nan Tzu* book, where the lengths of the pitch-pipes with their complicated standard fractions have been simplified into round numbers.[c] This was clearly done with some reference to the ear, and not merely as a mathematical convenience, for whereas the correct length of the pitch-pipe Chia-chung expressed in hundredths of an inch was 674·23, the *Huai Nan Tzu* gives it as 680, though 670 would have been an approximation mathematically truer.[d] This temperament does not, however, differ fundamentally from the cycle-of-fifths tuning.

This first approximation recorded in the *Huai Nan Tzu* dates from the −2nd century. From then on for some seventeen centuries there was an almost continuous succession of experimentalists, not all of whom can be mentioned here.[e] Developments oscillated between two extremes, one of which retained the purity of the tuning by reducing the number of mode-keys, while the other sacrificed the purity of tuning in attempts to embrace as many modes as possible. The former tendency reached its logical fulfilment during the Sui period (+581 to +618), when apparently only one mode-key was used, that of the *kung* mode in the Huang-chung key, for ceremonial music. Seven bells were used for giving a heptatonic scale in this mode-key, the other five bells of the gamut being held in abeyance and called 'dumb bells' (*ya chung* [1]).[f] Attempts to avoid this impoverishment took different channels, for which close parallels may be found in the history of European musical development. Only two forms of solution are possible: either to increase the available choice of notes so that every key may be rendered in perfect intonation, regardless of the difficulties of the performer and the complexities of the instruments; or to sacrifice purity of sound deliberately for the sake of a manageable compromise.

The most famous exponent of the former solution was Ching Fang [2] (*fl.* −45),[g] whose system continued the never-ending spiral of fifths calculated on a 10 ft. wooden tuner with thirteen strings starting at Huang-chung and working five times

---

[a] The Chinese text refers, as is normal, to 'mode-keys', by the method described above, p. 169. If the modes are isolated it will be found that the winter sacrifice uses modes III, IV and V, while the summer sacrifice uses modes I, II, III and IV. The ancestral sacrifice uses only modes I, III and IV.

[b] But the entire group of 'mode-keys' deriving in tonality from the *shang* note was excluded, since *shang* was considered to be 'hard' and therefore not suitable for the ritual music which purposed to lure to man's aid the spirits of heaven and earth by its sweetness.

[c] Ch. 3, p. 13 a (Chatley (1), p. 27).

[d] P. 12 a (Chatley (1), p. 25).

[e] Cf. Courant (2); Robinson (1).

[f] *Chhü Wei Chiu Wên,*[3] by Chu Pien[4] of the Sung period, ch. 5, p. 9 a.

[g] We have met with him before (cf. Vol. 2, pp. 247, 329, 350; Vol. 3, pp. 227, 433, 470, 483) in connection with mutationism, astronomy, meteorology, and other things.

[1] 啞鐘   [2] 京房   [3] 曲洧舊聞   [4] 朱弁

round the cycle to the 60th note.[a] His work may be compared with that of Nicolas Mercator in Europe some seventeen centuries later, who arrived at a system of temperament having 53 degrees.[b] Ching Fang's microtonic experiments were taken up five centuries after his death by another naturalist, Chhien Lo-Chih[1] (*fl. c.* +450),[c] who continued the calculation of the spiral to the 360th degree.[d] Such a system would have been quite unworkable in practice.

Of the many other experimenters through the centuries much has been written, notably by Courant[e] who is an indispensable authority on this subject, and by Yang Yin-Liu (*1*) who has written a detailed account of the efforts of Chinese scholars to attain equal temperament. Chu Tsai-Yü himself, when writing of his own experiments,[f] focuses attention on four pioneers who had all made use of the stringed tuner described above[g] as the *chün*, or referred to it. Of these, the first—Lingchou Chiu[2]— only appears in the *Kuo Yü* as one of the interlocutors of the High King Ching of Chou (*c.* −520) in the discussion about the gamut of bells, the function of the 'tuner' and its relation to good government.[h] The second is Ching Fang (*d.* −37). The third is Chhen Chung-Ju[3] (*fl.* +516), who combined certain of Ching Fang's ideas with others of his own in a way which Chu Tsai-Yü says could not have been successful; and the fourth is the famous Taoist scientist and engineer Wang Pho[4] (*fl.* +959).[i]

Wang Pho worked out his system on a thirteen-stringed tuner,[j] but is also said to have tuned sets of bells to this temperament.[k] In common with the astronomer Ho Chhêng-Thien[5] (+370 to +447) he realised that it was hopeless to attempt to reach a workable solution by extending the spiral of fifths as Ching Fang had done, and that the perfect octave must be accepted as the framework within which subdivision is to take place. Ho Chhêng-Thien[1] simply measured the difference between a perfect octave and that interval, the sharpened octave, which is produced in its stead as the thirteenth note in the cycle of fifths, the difference being the so-called Pythagorean

---

[a] It is described at length in the *Hou Han Shu*, ch. 11, pp. 3*a*–16*a*; and briefly by Chu Tsai-Yü, *Lü Hsüeh Hsin Shuo*, ch. 1, p. 23*b*. Cf. Robinson (*1*), p. 101; Wu Nan-Hsün (*1*), pp. 132 ff.

[b] Courant (*2*), p. 89. Christopher Simpson (*d.* +1669) in his *Division Violinist* advocated quarter-tones (Scholes (*1*), p. 575), as Tshai Yuan-Ting[6] had done before him in the +12th century when he inserted a *pien lü*[7] between each semitone (*Sung Shih*, ch. 131, pp. 11*a*ff., esp. pp. 12*b*, 13*a*). But the Moravian Aloys Hába (*b.* 1893) seems to have been the first European to have elaborated a scale of sixty notes to the octave like Ching Fang. According to Scholes, he could actually sing them all accurately.

[c] Well known to us as an astronomer and constructor of astronomical instruments, cf. Vol. 3, pp. 346, 384, etc.

[d] *Sui Shu*, ch. 16, pp. 4*b*ff. The names of the notes may be found in Shen Chung's[8] *Yo Lü I*[9] of about +570; in *YHSF*, ch. 31, pp. 31*a*ff.

[e] (*2*), pp. 88 ff.   [f] *Lü Hsüeh Hsin Shuo*, ch. 1, p. 22*a*.

[g] P. 185.   [h] See above, pp. 170, 204.

[i] Biography in *Chiu Wu Tai Shih*, ch. 128, pp. 1*a*ff.

[j] See *Chiu Wu Tai Shih*, ch. 145, pp. 3*a*ff. His form of temperament was somewhat analogous to that of Ramos de Pareja five centuries later in that it was worked out for a stringed instrument with movable bridges.

[k] In the *Chi Ku Lu Pa Wei*,[10] ch. 1, by Ouyang Hsiu (+1007 to +1072); quoted by Wei Chü-Hsien (*1*), p. 68.

[1] +370 to +447. Cf. Vol. 3, pp. 287, 292, 384, 392, etc.; *Sui Shu*, ch. 16, pp. 4*b*ff.

---

[1] 錢樂之   [2] 伶州鳩   [3] 陳仲儒   [4] 王朴   [5] 何承天

[6] 蔡元定   [7] 變律   [8] 沈重   [9] 樂律義   [10] 集古錄跋尾

comma. This difference he divided by twelve and distributed equally over all the thirteen notes except the fundamental. In this way he obtained a gamut which had the characteristics of a cycle-of-fifths scale and a true octave. It was not equal-tempered, however, for the original irregularities of the cycle-of-fifths tuning remained, and were in no way removed by the addition to each note of 1/12 of a comma.[a]

Wang Pho not only used the perfect octave as the basis of his calculations, but to a considerable extent broke away from the values of the cycle of fifths. His octave, fifth and major tone had the same values as for just intonation; but equal temperament would have required all his intervals to be sharpened except the octave. Ho Chhêng-Thien had taken a great step forward in establishing that the octave must be accepted as the framework for an equal-tempered system; Wang Pho departed further still from the orthodox tuning of the cycle of fifths. But by what calculation the twelve notes of the gamut could be so spaced that every semitone would be equal was still an unsolved mystery.

(iii) *The princely gift of Chu Tsai-Yü*

In +1536 was born one of China's most distinguished mathematical and musico-logical scholars. He was the son of Chu Hou-Huan[1] and a descendant of the fourth Ming emperor Chao. When[b] his father was unjustly reduced in rank by the emperor, he signified his filial grief by living in an earth-walled cottage for nineteen years. This time was spent in research into mathematical, musical and calendrical matters, the results of which were published at intervals and finally as a collected work.[c] His gift to mankind was the discovery of the mathematical means of tempering the scale in equal intervals, a system of such fundamental utility that people in all Western countries today take it for granted and are unaware of its existence.[d]

In the *Lü Hsüeh Hsin Shuo*[2] (A New Account of the Science of the Pitch-pipes), published in +1584, Chu Tsai-Yü[3] describes previous attempts at tempering the scale and shows their shortcomings before discussing his own 'new method', in which he 'used numbers for seeking harmony in the notes, and did not make the notes

---

[a] The work of Ho Chhêng-Thien was continued further by Hsiao Yen,[4] who ruled as Liang Wu Ti from +502 to +549. His book entitled *Chung Lü Wei*[5] (Apocryphal Treatise on Bells and Pipes) is still extant in fragmentary form. An excellent account of his interesting work is given by Wu Nan-Hsün (1), pp. 159 ff.

[b] His title in this capacity was Chêng Shih Tzu;[6] he is often so named in the dynastic history and similar official texts.

[c] His works, bound in four large volumes, are generally known by the name *Yo Lü Chhüan Shu*[7] (Collected Works on Music and the Pitch-pipes); but this title really refers to the earlier two of his four works on pitch-pipe theory, in addition to which there is one book on a perpetual calendar and another on the orchestration of ancient songs (cf. Robinson, 1). Details of Chu Tsai-Yü's life are to be found in the *Ming Shih*, ch. 119, pp. 1 aff. esp. 3 b, 4 a.

[d] The story of his discovery and its probable transmission to the West has been fairly well known in China (cf. Liu Fu (3), Chang Chhi-Yün (1, preface), Wu Nan-Hsün (1), pp. 190 ff. etc.) but has not before been recounted in a Western-language publication. For a much fuller survey, however, see Robinson (1).

[1] 朱厚烷   [2] 律學新說   朱載堉   [4] 蕭衍   [5] 鐘律緯
[6] 鄭世子   [7] 樂律全書

submit to (natural series of) numbers'.[a] That is to say, he found a true mathematician's solution. But fearing that mathematics alone might not guarantee the success of his system with posterity, he also made a deliberate study of ancient tuning instruments and then constructed one of his own (Fig. 320). In this the proportions of the scale were made manifest by studs placed at the appropriate tempered intervals to indicate the correct positions for the placing of the fingers after the manner of frets.

It was suggested at an earlier stage in this Section that the mathematical formula by which the lengths of the pitch-pipes were calculated was a foreign importation grafted on to the indigenous system. If the indigenous system had survived one would expect to find it as a still living tradition among the craftsmen and practising musicians, who were not concerned with the theories of scholars, naturalists and court ritualists. It was precisely a contradiction between these two different systems which focused Chu Tsai-Yü's attention on that aspect of the problem which gave him the solution.

After describing the musical theories of the great Sung philosopher Chu Hsi, who had advocated the orthodox pitch-pipe dimensions, Chu Tsai-Yü writes as follows:[b]

I had made an attempt with the theory of the Sung (scholar) Chu Hsi, based on the ancient up-and-down principle, and using this tried to get the positions for the standard pitches on the zither (*i chhiu chhin chih lu wei*[1]). But I noticed that the (normal) notes of the zither were not in consonance with (those produced from) the positions of the standard pitches, and suspicions therefore arose in my mind.

Night and day I searched for a solution and studied exhaustively this pattern-principle. Suddenly early one morning I reached a perfect understanding of it and realised for the first time that the four ancient sorts of standard pitches all gave mere approximations to the notes. This moreover was something which pitch-pipe exponents had not been conscious of for a period of two thousand years.

Only the makers of the zither (*chhin*[2]) in their method of placing the markers at three-quarters or two-thirds (etc. of the length of the strings) had as common artisans transmitted by word of mouth (the way of making the instrument) from an unknown source. I think that probably the men of old handed down the system in this way, only it is not recorded in literary works.

From this statement one might infer that Chu Tsai-Yü had recovered the secret of equal temperament from the remotest antiquity, but in fact he does not say so.[c] His elation was due to the fact that as a conscientious antiquarian he had discovered in this living tradition a moral justification for defying the cycle of fifths which had been

[a] Note this very conscious repudiation of the numerological games and number-mysticism which had become so hallowed in China by centuries of transmission from the classics. Cf. pp. 134ff. above, and our comments in Vol. 2, pp. 287ff. Though so far from Europe, Chu Tsai-Yü was 'a man of the Renaissance', contrasting as much with Chhen Thuan (cf. Vol. 2, pp. 442ff.) or Shao Yung (pp. 455ff.) as a Joseph Glanville with an Agrippa of Nettesheim.
[b] *Lü Hsüeh Hsin Shuo*, ch. 1, p. 5a, tr. auct.
[c] He was careful to give his innovations an appearance of respectability (*Lü Hsüeh Hsin Shuo*, ch. 2, pp. 7aff.) by taking as his unit of measurement an inch of fictitious antiquity which he described as the Hsia inch, the inch of the most ancient Chinese dynasty. Of course, no scholar of his time would have been misled by this. What he says about the handing down of oral tradition among artisans is interesting (cf. Sect. 29 below in connection with shipbuilding).

[1] 以求琴之律位   [2] 琴

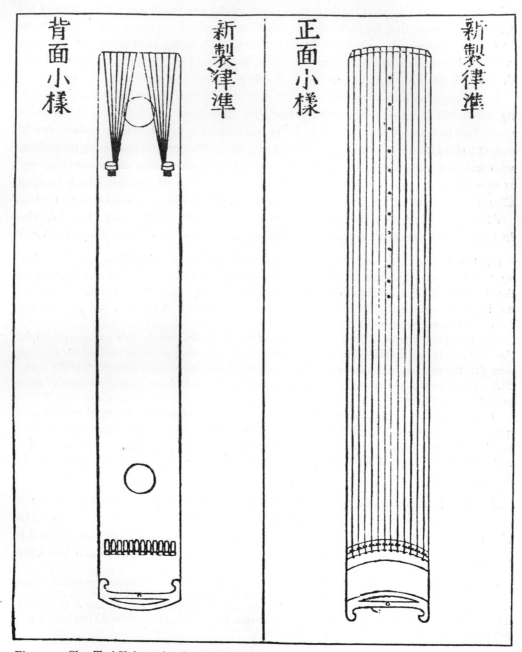

Fig. 320. Chu Tsai-Yü's tuning instrument (*chun*). From his *Lü Hsüeh Hsin Shuo*, ch. 1, p. 28*a* (+1584). The legends at the top read 'Sketch of the New *Chun*'; on the right the front, on the left the back of the instrument.

hallowed by two thousand years of history. Possibly the answer to his mathematical problem also occurred to him at the same instant, but this, which is for us the most interesting part of the story, he dismisses in a few words:[a]

> I have founded a new system. I establish one foot as the number from which the others are to be extracted, and using (square and cube root) proportions I extract them. Altogether one has to find the exact figures for the pitch-pipes in twelve operations.[b]

Applying this new principle he gives tables showing the lengths of the standard pitch-pipes and also of half-length and double-length pipes, giving in all a compass of three octaves. The principle which he had discovered was that to divide the octave into twelve equal semitones, the length of the fundamental string (or pipe if one ignores end-effect), and thereafter each successive length obtained, must be divided by the 12th root of 2. This is a very different matter from merely dividing the string into twelve equal parts, i.e. the whole string, $11/12$, $10/12$, $9/12$, etc., for these proportions produce a very unequal temperament. If all intervals are to be equal, the ratio for each semitone must be altered by an equal amount. The ratio of the octave is $1:2$. This can be expressed in the form $1:2^{12/12}$, since two to the power of twelve twelfths is two. To alter the ratio of each semitone by an equal amount, it was merely necessary to express each proportion thereafter as $1:2^{11/12}$, $1:2^{10/12}$, $1:2^{9/12}$, etc., and to know the exact length of each string, it was simply a matter of dividing the 1 ft. Huang-chung string by $\sqrt[12]{2}$, which is the same as $2^{1/12}$, and then dividing the length of each successive string so obtained by $\sqrt[12]{2}$. Thus a perfectly tempered scale was achieved.

This is the simplest method of calculating the lengths of the strings, and may have been the one used by Chu Tsai-Yü in the first place. Sung algebraists such as Chu Shih-Chieh[1] (+13th century) could certainly handle roots of high powers, but their books had at that time disappeared, and it is doubtful whether Chu Tsai-Yü could have been acquainted with their methods.[c] The method he shows in his published calculations achieves the same results by the use only of square and cube roots. Briefly it was as follows. Of the thirteen strings the lengths of which required calculation, two were already known, no. 1 being 1 ft., and no. 13 half a foot. He then found the square root of the product of these two lengths, which gave him the length of the middle string no. 7. The square root of the products of nos. 1 and 7 and nos. 13 and 7 then gave him the lengths of the two strings intermediate between these three, namely nos. 4 and 10 respectively. The lengths so far discovered may be expressed thus: 1 ** 4 ** 7 ** 10 ** 13.

In order to find the lengths of the remaining strings (nos. 11 and 12 may be taken as examples), he evidently had it clearly in his mind that the length of no. 10 was

---

[a] *Lü Hsüeh Hsin Shuo*, ch. 1, p. 5b, tr. auct.
[b] *Chhuang li hsin fa. Chih i chhih wei shih i mi li chhu chih. Fan shih-erh pien so chhiu lü lü chen shu.*
[c] Cf. Vol. 3, pp. 126ff. above.

[1] 朱世傑   [2] 創立新法置一尺爲實以密率除之凡十二遍所求律呂眞數

1 ft. $\times 2^{3/12}$; of no. 11, 1 ft. $\times 2^{2/12}$; and of no. 12, 1 ft. $\times 2^{1/12}$; and that as $2^{1/12}$ is the cube root of $2^{3/12}$, to find the length of string no. 12 he had merely to find the cube root of no. 10, and multiply one foot by it. Similarly 1 ft. multiplied by the square root of the length of no. 10 gave the length of no. 11, and like results may be got for the other unknown lengths by a similar process.

This rather complicated procedure was perhaps used as a check on the simpler method first mentioned, namely dividing the fundamental string by $^{12}\sqrt{2}$, which is 1·05946, and repeating the process for each string or pipe as described above. The figure 1·05946 was of course obtained by Chu Tsai-Yü for the note immediately below his standard Huang-chung length. Double-length Huang-chung being 2 ft., the 12th note above it, Ying-chung, was 1·05946 ft. long. Chu says quite explicitly[a] that each string in turn must be 'divided by the figure for double-length Ying-chung... which is a way of getting the pitches in their serial order'.[b]

This statement leaves no doubt that Chu Tsai-Yü invented the formula for equal temperament just as it was known in Europe at a later date. It is particularly important to note how little had to be memorised by any traveller in touch with Chinese ideas for him to be able to transmit the idea to the mathematicians and musicians of Europe. Such a traveller would only have to say: 'I understand that the Chinese temper their viols with great accuracy. They simply divide the length of their first string by $^{12}\sqrt{2}$ to get the length of the string for the second note, and then do the same again for the third note, and so on, till they reach the 13th which is a perfect octave.' Not a book but a sentence only was required for the diffusion of this great idea.

Although this temperament was worked out for strings Chu Tsai-Yü also applied it to pipes.[c] The pipes were the same lengths as the strings and if left uncorrected the equalness of their temperament would have been distorted by end-effect, but Chu Tsai-Yü compensated for this by also tempering their diameters, dividing each successive diameter by the 24th root of 2. Considering how little he can have known of the physics of end-effect his success is quite remarkable, remaining distortions being imperceptible to the human ear.

(iv) *Equal temperament in East and West*

Chu Tsai-Yü's formulation of equal temperament may justly be regarded as the crowning achievement of China's two millennia of acoustic experiment and research. One great question yet remains. Was it discovered independently or could it have been transmitted from China to Europe at the end of the +16th century? Certainly a time was coming when Europe could profit from such an invention, and there were many mathematicians in Europe capable of calculating just as Chu Tsai-Yü had done. Pacioli (+1494), for example, and Cattani (+1546), were able to handle roots of

[a] *Lü Hsüeh Hsin Shuo*, ch. 1, pp. 10 a ff. Cf. Robinson (1), p. 156.
[b] *Chieh i Ying-chung pei shu...wei fa chhu chih chi tê chhi tzhu lü yeh.*[1]
[c] Tables and illustrations showing lengths and internal and external diameters of his pitch-pipes are given in the same book, pp. 15 b ff. *Lü Lü Ching I*, ch. 8, pp. 4 b ff., 6 a ff.; *Lü Shu*, ch. 1, pp. 31 ff.

[1] 皆 以 應 鐘 倍 數 … 爲 法 除 之 卽 得 其 次 律 也

high powers.[a] Nevertheless, it would be a remarkable coincidence if the same solution was found independently at opposite ends of the earth within a few years. And it is striking that so little can be ascertained about its European origin when everything is known about its invention in China.

The opening pages of the *Lü Hsüeh Hsin Shuo* contain the date +1584.[b] There is no doubt that by this time Chinese books were beginning to make their way into Europe in some quantity. From Paolo Giovio's reference in his *Historia Sui Temporis* (+1550) to the gift of a Chinese book by the King of Portugal to the Pope, down to Amiot's detailed *Mémoire* on Chinese music (+1776), there were innumerable literary contacts of Europe with China. But it is worth noting that both the friars and the Jesuit missionaries were almost invariably interested in Chinese music; not only because they lived in a musically educated age, but because one of the problems of their work was either how to adapt European music for the use of Chinese congregations, or how to teach European music to Christians brought up in a different musical tradition. In +1294, when John of Monte Corvino was singing masses in Cambaluc,[c] European music was not so different from East Asian music that it could not be appreciated there. But during the following centuries it underwent so marked and rapid a development on the road to harmony that by the time Amiot, some five hundred years later, came to play contemporary European music to gentlemen of the Chhing court, they found it so meaningless that they were scarcely affected by it.

Gaspar da Cruz in +1556 stood midway between these two points, and was able to give an accurate description of the sort of music he heard.[d] Twenty years later (+1575) the Austin friars Martín de Rada (Herrada) and Jerónimo Marín, who spent three months in China (Fukien), brought away many books some of which they afterwards caused to be translated in the Philippines. These included some 'Of musicke and songs, and who were the inventors thereof'.[e] But already ten years before (+1565)

[a] Cf. D. E. Smith (1), vol. 2, pp. 471 ff., and above, Vol. 3, p. 128.

[b] This, the 12th year of the Wan-Li reign-period, was the very year which saw the first establishment of Matteo Ricci at Chao-chhing.

[c] Cf. Vol. 1, pp. 169, 230.   [d] Cf. p. 217 above. On him see Boxer (1), pp. lviii ff.

[e] See Gonzales de Mendoza (1), Parke tr., pp. 103 ff., 134, 140 and 250; Boxer (1), pp. lxxxiv ff., 243 ff.

As we have not mentioned this remarkably well-attested transmission elsewhere, the following other topics may also be noted:

'Of the mathematicall sciences, and of arithmeticke, and rules how to use the same.'

'Of the number, and moovings of the Heavens; of the planets and stars, and of their operations and particular influences.'

'Of the properties of stones and mettals, and of things natural that have vertue of themselves....'

'For the making of ships of all sorts, and the order of navigation, with the altitudes of every port, and the qualitie of every one in particular.'

'Of architecture and all manner of buildings, with the bredth and length that everie edifice ought to have for his proportion.'

'Manie herbals, or bookes of herbes, for phisitions, shewing how they should be applied to heale infirmities.'

Boxer adds several interesting speculations about the fate of the books collected by Martín de Rada; some of them may still exist in European libraries. There are, for example, about half a dozen Chinese works in early +16th-century editions in the Library of the Escorial. Apprised of their existence by Professor Donald Lach, Dr Lu Gwei-Djen and I had the pleasure of examining them in September 1960. Since this house is Augustinian it seems quite likely that some of these books were brought back by de Rada—but today nothing of scientific interest remains except a small medical treatise and a calendar, and no book on music or acoustics.

the Jesuits had opened their house in Macao for training missionaries,[a] teaching them to read Chinese books, and from this college there soon came a flow of letters informing the Western world of the nature of Chinese civilisation. In +1582 the great Matteo Ricci (Li Ma-Tou[1]) began his Chinese studies in Macao,[b] and the Franciscan friars Jerónimo de Burgos and Martín Ignacio de Loyola landed at Canton.[c] Ignacio was one of the informants of Juan Gonzales de Mendoza whose book[d] was first published in Spanish in +1585. In +1588 Cavendish returned to England from his first voyage round the world. As was customary in those days he had his private musicians on board, and it is interesting to note that when the 'Great St Anna' was captured, one of the prisoners was a certain Nicholas Roderigo 'a Portugall, who hath not onely beene in Canton and other parts of China, but also in the islands of Iapon. . .'.[e] The following year the Jesuit Edouart de Sande (Mêng San-Tê[2]), writing of his travels in China, described how the officials discovered in his possession several books in the Chinese language, 'de quoy ils monstrérent estre bien aisés'.[f] The last two decades of the century were the golden age of Macao when relations with the Chinese were becoming stabilised, and interchange of ideas relatively easy. For a period beginning in +1580 the Viceroy of Kuangtung opened bi-annual 'fairs' at Canton which lasted for several weeks, during which there was an opportunity for the interchange of Chinese and Western goods as well as ideas.[g] In +1595 Ricci was in Nanking discussing amongst other things mathematics with Chinese scholars, and in +1601 he succeeded in making his home in Peking. From that time forward knowledge of Chinese civilisation spread in Europe with great rapidity.

Thus by the beginning of the 17th century Europeans were interested in Chinese music, and had some access to Chinese books. It cannot be proved that a copy of Chu Tsai-Yü's *Lü Hsüeh Hsin Shuo* or of his *Lü Shu* made its way to Europe and was there acted upon, but it is reasonable to say that there was ample opportunity between +1585 and +1635 for this to happen. The matter is worth pressing a little further. Between +1597 and the year of his death in +1610 Matteo Ricci became increasingly conscious of the part which his mission could play in the rectifying of the Chinese calendar. He would naturally have studied Chinese books on the subject, as did his successor in the task, Sabbathin de Ursis (Hsiung San-Pa[3]).[h] Among these would have been Chu Tsai-Yü's *Shêng Shou Wan Nien Li*, which even Wieger could do no less than describe as 'a complete treatise on the calculation of time with a perpetual calendar, a masterpiece which...magisterially sums up all the previous works on the

[a] Pfister (1), pp. 5, 10.   [b] Pfister (1), p. 23.
[c] See Boxer (1), p. lxxxix; and in full detail Pelliot (45).
[d] *Historia de las Cosas mas notables, Ritos e Costumbres del Gran Reyno de la China....*
[e] Hakluyt, *Voyages*, vol. 3, p. 817.
[f] Letter from Father Edouart de Sande in Macao, 28 September 1589, to R.P. General (of the Society of Jesus), in *Sommaire des Lettres du Japon et de la Chine de l'an MDLXXXIX et MDXC*, p. 127. Cf. Pfister (1), p. 44.
[g] Cf. A. Kammerer (1).
[h] On the work of de Ursis see e.g. Bernard-Maître (1), p. 76; Pfister (1), p. 104. The early Jesuits must certainly have known of Chu Tsai-Yü's writings.

[1] 利瑪竇   [2] 孟三德   [3] 熊三拔

subject'.[a] Now pitch-pipe lore is so intermingled with calendrical science in Chu Tsai-Yü's writings, and the two were in fact so closely connected in Chinese thinking, that it would hardly have been possible to study the calendrical ideas without becoming acquainted with the pitch-pipe theories. Europeans as intelligent as Ricci and de Ursis discussing books with educated Chinese at the close of the +16th century could scarcely have avoided hearing of Chu's books so recently published.

An independent invention of equal temperament in Europe must therefore raise grave doubt. This doubt is increased when one finds that some sixteen years before Mersenne, the great Flemish mathematician and engineer Simon Stevin (+1548 to +1620)[b] left figures for the calculation of the scale in equal temperament among his unpublished papers.[c] Many of Stevin's papers had been circulated among his friends and were never returned. His son Hendrik gathered as many as possible intending to publish them, but only two volumes left the press.[d] The vital paper on equal temperament was not rescued from oblivion till it was found and published by Bierens de Haan (1) in 1884.[e] From this it can be seen that Stevin calculated 12 equal degrees within the octave represented by the figures 1 and $\frac{1}{2}$. The method is interesting, for just as Chu Tsai-Yü first computed the length of his middle string Jui-pin, which was the square root of the product of the two octave lengths, i.e. $^2\sqrt{2} \times 1$ or $^2\sqrt{1 \times \frac{1}{2}}$, so Stevin took $^2\sqrt{\frac{1}{2}}$ as the ratio of the middle note of his octave, and expressed the other ratios in comparable fashion as follows:

$$1 : \begin{cases} 1 \\ \sqrt{(12)}\frac{1}{2} \\ \sqrt{(6)}\frac{1}{2} \\ \sqrt{(4)}\frac{1}{2} \\ \sqrt{(3)}\frac{1}{2} \\ \sqrt{(12)}\frac{1}{32} \\ \sqrt{\frac{1}{2}} \\ \sqrt{(12)}\frac{1}{128} \\ \sqrt{(3)}\frac{1}{4} \\ \sqrt{(4)}\frac{1}{8} \\ \sqrt{(6)}\frac{1}{32} \\ \sqrt{(12)}\frac{1}{2048} \\ \frac{1}{2} \end{cases}$$

Perhaps the most striking fact about all this is that if Stevin discovered these formulae uninfluenced by Chinese work on the subject, it was the second remarkable invention of his which had previously appeared in China, the first being his celebrated

[a] (3), p. 249. Chu Tsai-Yü presented his calendar to the throne in +1595, the year of Ricci's first journey to Nanking. In doing so Chu drew attention to the deficiencies of the current calendar (see *Jih Chih Lu*, ch. 30, p. 1a; cf. *Ming Shih*, ch. 31, p. 33a), envisaging doubtless, however, a reform along the traditional and characteristic Chinese lines, not a wholesale adoption of Greek conceptions such as the Jesuits were beginning to urge. But there is no doubt that the prospect of their aid was one of the factors which facilitated the journey of Ricci and his companions through Nanking to Peking in 1598 (see d'Elia (2), vol. 2, p. 8; Trigault (Gallagher tr.), p. 297).

[b] Cf. Vol. 3, p. 89, and in detail Sarton (2); Dijksterhuis (1).

[c] Fokker (1), p. 18; Dijksterhuis (1), pp. 276ff.

[d] Sarton (2), p. 243.   [e] See esp. pp. 54ff.

sailing carriage.[a] Although it could be a coincidence that he happened to design a machine, the idea of which is known (for example, from Ortelius' +1584 map of China) to have reached Europe from there, without ever having heard about it, yet as Duyvendak has said, this is generally not the way in which things happen. But it would be a still more remarkable coincidence if, after a large crowd of distinguished people had witnessed the trials of his 'sailing chariot' on the sands at Scheveningen (c. +1600), and had discussed this and other Chinese inventions, Stevin were then able to invent the formula for the equal-tempered scale a few years later[b] without being influenced at least through hearsay by the work of his distinguished contemporary in that distant land which had so roused the interest of Europeans that Mendoza's *History . . . of China* had run into eleven editions in six languages in as many years.

It is a strange irony that though Chu Tsai-Yü's work was held in high esteem, his theory was put into practice but little in his own country; while Stevin's theory seems from Mersenne's account to have been widely adopted and utilised in Europe.[c] In any case it is fair to say that the European and modern music of the last three centuries may well have been powerfully influenced by a masterpiece of Chinese mathematics, though proof of transmission be not yet available. The name of the inventor is of less importance than the fact of the invention, and Chu Tsai-Yü himself would certainly have been the first to give another investigator his due, and the last to quarrel over claims of precedence. To China must certainly be accorded the honour of first mathematically formulating equal temperament. A less obvious but more precious gift may lie concealed in the example of this retiring scholar who declined the princely rank to which he was heir in order that he might carry on his researches, believing that for him who understands the meaning of the Rites and Music all things are possible. Such was the faith which animated Chinese students of sound for more than two millennia.

[a] This idea, and the probable Chinese sources of it, will be fully discussed in Sect. 27e below. Meanwhile, see Duyvendak (14).

[b] If his calculation of equal temperament may be placed in the same period as his *Hypomnemata mathematica*, it would be datable between about +1605 and +1608.

[c] It is true that at first his words were held in such slight regard that, to quote his son Hendrik, 'the erudite persons to whom the manuscripts had been entrusted detached several portions and left the rest scattered pell-mell in total confusion'.

(*i*) MAGNETISM AND ELECTRICITY

(1) INTRODUCTION

With this part of the Section we reach the discussion of what was the greatest Chinese contribution to physics. Unfortunately, no subject treated in this book has aroused more controversy or given rise to a more voluminous literature.[a] That the attractive power of the lodestone was known, both in China and the West, from an early time, about the middle of the −1st millennium, there is no dispute. Differences of opinion have centred round the discovery of the directive property, both of the lodestone itself, and of pieces of iron magnetised by contact with it. Knowledge of this appears rather suddenly in Europe at the very end of the +12th century, and the search for immediate antecedents in Arabic and Indian spheres has still not proved successful.

That the Chinese were the first to understand and utilise the directive property of the lodestone has traditionally been admitted, but strangely enough for entirely wrong reasons. From the Han onwards, Chinese texts, which we shall later examine, speak of the 'south-pointing carriage' (*ting nan chhê*[1] or *chih nan chhê*[2]), the art of making which was continually being lost and revived. From the time of the Jesuit missionaries onwards, it was assumed that these were references to some form of magnetic compass,[b] but it may now be considered solidly established that the south-pointing carriage had nothing whatever to do with magnetic directivity. It was in fact essentially a self-regulating device, involving a system of gear-wheels such that a pointer would maintain an originally fixed direction by continually compensating for any excursions of the vehicle away from that direction. It therefore falls under the head of engineering, and it is in that context that we shall deal with it (Sect. 27*e* below). Much of the dissatisfaction of modern historians of science with what have sometimes been called 'the Chinese claims' has arisen from an inability to recognise the legendary component in the early Chinese references to the south-pointing carriage. Contributors to the subject otherwise deserving much credit[c] have been completely at sea in approaching the Chinese literature owing to their lack of sinological competence, and the history of magnetism has therefore fallen into great confusion. There has also been the usual tendency to presuppose that nothing of real importance could have started outside Europe; thus in 1847 Whewell[d] opened his discussion by saying, patronisingly, 'Passing over certain legends of the Chinese, as at any rate not bearing upon the

---

[a] So much of this is unsound that I shall make no attempt to give a complete bibliography here; recourse may be had to Mitchell (1, 2) by anyone interested in probing it. A review of the European opinions during the 18th and 19th centuries was given by Schück (1), vol. 2, thirty years ago. His book, however, is rather rare; the only copy we could find is in the Hamburg Staatsbibliothek.

[b] It is probable that the confusion started much earlier than the Jesuits. Perhaps Chin Li-Hsiang,[3] in his historical work *Thung Chien Chhien Pien*,[4] about +1275, was the first to make it. Living under the Mongols in the north, he was not perhaps conversant with what had been done under the southern Sung. Hence Chhen Yin[5] in *TCKM*, ch. 1, 9*a*, *b*, followed by de Guignes (1) in +1784.

[c] Such as v. Lippmann (2) and Crichton Mitchell (1, 2).

[d] (1), vol. 3, p. 50. So also Gerland (2).

---

[1] 定南車　　　[2] 指南車　　　[3] 金履祥　　　[4] 通鑑前編　　　[5] 陳殷

progress of European science...'; though neither he nor anyone else succeeded in finding any precursors of European knowledge of the lodestone's directivity before the turning-point of +1190.[a]

The story of the magnetic compass in China[b] has recently been revolutionised by the contributions of Wang Chen-To (2, 4, 5),[c] who has been able to explain a fundamental text in the *Lun Hêng* (+1st century) and to reveal a probable connection between the magnetic compass and the diviner's board of the Han people.[d] In what follows we shall seek to show (a) that the first text clearly describing the magnetic needle compass is undeniably of about +1080, i.e. a century earlier than the first European mention of this instrument, (b) that the declination (i.e. the failure of the magnetic needle to point to the astronomical north), as well as the directivity, is there mentioned, (c) that the declination was discovered in China some time between the +7th and the +10th centuries, (d) that the use of the needle, which alone permitted the construction of an accurate pointer-reading instrument, was the limiting factor for this discovery and belongs to the beginning of this period; and (e) that the original Chinese compass was probably a kind of spoon carefully carved from lodestone and revolving on the smooth surface of a diviner's board. Lastly, we shall suggest that it has a detectable connection not only with divination practices but with games such as chess. This original form was certainly known and used in the +1st century, and may go back, as a secret of court magicians, to the −2nd. The failure to elucidate this before has been partly due to the fact that scholars were searching for traces of the south-pointing vehicle when they ought to have been looking for the 'south-controlling spoon' (*ssu nan shao*[1]).

The case of the earliest use of the magnetic compass in navigation is somewhat similar. It seems certain that by +1190 it was in use in the Mediterranean, but its use is also spoken of in a Chinese text just under a century previously. A mistranslation of this text by earlier sinologists led to the persistent statement that it was then found only on foreign (Arab) ships trading to Canton, but this idea, as we shall see, has no basis. Something will later be said as to possible means of transmission; William Gilbert himself[e] thought that Marco Polo or a man of his time brought

---

[a] The tide having now turned, it is possible for Mahdihassan (10) to seek to derive not only the Arabic word for magnet, but also the Greek word itself, from Chinese sources. But as his argument depends upon a Chinese term for the magnet which the Chinese themselves never used, and which indeed was invented *ad hoc*, it is singularly unconvincing.

[b] There most commonly called *chih nan chen*,[2] the south-pointing needle.

[c] Well summarised by Hsiang Ta (2) and Li Shu-Hua (1). We have not been able to see the papers by Chang Thai-Yen (1) and Chin Ching-Chen (1). The account by Chhêng Su-Lo (1) is reliable though very brief. Wu Nan-Hsün's book (1) has a good deal on magnetism, but it is not as original and valuable as his discussions of acoustics and optics. The only up-to-date accounts which have appeared in Western languages are those of Li Shu-Hua (2, 3). The English version is preferable to the French because slightly amplified and provided with the Chinese characters, but the references in both are somewhat imperfect, and both appeared long after the present Section was written. Cf. Needham (39).

[d] He has also succeeded in finding a number of important texts which take the knowledge of the developed magnetic compass in China back to the middle of the +10th century, and has placed us in his debt by many other discoveries of the greatest value.

[e] Followed by Mark Ridley and George Hakewill.

[1] 司南勺   [2] 指南針

it,[a] but this would have been a century too late. 'If the Chinese,' wrote Gibbon, 'with the knowledge of the compass, had possessed the genius of the Greeks and Phoenicians, they might have spread their discoveries over the southern hemisphere.'[b] In fact, that is exactly what they did.[c]

(2) MAGNETIC ATTRACTION

Let us first indicate briefly what was known in European antiquity[d] of the attractive power of the magnet.[e] By the beginning of the Middle Ages it had been established that (a) the lodestone attracts pieces of iron, (b) it does so across a distance, (c) the attracted iron adheres to the magnet, (d) the magnet induces a power of attraction in the attracted iron, which (e) it retains for some time. It had also been observed that (f) the magnetic influence would act through substances other than iron, and that (g) some magnets would repel some pieces of iron as well as attract them. The earliest observations on the magnet are supposed to have been made by Thales (− 6th century), who explained them animistically, but this has come down to us only through a quotation in Aristotle (− 4th).[f] In the − 5th century Empedocles[g] and Diogenes of Apollonia[h] also mentioned the magnet; but here again we have to rely on traditions relayed by Alexander of Aphrodisias, who lived at the beginning of the + 3rd century.[i] However, there is not much reason for doubting that these men, and Democritus their contemporary,[j] did indeed know of and discuss the magnet; the atomists seem to have been particularly interested in it. At any rate, all the fundamental properties mentioned above are described by Lucretius in the − 1st century,[k] while the earlier philosophers state no more than (a) and (b) above. In illustration of the power of transmission, he describes a chain of rings suspended in contact. His theory of attraction was that what we should call a vacuum was established between the magnet and the iron.

---

[a] 'Scientia Nauticae pyxidulae traducta videtur in Italiam, per Paulum Venetum, qui circa annum MCCLX apud Chinas artem pyxidis didicit; nolim tamen Melphitanos tanto honore privari, quod ab iis in mari mediterraneo, primum vulgariter fabricata fuerit' (*De Magnete*, I, I, p. 4). He is alluding in the second half of the sentence to the claim of Flavio Gioja of Amalfi; see on, pp. 249, 289.

[b] *Decline and Fall*, vol. 7, p. 95. He goes on to add: 'I am not qualified to examine, and I am not disposed to believe, their distant voyages to the Persian Gulf, or the Cape of Good Hope....'

[c] Cf. Duyvendak (8) on the Chinese discovery of Africa. Gibbon was writing a little rhetorically, since most of the southern hemisphere is water. The Chinese pushed at least as far as 10° lat. S., which left only Australasia and South Africa. On Australia see Sect. 29e below. Cf. also Vol. 3, p. 274.

[d] A note here on the books most helpful in this study. Falconet (1) and T. H. Martin (1) assembled all the references to magnetism in ancient writers, and sketches of the growth of knowledge about such phenomena have been given by Kramer (1); Mitchell (4); and Hoppe (2). Particularly useful is the book of Daujat (1) on the history of theories of magnetism and electricity. I have not been able to see the paper of Kuwaki (1). Humanists will like the introduction of Bitter (1).

[e] Review of ancient European names for the magnet in Martin (1), ch. 1.

[f] Freeman, p. 49; Aristotle, *De Anima*, 405a19.

[g] Freeman, p. 190.

[h] Freeman, p. 284; Martin (1), p. 53.

[i] I cannot avoid remarking again that sinologists seem to be much more sceptical about traditions handed down after such long intervals than are historians of ancient occidental thought.

[j] Freeman, p. 309.

[k] *De Rer. Nat.*, VI, 998−1088.

As in Europe, the magnet went by many names in China.[a] Its most usual one was *tzhu shih*,[1] the 'loving stone' (cf. *aimant*), generally combined in *tzhu*[2].[b] In spite of the usual austere philological viewpoint, one can hardly believe that in this case the phonetic was not chosen with semantic significance in mind. For example, another derivative, *tzhu*,[3] means copulation or breeding, and those who first observed magnetic attraction in China must have thought of it, like Thales, in animistic terms. *Hsüan shih*,[4] the 'mysterious stone', later signified non-magnetic iron ore, but one cannot help thinking that originally it was a name for the lodestone.[c] The attractive property is explicit in terms such as *hsieh thieh shih*[5] and *hsi thieh shih*;[6] while *nieh shih*[7] meant the stone which picks up pieces of iron as if with pincers. Most of these names (and there were others) go back to the Chin or at least the Thang.

The Chinese literature, between the −3rd and the +6th century, is as full of references to the attractive power of the magnet as the European.[d] There is nothing as far back as Thales, and the only one which might be contemporary with Aristotle is the statement in *Kuei Ku Tzu*, if it is genuinely Chou, which is perhaps unlikely. However, that in the *Lü Shih Chhun Chhiu* would be of the late −3rd century, about the same time as Archimedes.[e] Shortly before Lucretius was writing, the *Huai Nan Tzu* speaks as follows:[f]

If you think that because the lodestone can attract iron you can also make it attract pieces of pottery, you will find yourself mistaken. Things cannot be judged merely in terms of weight, (they have special and peculiar properties). Fire is obtained from the sun by the burning-mirror, the lodestone attracts iron, crabs spoil lacquer,[g] the mallow (*khuei*[8])[h] turns its face to the sun. Such effects are very hard to understand.

[a] Cf. Klaproth (1), p. 35.
[b] In later times this word came to be used for porcelain, but Wang Chen-To (2) shows that this was never the case before the Sung.
[c] Indeed Su Sung in the +11th century says just this (*Pên Tshao Thu Ching*, cit. in *Thu Ching Yen I Pên Tshao*, ch. 4, p. 2a), following *Shen Nung Pên Tshao Ching*.
[d] For example: in Chhin or before: *Kuei Ku Tzu*, ch. 2, p. 12a; *Lü Shih Chhun Chhiu*, ch. 45 (vol. 1, p. 88); *Kuan Tzu*, ch. 77, p. 2b. In Han: *Huai Nan Tzu*, ch. 4, p. 5b; ch. 6, p. 4a; ch. 16, p. 5a; *Shih Chi*, ch. 28, p. 27b; *Chhun Chhiu Fan Lu*, ch. 65, p. 4a; *Huai Nan Wan Pi Shu* (*TPYL*, ch. 736); *Chhien Han Shu* (I Wên Chih), ch. 30, p. 51a; *Lun Hêng*, ch. 47 (twice). Cf. *Shen Nung Pên Tshao Ching*, quoted in *Hsü Po Wu Chih*, ch. 9, p. 5a; cf. Mori ed. ch. 2, p. 57. In San Kuo: *Wên Chi Chiao Chih*[9] of Tshao Tzu-Chien.[10] In Chin: Kao Yu's commentary on the *Lü Shih Chhun Chhiu, ad loc*. Kuo Pho's note *Tzhu Shih Tsan*[11] (*CSHK*, Chin sect. ch. 122, p. 9b) and his commentary on the *Shan Hai Ching*, ch. 3, p. 6a, which mentions locations of lodestone. *Pao Phu Tzu* (*Nei Phien*), ch. 15, p. 5b and elsewhere (cf. Wang Chen-To (2), p. 151), *Wu Shu*, cit. in *San Kuo Chih*, ch. 57, p. 1a; *Nan Chou I Wu Chih* (in *TPYL*, ch. 988). In Liu Sung: *Lei Kung Phao Chih Lun*[12] by Lei Hsiao,[13] p. 42. In Liang and Northern Wei: *Shui Ching Chu*, ch. 19, p. 5a; and Thao Hung-Ching's *Ming I Pieh Lu*, in *PTKM*, ch. 10, p. 4a.
[e] It says: '*tzhu shih chao thieh, huo yin chih yeh*',[14] the lodestone calls the iron to itself, or attracts it. The word *yin*, however, is ambiguous and may be significant of something more; see on, p. 256. The Chin commentator, Kao Yu, adds that mineral is the mother of iron, and the attraction is like that of mother and son.
[f] Ch. 6, p. 4a, tr. auct.   [g] Cf. below, Sects. 33 and 40.
[h] General term for malvaceous plants (B11, 368).

[1] 慈石	[2] 磁	[3] 瑴	[4] 玄石	[5] 熠鐵石	[6] 吸鐵石
[7] 鑷石	[8] 葵	[9] 文集鑰志	[10] 曹子建	[11] 磁石贊	
[12] 雷公炮炙論		[13] 雷斅	[14] 慈石召鐵或引之也		

Elsewhere the writer also says [a] that 'the lodestone flies upwards' (*tzhu shih shang fei* [1]), meaning that a small piece of magnetite could be attracted by iron held above it. The Greeks, on the other hand, thought that only the iron could move to the lodestone, and not vice versa.[b] In a third place, Liu An and his colleagues say:[c]

Some effects are more pronounced at short range and others at long range. Rice grows in water but not in running water. The purple fungus grows on mountains but not in stony valleys. The lodestone can attract iron but has no effect on copper.[d] Such is the motion (of the Tao).

The *Lun Hêng* (+83) mentions the lodestone twice in ch. 47, and both passages link it with amber.[e] First:

Amber picks up mustard-seeds and the lodestone attracts needles.[f] This is because of their genuineness, for such a power cannot be conferred on other things; other things may resemble them but they will have no power of attraction. Why? Because when the nature of the *chhi* is different, things cannot mutually influence one another.[g]

Secondly:

Amber picks up mustard-seeds. (Of course) the lodestone, the 'image-hooking stone' [h] (*kou hsiang chih shih* [2]) is not amber, but it can also attract small things. Clay dragons (used in magic rain-making) are not genuine (dragons) either, yet they (have their effect because they) belong to the same category (of sympathetically attracting things).[i]

Then in the San Kuo period, Yü Fan,[3] later a Taoist official and *I Ching* expert, said that he had heard that amber will not attract 'rotten' mustard, nor the lodestone 'crooked' needles.[j]

The significance of Wang Chhung's remarks about mutual influence and sympathetic attraction in relation to the general Chinese concepts of 'resonance' and action at a distance will not be overlooked.[k] Thus it is interesting to find that about +300, Kuo Pho, in his note mentioned opposite, used the words

The lodestone attracts (lit. breathes in, *hsi* [4]) iron, and amber collects mustard-seeds. The *chhi* (of these things) has an invisible penetratingness, rapidly effecting a mysterious contact, according to the mutual responses of (natural) things (*chhi yu chhien thung, shu i ming hui, wu chih hsiang kan* [5]). This really goes beyond any conceptions that we can form.[l]

---

[a] Ch. 4, p. 5*b* (Erkes (1), p. 59).
[b] Martin (1), p. 27.
[c] Ch. 16, p. 5*a*, tr. auct.
[d] So also Tshao Tzu-Chien says that it has no effect on gold.
[e] The discussions include some remarks of Huan Than (*d.* +25) on the subject (cf. *CSHK*, Hou Han sect., ch. 15, p. 3*b*).
[f] This word is worth noting; cf. p. 278 below.   [g] Tr. auct. adjuv. Forke (4), p. 350.
[h] This is a very peculiar expression. I believe it may refer to the magnetised chess-men of Luan Ta, which we shall discuss below, p. 316.   [i] Tr. auct. adjuv. Forke (4), p. 352.
[j] *San Kuo Chih*, ch. 57, p. 1*a*, quoting *Wu Shu*. This was not very inspired, but again what is worth noticing is that needles are spoken of. Yü Fan was a schoolboy of twelve at the time.
[k] Cf. Vol. 2, index entries; also Ho & Needham (2).   [l] Tr. auct. adjuv. Klaproth (1), p. 125.

[1] 磁石上飛   [2] 鉤象之石   [3] 虞翻   [4] 吸
[5] 氣有潛通數亦冥會物之相感

The words *hsiang kan* are reminiscent of discussions on resonance in acoustics,[a] and reactivity in chemistry.[b]

In the Liang, Thao Hung-Ching said that the best lodestone came from the south, and that it would support a string of three or four needles hanging end to end.[c] Elsewhere he is quoted as saying that the highest quality, such as was annually offered to the court, could suspend a chain of more than ten needles and hold one or two catties (*chin* [1]) of iron knives, which might turn upon one another but could not fall down.[d] One would hardly expect to find the beginnings of quantitative measurement of magnetic force in pre-Renaissance times anywhere, yet China in the +5th century provides such an example, if what is reported of Lei Kung [2] really goes back to Lei Hsiao [3] of the Liu Sung himself.[e] In medieval times the therapeutic properties of non-magnetic iron ore were considered to be different from those of magnetite, and it was desired to distinguish the latter from *hsüan chung shih* [4] and *chung ma shih* [5] which looked like it but had no magnetic properties and were sometimes rather toxic. Lei Kung said:[f]

If you want to make a test take one catty of the stone and see whether using all four sides it can attract an equal weight of pieces of iron—if so this is the best, and may be called *yen nien sha* [6].[g] The sort which (in the same conditions) on all four faces attracts eight ounces is called *hsü tshai shih* [7].[h] Again, that which will only attract about four or five ounces is termed (ordinary) lodestone (*tzhu shih* [8]).

And presumably stones of less power graded into the category of non-magnetic ore. This mode of estimation, involving as it did the use of the balance, cannot be later than the Sung period (+12th century), texts of which often quote it, and may well be as early as the Liu Sung, more than five hundred years before.

In connection with this we may take note of the interesting circumstance that the first statement of the inverse square law of magnetic force was made by J. A. Dallabella in Lisbon in 1797, using an exceptionally large lodestone presented to the King of Portugal over a hundred years earlier by the Emperor of China.[i]

Another parallel between the ancient West and China was the growth of numerous

---

[a] Pp. 130, 161, 184, 186 above.   [b] Section 33 below.

[c] As cited in the +7th-century Thang *Hsin Hsiu Pên Tshao*, ch. 4, p. 9b, and in *Thu Ching Yen I Pên Tshao*, ch. 4, p. 2b.

[d] *Chêng Lei Pên Tshao*, +1249 ed., ch. 4 (p. 111.2); +1468 ed. ch. 4, p. 25b. Su Sung used the same words, ch. 4 (p. 112.1), ch. 4, p. 27b respectively.

[e] *Fl.* +450 to +470.

[f] *Lei Kung Phao Chih Yao Hsing Fu Chieh*, ch. 5 (p. 100), also quoted in *Chêng Lei Pên Tshao*, +1249 ed. ch. 4 (p. 111.2); +1458 ed. ch. 4, pp. 26aff.; *PTKM*, ch. 10, p. 4a.

[g] At first sight this only means 'prolongation-of-life powder'. But there were at least two famous Han technicians with the name Yen Nien in the −1st century, so possibly the term derived from one or other of them. Cf. Sect. 28f below.

[h] Textual variants read both *wei* [9] and *mo* [10] instead of *tshai*. We cannot say which is right, and forbear from interpreting.

[i] For this reference my thanks are due to Mr E. S. Shire of King's College.

[1] 斤   [2] 雷公   [3] 雷斅   [4] 玄中石   [5] 中麻石   [6] 延年沙
[7] 續采石   [8] 磁石   [9] 未   [10] 末

legends about the lodestone. These took various forms, e.g. that there were certain islands which ships could not pass if they were constructed with iron nails, or gates which men could not go through if they were armed with iron weapons; alternatively it was thought that somewhere or other statues of iron floated in mid-air, suspended by magnetic attraction.[a] In the +2nd century Ptolemy wrote of these magnetic islands,[b] which (interestingly enough) he placed between Ceylon and Malaya, and we find exactly the same story in the *Nan Chou I Wu Chih*[c] two centuries later. Afterwards it was often repeated.[d] But probably the idea had its own forms of purely Chinese origin,[e] for the early descriptions of the glories of Chhang-an (Sian) say that the A Fang Kung,[1] one of the palaces of Chhin Shih Huang Ti, had whole gates made of lodestone in order to arrest those who tried to enter with concealed weapons.[f] The use of such a device in the defence of a pass is referred to in the biography[g] of a Chin officer, Ma Lung.[2] The general notion of such gates seems to be connected with wider mythological ideas about ordeals and the escape from the mundane world.[h] Similar stories of magnetic mountains, islands and suspended images[i] occur in Arabic texts.[j]

It was natural that the lodestone should find application both in alchemy and medicine. Sung medical books[k] frequently speak of the opening of blocked passages

[a] T. H. Martin (1), p. 32; Daujat (1), p. 36.

[b] *Geogr.* VII, 2, 30. Also Palladius (+365 to +430), *De Brachmanibus*, in Pseudo-Callisthenes, III, 7; cf. Coèdes (1), pp. xxvii, 99ff.

[c] By Wan Chen. The passage is cited in *TPYL*, ch. 988, p. 3a, and many other places, e.g. *Pên Tshao Thu Ching*, in *Thu Ching Yen I Pên Tshao*, ch. 4, p. 2a.

[d] As in the *Hai Tao I Chih Chai Lüeh*[3] (Brief Selection of Lost Records of the Isles of the Sea) by Wang Ta-Hai,[4] in +1791; the passage has been translated in Anon. (37), p. 44.

[e] We are not convinced, however, by the argument of Hennig (3, 6, 7) who thought that the legend could only have originated in a land (i.e. China) where some form of the magnetic compass was known. It seems more likely that early European knowledge of Indian boats built with stitching or wooden pegs (cf. Sect. 29 below) may have had something to do with the matter. Cf. Peschel (1), vol. 1, p. 44.

[f] *San Fu Chiu Shih*[5] (Stories of the Three Districts in the Capital), and *San Fu Huang Thu*[6] (Illustrated Description etc.), ch. 4 (Kung); the latter Chin, and the former at any rate pre-Thang. Further in *Shui Ching Chu*, ch. 19, p. 5a, and in the Thang geographical encyclopaedia of +806, *Yuan-Ho Chün Hsien Thu Chih*.

[g] *Chin Shu*, ch. 57, p. 3a.   [h] Cf. Coomaraswamy (1).

[i] The suspension of metal objects in the air, and even the melting of them in that condition by high-frequency currents, have become practical possibilities in modern times, and even find industrial applications (see Anon. 12).

[j] Cf. Ferrand (1). Some even locate the magnetic mountains or islands in China; for instance, the *'Ajā'ib al-Hind* (Book of the Marvels of India), written by the sea-captain Buzurj ibn Shahriyār al-Rāmhurmuzī in +953 (cf. Mieli (1), p. 117). The passage occurs on p. 92 of the translation of van der Lith & Devic (1), and was noticed by Wiedemann (18). Buzurj was particularly interested in magnets, it seems. He met a man (p. 169) from China who told him that in that part of the world there were lodestones for lead, copper and gold, as well as iron, and that they exerted their influence just as effectively across the wall of a pottery vessel. It is strange that time's mutations have brought us to the knowledge of non-ferrous magnets such as the Heusler alloys, now rather more than fifty years old, which contain mixtures of copper, manganese and aluminium.

[k] E.g. the *Jih Hua Chu Chia Pên Tshao* (+970) by Ta Ming; the *Shêng Hui Hsüan Fang*[7] (+1046) by Ho Hsi-Ying;[8] the *Yen Chai Chih Chih Fang (Lun)*[9] (+1264) by Yang Shih-Ying[10] and the *Chi Sêng Fang*[11] (+1267) by Yen Yung-Ho.[12] Cf. *Shêng Chi Tsung Lu*, ch. 181, p. 4a.

[1] 阿房宮   [2] 馬隆   [3] 海島逸誌摘畧   [4] 王大海   [5] 三輔舊事
[6] 三輔皇圖   [7] 聖惠選方   [8] 何希影   [9] 仁齋直指方論
[10] 楊士瀛   [11] 濟生方   [12] 嚴用和

or the extraction of foreign bodies such as needles or arrow fragments by the use of lodestones—which suggests at any rate that even if the processes were more imaginative than successful, it was clearly realised that magnetic attraction acted through intervening substances other than iron.[a] It may be noted that magnets were much used in European medicine of the +17th and +18th centuries, though the background of this was distinctly mystical;[b] the medieval Chinese physicians here were simply trying to make use of the lodestone's attractive power for iron. Needless to say, it also played a part in the stock-in-trade of magicians and conjurers of all periods.[c] The *Wu Tsa Tsu*[1] (Five Assorted Offerings) by Hsieh Tsai-Hang[2] (+1573 to +1619), describes[d] an itinerant drug-seller who went about with a statue of the Goddess of Mercy, the hands of which were magnets; drugs mixed with iron powder adhered to them while others did not.

On the whole it may be said that between ancient and medieval knowledge of attraction in Europe and China there was nothing to choose. One finds less theory about it in China, perhaps because action at a distance was more congruent with the general Chinese world-view than with the Greek. Not having any theories as to the 'natural' motion of bodies seeking their 'natural' places, the Chinese were not bothered by the kind of reconciliation for which Peripatetics had to seek. As Daujat (1) has shown, there was great difficulty in fitting magnetic attraction into the Aristotelian distinction between 'natural' and 'violent' motions, so much so that Aristotle said as little about the magnet as possible, and Alexander of Aphrodisias in the +3rd century could solve the matter only by perpetuating a form of animistic theory.

A particularly Chinese conception was due to the heresiarch Hermogenes, known because of Tertullian's writings against him.[e] He believed that God had not created the world from nothing, but had organised all matter by acting upon it like a magnet. If this patterning principle had been immanent rather than transcendent there would have been no difference between it and the Tao.

The nearest approach to a discovery of magnetic polarity in Europe was the observation that in some cases a lodestone would repel iron. This was known to Pliny, to Marcellus (early +5th century) and to Joannes Philoponus (early +6th),[f] whom we have already encountered as the first proponent of the impetus theory in dynamics. But it was not carefully observed, and none of these writers realised that their effects were due to the approaching of two like poles. A proper understanding had to await

---

[a] Among the more practical procedures were those which sought to recover needles or other small iron objects mistakenly swallowed by children. A piece of lodestone the size of a date kernel was wrapped in meat and collected the iron as it passed through the digestive tract. Or a piece of similar size, threaded on a string, would ease the object out of the throat. Besides the authorities just given, there was the *Chhieh Chung Fang*[3] of Chhien Wei-Yen[4] (the statesman, G372, d. +1029). All are quoted in *CLPT*, +1249 ed., ch. 4 (p. 111.2); +1468 ed. ch. 4, p. 26a; *PTKM*, ch. 10, p. 6b.

[b] Wootton (1), vol. 1, p. 199; Beckmann (1), vol. 1, p. 43.

[c] Martin (1), p. 50.

[d] Ch. 6.

[e] *Contra Hermogenem*, ch. 54; Daujat (1), p. 36; Martin (1), p. 64. Cf. Vol. 2, p. 293.

[f] Daujat (1), pp. 12, 46; Martin (1), p. 39.

[1] 五雜俎   [2] 謝在杭   [3] 篋中方   [4] 錢惟演

the coming of the magnetic compass to Europe late in the +12th century. Similar observations may have been made in China. The curious story of the magnetised chess-men (see on, p. 315) has come down to us in several versions. The *Huai Nan Wan Pi Shu*, in *Thai-Phing Yü Lan*, ch. 988, uses the word *chü*,[1] to push away; and as preserved in Ssuma Chên's[2] Thang commentary on the *Shih Chi*, has the word *ti*,[3] to repel. The description in the text itself, both in *Shih Chi* and *Chhien Han Shu*, says *hsiang chhu chi*,[4] i.e. the pieces 'mutually hit' each other. The word used in most texts for attraction is *yin*.[5][a] We have therefore a hint that in China also repulsion was observed.

(3) ELECTROSTATIC PHENOMENA

Just as ancient and medieval Chinese knowledge about the lodestone paralleled that of the West, so also the fact that certain substances such as amber will, when rubbed, acquire an electric charge and attract small objects such as dried plant fragments or pieces of paper was known in both regions.[b] It is again Thales who is credited with the first of such observations, but this depends on a tradition transmitted by Diogenes Laertius as late as the +3rd century.[c] It is certain that the electrum of Homer was an alloy of gold and silver, but after the time of Herodotus (−5th century) the word generally refers to yellow amber, and our 'electricity' derives from it. Plato, in the *Timaeus*, is the first to speak of the attractive power of amber in a text still extant, but of all the many authors who also do this, only Plutarch and Pliny mention that it must be rubbed beforehand.

Greek amber was probably of Baltic origin, but most Chinese amber came from deposits of burmite in Upper Burma.[d] The Han histories say[e] that it was brought from Chi-Pin (Gandhāra) and Ai-Lao (the Shan region of north Burma), not mentioning Persia or sources further west. Wang Chhung, the contemporary of Plutarch and Pliny, is one of the first to mention it,[f] and the term he used, *tun mou*,[6] is likely to have been formed as a loan-word from some Shan or Thai language. Such is also supposed to

---

[a] But see below, p. 257.

[b] Martin (1), pp. 95, 139, gives a history of the statements of Greek and Roman writers on amber and its electrostatic properties.

[c] I, 24.

[d] See Laufer (17), who took due account of older work such as that of Jacob; and also Laufer (1), p. 521. Small amounts of European amber no doubt travelled eastwards from time to time over the Old Silk Road, and in late times Formosa was an entrepôt for Dutch trade to China in the Baltic product. Cf. Baker (1). Burmite has slightly different physico-chemical properties from succinite, the common amber, but is equally the fossil gum of conifers. Malaya also contributed.

[e] *Chhien Han Shu*, ch. 96A, p. 11a; *Hou Han Shu*, ch. 115, p. 18b. Cf. Parker (2, 3).

[f] *Lun Hêng*, ch. 47. A reference purporting to be from the *Shen Nung Pên Tshao Ching*, which may be of the Former Han, is preserved in *Hsü Po Wu Chih*, ch. 9, p. 5a, but the genuineness of this is doubtful for reasons given immediately below. The *Shan Hai Ching*, ch. 1, mentions amber under the name of *yü phei*,[7] but the passage would be hard to date. Cf. Chang Hung-Chao (1), p. 62, on this and other words which seem to have meant amber in this text and in Kuo Pho's commentary on it.

[1] 拒   [2] 司馬貞   [3] 抵   [4] 相觸擊   [5] 引   [6] 頓牟
[7] 育沛

be the origin of the more usual term *hu pho*[1]—the derivation from *hu pho*,[2] the 'concreted soul of the tiger' being fanciful and late. About +500, Thao Hung-Ching, in his commentary on the *Ming I Pieh Lu*, wrote that amber was fir-tree resin which had been buried for a thousand years, and in which entrapped insects might often be seen,[a] adding that there was a method of imitating it by heating hen's eggs with dark fish roe.[b] He goes on to say:[c]

But only that kind which, when rubbed with the palm of the hand, and thus made warm, attracts mustard-seeds (*chieh*[3]), is genuine. Nowadays amber comes from foreign countries, and is produced in those places where the *fu-ling*[4] grows;[d] on the other hand everyone knows that amber may occur in any place, whether the *fu-ling* is there or not.

The electrostatic test for genuine amber is still used.[e] Wang Chhung and Yü Fan, who had mentioned the phenomenon, had not stated that the amber had to be rubbed. In Europe, Alexander of Aphrodisias emphasised that while the lodestone would attract only iron, amber would attract any small or light bodies.[f] So also Li Shih-Chen, much later, said that straw fragments or any small bits of vegetable material would do;[g] indeed, the word *chieh*[3] had always had this wider meaning. After the Han, practically all the pharmacopoeias mention amber and its properties.[h] But there was no real advance, any more than in Europe, until the study of electricity really began as a characteristically post-Renaissance science in the 18th century.[i]

[a] Cf. Kirchner (1).

[b] No doubt a yellowish-brown translucent resinous mass could be produced in this way. Late encyclopaedias, such as *KCCY*, ch. 33, p. 12*b*, quoting *Pai Shih Hui Pien*,[5] a collection of excerpts assembled by Wang Chhi[6] about +1590, attribute the technique to the *Shen Nung Pên Tshao Ching*. But none of the modern reconstructions of this ancient work have an entry for amber at all. According to Li Shih-Chen (*PTKM*, ch. 37, pp. 9*b*) the *Ming I Pieh Lu* is the first to mention it, and on the counterfeiting method there is nothing earlier than the words of Thao Hung-Ching himself.

[c] *PTKM*, ch. 37, p. 10*a*, tr. Laufer (17).

[d] This is the fungus *Pachyma cocos*, parasitic on the roots of pine-trees; cf. p. 31 above. Understanding of the true nature of amber, and at the same time a belief in its association with plants parasitic on pine-trees, was quite general in medieval China; cf. the poem of Wei Ying-Wu[7] (G 2299) mentioned in Hui-Hung's[8] (Sung) *Lêng Chai Yeh Hua*[9] (Night Talks in a Cool Library), ch. 4, p. 4*b*. Cf. Minakata (2).

[e] See Farrington (1).   [f] Daujat (1), p. 23.

[g] *PTKM*, ch. 37, p 10*b*.

[h] Reference has already been made above (p. 74) to sparks of static electricity. Crackling sounds were also reported as coming from amber in various Han and later passages; cf. Chang Hung-Chao (*1*), p. 62.

[i] The introduction of electrical knowledge into modern China and Japan would make an interesting story but it has not, so far as we know, been written. Some of the pioneer works of the 18th century, such as Hashimoto Donsai's *Erekiteru Yakusetsu*, based mainly upon Dutch sources, have recently been reprinted in historical collections.

[1] 琥珀   [2] 虎魄   [3] 芥   [4] 茯苓   [5] 稗史彙編   [6] 王圻

[7] 韋應物   [8] 惠洪   [9] 冷齋夜話

(4) MAGNETIC DIRECTIVITY AND POLARITY

In entering upon this subject it may be well to pause for a moment to consider the incalculable importance of the discovery of the magnetic compass, as the first and oldest representative of all those dials and pointer readings which play so great a part in modern scientific observation. The sundial was of course far older, but there it was only a shadow which moved, and not a part of the instrument itself. The wind-vane was also older (cf. what has been said in the meteorological Section, Vol. 3, pp. 477 ff. above), but there the possibility of precise readings on a circular graduated scale was absent in all ancient forms. In the armillary sphere, the sighting-tube or alidade was moved to a position which was then read off on the graduated circles, but here the movement was made by hand, and the instrument was not self-registering. As we shall see, the limiting factor for making accurate readings was the introduction of the use of a needle as against the lodestone itself, and this implied another discovery of fundamental importance for science, namely that of an induction process. No apology is needed, therefore, for an attempt as thorough as possible to ascertain what was the oldest form of compass developed by the Chinese, and when its successive developments were introduced. That so fundamental an instrument did in fact spread so slowly is not difficult to understand once we realise that its original discovery took place in connection with the divination processes of imperial magicians; and that since it developed in an agrarian–terrestrial rather than in a primarily maritime civilisation, its use was for centuries limited to a specifically Chinese pseudo-science, namely Taoist geomancy, the minutiae of which were carried to a high level of refinement. The adoption of the compass by Chinese sailors was probably long retarded by the fact that all through the Middle Ages river and canal traffic predominated over ocean voyages.[a]

Since the geomantic art necessarily enters this Section as a kind of intercurrent refrain[b] it may also be well to remind ourselves of what was said on the subject in the Section on the pseudo-sciences.[c] The term geomancy has other meanings in other civilisations, but for the Chinese it meant 'the art of adapting the residences of the living and the tombs of the dead so as to cooperate and harmonise with the local currents of the cosmic breath'.[d] Known as the science of 'winds and waters' (*fêng shui*[1]), it did not mean merely the winds of everyday life, but rather the *chhi*[2] or *pneuma* of the earth circulating through the veins and vessels of the earthly macrocosm. The waters too were not only the visible streams and rivers but also those passing to and fro out of sight, removing impurities, depositing minerals,[e] and like the *chhi* affecting for good or evil the houses and families of the living, as also the descendants of those who lay in the tombs. The history of the magnetic compass is only under-

---

[a] For a brief account of the history of Chinese long-distance navigation, see Vol. 1, p. 179, and further in Sect. 29 below.

[b] Pp. 240 ff., 293 ff., 307 ff.

[c] Sect. 14a (6), in Vol. 2, pp. 359 ff.

[d] The words are those of Chatley (7).

[e] Cf. especially Vol. 3, pp. 637, 650.

[1] 風水   [2] 氣

standable in the context of this system of ideas, for this was the matrix in which it was generated.

Of all the forms of divination, geomancy was perhaps that which became most deeply rooted in Chinese culture throughout the traditional period. It led to a minute appreciation of the topographical features of any locality, the forms of the hills and the directions and windings of the streams, the presence of woods and flooded rice-fields, the building of pagodas on conspicuous eminences and the contours of city walls. A wealth of technical terms, as yet very imperfectly understood,[a] was applied to the configurations of terrain, connecting together in many varying ways the Yang and Yin, the dragon and tiger, the earth, planets and stars. The protection of a site from harmful influences was always a matter of great importance, and the achievement of a balance of Yang and Yin forces, high cliffs and rocky masses setting off thickets of bamboos, rounded hills and placid lakes. A whole book could be written on Chinese landscape paintings in relation to *fêng-shui* principles. Purely superstitious though in many respects they sometimes became, the system of ideas as a whole undoubtedly contributed to the exceptional beauty of positioning of farmhouses, manors, villages and cities throughout the realm of Chinese culture. Anyone who has visited the tomb-temples of the Ming emperors in their group of exquisite valleys north of Peking will know something of what the geomancers, at their best, could do.[b]

Thus the background history of geomancy is of some importance for that of the magnetic compass itself. There can be little doubt but that it was something which developed during the Warring States period at the time of the schools of philosophical magic when Tsou Yen was flourishing. The *Kuan Tzu* book, perhaps the product of the Chi-Hsia Academy in the late −4th century, speaks of water as the blood and breath of the earth, 'flowing and communicating within its body as if in sinews and veins'.[c] An early reference of importance is the remark of Mêng Thien,[1] the builder of the Chhin Great Wall, just before his death in −210: 'I could not make the Great Wall without cutting through the veins of the earth.'[d] The *Shih Chi* (−90) also mentions[e] a class of diviners called *khan yü chia*[2] (diviners by the canopy of Heaven and the chariot of Earth). Then in the Han the system was well under construction, as we know from the diatribes of Wang Chhung against it[f] towards the end of the +1st century. Its consolidation took place in the San Kuo period.[g] We shall see

---

[a] And indeed in danger of total oblivion, since the tradition is now dying so fast in contemporary China.

[b] The series of tumuli and temples was begun in +1409 for the Yung-Lo emperor who moved up from Nanking. They stand on the southern-facing slopes and bluffs of a beautiful range of hills, over-looking the streams which flow southwards to the neighbourhood of the capital. A dam built in 1958 is converting the main valley into a lake in which they will be mirrored. Two geomancers have traditionally the credit of the siting, Wang Hsien[3] from Shantung, and Liao Chhiung-Ching[4] from Chiangsi. Cf. Grantham (1).

[c] Ch. 39, p. 1a. The whole chapter has been given in translation in Vol. 2, pp. 42ff.

[d] *Shih Chi*, ch. 88, p. 5b. Cf. Bodde (15), pp. 61, 64, 65.

[e] Ch. 127, p. 7b.

[f] Cf. Forke (4), vol. 1, p. 531, as also Vol. 2, p. 377 above.

[g] See the brief survey of the literature in Vol. 2, p. 360 above, and further, pp. 268ff., 302ff. below.

[1] 蒙恬   [2] 堪輿家   [3] 王顯   [4] 廖瑰靜

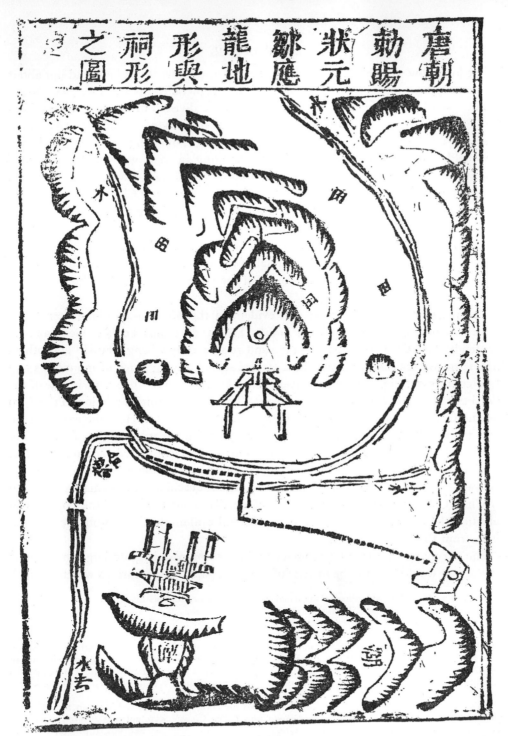

Fig. 321. The geomantic background to magnetical science: a selected topographic disposition of the early +13th century. The caption says: 'Map of the Grounds and Family Temple donated by Imperial Rescript to Tsou Ying-Lung, Optimus Graduate in the Palace Examinations.' In a region of hills three streams join together at a point left of centre to form a small river which flows away at the left bottom corner. The temple, backed by two small lakes, faces the upper part of the plan, with a view of the auspicious point (marked by a pavilion) at the tip of the hilly ridge separating the two upper valleys, each of which contains numerous rice-fields. A similar auspicious situation was shown and described in Fig. 45 (Vol. 2). Two bridges and a gateway in the hills are also marked. From *Ti Li Cho Yü Fu*, ch. 3, p. 13*b*. The relation of this kind of drawing to physiographic map-making (cf. Vol. 3, p. 546) is evident. Although the caption ascribes him to the Thang, Tsou Ying-Lung gained his high honours in the Sung, between +1195 and +1200.

before long how well this sequence of events fits in with what can be found out about the development of the successive forms of the magnetic compass.

The rise of the compass seems to have led to the division of the geomancers from the Thang onwards into two schools. The 'gentlemen of Ganchow' (*Kan-chou hsien-sêng*) stemmed from Chiangsi, following Yang Yün-Sung[1] of that province;[a] they held mainly to the older principles, reasoning in their way on the shapes of mountains and the courses of rivers, much no doubt as their Han predecessors had done. The men of Fukien on the other hand, following Wang Chi[2] of that maritime region,[b] regarded the compass as all-important for determining the indications of the topography, but besides this they made more use of the *kua* of the Book of Changes, and the astrological element was more prominent in their ideas. Yang Yün-Sung's chief work, the *Han Lung Ching*[3] (Manual of the Moving Dragon) certainly contains some astrology, but nothing about the use of the compass. Wang Chi's books appear to be all lost.

The marks of this division are still evident in the Ming and Chhing literature.[c] If we take up a book such as the *Ti Li Cho Yü Fu*[4] (Precious Tools of Geomancy),[d] written by Hsü Chih-Mo[5] about +1570 and re-issued by Chang Chiu-I[6] and others in +1716, we see that it follows the Chiangsi school, saying much of mountains and watercourses (cf. Fig. 321) and relatively little of the compass.[e] The same is true of the *Yin Yang Erh Chai Chhüan Shu*[7] (Complete Treatise on Siting in relation to the Two Geodic Currents), due to Yao Chan-Chhi[8] and dated +1744. Here the emphasis on pure topography includes an interesting variety of contour mapping (cf. Fig. 322) in a style which goes back to Thang delineations of the five sacred mountains.[f] But again there is little on the magnetic compass and orientation by its bearings.[g] By contrast the *Ti Li Wu Chüeh*[9] (Five Transmitted Teachings in Geomancy), written by Chao Chiu-Fêng[10] in +1786, is full of details about the use of the compass in selecting the most auspicious sites for tombs and buildings. Chao probably derived, then, from the Fukien school.

The surviving literature on geomancy and the compass is quite large, but not as large as we would like it to be in view of the great gaps and obscurities which we shall

---

[a] *Fl.* +874 to +888, according to the usual view. But some, such as his editor of 1892, Li Wên-Thien, place him in the Later Thang dynasty, *c.* +923 to +936.

[b] Probably born *c.* +990, as we seek to show on p. 305 below.

[c] Further on this late geomantic literature see p. 300 below.

[d] The expression *Ti Li*, more commonly used of geography, is also characteristic of books on geomancy. Dr Ho Kuang-Chung at the University of Malaya, Singapore, has made a collection of *fêng-shui* books with this type of title, and promises to provide a bibliography of them if requested. The largest published catalogue of geomantic books is that of Chhien Wên-Hsüan (*1*); cf. Wang Chen-To (*5*), pp. 110ff.

[e] This does not mean that Hsü Chih-Mo neglected the compass in other writings, on the contrary, his *Lo Ching Ting Mên Chen*[11] and his *Lo Ching Chien I Thu Chieh*[12] were entirely devoted to it. But the two methods were kept somewhat separate.   [f] See Vol. 3, p. 546.

[g] On the possibility that compass bearings were used for geographical map-making in the time of Shen Kua (late +11th century), cf. Vol. 3, p. 576.

¹ 楊筠松   ² 王伋   ³ 撼龍經   ⁴ 地理琢玉斧   ⁵ 徐之鎮

⁶ 張九儀   ⁷ 陰陽二宅全書   ⁸ 姚瞻旂   ⁹ 地理五訣

¹⁰ 趙九峯   ¹¹ 羅經頂門鍼   ¹² 羅經簡易圖解

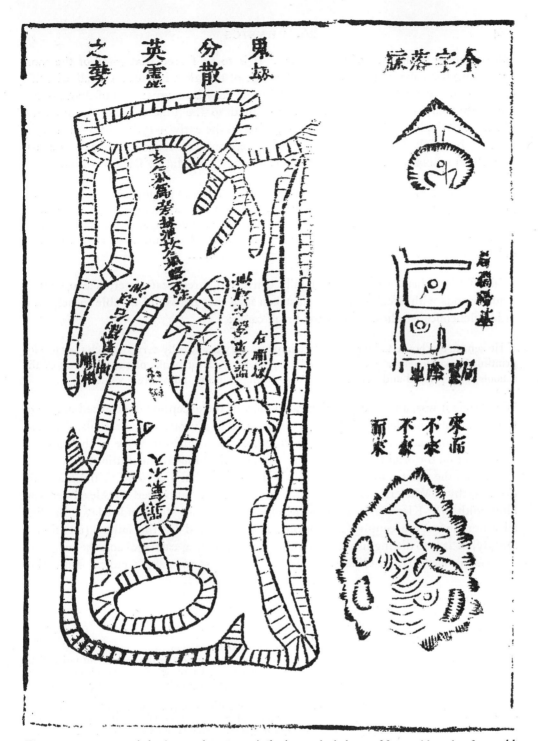

Fig. 322. The geomantic background to magnetical science: the balance of favourable and unfavourable influences in a hilly region. The caption says: 'How the Mysterious Geodic Influences are distributed according to the Contours.' Again several small streams combine to flow out through the gap on the left. The inscriptions within the diagram mark the presence and effectiveness of various kinds of *chhi* classified according to the *kua* of the *I Ching* (cf. Vol. 2, pp. 304ff.). The basin in the upper half of the chart seems thus to enjoy a favourable balance. The cartographical convention may be compared with that in Vol. 3, Fig. 224. From *Yin Yang Erh Chai Chhüan Shu*, ch. 1, p. 12b.

still be facing at the end of this Section. It is to be feared that some of the most interesting facts about the development of one of the most important of all scientific instruments have perished for ever. While there may be a great deal of doubt about the traditional burning of the books attributed to the first emperor Chhin Shih Huang Ti,[a] the bonfires inspired by the Jesuits in the early +17th century can enjoy no such benefit.[b] Li Ying-Shih,[1] whose conversion occurred in +1602, was a distinguished scholar particularly skilled in geomancy. In Matteo Ricci's words:[c]

He had a rather good library and it took him three full days to purge it of books on subjects prohibited by our (churchly) laws, books which were very numerous, especially on divinatory arts, and the most part in manuscript, collected with the greatest assiduity and expense. So at this time, all of these, amounting to three trunks full, were committed to the flames, either in his own courtyards or publicly outside our own house....

Again, Chhü Thai-Su,[2] who was finally converted three years later, had originally made friends with Ricci under the mistaken idea that he would be able to help him in his alchemical studies.[d] When the break came:

He sent to our house to be burned certain plates very beautifully carved which they use for printing books, together with three or four porters' loads of books of the doctrines of the schools, part printed and part manuscript awaiting printing, all of great value.[e]

How tragic it was, and how paradoxical in view of the exceptionally learned nature of the Jesuit mission, that one of the most shameful of European aberrations, the ideal of 'holy ignorance', should thus have closed, perhaps for ever, some of the doors of knowledge on the origins of one of the greatest of all Chinese contributions to science.[f]

Perhaps the most remarkable work which has ever been devoted to the magnetic compass by a Western scholar was the erudite letter addressed to Alexander von Humboldt by J. Klaproth in 1834; many of his conclusions still stand today.[g] But like all other nineteenth-century discussions, it was vitiated by the failure (mentioned already, p. 229) to distinguish between the 'south-pointing carriage' and the magnetic compass proper. H. A. Giles[h] was probably the first to make a radical distinction

---

[a] Cf. Vol. 1, p. 101.
[b] Yet they seem still to have the approbation of Cronin (1), p. 203.
[c] From d'Elia (2), vol. 2, p. 262. Cf. Trigault (Gallagher tr.), p. 434.
[d] See d'Elia (2), vol. 1, p. 297.
[e] From d'Elia (2), vol. 2, p. 342. Cf. Trigault (Gallagher tr.), p. 468.
[f] It will not do to say that the Jesuits were combating superstition in the capacity of rationalists. They did not disbelieve in the geomantic art; they considered it diabolical. Sixty years later the Royal Society was still taking such things seriously, but did not propose to destroy the written word, however untrustworthy scientifically.
[g] Von Humboldt drew attention to them in many publications, e.g. (1), vol. 1, p. 187, (2), p. xxxvii, (3), vol. 3, p. 36. Following in the steps of Leibniz (cf. Vol. 2, p. 497), von Humboldt was the first who succeeded in setting in motion a systematic world magnetic survey 'from China to Peru'. The story has been told by Kellner (1).
[h] (5), vol. 1, pp. 107, 219, 274.

[1] 李應試   [2] 瞿太素

between them, and to demonstrate that the former was a purely mechanical device;[a] this has been followed up by Hashimoto (3, 1), Moule (7) and other workers. But understanding of the matter has still not penetrated to historians of science, and the reviews of Mitchell (1, 2), which are the authority behind such widely trusted works as those of Chapman & Bartels (1) or Stoner,[b] have not only perpetuated the confusion but allowed it to prejudice the unbiased consideration of texts which undoubtedly do refer to the magnetic compass proper.[c] Others, such as Hoppe (1), grant the Chinese the compass in the + 1st century, but for the wrong reasons, since at the time he wrote the new archaeological discoveries of Wang Chen-To (2) had not been made.

(i) *Appearance of the magnetic compass in Europe and Islam*

The first thing to do is to note the exact dates at which knowledge of the magnetic compass first appears among the Europeans and Arabs. These dates were not clearly known to Samuel Purchas, whose remarks in the *Pilgrimes*,[d] however, are worth quoting:

The Vertue of the Loadstone, to be the Lead-Stone and Way-directing Mercurie thorow the World; Plato, Aristotle, Theophrastus, Dioscorides, Galen, Lucretius, Plinie his Solinus, and Ptolomee, Scholers of the highest Formes in Natures Schoole, knew not, though the Magneticall attraction of Iron be mentioned in their Workes....

This invention some ascribe to Salomon; which I would beleeve, if he had written of Stones, as he did of Plants; or if the Tyrians, which were almost the engrossers of Navigation in that Age, and were the Sea-men which Salomon used in his Ophyrian Discoveries...had left any Tradition or Monument thereof to Posteritie; which could no more have been lost than Sayling itself; which the Greeks, Carthaginians, and other Nations, successively derived from them. Others therefore looke further unto the East, whence the Light of the Sunne, and Arts, have seemed first to arise to our World; and will have Marco Polo the Venetian above three hundred yeeres since to have brought it out of Mangi (which wee now call China) into Italy. True it is, that the most magnified Arts have there first been borne, Printing, Gunnes, and perhaps this also of the Compasse, which the Portugals at their first entry of the Indian Seas found amongst the Mores, together with Cards and Quadrants to observe both the Heavens and Earth.[e]

Modern historians, however, have been able to draw the dividing line with much greater precision.[f] Their results may be summarised by saying that among the

---

[a] It is true that Chalmers (2) had preceded him, some fifteen years earlier, but Chalmers' views were based not so much on an open-minded study of texts as on a conviction that the Chinese could never have had the intelligence to discover the directive properties of magnets.

[b] Or the historical accounts of Hitchins & May (1) and May (4).

[c] Even Sarton (vol. 1, p. 764; vol. 2, pp. 24, 509, 629; vol. 3, p. 714) did not speak clearly on this point, though giving to Shen Kua the credit he deserves (see on, p. 249). In view of what we shall see, Sarton's statements that the Chinese did not apply the compass to any rational purpose, and that the first use in navigation was by foreign sailors, rest on mistakes and cannot be retained.

[d] Pt. 1, Bk. ii, ch. 1 (i), p. 2.

[e] In fact, there is no mention of the compass in Marco Polo.

[f] Poggendorff (1); Libri-Carrucci (1); Libes (1); von Lippmann (2); Mitchell (1); Hennig (5); Schück (2, 3, 5, 6, etc.). Schück (1), vol. 2, gives a long and elaborate account of the earliest European mentions of the compass.

Europeans the directive property of the lodestone was unknown to Adelard of Bath[a] in +1117, or any previous lapidary such as Marbodus, but is mentioned by Alexander Neckam[b] in +1190—and then by many others, such as Guyot de Provins[c] in +1205 and Jacques de Vitry[d] in +1218.[e] After that comes the important treatise on the magnet by Petrus Peregrinus (Peter the Wayfarer, or Pierre de Maricourt)[f] in +1269; one of the finest contributions to physics of the whole medieval period.[g]

What Neckam said was this:[h]

The sailors, moreover, as they sail over the sea, when in cloudy weather they can no longer profit by the light of the sun, or when the world is wrapped up in the darkness of the shades of night, and they are ignorant to what point of the compass their ship's course is directed, they touch the magnet with a needle. This then whirls round in a circle until, when its motion ceases, its point looks direct to the north.[i]

And Guyot de Provins, in his satire, *La Bible*:[j]

De nostre père l'apostoile[k]
Vousisse qu'il semblast l'estoile
Qui ne se meut; mout bien la voient,
Li marinier qui se navoient.
Par cele estoile vont et viennent
Et lor sens et lor voie tienent;
Ils l'appellent la Tresmontaigne[l]
Celle est atachie et certaine;
Toutes les autres se removent,
Et lor leus eschangent et muevent
Mais cele estoile ne se meut.
　　Un art font qui mentir ne peut,
Par la vertu de la magnette.
Une pierre laide et brunette
Où li fers volontiers se joint
Ainsi regardent le droict point;
Puis, qu'une aiguile l'ait touchie
Et en un festu l'ont fichie
En l'eaue la mettent sans plus
Et le festus la tient desus.

[a] Sarton (1), vol. 2, p. 167.　　　　　　　　　　　　[b] Sarton (1), vol. 2, p. 385; May (5).
　[c] Sarton (1), vol. 2, p. 589. One of the first to draw attention to Guyot was the great naval archaeologist Jal (1), vol. 1, p. 205. He has sometimes been identified with Hugues de Bercy (cf. May, 3); the idea that he was a Provençal seems to have been due to a mistake of von Humboldt's (Schück (1), vol. 2, p. 27). But he had been to the Holy Land and to Byzantium.
　[d] *Historiae Hierosolimitanae*, ch. 89; Klaproth (1), p. 14; Sarton (1), vol. 2, p. 671.
　[e] Another early reference is in a lapidary written by a French Jew, Berakya ha-Naqdan, c. +1195 (Sarton (1), vol. 2, p. 349). According to Schück (1), vol. 2, p. 30, there is evidence for the use of the compass in mining near Massa in North Italy shortly after +1200.
　[f] Sarton (1), vol. 2, p. 1030.
　[g] His contribution is excellently explained by Daujat (1), pp. 79ff. The older authority on him was Bertelli(1), summarised in English by Hazard. Cf. Winter (4); Chapman & Harradon (1); Thompson (4).
　[h] *De Naturis Rerum*, II, xcviii; ed. Wright, p. 183.
　[i] Tr. Bromehead (4).　　　　　　　　　　　　　　[j] Cf. Klaproth (1), p. 41; Michel (6).
　[k] The Pope.　　　　　　　　　　　　　　　　　[l] The pole-star.

Puis se tourne la pointe toute
Contre l'estoile, si sans doute
Que jà por rien ne faussera
Et mariniers nul dotera.
Quant la mers est obscure et brune,
Qu'on ne voit estoile né lune,
Dont font à l'aiguile alumer;
Puis, n'ont-il garde d'esgarer.
Contre l'estoile va la pointe,
Por ce, sont li marinier cointe
De la droite voie tenir,
C'est un ars qui ne peut fallir.
Mout est l'estoile bele et clère;
Tel devroit estre nostre père.[a]

It is not necessary to quote later texts, which will be found in the literature cited.[b]

The earliest Arabic references to the magnetic compass are all somewhat later than the European ones.[c] In the *Jāmiʿ al-Ḥikāyāt*, a collection of anecdotes compiled in Persian about +1232, by Muḥammad al-ʿAwfī,[d] there is mention of sailors finding their way by means of a fish-shaped piece of iron rubbed with a magnet. Then, in the *Kanz al-Tijār*, a lapidary written by Bailak al-Qabajaqī[e] in +1282, the writer describes the use of a floating compass-needle witnessed by himself in +1242,[f] and adds that the captains who sail the Indian seas employ a floating fish-shaped iron leaf. There is no mention of the compass by astronomers such as Ibn Yūnus (c. +1007) nor by +10th-century geographers such as al-Masʿūdī. Even in later Arabic literature references to it are not numerous; thus the lapidary of al-Tīfāshī (+1242) does not speak of it, nor the encyclopaedia of al-Qazwīnī (c. +1250). No Indian reference of any importance has been discovered.[g]

Appropriately enough, Ibn Ḥazm discoursed on magnetism in his treatise on love, the *Ṭauq al-Ḥamāma*, and it has been thought that one of the poems which it contains

---

[a] Perhaps a brief modernisation is necessary. 'Our apostolic father ought to be like the Pole-Star, which never moves, and can well be seen by every navigating mariner. By this they go and come, and hold their right course; it is called the "tramontana" and is fixed while all the others move. Moreover, there is an art which the sailors have, which cannot deceive. They take an ugly brown stone, the magnet, to which iron willingly attaches itself, and touching a needle with it, they fix the needle in a straw, and float it on the surface of water, whereupon it turns infallibly to the Pole-star. It makes no mistake about this and no mariner doubts it. And when the sea is all dark, and no stars nor moon can be seen, they light a lamp with which to see the needle, and thus they go not astray from their course. This art is unfailing, bright and clear is the Star; so also ought our Father to be.'

[b] One of the earliest pictures of a sailor using the compass is claimed for the *Livre des Merveilles* of +1385 (figured by Feldhaus (2), p. 288). But it is more probably an astrolabe (cf. Sect. 29f).

[c] Cf. the review of Vernet (1).

[d] Mieli (1), pp. 159, 263. A translation of the passage is given in Balmer (1), p. 54.

[e] Mieli (1), p. 159.

[f] Sarton (1), vol. 2, p. 630.

[g] Sardar K. M. Panikkar tells me, however, of mention of the compass in Tamil nautical books attributed to the +4th century. This should be thoroughly investigated. According to Mukerji (1), p. 48, an old name for the compass in India was the *maccha-yantra* (fish machine), the magnet being in the shape of a fish and floating on oil. This is evidently akin to both Chinese and early European techniques, but no dating can be given to it.

may refer to the polarity of the magnet.[a] This would be important, if true, since Abū Muḥammad 'Ali ibn Ḥazm al-Andalusī lived from +994 to +1064. The second chapter of his 'Book of the Ring of the Dove' has the following verse:[b]

> My eyes find nowhere else to look but at you,
> Like what happens with the lodestone (and the pieces of iron),
> Changing direction to right and left, in accordance with where you are,
> Or like as in grammar adjectives and nouns must accord.

At first sight this might be interpreted as meaning that the girl was polar and her lover the lodestone always turning to her, but it is far more likely that in the writer's thought the girl was the lodestone, the attractor, and that his eyes were like pieces of iron drawn towards it wherever it moved. Thus it is most probable that Ibn Ḥazm knew of attraction but not of polarity.

There have, of course, been claims for earlier European references, but all have been discredited (Mitchell, 1). One of them, however, is worth looking at, namely a manuscript[c] of the *Cosmography* (so-called) of Asaph ha-Yehūdī,[d] described and figured by Winter (1). If the statement about the compass was really in the original, it would be of the +9th century, but the MS. is in a +14th-century hand, and the passage is undoubtedly a later interpolation.

> And thus at all times days and nights pass in order just as the heavens go round unceasingly from east to west under the two signs which are one in the south and the other in the north. And these never move from their places as the others do. And so sailors navigate to the signs of the stars which are there, which they call pole-stars (*tramuntanam*), and the people in Europe and those parts navigate by the sign of the south. I can show that this is true. Take a diamond stone (lodestone); you will find that it has two aspects, of which one lies towards one pole and the other towards the other; by either of them the needle can be bound, and will point to that pole towards which the lodestone's aspect pointed. Sailors might have been deceived by this if they had not taken due care. And that these two stars do not move happens because the other stars which are nearer to the equator revolve in various circles, some shorter, some longer.

Assuming that this passage dates from just after +1300, it still has considerable interest since it suggests that some needles pointed to the south while others pointed to the north.[e] This is exactly what Shen Kua had said in the +11th century, as we shall shortly see. The expression 'locum tramontane septentrionalis', often found, shows that although the word 'ultramontane' had originally been used to designate the north pole, it could mean the south pole also. Haskins has drawn attention to the

---

[a] For this reference I am indebted to Professor Juan Vernet of Barcelona.

[b] Cf. Arberry (2), p. 33. A quite different discussion of the attractive power of the lodestone occurs in the same book, p. 26. It was this to which Wiedemann (17) drew attention.

[c] Paris, Bib. Nat. 6556; tr. auct. I am indebted to Professor Christopher Brooke for checking this translation.

[d] Sarton (1), vol. 1, p. 614.

[e] A closely similar passage is found in the works of Brunetto Latini, a celebrated Florentine grammarian, written about +1260 (text in Klaproth (1), p. 45).

words of Michael Scott, in his *Liber Particularis* (+1227 to +1236), where he says that there are two kinds of lodestones, one of which points north and the other south.[a] This is exactly in the Chinese tradition. Sarton, moreover, tells us[b] that in most of the Arabic accounts, the south-pointing property of the lodestone or needle is regarded as more important than a north-pointing property; which again is suspicious of a Chinese origin. And, as is well known, the Persian and Turkish names for the compass mean 'south-pointer'.[c] Finally, it has been demonstrated in detail by Taylor (6) that compasses made for astronomers rather than navigators pointed south, not north, even as late as +1670.

Perhaps a word should be said about certain other relevant views. Winter (2, 3) and others have attempted to show that the Norsemen or Vikings had the magnetic compass at an earlier date than +1190, but this has been invalidated as the argument rested on interpolated texts.[d] Still, there is the evidence of the interpolator that they had it as early as +1225. Another story widespread in the literature is that the compass was invented by Flavio Gioja of Amalfi about +1300, but this also has no basis, except that he or some other Italian may well have been responsible for the amalgamation of the wind-rose, or some improvement of it, with the magnet to form the compass-card.[e] There have also been suggestions that medieval European churches were oriented by means of the compass, but this is a highly controversial subject, and the fact is as yet far from being established.[f]

(ii) *Development of the magnetic compass in China*

What, then, happened in China? We propose now to take the basic text of Shen Kua, in the *Mêng Chhi Pi Than*, written about +1088, i.e. a century before the earliest European mention of the magnetic compass, as our fixed point, and to work back from that, considering all earlier references to the directivity of iron magnets, as well as other Sung texts. This important passage[g] runs as follows:

Magicians rub the point of a needle with the lodestone; then it is able to point to the south. But it always inclines slightly to the east, and does not point directly at the south (*jan chhang*

---

[a] 'Item est lapis qui sua virtute trahit ferrum ad se ut calamita et ostendit locum tramontane septentrionalis. Et est alius lapis generis calamite qui depellit ferrum a se et demonstrat partem tramontane austri.' Haskins (1), p. 294. [b] (1), vol. 2, p. 630.

[c] Klaproth (1) pointed out other suspicious facts. Both Albertus Magnus and Vincent of Beauvais, in describing the compass, used the 'Arabic' words *aphron* and *zohron* for north and south. Jacques de Vitry said the lodestone came from India. [d] See Sølver (1); von Lippmann (2) and Marcus (1, 4).

[e] Sarton (1), vol. 3, p. 715. The subject was exhaustively discussed by Schück (1), vol. 2, pp. 10ff. and Breusing (1). Yet in spite of crucial Chinese evidence which has been available in Europe for more than a century, many writers, e.g. Motzo (1) and Taylor (9), have continued to favour the Amalfitani as the first discoverers of magnetic directivity.

[f] The most recent study, that of Cave (1), who measured 642 churches, yielded a Gaussian distribution curve with two peaks, one due east and another a few degrees north of it. This was thought to be probably due to orienting to sunrise at particular times of the year (other than the equinoxes) when the foundations were first marked out in each case. Compass declination was hardly possible since the range of the distribution was so wide. See, however, the different problem of Chinese city plans, p. 312 below.

[g] The first sentence of it was given by Klaproth (1, p. 68) in 1834, and the whole by Biot (14) ten years later. On Shen Kua himself see Vol. 1, pp. 135ff. and many references in Vol. 3. Biographical information will be found in Holzman (1).

wei phien tung, pu chhüan nan yeh[1]).[a] (It may be made to) float on the surface of water, but it is then rather unsteady. It may be balanced on the finger-nail, or on the rim of a cup, where it can be made to turn more easily, but these supports being hard and smooth, it is liable to fall off. It is best to suspend it by a single cocoon fibre of new silk attached to the centre of the needle by a piece of wax the size of a mustard-seed—then, hanging in a windless place, it will always point to the south.

Among such needles there are some which, after being rubbed, point to the north. I have needles of both kinds by me. The south-pointing property of the lodestone is like the habit of cypress-trees of always pointing to the west. No one can explain the principles of these things.[b]

Moreover, the same book contains another, less well known, passage:

When the point of a needle is rubbed with the lodestone, then the sharp end always points south, but some needles point to the north. I suppose that the natures of the stones are not all alike. Just so, at the summer solstice the deer shed their horns, and at the winter solstice the elks do so. Since the south and the north are two opposites, there must be a fundamental difference between them. This has not yet been investigated deeply enough.[c]

Here, then, we have not only the undeniably earliest clear description of the magnetic needle compass in any language, but also a clear statement of the magnetic declination. Some Western historians of science have been willing to recognise this, for example Cajori (5), although it greatly antedates the traditional time of the discovery of declination, by Columbus in +1492.[d] The two kinds of needles mentioned by Shen Kua may of course have been magnetised at different poles of the lodestone, but there may also have been another origin for this traditional idea (see on, p. 267). The analogy with cypress-trees can only have rested on local observations of curvature due to prevailing winds.[e]

Wang Chen-To (4) has pointed out that some of Shen Kua's experimental conditions indicate a considerable amount of careful investigation. The use of a single thread only for the suspension would avoid twisting effects. That the thread should be of silk meant that it would be a continuous fibre, unlike a thread of hempen yarn (cotton was almost certainly not known in China in his time), in which shorter fibres would be spun together under tension. That the silk thread should be new would imply an evenly distributed elasticity.

[a] I shall never forget the excitement which I experienced when I first read these words. If any one text stimulated the writing of this book more than any other, this was it.

[b] Ch. 24, para. 18; tr. auct. adjuv. Klaproth (1); Biot (14); Wylie (11); Hirth (3). Cf. Hu Tao-Ching (1), vol. 2, pp. 768 ff.

[c] Appended chapters, ch. 3, para. 19, tr. auct.

[d] Not so Mitchell (2), however, who urges that the tendency of the needle to point to the east of south, described by Shen Kua, was merely 'an accidental result due to imperfect support or suspension'. The spirit in which some Europeans approach Chinese texts is exquisitely shown by his comment on the whole Shen Kua passage—'There is no immediately apparent ground on which this can be discredited' (Mitchell, 1).

[e] The directivity of the cypress must have been widely believed at the time since it is mentioned in other books, e.g. the *Phi Ya*[2] (A Heap of Elegances) by Lu Tien,[3] written just ten years after the *Mêng Chhi Pi Than*.

[1] 然常微偏東不全南也   [2] 埤雅   [3] 陸佃

Shen Kua's description of the compass, and of the needle's declination, was written just over a hundred years earlier than the first occidental mention. It will be convenient to divide the other noteworthy Chinese texts into several groups. First we must consider all texts down to the end of the Sung which describe or hint at the use of iron magnets, whether in needle shape or not. It will be found that this subject cannot be separated from the persistent use of the lodestone itself in various developed forms. Pushing back the analysis, we shall come, secondly, to the earliest techniques which made use of the directive properties of pieces of magnetite. In a third group of texts, we shall be concerned with the first date at which the use of the compass in navigation by Chinese sailors is mentioned. In yet a fourth group, we must study those which bear upon the problem of the discovery of the declination.

Before proceeding upon our backward course, however, there is one statement of much interest to detain us, though written a little later than the remarks of Shen Kua. It comes from one of the *Pên Tshao* compendia—still many decades before the first European mention.

(iii) *Sung compasses, wet and dry*

In the *Pên Tshao Yen I*[1] (The Meaning of the Pharmacopoeia Elucidated), a work which dates from +1116, Khou Tsung-Shih[2] says:[a]

Lodestone (*tzhu shih*[3]) is somewhat purplish in colour,[b] and its surface is rather rough. It 'inhales' needles or (small pieces of) iron, which may adhere to one another in serial attachment—hence it is commonly called 'the stone which attracts (or compels) iron (*hsieh thieh shih*[4]).' *Hsüan shih*[5] is another kind of lodestone dark or black in colour, but smoother; its medical use is more or less the same though with slight differences, so one must know that there are two kinds (and how to distinguish them). When one rubs a pointed (iron) needle (upon the lodestone) (proper), it acquires the property of pointing to the south, yet it inclines always towards the east (*jan chhang phien tung*[6]), and does not point due south. The (best) way is to attach a single fibre of new silk (fresh from the cocoon) to the middle of the needle with a very tiny quantity of wax (lit. a piece half the size of a mustard-seed). When this is hung up in a windless place it always points to the south. Again, if one pierces a small piece of wick (pith or rush)[c] transversely with this needle, and floats it on water, it will also point to the south, but will always incline (to the east) towards the compass-point Ping[7] (i.e. S. 15° E.). This is because Ping belongs to the principle of Fire, and the points Kêng[8] and Hsin[9] (in the west), which belong to Metal (the needle being of metal), are controlled by it.[d] Thus its (declination) is quite in accord with the mutual influences of things (*wu li hsiang kan*[10]).

ᵃ Ch. 5, p. 5*a*, tr. auct.; adjuv. Klaproth (1), p. 68.

ᵇ Li Shu-Hua (2) is right in reading *sê*[11] instead of *mao*[12] here, with the *Tao Tsang* edition of the text. But the mistake began at least in the +15th century and not with Li Shih-Chen, as may be seen from the +1468 edition of the *Chêng Lei Pên Tshao*.

ᶜ Rush wicks were common in China, especially *Juncus effusus* (R696), the *têng hsin tshao*.[13]

ᵈ According to the principle of Mutual Conquest in five-element theory (cf. Vol. 2, p. 257) Fire dominated Metal, since metal could be melted by fire. Khou Tsung-Shih's thought was that while a metal needle should naturally point to the west, the preponderant influence of Fire, associated with the south, drew it off to that direction.

¹ 本草衍義   ² 寇宗奭   ³ 磁石   ⁴ 熁鐵石   ⁵ 玄石
⁶ 然常偏東   ⁷ 丙   ⁸ 庚   ⁹ 辛   ¹⁰ 物理相感   ¹¹ 色
¹² 毛   ¹³ 燈心草

At first sight this passage[a] seems to be but a repetition of what Shen Kua had said thirty years earlier, but two things were added. Khou Tsung-Shih gives what was long supposed to be the earliest known description of the water compass,[b] so characteristic of all the oldest (but later) European accounts. Secondly, he not only gives a quite precise measure of the declination, but adds an attempted explanation for it.[c]

Actually, Khou Tsung-Shih was far from being the first to describe the water, or floating, compass. Until Wang Chen-To (4) noted it, everyone had overlooked a passage in the great compendium of military technology, the *Wu Ching Tsung Yao*,[1] edited by Tsêng Kung-Liang[2] and finished in +1044. Here the writer says:[d]

When troops encountered gloomy weather or dark nights, and the directions of space could not be distinguished, they let an old horse go on before to lead them, or else they made use of the south-pointing carriage, or the south-pointing fish (*chih nan yü*[3]) to identify the directions. Now the carriage method has not been handed down, but in the fish method a thin leaf of iron is cut into the shape of a fish two inches long and half an inch broad, having a pointed head and tail. This is then heated in a charcoal fire, and when it has become thoroughly red-hot, it is taken out by the head with iron tongs and placed so that its tail points due north (lit. in the Tzu[4] direction). In this position it is quenched (*chan shu*[5]) with water in a basin, so that its tail is submerged for several tenths of an inch. It is then kept in a tightly closed box (*i mi chhi shou chih*[6]).[e] To use it, a small bowl filled with water is set up in a windless place, and the fish is laid as flat as possible upon the water-surface so that it floats, whereupon its head will point south (lit. in the Wu[7] direction).

This statement is full of interest. The fish-shaped iron leaf must have been slightly concave so that it would swim like a boat (see Fig. 323 showing Wang's reconstruction). Though Wang Chen-To supposed that the writer's failure to mention rubbing the iron on a lodestone was due to the desire to keep a military secret, such a suggestion is unnecessary. When a piece of iron is allowed to cool rapidly through the Curie point (600–700°) while oriented in the earth's magnetic field, the metal (especially if steel) will be magnetised. It seems extremely probable that this phenomenon, now called thermo-remanence, was known to these Sung technicians; and though the magnetisation might be weak, it had the great advantage that it could be induced in the absence of a lodestone.[f]

---

[a] Repeated almost verbally about +1280 by Chhêng Chhi[8] in his *San Liu Hsüan Tsa Chih*[9] (Three Willows Miscellany), who adds that such compasses were widely used by geomancers.

[b] As de Saussure (35), p. 39, rightly pointed out.

[c] This further disproves the belief of Mitchell that the declination was not understood to be a constant and distinct phenomenon. Whether Shen and Khou really thought that it was not seen when the needle was suspended, instead of being supported or floated, as would follow from an absolutely literal interpretation of their texts, is not certain, and seems to me doubtful.

[d] *WCTY/CC*, ch. 15, p. 15 b; tr. auct.

[e] This direction might mean that the magnetised 'fishes' should be kept in a small box with a lodestone floor; such, at any rate, was a method which has continued in use until the present time. It is known as 'nourishing the needle' (*yang chen fa*[10]), and Wang Chen-To himself saw it in use in Anhui province.

[f] Thermo-remanence was of course known to William Gilbert, who illustrated the forging of a steel rod in the direction of the earth's field (cf. Andrade, 2).

| [1] 武經總要 | [2] 曾公亮 | [3] 指南魚 | [4] 子 | [5] 蘸水 |
| [6] 以密器收之 | [7] 午 | [8] 程棨 | [9] 三柳軒雜識 | [10] 養針法 |

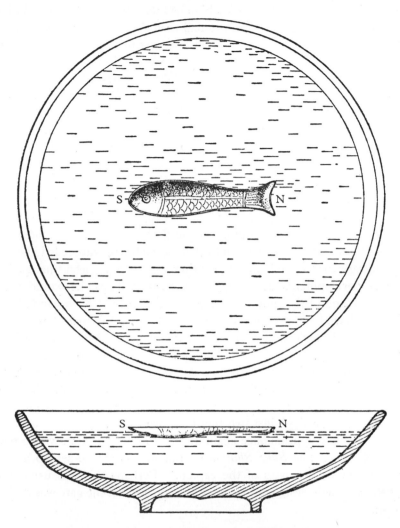

Fig. 323. Early forms of magnetic compass: the floating fish-shaped iron leaf described in the *Wu Ching Tsung Yao* (*Chhien Chi*), ch. 15, p. 15*b* (+1044). After Wang Chen-To (*4*).

That heating destroys magnetisation must have been long known in China; it is mentioned in the *Wu Tsa Tsu* of +1600.[a] W. A. Harland (1) described a Cantonese formula of 1816 for making needles for magnetisation,[b] which consisted in heating them to the highest degree of a charcoal fire for seven days and nights together with

[a] Klaproth (1), p. 97. And in the *Wakan Sanzai Zue* passage quoted below, p. 275.

[b] It was taken from the *Thung Thien Hsiao*[1] (Book of General Information) compiled by Wang Hsiang-Thang.[2] The content of this dates from the last years of the 18th century, but it was often reprinted (1816, 1837, 1856).

[1] 通天曉   [2] 王纕堂

vermilion, orpiment, iron filings, cock's blood,[a] etc. Heating alone would have done just as well for destroying spontaneous small magnetic fields and so improving later overall magnetisation. Perhaps the method was part of a steel-making process,[b] and as no lodestone is mentioned it may well be that the old discovery of cooling in the earth's magnetic field was still being used. Charnock,[c] the historian of naval architecture, repeated in 1800 rumours that the Chinese were 'not tormented by the variation of the needle', and that 'their needles do not receive their virtue from the loadstone', but by heating according to the formula just mentioned. Some thought that the former property was the result of the latter treatment, but Charnock justly agreed with Barrow (see p. 290 below) that it was due more likely to the excellent method of suspension.

A word must be said on the date of the passage about the hollow fish. Though the preface of the *Wu Ching Tsung Yao* is dated +1040, and the imperial preface +1044, it seems very likely that the description of the floating-compass cannot be later than +1027. For it was in that year that Yen Su,[1] as we shall see,[d] succeeded in constructing a south-pointing carriage, and since our writer cannot have been unacquainted with technological developments in the government workshops, he would hardly have said that the method of the south-pointing carriage had not been handed down. He may, of course, have been copying from some source earlier than +1027, but in any case the existence of the floating-compass seems clearly established for the first decades of the +11th century, and in all probability would go back to the later part of the +10th.[e] The fish form is particularly to be noticed in connection with what we shall shortly see regarding the first beginnings of the compass as a spoon. That the floating fish-shaped iron leaf spread outside China as a technique we know from the description of Muḥammad al-'Awfī just two hundred years later.[f]

Tsêng Kung-Liang appears once again in history in connection with a fish, in this case a symbolic one. The *Sung Shih* says[g] that in +1072 he set up a bronze fish talisman over one of the gates of the capital.[h] Perhaps this was using the south-pointing fish as a magic design or ornament to indicate that the gate was directed truly to the south.

---

[a] This speaks for the antiquity of the process, since the ingredient was so prominent in the *Huai Nan Wan Pi Shu* (cf. p. 278).

[b] See Sect. 30d, and Needham (32).

[c] (1), vol. 3, p. 299.

[d] In the Section on mechanical engineering, Sect. 27e below.

[e] This is the date of the important figure of Wang Chi[2] (p. 305) and the editors of the *Thai-Phing Yü Lan* encyclopaedia, to whom we must shortly pay attention (p. 274 below). They attest the use of the needle itself just at this time.

[f] Al-'Awfī gave a first-hand eye-witness account of a captain with whom he sailed taking a hollow piece of iron shaped like a floating fish, and laying it on water in a basin. It pointed in the *Qiblah* direction (south). The captain explained that when rubbed with the lodestone the iron naturally had this property (see Wiedemann, 8). Cf. above, p. 247 for the similar mention of Bailak (+1242).

[g] Ch. 15, p. 7b.

[h] *Ching chhêng mên thung yü fu.*[3]

[1] 燕肅   [2] 王伋   [3] 京城門銅魚符

It seems certain, however, that while these things were going on, the use of magnetite itself persisted in other forms of compass. These are described[a] in the *Shih Lin Kuang Chi*[1] encyclopaedia (Guide through the Forest of Affairs), written by Chhen Yuan-Ching.[2] This book cannot be exactly dated; though first printed only in +1325, it was compiled at some time between +1100 and +1250, probably after the move of the capital to the south in +1135. In his section on the magical arts of the Taoist immortals (*shen hsien huan shu*[3]), he wrote:[b]

They (the magicians) cut a piece of wood into the shape of a fish, as big as one's thumb, and make a hole in its belly, into which they neatly fit a piece of lodestone, filling up the cavity with wax. Into this wax a needle bent like a hook is fixed. Then when the fish is put in the water it will of its nature point to the south, and if it is moved with the finger it will return again to its original position.

Fig. 324 shows Wang Chen-To's reconstruction of this little object.[c] The text goes on to say:

They also cut a piece of wood into the shape of a turtle, and arrange it in the same way as before, only that the needle is fixed at the tail end. A bamboo pin about as thick as the end of a chopstick is set up on a small board, and sustains the turtle by the concave under-surface of its body, where there is a small hole. Then when the turtle is rotated, it will always point to the north, which must be due to the needle at the tail.

This is of great interest, because it indicates that the dry suspension from below was known during the Sung period. The reconstruction is seen in Fig. 325. We shall later find (pp. 290 ff. below) that the floating compass remained more popular among the Chinese during the following centuries, however, and that its supersession by the dry suspension was connected with European maritime influence. But nevertheless the first dry suspension was Chinese. In both the types just described, we should note again the significance of fish and turtle shapes in relation to the spoon shape. What is particularly interesting is the association of a needle with the lodestone, not as a 'detached' pointer, but for making more precise, as it were, the orientation of the lodestone-carrying object.

---

[a] As Wang Chen-To (*4*) has pointed out.

[b] Ch. 10, tr. auct. Only in the +1325 edition, not in those of +1340 and +1478.

[c] Something very similar was described by al-Zarkhūrī al-Miṣrī in a MS. dating from +1399 summarised by Wiedemann (8). It was a 'fish' of wood carrying a magnetised rod or needle of iron, and so floating. If any Muslim objected to images of animals, the lacquered wood was painted with the form of a *miḥrāb* (prayer apse). Here again we see the spread of a precisely identifiable Chinese technique. Experimental verification of it was carried out successfully by Schück, Ritter & Vollmer (in Schück (1), vol. 2, p. 54). Schück found some of the procedures of Petrus Peregrinus more difficult to repeat. Another late Arabic work which needs further investigation is the *Risālah fī Maʿrifah Bait al-Ibrah* (Treatise on the Knowledge of the House of the Needle) by Abū Zaid ʿAbd al-Raḥmān al-Tājūrī (*d.* +1590). A MS. is in the John Rylands Library at Manchester.

[1] 事林廣記   [2] 陳元靚   [3] 神仙幻術

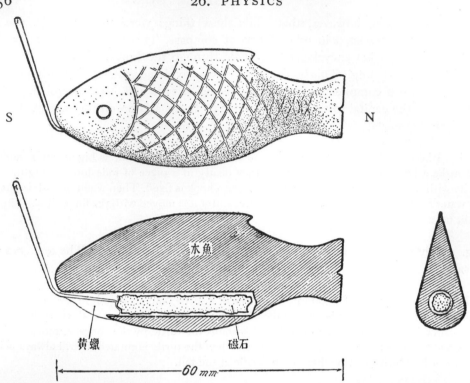

S · · · · · · · · · · N

木魚

黄蠟   磁石

|← ————————— 60 mm ————————— →|

Fig. 324. Early forms of magnetic compass: the floating wooden fish containing lodestone and needle described in the *Shih Lin Kuang Chi*, ch. 10 (*c.* +1150). After Wang Chen-To (*4*). The lodestone and the projecting needle pointer are fixed in place by wax.

(iv) *References in the Thang and earlier*

Of reliably Thang references to the magnetic needle we know none, but there are many in texts which were afterwards ascribed to the Thang,[a] for instance, the *Chhing Nang Ao Chih*[1] (Mysterious Principles of the Blue Bag, i.e. the Universe) by the famous geomancer Yang Yün-Sung.[2] It is much to be wished that some scholars would thoroughly examine the early geomantic texts so that literary evidence could be compared with internal evidence of a scientific nature, and approximate datings established.

Another research which should be undertaken is the systematic examination of Sung and Thang encyclopaedias and dictionaries for references to the lodestone and the magnetic needle. Klaproth stated long ago[b] that a reference to the magnet was to be found in the *Shuo Wên*, the great dictionary completed in +121 by Hsü Shen. The words he found were *Shih ming kho i yin chen*[3] (The name of a stone which will '*yin*' the needle). Many have subsequently pointed out quite correctly that this phrase does not occur in the original *Shuo Wên*, as reconstituted by later scholars. De Saus-

---

[a] And may really be of that date.   [b] (*1*), p. 66.

[1] 青囊奧旨   [2] 楊筠松   [3] 石名可以引針

sure, however, attributed[a] much significance to the fact that it does occur in the early Sung *editio princeps* edited by Hsü Chhuan[1] in +986, from which it was copied by most later dictionaries; he laid weight on this because he accepted the translation which Klaproth gave for *yin*, i.e. 'the stone which will give directivity to the needle'. This raises the question of the senses which the word *yin* can bear. As we saw above,[b] it was

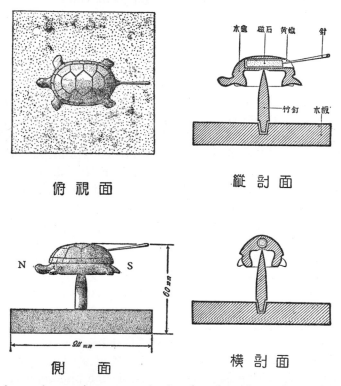

俯 視 面   縱 剖 面

側 面   橫 剖 面

Fig. 325. Early forms of magnetic compass: the dry-pivoted wooden turtle containing lodestone and needle described in the *Shih Lin Kuang Chi*, ch. 10 (c. +1150). After Wang Chen-To (4). Above left, seen from above; below left, from the side; above right, median sagittal section; below right, cross-section. The lodestone and the projecting needle pointer are again fixed in place by wax, and the object is free to rotate upon a sharpened bamboo pin.

perhaps the commonest expression used in ancient texts (such as the *Lü Shih Chhun Chhiu*, the *Huai Nan Tzu*, the *Lun Hêng*, and Tshao Tzu-Chien's note,[c] all of which have it) for what have generally been regarded as statements of attractive power only. But the word is a little ambiguous. Its original meaning was to pull or draw a bow, hence to pull up a fish or to attract, or to light a fire. But it also came to mean to lead, conduct, introduce or present, to derive, to propagate, to induce or to extend, even to prolong. There seems therefore no reason why it might not have meant, at some periods, the extension of directivity to the magnetised needle, as well as the attraction

---

[a] (35), p. 23.   [b] P. 237.   [c] P. 232.

[1] 徐 鍇

of the needle to the lodestone. One way of elucidating the matter further would be to draw up a statistical table of all those texts which speak of iron, as against those which say specifically needle. This we shall present below. Of course, there is also the possibility that some Sung or Thang dictionaries may be more explicit, and so justify more fully a Thang date.

It has often been pointed out that if the magnetic compass had been used for navigation in or before the Thang, there should have been some mention of it in the numerous accounts of the pilgrimages of Buddhist monks to India.[a] No such mention has yet been found.[b] At that time the compass may well have been purely geomantic and not nautical. However, it must be observed that in this literature there are references to the 28 *hsiu* (lunar mansions) as azimuth points—thus I-Ching, leaving Canton in +671, sailed in the direction of the *hsiu* I[1] and Chen[2] (cf. Table 24 above).[c] These are still placed on Chinese compasses (Fig. 326 below) and would indicate the region from S. 15° E. to S. 45° E. The implication is that the 'wind-rose' of the sea-captains of Thang times followed closely the cosmological principles embodied in that ancestor of the magnetic compass, the *shih* or diviner's board. Indeed this is so old that in all probability the mariners of the Han had also been acquainted with it.

In this connection it is also to be noted that no mention of the mariner's compass has so far been found in the accounts of China during the Thang dynasty written by Arab travellers. Such negative evidence can carry but limited weight, for in all ages Chinese ships carried Taoist priests, and the use of the floating needle may well have been regarded as a numinous secret which would not readily be revealed to foreign merchants. It is interesting to remember that this negative evidence played a part in the Jesuit controversies of the 17th and 18th centuries.[d] One of their opponents, Eusebius Renaudot, published in +1718 a translation of the voyages of Sulaimān al-Tājir (in China around +851)[e] and Ibn Wahb al-Baṣrī (in China around +870),[f] as completed in the *Silsilat al-Tawārīkh* of Abū Zaid al-Ḥasan al-Sīrāfī[g] (+920); with the object of minimising as much as possible the philosophical and technical achievements of the Chinese. Naturally he laid much weight on the absence of mention of the magnetic compass in these writings.[h] Yet Gaubil, whose opinion

[a] Cf. Vol. 1, pp. 207ff.

[b] Cf. p. 281 below. The alarming narrative of the voyage of Fa-Hsien about +414 from India to China has often been quoted in this connection; cf. Hirth & Rockhill (1), p. 27; Schück (1), vol. 2, p. 9; Li Shu-Hua (2), p. 191. But it is the considered opinion of some experts in navigation and nautical technology (such as Cdr. D. W. Waters at the International Congress of the Discoveries at Lisbon in 1960) that the voyages of the Chinese to the Persian Gulf in the +8th century (cf. Vol. 1, p. 179) could not have been accomplished without the aid of some form of magnetic compass.

[c] Chavannes (4), p. 117; de Saussure (35), p. 36. The curious process whereby an essentially equatorial cycle, that of the *hsiu*, was first immobilised thereon, and then transferred *en bloc* to the horizon to form an immobile azimuth cycle, will be explained shortly (p. 265).

[d] Cf. Pinot (1), pp. 109, 160, 229, 237.

[e] Cf. Hitti (1), pp. 343, 383; Mieli (1), pp. 13, 79, 81. Latest translation: Sauvaget (2).

[f] Cf. Mieli (1), p. 302.

[g] Cf. al-Jalil (1), p. 138; Mieli (1), p. 115.

[h] Note the section 'Éclaircissements sur les Sciences des Chinois', p. 340.

[1] 翼   [2] 軫

should never be despised, stated[a] in one of his last writings[b] that he believed the compass had taken its definitive form about the time of the Emperor Hsien Tsung, c. +800. Those who read this Section to the end may feel that he was not far off the truth.

The next three passages have not previously been noticed in the study of this subject.[c] It may be remembered that in the Section on mathematics we had occasion to say something of the *Shu Shu Chi I* (Memoir on some Traditions of Mathematical Art), a book attributed (not without justification) to Hsü Yo (*fl.* +190) but also suspected of having been in fact written by its own commentator, the Northern Chou mathematician, Chen Luan (*fl.* +570). Here we need adopt no decision as to whether the work is genuinely of the Later Han or not, for any mention or hint of the magnetic compass would be quite remarkable enough (on conventional views) for the mid +6th century. In a passage quoted already in another connection,[d] we have something which at first sight seems relevant; it runs:[e]

Those who do not recognise the 'three', yet boastfully claim to know the 'ten' (i.e. ignorant mathematicians), are like the River People who lost the direction of their return destination, and blamed it on the unskilful hand of the direction-finder (*yu Chhuan-Jen shih mi chhi chih kuei, nai hên ssu-fang chih shou shuang*[1]).

An unwary interpretation might take the 'river people' as sailors, and the direction-finder as the compass, but Chen Luan's commentary explains that the remark refers to a legendary story about Huang Ti going into the mountains with people from a land of rivers,[f] who blamed the figure on the south-pointing carriage when they lost their way, and were then taught the use of the gnomon by Jung Chhêng.[2] We agree, therefore, with Wang Chen-To (5), that this passage does not concern the magnetic compass.

Nevertheless, the book of Hsü Yo proves to contain something valuable for our quest. In an adjacent passage, the text says:[g]

In the Eight Kua method of calculation, a needle points at the eight directions. When a position is lacking, it points at Heaven. (*Pa Kua suan, chen tzhu pa fang; wei chhüeh tsung Thien.*[3])

[a] (11), p. 179.
[b] In another place he wrote (2), p. 95: 'Je viens de trouver dans un Livre fait sur la fin des *Han* l'usage de la Boussole marqué distinctement pour connôitre le Nord et le Sud. On y parle expressement de l'aiguille aimantée.' And in a footnote he added, 'Dans ce même Livre on parle d'un Indice au-devant des Chariots pour marquer le nombre des *Lys* qu'on avait fait en chemin'. As Li Shu-Hua (2) remarks, this book has never been identified, but we feel sure that it must have been the *Ku Chin Chu*, which does in fact speak of the south-pointing carriage (see Sect. 27e), of the hodometer (see also Sect. 27e), and of the 'mysterious needle' (see p. 273). Gaubil might easily have dated this a couple of centuries too early.
[c] Except by Wang Chen-To, whose later work became known to us only after the writing of this Section.
[d] Vol. 3, p. 600.   [e] P. 3b, tr. auct.   [f] Szechuan, perhaps.
[g] P. 7a, tr. auct. Still valuable even if, as some believe, no earlier than the +10th century.

[1] 猶川人事迷其指歸乃恨司方之手爽   [2] 容成
[3] 八卦算針刺八方位闕從天

Then comes Chen Luan's commentary:

In this kind of computing, the digits are indicated by the pointing of the sharp end of a needle (*yung i chen fêng so chih i ting suan wei shu*[1]). The first digit occupies the *Li*[2] position, i.e. pointing full south; second, or two, is at *Khun*,[3] south-west; the third, or three, is at *Tui*,[4] full west. Then come four at *Chhien*[5] in the north-west, five at *Khan*[6] full north, six at *Kên*[7] in the north-east, seven at *Chen*[8] in the east, and eight at *Sun*[9] in the south-east. Thus there is no place for the ninth digit, and so it is called 'perpendicular', (as if) the needle were pointing at Heaven.

This technique, which would seem to have been some simple sort of abacus-like device,[a] arising out of the old diviner's board, is elsewhere attributed to, or associated with, the name of Chao Ta,[10] a famous diviner of the San Kuo period.[b] But the remarkable thing is that a needle is said to be used as pointer, and the series starts from full south. It seems hard to believe that all this can have had no connection with the magnetic compass, and it must be at least of +570 if not earlier.

[a] Chen Luan explains (p. 10b) that a digit to be multiplied is indicated by the point of the needle, while a digit to be divided is indicated by the needle's tail, hence the expression *thou chhêng wei chhu*.[11] The fact that the two ends of the needle were clearly different in shape may have special significance, as we shall shortly see. On the *kua* positions, see p. 296 below.

[b] Chao Ta flourished between +225 and +245, gaining fame not only as the adviser and minister of the first Wu emperor, Sun Chhüan,[12] but even more as a mathematician and diviner. His association with a dial-and-pointer system makes him especially relevant in the history of the compass needle, and further research on him would be very useful. His biography (*San Kuo Chih*, ch. 63, pp. 4aff.) tells us that when young he mastered the 'nine palaces' (*chiu kung*[13]) method of computation (possibly permutations and combinations, cf. Vol. 3, pp. 139, 542). His reputation as a diviner was extraordinary—he numbered locust swarms, shot at men in ambush, and using chopsticks as counting-rods reckoned after a very poor entertainment that there was plenty more wine and venison in the house. Later he taught part of his knowledge to the Confucian scholars Khan Tsê[14] and Yin Li,[15] both high officials. Another man of the same kind, Kungsun Thêng,[16] studied under him for a long time, hoping that he would reveal all his secrets, but failing in this, offered wine ceremonially and begged to be taught them as his intellectual heir. Chao Ta said: 'My family has been ambitious to teach kings and emperors, but none of us has ever risen above the rank of Astronomer-Royal, so now we have no wish to hand down our arts....Some are very delicate and precise, such as the method of multiplying at the head and dividing at the tail.' This may well have been a reference to some early form of abacus (cf. Vol. 3, p. 75). 'Father and son', said Chao Ta, 'rarely spoke, but it was handed down in the family.' Finally, after a little more wine, he said that he had a secret book entitled *Su Shu*,[17] as thick as a man's fist, which he would give to Kungsun Thêng. This would explain everything. But he added that he was out of practice in these matters, and needed time to remind himself of them. However, when Kungsun Thêng returned, the book proved to be lost, and though Chao Ta thought his son-in-law might have taken it, it never came to light. Here we see strongly at work the inhibitory factor of family property in ancient Chinese mathematics and science. Chen Luan himself in the +6th century did not believe the general view that the dial and needle-pointer method started with Chao Ta (*Shu Shu Chi I*, p. 10a). 'Someone asked him, saying, "Formerly there was Chao Ta of Wu, who had a method of computing by ranks (*têng*[18]). Was not this (dial-and-pointer) method something to do with his multiplying at the head and dividing at the tail?" Chao Ta replied "That is the tradition, but it has no substance. It is like the musician Khuei having but one foot, or Mr Ting digging a well and finding a man at the bottom of it...."' And he went on to apply the famous phrase to the dial-and-pointer method, leaving us in the dark as to what Chao Ta had meant by it. One cannot help believing that more light on this whole subject would be of great value for the early history of dial-and-pointer readings.

[1] 用一針鋒所指以定算位數   [2] 離   [3] 坤   [4] 兌   [5] 乾
[6] 坎   [7] 艮   [8] 震   [9] 巽   [10] 趙達   [11] 頭乘尾除
[12] 孫權   [13] 九宮   [14] 闞澤   [15] 殷禮   [16] 公孫滕
[17] 素書   [18] 等

A passage which cannot be found in the present text of the *Tshan Thung Chhi* of Wei Po-Yang, the earliest of the alchemical books (*c.* +145), is preserved as if it were due to Wei himself in the *Phei Wên Yün Fu* encyclopaedia.[a]

Bright like the candle glowing in the depths of the darkness; like the needle shining in the midst of the bewildering sea (*tshan jan jo hun chhü chih chu; chao jan jo mi hai chih chen*[1]).

But this has been identified[b] as coming from a commentary of the late +13th century, quoting Yü Yen,[2] who applied these words to the book itself.

Lastly, we may note a strange passage, typical of Ko Hung, in the *Pao Phu Tzu*, *c.* +300:[c]

Those who are prejudiced by affections do not bear criticism. Those who are envious of the beauty of others—their needle is not bright and straight. Mugwort (i.e. ordinary) people are bewildered (by these affections) and have no 'south-pointer' to bring them back.

Here it is difficult to believe that the writer did not have some kind of magnetic compass in mind.[d] We are not bound, of course, to believe that this was in the original text, but if an interpolation, it is not likely to be later than Chen Luan.

(v) *The Han diviners and the lodestone spoon*

We come now, pushing back a few centuries further, to the *Lun Hêng* of Wang Chhung (+83), and to the passage which Wang Chen-To (2) has so interpreted as to throw a flood of light on the origin of the compass.[e] It runs as follows:[f]

As for the *chhü-yi*[3] (the 'indicator-plant')[g] it is probable that there was never any such thing, and that it is just a fable. Or even if there was such a herb, it was probably only a fable that it had the faculty of pointing (at people). Or even if it could point, it was probably the nature of that herb to move when it felt the presence of men. Ancient people, being rather simple-minded, probably thought that it was pointing when they really only saw it moving. And so they imagined that it pointed to deceitful persons.

---

[a] Ch. 27, under 'needle' (p. 1403.2); tr. auct.

[b] Wang Chen-To (4).

[c] *Wai Phien*, ch. 25, p. 2*b*, tr. auct.

[d] We shall shortly review a number of other texts which speak of a 'south-pointer' without the word needle in the context.

[e] The first suggestion of such an interpretation is due perhaps to Chang Yin-Lin (2) in 1928.

[f] Ch. 52 (ch. 17, p. 4*a*), tr. auct.

[g] This was an old legend to the effect that in the palace courtyards of the Chou emperors there was a plant which pointed accusingly at potential traitors, cf. the *Ti Wang Shih Chi*[4] (Stories of the Ancient Monarchs) by Huangfu Mi (+3rd century). It attracted the attention of the early Portuguese; see de Mendoza (Parke tr.), p. 70. De Lacouperie (4), with much eccentric learning, sought an origin for it in Babylonia ('the calendar plant'). Cf. Sect. 27*j*, and Needham, Wang & Price (1), p. 103.

[1] 燦 然 如 昏 衢 之 燭 照 然 如 迷 海 之 鍼   [2] 俞 琰   [3] 屈 軼

[4] 帝 王 世 紀

But when the south-controlling spoon is thrown upon the ground, it comes to rest pointing at the south. (*Ssu nan chih shao, thou chih yü ti, chhi ti chih nan*[1]).[a]

So also certain maggots which arise from fish and meat, placed on the ground, move northward. This is the nature of these maggots. If indeed the 'indicator-plant' moved or pointed, that also was its nature given to it by Heaven.[b]

Here Wang Chhung is contrasting a fabled phenomenon in which he did not believe, with actual phenomena which he had seen with his own eyes. The passage was garbled in some later editions[c] so that *ssu-nan* was replaced by *ssu-ma*,[2] and there was no explanation of what 'the Marshal's spoon' could mean. *Ssu-nan* may be translated here by 'south-pointing' instead of 'south-controlling', and *shao* may be taken to mean specifically the handle of the spoon. The essence of Wang Chen-To's proposal may be summarised by saying that the phrase *thou chih yü ti* should be interpreted as 'made to revolve upon the ground-plate of the diviner's board',[d] the spoon itself being nothing else than lodestone worked into that shape by the jade-cutters, in conscious imitation of the form of the Great Bear (Pei Tou[3] or Northern Dipper). As for the description of the maggots, it may be regarded as the earliest mention of a tropism; what Wang Chhung had seen was probably some insect larvae of a species with strongly phototropic tendency.

What, then, was the diviner's board (*shih*[4] or *shih*[5])? It was composed of two boards or plates, the lower one being square (to symbolise the earth, hence called the *ti phan*[6]); and the upper one being round (to symbolise heaven, hence called the *thien phan*[7]). The latter revolved on a central pivot and had engraved upon it the 24 compass-points, composed, just as in the later traditional compass (see on p. 297), of the denary and duodenary cyclical characters, *wu*[8] and *chi*[9] (which symbolised the earth) being repeated in order to make up the full number. It always bore, engraved at the centre, a representation of the Great Bear. The 'ground-plate' was marked all about its edge with the names of the 28 *hsiu* (equatorial divisions or constellations; cf. Vol. 3, p. 231 above), and the 24 directions were repeated along its inner gradations. Moreover, it carried the eight chief *kua*[10] (trigrams; cf. Vol. 2, p. 313 above) arranged according to the Hou Thien[11] system so that *Chhien*[12] occupied the north-west and *Khun*[13] the south-east. This is the same arrangement which we met with a moment ago in the

---

[a] This interpretation of the second *ti* is Wang Chen-To's. A more obvious rendering would be 'its handle (lit. root) points to the south'. If, however, this *ti* is regarded as equivalent to *ti*[14] or *chih*[15] then one might say 'it is knocked or tapped until it points to the south'. As the grammar seems to require a noun, the former of these possibilities may be the best. Moreover, quotations of the passage in encyclopaedias are liable to replace *ti* by *ping*,[16] unequivocally 'handle', as e.g. *TPYL*, ch. 762, p. 2b, in the +10th century.

[b] Forke (4), vol. 2, p. 320, translated the essential words correctly, but passed them over in complete silence. He suggested that the 'indicator-plant' was some sensitive species such as *Mimosa pudica*.

[c] As in *TPYL*, ch. 762.

[d] This does no violence to the text, as may be proved by ancient dictionaries. *Thou* can be equivalent to *thi*,[17] to move gently.

[1] 司南之杓投之於地其柢指南   [2] 司馬   [3] 北斗   [4] 式   [5] 栻
[6] 地盤   [7] 天盤   [8] 戊   [9] 已   [10] 卦   [11] 後天   [12] 乾
[13] 坤   [14] 抵   [15] 抵   [16] 柄   [17] 攡

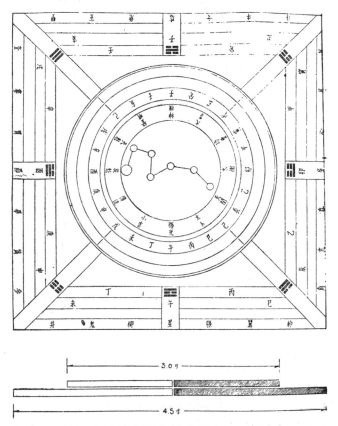

Fig. 326. The diviner's board (*shih*) of the Han period: diagrammatic reconstruction by Wang Chen-To (2). The apparatus was a double-decked cosmical diagram having a square earth-plate surmounted by a rotatable discoidal heaven-plate (see cross-section below the plan). Both plates were marked with cyclical and astronomical signs (compass-points, *hsiu*, etc.) as well as *kua*, and certain technical terms used only in divination. The round heaven-plate carried centrally a representation of the Great Bear. The apparatus was of bronze or painted wood. North is at the top. Cf. Fig. 223 in Vol. 3.

curious computing-machine of the *Shu Shu Chi I*, but differs from that found on all later geomantic compasses, where *Chhien* is the south and *Khun* the north. We will defer consideration of this difference until the description of the later standard compass (p. 296 below). Fig. 326 shows a reconstruction of what the *shih* looked like from above, and in elevation.

There is no dispute about the construction of the *shih*. Fragments of actual examples have been found in Han tombs,[a] notably the tomb of Wang Hsü[1] at Lo-lang[2] in Korea, excavated by Harada & Tazawa (1),[b] and in another Korean Han

---

[a] It should be noted that Wang Chen-To's theory of the lodestone spoon still lacks the caucial confirmation which would be afforded by the discovery of an actual example from a Han tomb.
[b] Pl. cxii.

[1] 王盱   [2] 樂浪

tomb,[a] that of the 'Painted Basket' (Koizumi & Hamada, 1). These were flat pieces of wood, cut to the right shape and lacquered. One of the 'heaven-plates'[b] is illustrated here (Figs. 327, 328). Since several objects in Wang Hsü's tomb were dated, and the latest date was +69, the burial can certainly not have been earlier than that, and probably took place about that time. It is of much interest that this was just

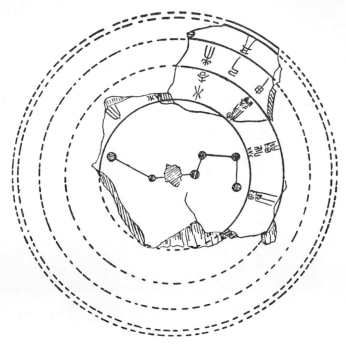

Fig. 327. Drawing of a fragment of a heaven-plate to show the characters identifiable (Wang Chen-To, 2).

contemporary with Wang Chhung's description of the use of the lodestone spoon on the *shih*. The other, 'Painted Basket' (*tshai hsia*[1]), tomb was rather earlier, containing objects dated +4 and +8; moreover, the style of their inscriptions pointed to the time of Wang Mang. Among them were spoons (though none of lodestone) the shape of which permitted easy rotation when balanced on their bowls.[c]

Not all the 'ground-plates' were made of wood, however; some bronze pieces of this kind may have been confounded with mirrors.[d] A Chhing archaeologist, Liu

[a] Korea at that time was partly a province of the Han empire. Further on the archaeology of the Lo-lang region see Sekino, Yatsui *et al.* (1).

[b] It was figured also by Rufus (2) shortly after its discovery, and is seen in Koizumi & Hamada's pl. CVII (cf. their p. 23). Curiously, the drawing of the Great Bear is inverted, as if a mirror image, or as if seen from 'outside the world'. Our attention was drawn to this by Dr Crone of the Netherlands Geographical Society. It would be interesting to know whether the inversion was a general practice among the Han makers of *shih*, and if so, why they adopted it.

[c] These were taken as models by Wang Chen-To in his reconstruction of the lodestone spoon.

[d] Cf. Harada & Tazawa (1), fig. 27.

[1] 彩篋塚

PLATE CXIII

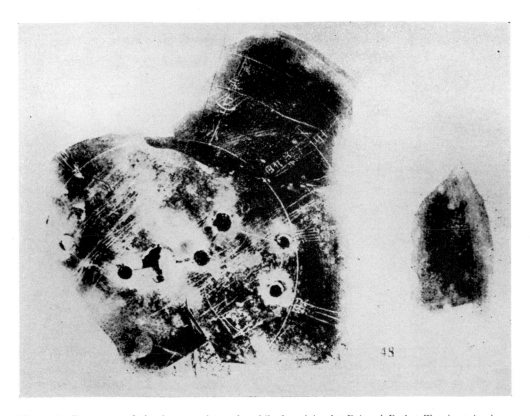

Fig. 328. Fragment of the heaven-plate of a *shih*, found in the Painted Basket Tomb at Lo-lang, Korea, +1st century. At that time this region was a Han commandery. After Koizumi & Hamada (*1*); Rufus (2).

Hsin-Yuan,[1] described such a bronze 'ground-plate', calling it a 'Four-Direction Mirror' (ssu mên fang ching[2]); it was certainly a Han piece.[a] The most complete example extant is of bronze, dating from the Sui period or somewhat earlier; it has now been described by Yen Tun-Chieh (15). Something is known of the wood from which the shih were usually made in later times. The Thang Liu Tien[3] (Institutions of the Thang Dynasty)[b] from between +713 and +755 states that the round 'heaven-plate' was made of maple wood, and the square 'ground-plate' of selected jujube wood.[c] It may be significant that a particularly hard wood was chosen for the lower board, for that would be the surface on which the spoon would have to rotate if, as Wang Chen-To suggests, the upper board was removed and the spoon substituted for it.

There is a close correspondence between the markings on the shih and the cosmological theories in the Huai Nan Tzu book (−120). The division of the hsiu into nine spaces (chiu yeh[4]), and the use of the denary and duodenary series of characters as compass-points, follow closely its expositions.[d] Moreover, an important section of that book is devoted to a survey of the characteristics and qualities of the seasons of the year as indicated by the tail of the Great Bear.[e] This is based on the annual displacement of the tail which, observed at a given hour of the night, shifts round a complete circuit in the circumpolar region,[f] pointing successively to different equatorial positions. By what de Saussure (16a) called a 'physico-astronomical association of ideas', the four sidereal palaces (separated by hour-circles) into which the hsiu segments were divided by sevens (cf. Vol. 3, p. 240 above), and each of which was connected with a season, were transferred to the horizon; so that the Great Bear's tail was considered to make the round of the azimuthal compass-points. In the passage under discussion,[e] the tail is said to indicate successively these 24 points.[g] Since the hsiu, or equatorial constellations, were also brought down and immobilised round the horizon, they also participated in these recurring attentions, though in fact in the apparently circling heavens, the tail invariably points to the second hsiu Khang[5].[h] It will also be remembered (Vol. 3, p. 250 above) that ch. 5 of the Huai Nan Tzu book is entirely devoted to the annual peregrination of a star, γ-Bootis (Chao yao,[6] the 'Twinkling

[a] Chhi Ku Shih Chi Chin Wên Shu[7] (Bronze Inscriptions from the Chhi-Ku Studio).

[b] Ch. 14, p. 27b.

[c] The same statement is found in the earlier Kuang Ya dictionary. The jujube wood is from Zizyphus jujuba or vulgaris (R292, 293; B11, 484; 111, 275).

[d] Ch. 3, pp. 2b, 3a, 4b. Much elaboration of these parallels in Wang Chen-To (2).

[e] Ch. 3, pp. 10b–11b.

[f] Owing to the proximity of the Great Bear to the northern horizon, the azimuths 'pointed to' at monthly intervals do not proceed by equal differences, but Liu An and his friends did not worry about this inequality, and assumed a uniform rate (Chatley, 1).

[g] Actually only the 12 duodenary characters are mentioned by name.

[h] This Chinese system is to be sharply distinguished from the azimuth stars of the Arabs, as de Saussure (35), pp. 50ff., well explains. The Arabs chose stars to represent the compass-points according to the positions of their rising and setting as seen from a position approximately on the terrestrial equator (perhaps about 10° lat. N.), i.e. according to their NPD's. See further Sect. 29 below.

[1] 劉心源   [2] 四門方鏡   [3] 唐六典   [4] 九野   [5] 亢
[6] 招搖   [7] 奇觚室吉金文述

indicator'), which may be considered as a prolongation of the tail of the Great Bear, round the azimuthal points.[a] In sum, therefore, the tail of the Great Bear may be unhesitatingly denominated the most ancient of all pointer-readings, and in its transference to the 'heaven-plate' of the diviner's board, we are witnessing the first step on the road to all dials and self-registering meters.

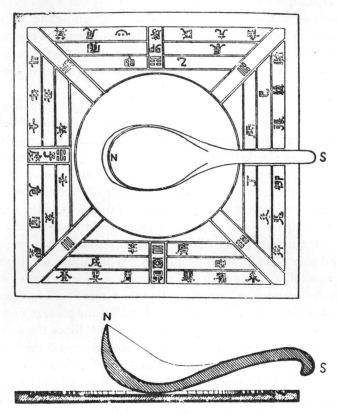

Fig. 329. Diagram (plan and elevation) of the earth-plate and the lodestone spoon, showing how the shape of the latter permits rotation upon its bowl in reponse to magnetic torque (Wang Chen-To, *2*).

The fascinating problem of exactly why and how this transference took place, and what it was that suggested the replacement of the 'heaven-plate' by an actual model of the Great Bear itself, would entail too long a digression at this stage and must be deferred until a later point (p. 315 below). On several grounds, Wang Chen-To (*2*) rejected the possibility that the spoon was made of magnetised iron, and resolved to explore the properties of a spoon carved from natural magnetite (lodestone). The only useful approach could be experimental, for in no other way would it be possible to predict whether a spoon of this shape would exert sufficient magnetic torque to overcome the friction between its bowl and the underlying smooth surface. One of the models which Wang constructed is shown in Figs. 329 and 330. He first made

[a] Cf. de Saussure (*11*), and Vol. 3, pp. 251 ff. above.

PLATE CXIV

Fig. 330. Model lodestone spoon (*shao*) and bronze earth-plate of the *shih* reconstructed by Wang Chen-To (*2*). The earliest form of the magnetic compass, if the true interpretation of Wang Chhung's words of +83 (see text, p. 262). North is to the left.

PLATE CXV

Fig. 331. Han stone relief dated +114 (van der Heydt Collection, Rietberg Museum, Zürich), a demonstration of art magic, sleight-craft and gramarye. Jugglers and acrobats are performing (on the significance of this cf. Vol. 1, p. 197), and music is being played (second row from the bottom, left to right: small pipe, large pipe, *shêng*, and pan-pipes). The figures beating upon the large drum in the centre may be mechanical jacks, for they closely resemble those which are seen in representations of Han hodometer carriages (see Sect. 27e below), and the whole structure, with its acrobats, may well be some kind of roundabout or carrousel automatically turning. The seated figures in the top row constitute perhaps the imperial audience. And still another wonder is presented to them at the extreme right top corner, where we see a large ladle of the same shape as the lodestone spoon, resting on the surface of a small square table (see inset drawing), and being attentively observed by a kneeling figure. It is possible that we have here, only a few decades after the words of Wang Chhung, a depiction of the earliest form of the magnetic compass.

a spoon from tungsten steel[a] and after magnetising it as strongly as possible, tested it on a 'ground-plate' made of highly polished bronze. It proved to be essentially south-pointing, with minor variations due to friction which the highest polishing could not overcome. Wood, even the hardest, lacquered or not, failed to provide a surface sufficiently smooth. He then proceeded to make further tests with lodestone from Tzhu Shan,[1] a source of iron ore and lodestone which was almost certainly available to the Han people, and obtained equally conclusive results.[b]

It might seem at first sight that for the success of the most ancient experiments there would have been one important condition, namely, that the natural magnetite block should be cut from its vein in accordance with the earth's magnetic field, i.e. in an approximately north–south direction. Had this not been done, one might think, the handles of the spoons might have pointed in any direction, not north or south. But in fact once a bar-shaped piece has been cut out from the rock, the free poles at the ends will cause the magnetic polarity to lie along the main axis of the bar. Such a polarity too would be intensified by thermo-remanence if the bar were heated and cooled, as in alchemical processes to which it may well have been subjected by the Han magicians. But since there was no way of testing the polarity beforehand the two ends of the bars may often have been confused, in which case some spoons would in fact have pointed to the north with their handles, while other specimens pointed south. This might conceivably account for Shen Kua's statement that there were two kinds of needles, some of which pointed north and others south. He may have been using two techniques which magnetised them differently, but on the other hand he may have been relaying an ancient tradition going back to the days of spoons, in which two kinds had been known.

It would seem almost impossible that any representation of the use of the lodestone spoon should have survived, but in this connection one Han relief is at least worth reproducing. Ladles in bowls are not at all uncommon, but in most cases they seem to be really ladles for serving soup or wine, as in four Han carvings figured by Rudolph & Wên.[c] Only their curved handles are visible over the edge of the bowls. But in a flat relief on a stone slab in the van der Heydt Collection,[d] dated +114, one sees (Fig. 331), in the extreme right-hand top corner, a personage apparently consulting what seems to be a large spoon of exactly the shape in question, not hidden by a bowl

[a] 15–20% Wo. O_2-content of magnetite would be some 30%.
[b] In contrast to certain failures to repeat the experiments of Pierre de Maricourt on floating lodestones (Michel, 6). It would be interesting to repeat Wang Chen-To's experiments in Europe, but I myself, in the summer of 1952, witnessed personally the performance of one of the lodestone spoons in the presence of Dr Wang and his assistant Mr Lu Shao-Shen in Peking. Although his best piece was not available, the spoon in question duly came to rest time after time pointing south if it was made to rotate with a mild rocking motion. High polish of the bronze base and (as much as possible) of the magnetite itself, is essential.
[c] No. 57, two winged deities playing the *liu-po* game (in the Szechuan Provincial Museum). Nos. 77, 78, 79, a banquet with dancers and jugglers (in a private collection at Chhêngtu).
[d] Now in the Rietberg Museum at Zürich.

[1] 磁 山

but resting almost fully in sight on the flat surface of a small square table.[a] The rest
of the relief, moreover, depicts, not a banquet, but some kind of ritual or entertainment accompanied by music and involving what may well be mechanical toys. Thus
the drum in the centre, with the jacks hitting it, is just like the drum of the hodometer
carriage (see Sect. 27e below), and the shaft rising above it carries what might
reasonably be interpreted as a roundabout or carrousel, the figures on which would
be mechanically operated.[b] It looks like a field-day of the emperor An Ti's magicians
and conjurers.

(vi) Literature on the diviner's board

We may now return to the consideration of further literary evidence. Two groups of
references are to be considered, first those which speak of the shih (diviner's board)
itself, and secondly those which speak of a ssu-nan (south-pointer or south-controller).

As to the first, there is no controversy. The diviner's board was well known at all
times from the Han onwards. Exactly how it was used has long been forgotten, and
all that we can be sure of was that the 'heaven-plate' was turned on its axis in imitation of the imaginary passage of the Great Bear's tail around the horizon according
to the seasons. One of the earliest mentions must be Tao Tê Ching, ch. 22.[c] The
reference to the shih in the Chou Li has already been studied[d] in connection with the
technical term lü[1] (standard pitch-pipes, also laws or regulations); here it may suffice
to repeat that it depicts the diviner's board in the charge of one of the highest officers
of state, who had to consult it for prognostication in time of war. The reference in the
Chhien Han Shu concerning Wang Mang requires more detailed consideration, and
will be postponed for a few pages. Ssuma Chhien says:[e]

Nowadays the diviners have a model of heaven and earth representing the four seasons
and following benevolence and justice[f] (jen i[2]); in this way they group the bamboo stalks and
decide upon what kua are involved. They turn the shih to adjust the kua to the right place
(hsüan shih chêng chhi[3]).[g] And then they are able to predict benefit or misfortune, success or
failure, in human affairs.

[a] We are indebted to Dr A. Bulling for calling our attention to this interesting piece and providing
a photograph of it. A closely similar representation is found on one of the reliefs in the Nan-yang Han
Hua-Hsiang Hui-Tshun collection (p. 66b), figured by Rudolph (3).

[b] Cf. Sect. 7h above, including Fig. 33, in Vol. 1, pp. 197ff., where emphasis was placed on the
hitherto unrecognised importance of 'jugglers and acrobats' in the history of technology.

[c] Cf. Vol. 2, p. 46 above.

[d] Biot (1), vol. 2, p. 108; see Sect. 18 on Laws of Nature, Vol. 2, p. 552.

[e] Shih Chi, ch. 127, p. 4a; tr. auct.

[f] I suspect that these adjectives refer to qualities of the natural phenomena of the heavens; cf. Vol. 2,
p. 43, in connection with the difficulty of inventing non-ethical technical terms for the properties of
minerals; and above, pp. 15ff., in connection with conceptions of 'justice' in Nature, with regard to
natural law.

[g] This is a strange word, for chhi means also 'chess pieces', though here, according to the Thang
commentator, it means the lower part of the board, the 'ground-plate'. See below, for the connection
between the compass and chess.

[1] 律   [2] 仁義   [3] 旋式正棊

This belongs to about −90. One can begin to see how natural it was that since the Great Bear revolved in the sky, the upper plate of the diviner's board should turn on the lower, and hence that the model Bear, or spoon, should also turn on it. Other mentions occur in the *Kuang Ya* (enlargement of the *Erh Ya* dictionary) which is perhaps +4th century (early Northern Wei); in the *Shih Shuo Hsin Yü* (late +5th or early +6th)[a] where the *shih* is associated with the Later Han scholar Ma Jung;[1] in the *Yen Shih Chia Hsün* (about +590)[b] whose author said that he had used it himself; and twice in the *Sung Shu* (written mid +6th)[c] where we find a certain Yen Ching[2] famous for using it.[d] Thang encyclopaedias record at least three forms of it. Also the bibliographies are interesting. That in the *Chhien Han Shu*[e] records two books on the use of the *shih*, one being entitled *Hsienmên Shih Fa*[3] (The Shaman Method of Using the Diviner's Board—or perhaps Hsienmên Kao's Method).[f] In about +1150 the bibliographer of the *Thung Chih Lüeh* is able to record[g] no less than 22 books on the *shih*, including about half-a-dozen *Shih Ching*,[4] attributed to various ancient men such as Wu Tzu-Hsü[5] and Fan Li.[6] Naturally enough, there was some connection with alchemy. In the *Pei Mêng So Yen*, written about +930, we read[h] of an alchemist called Tsung Hsiao-Tzu,[7] who was expert with the diviner's board, which he used to foretell the success of his chemical operations while in exile in Szechuan about +880. It is a pity that no detailed study of the *shih* has been attempted.[i] Scholars are still affected by the contempt with which modern scientists sometimes reject even historical investigation into the pit of the pseudo-sciences whence they themselves were digged.[j]

(vii) *References to the 'south-pointer'*

Of the texts which speak of the *ssu-nan* (south-pointer or south-controller), what might be the oldest is the reference in the *Kuei Ku Tzu* book. This says:[k]

When the people of Chêng[8] go out to collect jade, they carry a south-pointer with them so as not to lose their way. The assessment of abilities and probabilities is the south-pointer for human affairs.

Many editions of this book have slipped in the word *chhê*[9] (carriage) after *ssu-nan*, but Wang Chen-To regards this as a typical case where scribes or editors, unskilled in geomancy, had heard only of the carriage, and thought that every reference to south-pointing must refer to that. This did not always happen, however, for the

---

[a] Ch. 1B, p. 9*b*.
[b] Ch. 19, p. 8*b*.
[c] Ch. 97, pp. 8*b*, 21*a*.
[d] Ch. 57, p. 23*b*.
[e] Ch. 30, p. 45*b*.
[f] Cf. Vol. 2, p. 133 above.
[g] Ch. 44, pp. 31*b*ff.
[h] Ch. 11, p. 5*b*.
[i] Yen Tun-Chieh (*14*) has, however, given an account of the *liu-jen* method of divination (cf. Vol. 2, p. 363), which was associated with the *shih*.
[j] As to this, the excellent note which Neugebauer (4) entitled 'The Study of Wretched Subjects' is well worth reading. Compare 'The Vindication of Rubbish' by Pagel (5). Echoes of the *shih* in other parts of the world would be of interest. Cf. Thorndike (1), vol. 2, pp. 116ff. on the *Prenosticon Socratis Basilei* (perhaps +12th century).
[k] Ch. 10, p. 19*a*, tr. auct. adjuv. Forke (13), p. 490.

[1] 馬融　　[2] 顏敬　　[3] 羨門式法　　[4] 式經　　[5] 伍子胥
[6] 范蠡　　[7] 宗小子　　[8] 鄭　　[9] 車

original form of the text is sometimes preserved, as in Shen Yo's *Sung Shu*.[a] That the writer of the *Kuei Ku Tzu* cannot have had a carriage in mind is shown by the use of the word 'carry'. There is, of course, little probability that this passage really goes back to the early −4th century, when Wang Hsü[1] is supposed to have lived;[b] it is almost certainly Han, but that does not lessen its value for the present argument.

More authentically pre-Han must be the remark in *Han Fei Tzu*, which runs as follows:[c]

Subjects encroach upon the ruler and infringe his prerogatives like creeping dunes and piled-up slopes. This makes the prince to forget his position and to confuse west and east until he really does not know where he stands. So the ancient kings set up a south-pointer, in order to distinguish between the directions of dawn and sunset.

Now distinguishing between these directions was what the gnomon, the sun-shadow and the pole-star are supposed to do in the *Chou Li* (cf. Vol. 3, p. 231 above), so that if astronomical methods had been in Han Fei's mind, these would surely have been mentioned rather than a special 'south-pointer'. As Wang Chen-To has pointed out, it is in fact rather remarkable that the south-pointer never occurs among the lists of official astronomical instruments. It seems likely that this was because it took its origin among a group of technicians quite different from the astrologers, namely the geomancers. In any case, these words of Han Fei, part of a rather reliable text, carry the reference to the *ssu-nan* farthest back, into the time of the rise of Chhin just before the Han.[d]

Another instance of what was probably an insertion of the word 'carriage' by editors occurs in the *Chih Lin Hsin Shu*[2] (New Book of Miscellaneous Records) written about +325 by Yü Hsi[3] the astronomer.[e] After retailing the legend about the first construction of some kind of south-pointer by Fêng Hou[4] at the behest of Huang Ti to enable the army to find its way through the mists (a legend which later became firmly connected with the carriage) Yü Hsi goes on to say that he did this 'modelling it on the Great Bear' (*fa tou chi*[5]).[f] The passage must therefore originally have been a reference to the *shih*, and the word *chhê* was inserted afterwards.

Another indication that the south-pointer was an actual object occurs in the biography of Jen Fang[6] in the *Nan Shih*,[g] where someone laments that there is now no

---

[a] Ch. 18, p. 4*b*.

[b] A view gaining support recently is that the *Kuei Ku Tzu* book was written by Su Chhin[7] and other members of the School of Politicians (see Vol. 2, p. 206). Wang Hsü's historicity is rather uncertain.

[c] Ch. 6, p. 2*b*, tr. auct. adjuv. Liao (1), p. 44. Han Fei died in −233.

[d] Hirth (3) was the first Western sinologist who found these two passages, and de Saussure (33) recognised their interest, though not knowing the essential *Lun Hêng* passage.

[e] Cf. Vol. 3, p. 247.

[f] *YHSF*, ch. 68, p. 40*b*. Cf. an ode by the Thang scholar Chang Yen-Chen,[8] who says that the *ssu-nan* on earth is what the Great Bear is in the heavens (*TSCC*, *Khao kung tien*, ch. 175 in *pên* 794, p. 39*b*; also given in *YCLH*, ch. 387, p. 9*b*). The context shows rather clearly that the instrument and not the carriage was what Chang had in mind.

[g] Ch. 59, p. 8*b*. The reference will be of about +500.

[1] 王詡   [2] 志林新書   [3] 虞喜   [4] 風后   [5] 法斗機
[6] 任昉   [7] 蘇秦   [8] 張彥振

sage to whom the *ssu-nan* can be entrusted. There are a great number of metaphorical references to the south-pointer—we may instance only the following:

HAN +107   Chang Hêng's[1] ode, *Tung Ching Fu*:[2] 'You have been my south-pointer.'[a]

SAN KUO +205   Letter from Sung Chung-Tzu[3] to Wang Shang,[4] preserved in *San Kuo Chih*:[b] 'You, Sir, ought to take him (Hsü Ching[5]) as your south-pointer.'

SAN KUO *c.* +260   Tso Ssu's[6] ode, *Wu Tu Fu*:[7] 'The kobold Yü(-Erh)[8,c] rides in the van (of the emperor's hunting train) to lead the way, and the south-pointer gives orientation as to the four quarters. The carts (from the palace) rumble forward and the soldiers in armour march on with steady tramp.'[d]

CHIN *c.* +300   *Pao Phu Tzu* (*Wai Phien*), ch. 25, p. 2*b*: 'Mugwort (i.e. ordinary) people are confused and have no south-pointer to bring them back.'

LIANG +6th   *Wên Hsin Tiao Lung*[9] (On the Dragon Carving of the Literary Mind),[e] by Liu Hsieh:[10] 'Let these principles be your south-pointer.'

THANG *c.* +630   *Pien Chêng Lun*[11] (Discourse on Proper Distinctions), by the monk Fa-Lin:[12] 'Take Buddhist classics as your south-pointer.'

The implication of all this is that during these centuries (between early Han and late Thang) the 'south-pointer' was an actual thing, a compass-like instrument, known of as a famous device even by those who had not seen or used one, and indeed widely adopted as a literary metaphor. Since texts of this type never mention the carriage, they must refer to something else. If this was not the lodestone spoon on the ground-plate of the *shih*, what else could it have been? And if the beginning of it was not in the time of Wang Chhung and Chang Hêng, may it not perhaps go back beyond Luan Ta to the great school of Tsou Yen?

(viii) *The 'ladle of majesty'*

One of the most striking historical events connected with the *shih* was the latter end of Wang Mang, first and only emperor of the Hsin dynasty, between the Former and the Later Han.[f] Part of the account of the final capture of his palace in +23 by the Han people, in the course of which he was killed, runs thus:[g]

The conflagration then reached the Chhêng-Ming (Hall) in the lateral courts, where the Princess of the Yellow Imperial House (his daughter) dwelt. (Wang) Mang fled from there to

---

[a] *Wên Hsüan*, ch. 3, p. 21*b*; tr. von Zach (6), vol. 1, p. 37.

[b] Ch. 38 (*Shu Shu*, ch. 8), p. 4*b*.

[c] A mountain gnome in *Kuan Tzu*, who leads men to water and ore.

[d] *Wên Hsüan*, ch. 5, p. 11*a*, *CSHK* (Chin sect.), ch. 74, p. 8*b*; quoted in *YCLH*, ch. 387, p. 9*b*; tr. von Zach (6), vol. 1, p. 65.

[e] The oldest work in Chinese literature on literary criticism; see Chhen Shih-Hsiang (1). The passage occurs in ch. 27.

[f] Cf. Vol. 1, pp. 109 above.

[g] *Chhien Han Shu*, ch. 99C, p. 31*b*, tr. auct.

[1] 張衡   [2] 東京賦   [3] 宋仲子   [4] 王商   [5] 許靖   [6] 左思

[7] 吳都賦   [8] 俞兒   [9] 文心雕龍   [10] 劉勰   [11] 辨正論

[12] 法琳

the Hsüan Room, but the flames from the Front Hall immediately followed him. The palace maids and women wailed, saying 'What ought we to do?'. Meanwhile, (Wang) Mang, dressed all in deep purple[a] and wearing a silk belt with the imperial seals attached to it, held in his hand the spoon-headed dagger of the Emperor Shun. An astrological official placed a diviner's board (*shih*[1]) in front of him, adjusting it to correspond with the day and hour (*jih shih chia mou*[2]). And (the emperor) turned his seat, following the handle of the ladle, and so sat (*hsüan hsi sui tou ping erh tso*[3]). Then he said, 'Heaven has given the (imperial) virtue to me; how can the Han armies take it away?'

In this impressive scene, the chief question for us is what exactly Wang Mang was following as he turned his seat to face the magically right direction. Wieger[b] and Dubs[c] take *tou* to mean the Great Bear itself, and suppose therefore that the doomed emperor's movement was continuous,[d] but it seems surely more probable that his essential aim was to face truly to the south, the fundamental cosmo-centric ritual position of all Chinese rulers from earliest times. We are not even told that the Great Bear was visible; if it had been, he would not have needed the *shih*. The preferable suggestion is that the ladle or dipper in question was not the constellation but either its image engraved on the *shih*'s 'heaven-plate' or indeed its lodestone model, the handle of which indicated to him the true direction in which he must face to remind the universe of his imperial virtue, and so to defeat the Han rebels.

The great importance of objects shaped like spoons and ladles in Wang Mang's time is indicated also by another passage relating how he organised the making of a 'Ladle of Majesty'. It is not clear whether there was only one of these, or five, one for each of the five elements. The event occurred in +17.[e]

(In the fourth year of the Thien-Fêng reign-period) in the 8th month (Wang) Mang went in person to the place for the suburban sacrifice south of the capital to superintend the casting and making of the Ladle of Majesty (Wei Tou[4]). It was prepared from minerals of five colours and from copper. It was in shape like the Northern Dipper (Great Bear), and measured two feet five inches in length. (Wang) Mang intended (to use it) to conquer all rebel forces by means of conjurations and incantations. After the Ladle of Majesty was finished, he ordered the Directors of Mandates (from the Five Elements) to carry it solemnly on their shoulders whenever he went out, in front of him, and when he entered the palace, they waited upon him at his sides. On the day that the Ladle of Majesty was cast, the weather was exceedingly cold, so that some officials and horses from the government offices were frozen to death (while in attendance).

---

[a] This colour was appropriate to the North Polar region, symbolically imperial (cf. Vol. 3, p. 259).
[b] *TH*, pp. 634, 635.
[c] Personal communication. Cf. Dubs (2), vol. 3, p. 463, published later.
[d] There is every reason to believe that the emperor's seasonal progress through the cycle of throne-rooms in the Ming Thang (Cosmic Temple) followed a succession based upon the movement of the Great Bear (cf. Soothill (5), p. 93), but that was a slow process. Here we are dealing with the events of one agonising night.
[e] *Chhien Han Shu*, ch. 99c, p. 2b, tr. auct. Cf. Stange (1); Dubs (2), vol. 3, p. 372.

ⁱ 杙   ² 日 時 加 某   ³ 旋 席 隨 斗 柄 而 坐   ⁴ 威 斗

There can be no doubt that this magico-ritual object was a model representation of the Great Bear, and it has therefore much relevance to the lodestone spoon. Later commentators were somewhat perplexed as to the materials from which it was made (Yen Shih-Ku, for example, suggested brass), but the most probable suggestion, in the light of archaeological evidence already given (p. 111 above), is that it was of bronze encrusted with glass.[a] The name appears again in one of the Han apocryphal treatises, the *Li Wei Tou Wei I*[1] (System of the Majesty of the Ladle), which is full of astrological material.[b] Some of Wang Mang's ritual ladles were recovered about +430 near Nanking,[c] and one at least still existed in the Sung, for the *Yeh Kho Tshung Shu* of +1210 says[d] that the premier Han Yü[2] had in his house a bronze ritual *thung tou*[3] of Wang Mang's time, 1 ft. 3 in. long according to Sung scales, and inscribed with a date corresponding to +19. Wang Mou adds that though also called *tou*,[4] it was shaped like a spoon. Our conclusion therefore is that the 'Ladle of Majesty' was a ritual model of the Great Bear just as the 'south-pointer' was a practical lodestone model.[e]

(ix) *From the spoon to the needle*

Those who are acquainted with the literature on the mariner's compass in Europe are well aware of the fact that one of its earliest names was 'calamita'.[f] While some have suggested that this was derived from the Greek word for reed (*kalamos*, κάλαμος), and referred to the small piece of reed by which the needle was assisted to float, the more generally accepted view has been that the word meant a small frog or tadpole. It is thus used, for example, by Pliny.[g] Those who can also read Chinese are therefore liable to receive a considerable shock when they find, in the *Ku Chin Chu*, a +4th-century dictionary by Tshui Pao, the following remarks:[h]

Hsia-ma tzu,[5] the tadpole, is also called 'the mysterious needle' (*hsüan chen*[6]), or the 'mysterious fish' (*hsüan yü*[7]), and another name for it is the 'spoon-shaped beastie' (*kho tou*[8]). Its shape is round and it has a long tail. When it divests itself of its tail, its limbs grow out.

---

[a] Various lists of constituents of 'five coloured minerals' occur in Han and Chin alchemical works, e.g. *Tshan Thung Chhi*, ch. 12 (Wu & Davis (1), ch. 29); *Pao Phu Tzu*, ch. 4, pp. 32*b*, 33*a*; ch. 17, p. 12*a*; and the lodestone occurs in three of these. But glass is much more probable in this case. That would not preclude the addition of some magnetite to the melt by the artificers.

[b] *Ku Wei Shu*, ch. 19, p. 1*a*.

[c] *Nan Shih*, ch. 33, pp. 24*a* ff.

[d] Ch. 13, p. 8*a*. Repeated in other Sung sources.

[e] I am much indebted to Professor H. H. Dubs for discussions on this subject, and regret that our interpretations of the above passages could not be brought to coincide. He believes that the Wei Tou were measures of volume, but magic and metrology do not consort well together.

[f] Klaproth (1), pp. 15 ff.; de Saussure (35), p. 43, etc. In modern Italian, the word is used for a horseshoe magnet.

[g] *Hist. Nat.* XXII, 42. No doubt the extension of meaning occurred because amphibia live often among reeds.

[h] Ch. 5, tr. auct.

[1] 禮緯斗威儀   [2] 韓玉   [3] 銅科   [4] 鈄   [5] 蝦蟆子
[6] 玄鍼   [7] 玄魚   [8] 蝌蚪

The genuineness of the text of the *Ku Chin Chu* (Commentary on Things Old and New) has of course been contested, but a convincing defence of it was made by Hashimoto (1). In any case, for the present argument there is no need to insist that these strange words are of the +4th century, for we find an almost identical passage, with all the essential words, in the *Chung Hua Ku Chin Chu*[1] (Commentary on Things Old and New in China), which was written by Ma Kao[2] between +923 and +936. On the conventional view, this is quite bad enough. We see no other way of explaining them than to suppose that at some time between the +4th and the +10th century the south-pointing lodestone spoon was superseded by the south-pointing iron 'fish', 'tadpole', or needle which had been magnetised by being rubbed on the spoon (or on some other piece of magnetite), or by sudden cooling in the earth's field, and that by a natural association of ideas, the needle came to be called a 'frog' or 'tadpole', while the tadpole itself acquired the popular name of 'mysterious needle'. That the very character for 'tadpole' contains the spoon-ladle radical (*tou*[3]) as its phonetic is a point not to be missed. Tadpoles and Chinese spoons are strikingly similar in shape; but the association of tadpoles with needles becomes much more understandable when one has reason to think that iron needles derived from lodestone spoons.

Finally, we know (p. 257 above) that between +1040 and +1160, i.e. long before the first European references, the Chinese were using (*a*) floating wooden fish-shaped objects containing lodestone, (*b*) floating fish-shaped iron or steel magnets, and (*c*) pivoted wooden turtle-shaped objects containing lodestone.

Closer approximation to the date of origin of one of the Chinese methods of mounting a magnetic needle may be obtained by a critical study of textual variants. The *Han Wei Tshung-Shu* edition of the *Ku Chin Chu*[a] writes *yuan chen*[4] in the tadpole passage, but that *hsüan chen*[5] ('mysterious needle')[b] is correct is shown by reference to the *Pei-Chhüan Thu-Shu-Kuan Tshung-Shu* edition.[c] This is repeated by the *Chung Hua Ku Chin Chu* (both *Pai Chhuan Hsüeh Hai*[d] and *Han Wei Tshung-Shu*[e] editions). But when one looks at the citation of the passage in the *Thai-Phing Yü Lan* encyclopaedia[f] one is astonished to find that the editors wrote *hsüan chen*[6]— 'suspended needle'. Thus whether the change was made because the older term was not understood, or simply by an act of inadvertence, it betrays to us the fact that in +983, a full century before the book of Shen Kua (cf. p. 249 above), the system which he described of hanging the needle (presumably on a thread of silk) was known and used. The term employed (*hsüan*[6]) is the same as that in the language of Shen Kua himself.[g]

---

[a] Ch. 5, p. 9*a*.

[b] It will be remembered that the term *hsüan shih*,[7] 'mysterious stone', has already been discussed in connection with the names for magnetite. Another common meaning of *hsüan* is, of course, 'dark', but there the lodestone and the tadpole agreed also.

[c] Ch. 5, p. 6*a*.   [d] Ch. 3, p. 10*b*.   [e] Ch. 3, p. 7*a*.

[f] Ch. 949, p. 4*a*. The 1818 edition of Juan Yuan and Pao Chhung-Chhêng, based on a Sung edition.

[g] Further confusion took place in the Ming, when both MS. copies and printed editions wrote *hsüan kou*[8] ('suspended hook'), doubtless because magnetic needles were no longer hung (Peking Nat. Library Microfilm Collection, nos. 270, 1490).

[1] 中華古今注   [2] 馬縞   [3] 斗   [4] 元針   [5] 玄針

[6] 懸針   [7] 玄石   [8] 懸鉤

Certain other facts may here be relevant. First, the *Shu Shu Chi I*, the mathematical book already referred to (p. 259) dating from the +2nd to the +6th century, and in which references important for the history of the compass are found, has a near-by passage referring to a method of computing by the use of the character *liao* (*liao chih suan*[1]). This is like a tadpole in shape. Secondly, in the Thang (+766 to +779) Wei Chao[2] wrote an essay on scoops made from gourds and calabashes (*phiao*[3]), in which he said that if one spins such a spoon-shaped object round like a toy, it reminds one of the 'south-pointer'.[a] The lodestone spoon was evidently still known or remembered at that date. Then, in connection with the passage to Europe, we should not forget the earliest Arab references (al-'Awfī and Bailak al-Qabajaqī) both of which describe the floating magnet as a 'fish'. Finally, the tradition persisted as late as the early 18th-century *Wakan Sanzai Zue*,[4] where one may read:[b]

The *chhi* and influence of the lodestone is truly as if it were living. It has a head and a tail. Its head points towards the north and its tail towards the south. The force of the head is superior to that of the tail. If it is broken into pieces, all of them also have heads and tails. If it is fed with small pieces of iron it gets 'fat', if starved it gets 'thin'. If it is heated in the fire, it 'dies' and can no longer point to the south. It is also afraid of tobacco. The artisans who make magnetic needles rub their front ends on the head of the lodestone and their tips on the tail, then the front end indicates the north and the tip indicates the south. If one brings such a needle near a lodestone it will turn right round, the front end according with the head and the tip according with the tail. It is in this way that one can recognise the heads and tails of lodestones, which again is a wonderful thing.

Here we see the application of the ancient head-and-tail ideas originating from the lodestone spoons to the newer knowledge of magnetic polarity.

The term *hsüan chen*, whether in the form of 'suspended needle' or 'mysterious needle', deserves further pursuit, more indeed than we have been able to give to it. For example, in an alchemical work of the Sung, the *Thai Chi Chen-Jen Tsa Tan Yao Fang*[5] (Tractate of the Supreme-Pole Adept on Miscellaneous Elixir Recipes), which has illustrations of chemical apparatus,[c] we are told of a 'suspended needle aludel' (*hsüan chen kuei*[6]) used, no doubt, for digestions, sublimations or even reactions under pressure. It is one of a number of curiously named closed vessels. Although this tractate is not dated, its title, deriving as it must from the famous philosophical diagram of the Neo-Confucians,[d] would place it most probably about the middle of the +11th century, i.e. just at the time we would expect from all other evidence of the study of the magnetised needle. On the other hand, it might easily be a century earlier, since the Thai Chi doctrines are quite credibly thought to have come down

---

[a] *TSCC, Khao kung tien*, ch. 198, in *pên* 796, p. 25b.

[b] Ch. 61, pp. 19b, 20a, tr. auct. adjuv. de Mély (1), p. 108, whose translation, however, is wrong in several places. This encyclopaedia was compiled by Terashima Ryōan in +1712 taking the *San Tshai Thu Hui* of +1609 as a basis. We have left uncorrected the inversion in the penultimate sentence.

[c] See Vol. 5, pt. 2, and meanwhile Ho & Needham (3).   [d] Cf. Vol. 2, p. 461.

[1] 了知算   [2] 韋肇   [3] 瓢   [4] 和漢三才圖會
[5] 太極眞人雜丹藥方   [6] 懸針匱

Table 50. *References to 'iron' and 'needle' in Chinese texts on magnetism*[a]

Dynasty	IRON	NEEDLE
HAN or before	*Lü Shih Chhun Chhiu* *Huai Nan Tzu* *Shih Chi* *Chhien Han Shu (I wên chih)* *Chhun Chhiu Fan Lu* *Huai Nan Wan Pi Shu* *Chhun Chhiu Wei Khao I Yu*[1,c]	*Kuei Ku Tzu* *Lun Hêng* *Yen Thieh Lun*[b]
SAN KUO	*Wên Chi Chiao Chih*	

CHIN	Kao Yu's commentary on
	  Lü Shih Chhun Chhiu
	Nan Chou I Wu Chih
	Kuo Pho's *Tzhu Shih Tsan*
	Kuo Pho's commentary on
	  the *Shan Hai Ching*
		Pao Phu Tzu
		Wu Shu (in *San Kuo Chih*)

LIU SUNG	*Lei Kung Phao Chih Lun*	
LIANG and N/WEI	*Shui Ching Chu*	Thao Hung-Ching's *Ming I* *Pieh Lu*
THANG	*Pên Tshao Shih I* *Hua Shu*[e] *Kuan Yin Tzu*[g]	*Hsin Hsiu Pên Tshao*[d] (*Thang Pên Tshao*[f]) *Hsüan Chieh Lu*[h] *Kuan shih Ti Li Chih Mêng*

WU TAI		*Shu Pên Tshao*
		Chiu Thien Hsüan Nü Chhing
		  Nang Hai Chio Ching
		Wamyō Ruijushō[2] (Japanese
		  dictionary)[i] (+934)

SUNG	Sung commentaries on the
	  Tshan Thung Chhi by
	  Chhen Hsien-Wei and
	  others
		Pên Tshao Thu Ching[j]
		Pên Tshao Yen I
		Sung commentaries on the
		  Tshan Thung Chhi
		Mêng Chhi Pi Than
		Jih Hua Pên Tshao
		Shih Lin Kuang Chi[3]
		Phing-Chou Kho Than[4]
		Chu Fan Chih
		Hsüan-Ho Fêng Shih Kao-Li
		  Thu Ching
		Chhu I Shuo Tsuan
		Mêng Liang Lu
		Mo Chuang Man Lu[5,k]
		Ko Wu Tshu Than[6,1]
		Wu Lei Hsiang Kan Chih[7,m]
		Hsü Po Wu Chih[n]
		Lu Shih
		Lung Hu Huan Tan Chüeh[8]

YUAN		*Chen-La Fêng Thu Chi*
MING		*Shih Wu Kan Chu*[9] *Ko Chih Tshao*, etc.

to Chou Tun-I from Chhen Thuan (*d.* +989), if not from the late Thang.[a] This again takes us back to the +10th century, the Wu Tai period, on other evidence (pp. 259, 274, 305) a focal one for the development of this subject. Why the aludel got this name we cannot say, but since in later ages the vertical stroke in writing a character acquired the fancy name of 'suspended needle', this plumb-line association may have existed already in the Sung or before, and could have been chosen here because the apparatus was intended to stand perfectly upright in the furnace.[b]

The better one gets to know the medieval alchemical literature the more valuable are the hints which it is found to contain. A book of much earlier origin than the anonymous tractate just mentioned is the *Thai Chhing Shih Pi Chi*[1] (The Records in the Rock Chamber; a Thai-Chhing Scripture), revised and enlarged by Chhu Tsê hsien-sêng[2] *c.* +500, but based upon a text deriving from the alchemist Su Yuan-

---

[a] Cf. Vol. 2, p. 467.

[b] One should not of course exclude the possibility that the name might have come from substances subliming in needle-shaped crystals such as some of the salts of silver and lead. Even so, a punning reference need not be absent.

[1] 太清石壁記   [2] 楚澤先生

Notes to Table 50.

[a] Most of the detailed references to the earlier books in the Table will be found on p. 232.

[b] On the authority of *PWYF*, ch. 27 (p. 1402.3).

[c] In *YHSF*, ch. 55, p. 62b.

[d] Ch. 4, p. 9b. This book was of course not known to Li Shih-Chen, who simply quoted it as the *Thang Pên Tshao* from previous pharmacopoeias.

[e] P. 13a. A translation of the passage has already been given in Vol. 2, p. 447.

[f] In *Thu Ching Yen I Pên Tshao*, ch. 4, p. 5a; *PTKM*, ch. 10, p. 7a.

[g] *Wên Shih Chen Ching*, ch. 2, p. 19b. A translation of the passage has already been given in Vol. 2, p. 54. In this case neither iron nor needle is mentioned, simply attractive power as such.

[h] *TT* 921, also in *Yün Chi Chhi Chhien*, ch. 64, p. 6b. The passage occurs again in the *Yen Mên Kung Miao Chieh Lun* (The Venerable Yen Mên's Explanations of the Mysteries), *TT* 937, another alchemical work the text of which is substantially identical with that of the *Hsüan Chieh Lu*, and therefore certainly also Thang (p. 2b). These books will be discussed more fully in Vol. 5, pt. 2; here we wish to express our indebtedness to Dr Ho Ping-Yü for his collaboration in matters concerning alchemical literature.

[i] Ch. 2.

[j] In *Thu Ching Yen I Pên Tshao*, ch. 4, p. 5a.

[k] Ch. 1, p. 14a. Written by Chang Pang-Chi about +1131.

[l] P. 28. Written by Su Tung-Pho about +1080.

[m] P. 1. Written by Su Tung-Pho late in the +11th century. The phenomenon of magnetic force is referred to in the opening words of this tractate 'On the Mutual Influences of Things according to their Categories', indicating clearly the 'field' conception of natural philosophy so classical in Chinese thought (cf. Vol. 2, p. 293). See further Ho & Needham (2).

[n] Ch. 9, p. 5a; this is the quotation which purports to be from the Han *Shen Nung Pên Tshao Ching*. But as the best reconstructed versions of this book have no mention of magnetic attraction in the entry for lodestone, it is better to place this text in the Sung, for the *Hsü Po Wu Chih* was compiled in the +12th century.

[1] 春秋緯考異郵   [2] 和名類聚抄   [3] 事林廣記   [4] 萍州可談
[5] 墨莊漫錄   [6] 格物麤談   [7] 物類相感志   [8] 龍虎還丹訣
[9] 事物紺珠

Ming [1] who lived in the last decades of the +3rd century. Here we find a long list of alchemical synonyms,[a] and among them the following statement:[b]

The 'imperial mush' (*ti liu chiang* [2]) and the 'needle-attraction on a fixed platform' (*ting thai yin chen* [3]) are both (cover-names for) magnetite (*tzhu shih* [4]).

No doubt the first of these referred to a suspension of powdered magnetic iron ore in water. But it is the second which interests us here. For if the thought concerned only the attraction of the lodestone for iron filings or needles, there would be no mention of any arena for these activities—the very idea of a 'fixed platform' suggests that the writer had in mind a diviner's board or some other oriented dial-plate upon which a directive property was manifested.[c] We need not insist that he was Su Yuan-Ming, for Chhu Tsê was a contemporary of Thao Hung-Ching and there is much other evidence which points to the lifetime of that great naturalist as a crucial period in the development of the compass-needle from the lodestone spoon.

In order to ascertain more closely the time of introduction of the needle, which alone made possible precise determinations of azimuth directions by magnetic means, let us assemble as many texts as we can find in which there is mention of the lodestone, and separate them into two groups, according to whether the words 'iron' (*thieh* [5]) or 'needle' (*chen* [6]) are used. The earliest would be expected to refer to small bits of iron in general, while later ones would specify that needles were attracted or oriented. Table 50 shows that this expectation is fulfilled.

Broadly speaking, then, references to iron are more numerous in the earlier periods, while after the Liang the lodestone is almost universally said to attract (or orient) needles rather than simply iron. This result is presented, of course, with all reserve, since copyists' confusions between iron and needle would be quite easy. Such mistakes indeed probably explain the occurrence of the word needle in the *Kuei Ku Tzu* text, and in the *Lun Hêng*. If this be admitted, then the evidence suggests the first use of magnetised needles about the +4th century, i.e. in the time of Ko Hung or a little later. It will naturally be realised that this was the limiting factor for the discovery of the declination, which could not be revealed by the lodestone spoon with its blunt end and clumsy frictional drag. It will also be seen that an approximate +4th-century date agrees with the bulk of the evidence so far assembled, including the reference in the works of Thao Hung-Ching.

What first suggested the flotation of the needle on water is indeed an interesting question. It seems that there was an ancient method of divination which consisted in watching the shadow of a floating needle on the bottom of a bowl of water. This has actually been seen still in use among girls and young women of South China at certain seasonal festivals by modern observers.[d] We have good reason to place the technique

[a] The whole book is remarkable for the lucidity and frankness of its chemical and pharmaceutical descriptions. Written in a style designedly clear rather than obscure, it deserves study and translation.

[b] Ch. 2, p. 10*a*. The passage was brought to light by Dr Ho Ping-Yü.

[c] This could easily have been a saucer of water, but whatever it was the points of the compass would have had to be marked on it.

[d] Notably Przyłuski (3). And at Peking by Bodde (12), p. 59, where it is called 'losing the needle' (*tiu chen* [7]).

[1] 蘇元明   [2] 帝流漿   [3] 定臺引針   [4] 磁石   [5] 鐵   [6] 針   [7] 丟針

at least as early as the Han because there is a reference to it in the *Huai Nan Wan Pi Shu*,[a] which recommends that to make a needle float it should be greased, as with sweat or the natural oil of the hair. This accompanies an entry about the lodestone.[b]

(5) THE USE OF THE COMPASS IN NAVIGATION

Here it will be best to proceed as we did with the magnetic compass as such, and allow the story to develop around a central text. It so happens that this text is approximately contemporary with the words of Shen Kua in the *Mêng Chhi Pi Than* (above, p. 250). The *Phing-Chou Kho Than*[1] (Phingchow Table-Talk) was written between +1111 and +1117, but referred to events concerning that and other ports from +1086 onwards; its author was Chu Yü.[2] He knew what he was talking about, for his father, Chu Fu,[3] had been a high official of the Port of Canton from +1094 and Governor from +1099 to +1102. The essential passage runs as follows:[c]

According to the government regulations (*chia ling*[4]) concerning sea-going ships (*hai po*[5]), the larger ones can carry several hundred men, and the smaller ones may have more than a hundred men on board. One of the most important merchants is chosen to be Leader (Kang Shou[6]), another is Deputy Leader (Fu Kang Shou[7]) and a third is Business Manager (Tsa Shih[8]). The Superintendent of Merchant Shipping (Shih Po Ssu[9]) gives them an officially sealed red certificate (*chu chi*[10]) permitting them to use the light bamboo for punishing their company when necessary. Should anyone die at sea, his property becomes forfeit to the government....The ship's pilots are acquainted with the configuration of the coasts (*ti li*[11]); at night they steer by the stars, and in the day-time by the sun. In dark weather they look at the south-pointing needle (*chih nan chen*[12]). They also use a line a hundred feet long with a hook at the end, which they let down to take samples of mud from the sea-bottom; by its (appearance and) smell they can determine their whereabouts.

Here then is a very detailed statement of the use of the mariner's compass just about a century before its first mention in Europe.[d] In connection with this there has been a persistent theory that the account refers to foreign (Arab) ships trading to Canton, and that it was therefore Arabs who first saw the possibilities of the Chinese geomantic compass. This is quite erroneous; it originated because of a mistranslation by

[a] Cit. in *TPYL*, ch. 736, p. 8*b*, and the modern reconstructions of the text. There is a parallel passage in Fang I-Chih's[13] *Wu Li Hsiao Shih*[14] (+1664), ch. 12, p. 11*b*, given without indication of origin, though the *Wan Pi Shu* is quoted by name elsewhere in the same chapter (cf. Wang Chen-To (4), p. 195). On the history and bibliography of the *Huai Nan Wan Pi Shu*, which we so often have occasion to quote, see Kaltenmark (2), pp. 31 ff. A late Arabic example of floating-needle divination will be found in Wiedemann (8).

[b] See below, p. 316.

[c] Ch. 2, p. 2*a*, tr. auct. adjuv. Hirth & Rockhill; Kuwabara (1).

[d] This text has been known to Europeans for more than a century. Yet an eminent scholar such as Forbes can still write (1950) that 'it has been established that the compass is a Western invention', (2), pp. 101, 108, 132.

[1] 萍州可談   [2] 朱彧   [3] 朱服   [4] 甲令   [5] 海舶   [6] 綱首
[7] 副綱首   [8] 雜事   [9] 市舶司   [10] 朱記   [11] 地理
[12] 指南針   [13] 方以智   [14] 物理小識

Hirth (3) and Hirth & Rockhill,[a] who thought that *chia-ling* was the name of some foreign people, the 'Kling',[b] not realising that it was a technical term for 'government regulations'. We came to this conclusion independently, unaware of the fact that Kuwabara had pointed out the mistake more than twenty years ago, and since then others too have recognised it,[c] so perhaps we may dare to hope that the story will now be abandoned. In any case, one has only to read the whole passage to see that it would not fit, for foreign merchants would not have required an authorisation from the local Chinese authorities giving them disciplinary powers over their seamen, nor would the property of a foreign merchant dying at sea have been forfeit to the Crown. Any assertion that the Chinese were not voyaging far afield at that time collapses when adjacent passages are read, in which it is said that repairs were often carried out at places in Sumatra.

Before the first European mention (+1190) there are two further Chinese texts. After the fall of Khaifêng, the capital of Northern Sung, in +1126, and the move to Hangchow, Mêng Yuan-Lao[1] wrote the *Tung Ching Mêng Hua Lu*[2] (Dreams of the Glories of the Eastern Capital), in which he notes, regarding navigation:

> During dark or rainy days, and when the nights are overclouded, sailors rely on the compass (*chen phan*[3]). The Mate (Huo Chhang[4]) is in charge of this.[d]

It is interesting that the title of the ship's officer who looked after the compass remained unchanged down to the 18th century. Mêng Yuan-Lao's reference would refer to some time around +1110. Thirteen years later a diplomatic mission set out for Korea, and the account of it by Hsü Ching[5] we have already had occasion to mention more than once (Vol. 3, pp. 492, 511 above). His words on the compass are valuable because he states specifically that it was a floated needle. What he says is as follows:[e]

> During the night it is often not possible to stop (because of wind or current drift), so the pilot has to steer by the stars and the Great Bear. If the night is overcast then he uses the south-pointing floating needle (*chih nan fou chen*[6]) to determine south and north. When night came on we lit signal-fires (to transmit compass-readings?), and all the eight ships of our convoy responded.

These words, to which Edkins (13) first drew attention, demonstrate once again that it was Chinese ships, and not those of other peoples, which carried the mariner's compass during the century before the first occidental knowledge of it.

---

[a] P. 30.

[b] This is a real name; in modern times it referred to Tamils in Singapore, and derives from the Kalinga or Telugu coast of the Bay of Bengal (Yule & Burnell).

[c] Duyvendak (8); Wang Chen-To (4); Li Shu-Hua (2, 3). It is regrettable that Reischauer (3), p. 274, continues to give countenance to this mistake.

[d] Tr. auct. On the Mate see p. 292 below.

[e] *Hsüan-Ho Fêng Shih Kao-Li Thu Ching*, ch. 34, p. 9b, 10a; tr. auct.

[1] 孟元老　　[2] 東京夢華錄　　[3] 針盤　　[4] 火長　　[5] 徐兢
[6] 指南浮針

Pausing at this point to look backwards, we see first that no references to the compass of magnetised iron (as distinct from the lodestone compass) of the same degree of clarity as those of Tsêng Kung-Liang, Shen Kua and Chu Yü have so far come to light before the beginning of the +11th century. There are, however, earlier hints in abundance. Taken together, they build up a picture of its development indistinct but unmistakable. First there is the most ancient mention of the floating needle in the *Huai Nan Wan Pi Shu*. Then there are the obscure descriptions of a compass-like needle method of computing in the +2nd- or +6th-century *Shu Shu Chi I*. Ko Hung, in his +4th-century *Pao Phu Tzu*, makes a strange juxtaposition of the needle and the 'south-pointer', normally to be taken in his time as the lodestone spoon on the ground-plate of the *shih*. Next we have the frog-tadpole-gourd-needle nomenclature complex covering the +4th to the +10th century and paralleling that of the occidental *calamita*, though so much earlier in China. Comparative study of the terms employed in descriptions of magnetic attraction has given us a strong indication that magnetised needles were coming in from the +4th to the +6th centuries. Presently we shall have to consider a number of books which though very hard to date seem to come from the late +9th and the +10th century; these not only speak of needles but give values for the declination. So also does Wang Chi,[1] who must have been born not later than +990.[a] Thus the picture unfolds.

On the other hand, so far as sailors were concerned, there is the negative evidence. A mention of steering by the stars occurs in *Huai Nan Tzu* (−2nd century),[b] and in the account of the travels of Fa-Hsien (early +5th). None of the Buddhist pilgrim voyages, as we have seen, show anything else, and in +838 a rather circumstantial relation of the navigational difficulties of the Japanese monk Ennin,[2] in sailing from his country to Korea and China, indicates that if the compass was at that time applied to maritime use, it must have been the possession of very few pilots.[c]

It seems probable therefore that a rather long delay occurred between the first use of the magnetised needle in geomancy and its adoption by sailors for navigation. If our conclusion is plausible that the transference of polarity from lodestone to needle was first made use of about the +5th century, it may well be that it was not applied to navigation before the +10th. The most probable period would be between +850 and +1050. Making a guess in the light of all the evidence one might say that the magnetised needle was probably used for geomancy on an increasingly widespread scale during the Sui, Thang and Wu Tai periods, not finding application at sea until the beginning of the Sung. The subject calls urgently for further research, and it is possible that valuable references hitherto unknown will be discovered in Chinese literature.

[a] On all this, see pp. 302 ff. below. The title of one of these books includes the significant words *Hai Chio Ching*[3] (Sea Angle Manual).
[b] Ch. 11, p. 4*b*. Many similar statements could easily be found.
[c] Ennin's book is entitled *Nittō Guhō Junrei Gyōki*[4] (Diary of Travels to China in search of the (Buddhist) Law and for studying Good Customs). Tr. and annot. Reischauer (2, 3).

[1] 王伋   [2] 圓仁   [3] 海角經   [4] 入唐求法巡禮行記

In this connection an interesting point arises concerning the two schools of geomancy.[a] As we have noted already, there were two schools of this pseudo-science from the Thang onwards, that of Chiangsi and that of Fukien.[b] The former, centred on Ganchow, was founded by a Thang Geomancer-Royal, Yang Yün-Sung[1] (*fl.* +874 to +888), and his chief disciple, Tsêng Wên-Chhuan.[2] This school emphasised the shape of mountains and the directions of water-courses, concentrating on physiography. The Fukien school, on the other hand, associated with Chu Hsi, derived from Wang Chi[3] and his pupil Yeh Shu-Liang,[4] and laid much more emphasis on *kua*, compass-points and constellations, making particular use of the magnetic compass.[c] This is no coincidence, for the province of Fukien, forming as it does an amphitheatre of mountains facing the sea, has been for many centuries a nursery of Chinese sailors. In our own time the great majority of naval officers have been Fukienese. That the Fukien school of geomancy should have emphasised just those things which were most important for navigation would therefore be expected. I would suggest that any research on the earliest application of the magnetic compass to navigation would do well to concentrate its attention on this province, and especially on the Thang and early Sung periods.

Before coming to a few interesting records of the Chinese use of the compass after +1200, a point of substance raised by de Saussure (35), himself a practical sailor, should not be forgotten. He pointed out that the use of the compass in navigation depended to some extent upon metallurgical procedures for the production of steel. Soft iron does not retain its magnetism long, or show it strongly;[d] for extended voyages magnetised needles of good steel would have been desirable. De Saussure considered[e] that the Chinese narratives of deep-water sailing which we have from the +13th century, such as the embassy to Cambodia,[f] would not have been possible without steel needles. Now the history of steel itself is complicated enough, and the first use of it for needles still more obscure. Though known in Greek and Roman times, the two most important ancient sources of it were Asia Minor (the old Hittite centre) and Hyderabad in India, whence came the wootz steel worked up by the Arab smiths in Damascus. On this al-Kindī wrote in the +9th century. In a metallurgical Section later on, we shall bring forward evidence to show that it was being exported from India to China at least as early as the +5th century. There we shall also show that good steel was manufactured in China by remarkably modern methods at least from that time onwards also.[g] Fig. 332 shows a scene of wire-drawing and needle-

---

[a] Cf. Vol. 2, pp. 359ff., and p. 242 above.   [b] Cf. de Groot (2), vol. 3, p. 1007.

[c] For example, Chao Fang,[5] a Yuan expert on geomancy, describes how the compass-using adepts radiated from Fukien (*TSCC, I shu tien,* ch. 680, *i wên,* p. 9b).

[d] Schück & Vollmer (in Schück (1), vol. 2, p. 53) carried out experiments with iron needles of various kinds made by hand as they would have been in the +12th century. An interesting paper on the early history of permanent magnets has been written by Andrade (2).

[e] (35), p. 44. Cf. Michel (6).

[f] The descent of the Indo-Chinese coasts to Malaya requires two great short-cuts, crossing the openings of the Gulf of Tongking and the Gulf of Siam. Cf. p. 258 above.

[g] In the meantime Needham (31, 32) may be consulted.

[1] 楊筠松   [2] 曾文遄   [3] 王伋   [4] 葉叔亮   [5] 趙汸

抽線琢鍼圖

Fig. 332. Steel for magnetised needles: a scene of wire-drawing and needle-making
from the *Thien Kung Khai Wu* (+1637).

making from the *Thien Kung Khai Wu*[a] of +1637. It may well be, therefore, that good steel needles were available to the Chinese several centuries before Europe had them—but the investigation of this would be a research in itself.[b] Failing them, lodestones had to be carried on board every ship for remagnetisation, as Bromehead (5) has described, quoting from a book on navigation of +1597. Thus for 600 years the lodestone was an economic mineral.

Now for the later Chinese descriptions. Most famous is the *Chu Fan Chih* (Record of Foreign Peoples) written by Chao Ju-Kua in +1225. In this work he says:[c]

To the east (of Hainan Island) are the 'Thousand-Li Sand-Banks' and the 'Myriad-Li Rocks', and (beyond them) is the boundless ocean, where the sea and sky blend their colours, and the passing ships sail only by means of the south-pointing needle. This has to be watched closely by day and night, for life or death depend on the slightest fraction of error.

Though the nationality of these ships is not mentioned, the whole context, which deals with Hainan, concerns an island which had been a Chinese province since the Han, and mentions junks coming from Chhüanchow and other Chinese ports. Half a century later a similar account is found in the *Mêng Liang Lu*,[d] Wu Tzu-Mu's description of Hangchow.

As soon as the (merchant-ships) enter the gates of the ocean, there is the vast expanse of the sea without a shore, strong and very dangerous, the abode of mysterious dragons and marvellous serpents. At times of storm and darkness they travel trusting to the compass alone, and the pilot steering dares not make the smallest error since the lives of all in the ship depend upon him. I have often talked with great merchants who have told me all this in detail.... The water of the ocean is shallow near islands and reefs; if a reef is struck the ship is sure to be lost. It depends entirely on the compass, and if there is a small error you will be buried in the belly of a shark.

And he goes on to speak of other sailors' signs, the nature of the water, weather-forecasting, and so on, in terms remarkably like those used in books four or five centuries later.

Then comes, in Yuan times, the account of the embassy to Cambodia by Chou Ta-Kuan (the *Chen-La Fêng Thu Chi*[1]), which has been translated by Pelliot and others. By this time (+1296) not merely mentions of the compass, but actual compass-bearings, have got into the literature, as the following shows:[e]

Embarking at Wênchow (in Chekiang) and bearing S.S.W. (*hsing ting wei chen*[2]), one passes the ports of the coastal prefectures of Fukien and Kuangtung, as also those of the

---

[a] Ch. 10, pp. 4b, 7b. Hangchow about +1250 had a Street of the Iron-Wire Drawers.

[b] For example, we have noted a discussion of steel needles in Thao Ku's +10th-century *Chhing I Lu*, ch. 2, p. 23b. 'Seamstresses or medical men will tell you the merits and disadvantages of different kinds of needles in just as much detail as Confucian scholars talking about brush pens....' The ones which were most praised were those made of yellow steel and capped with gold heads (*chin thou huang kang hsiao phin*[3]).   [c] Ch. 2, p. 16a; tr. Hirth & Rockhill (1), p. 176.

[d] Ch. 12, p. 15aff., tr. Moule (5), p. 366, (15), p. 32.

[e] Ch. 1, p. 1a; tr. Pelliot (9, 33), eng. auct. Note that this was contemporary with Yü Yen's remark about the *Tshan Thung Chhi* (p. 261 above). Closer investigation of sources such as Liu Chi's[4] *Yu Li Tzu*[5] (c. +1360) might be useful here.

[1] 眞臘風土記   [2] 行丁未針   [3] 金頭黃鋼小品   [4] 劉基   [5] 郁離子

(four) overseas prefectures (of Hainan), then crossing the sea of the Seven Isles (Taya Islands) and the Sea of Annam, one arrives at Champa (somewhere near Qui-nhon). Thence with a good wind one can arrive in 15 days at Chen-phu (somewhere near Cape St James), which is the frontier of Cambodia. Thence bearing S. $52\frac{1}{2}°$ W. (*hsing khun shen chen*[1]) one crosses the Khun-Lun Sea (north of Pulo Condor Island) and enters the river mouths.

Still more is this the case for the numerous accounts of the voyages in the time of Chêng Ho (1400 to 1431).[a] There are such details, for example, in Huang Shêng-Tsêng's *Hsi-Yang Chhao Kung Tien Lu*[2] (+1520), already mentioned.[b] Liu Ming-Shu (2) and Wang Chen-To (5) have brought to light certain specialist books, which did not get into the official bibliographies, but must be of the +14th century, such as the *Hai Tao Ching*[3] (Manual of Sailing Directions), the *Hai Tao Chen Ching*[4] (Sea-ways Compass Manual), the *Hang Hai Chen Ching*[5] (Sailors' Compass Manual),[c] and the *Yuan Hai Yün Chih*[6] (Yuan Dynasty Sea Traffic Record).[d] At the beginning of the Ming there must have been quite an abundant literature of sailing directions recording compass-bearings. Huang Shêng-Tsêng named as one of his sources a *Chen Wei Pien*[7] (Collection of Needle Positions), which may or may not have been a specific printed book.[e] If it was, it must have been a 'routier' or 'rutter' like the *Yüeh Yang Chen Lu Chi*[8] (Record of Courses Set by the Needle in the Cantonese Seas), which is known to have still existed in the 18th century.[f] At an earlier point[g] a few words have already been said about the way in which the compass was used by the pilots of Chêng Ho's time. Some idea of their skill may be gained by the fact that in circumnavigating Malaya they laid their course through the present Singapore Main Strait, which was not discovered (or at least not used) by the Portuguese till +1615 when they had been in those waters for over a century.[h]

Oxford possesses a manuscript of remarkable value in this connection,[i] the *Shun Fêng Hsiang Sung*[9] (Fair Winds for Escort), which Duyvendak (1) recognised as stemming from about +1430, when the great series of expeditions led by Chêng Ho was ending. It contains a mass of general nautical information (tides, winds, stars, compass-points, etc.) and on the compass runs as follows:[j]

In olden times, the worthies of antiquity traversed the sea-routes, and everywhere they depended upon the twenty-four positions of the 'earth dial' (*ti lo ching*[10]),[k] modifying their

---

[a] Cf. Vol. 1, pp. 143ff. above; Vol. 3, pp. 556ff. See also Vol. 4, pt. 3.   [b] Vol. 3, p. 558.

[c] There are a dozen *Chen Ching* in the bibliographies, but all concern acupuncture.

[d] By Wei Su[11] of the Yuan or early Ming. Cf. Hu Ching's[12] edition of *Ta Yuan Hai Yün Chi*.

[e] Rockhill (1), p. 77, thought not, but Pelliot (2a), p. 345, (2b), p. 308, (33), p. 80, always believed that evidence of a specific book with this title would some day come to light.

[f] Pelliot (2b).   [g] Vol. 3, p. 560.   [h] See Mills (1); Duyvendak (1).

[i] Laud Orient. MSS. no. 145. Translation from it is rather difficult since it seems to have been written by some not very literate person, and the meaning is not always clear. Brief description in Hsiang Ta & Hughes (1), who think that the MS. itself may not be older than the second half of the +16th century.

[j] P. 4a, tr. auct. adjuv. Mills (5).   [k] The compass.

[1] 行坤申針　　[2] 西洋朝貢典錄　　[3] 海道經　　[4] 海道針經
[5] 航海針經　　[6] 元海運志　　[7] 針位編　　[8] 粵洋針路記
[9] 順風相送　　[10] 地羅經　　[11] 危素　　[12] 胡敬

practice according to the conditions. Whether going or returning, it is necessary to record the time and day, whether early or late, in passing islands and mountains; and to observe the wind, whether it be east, west, south or north. Now the wind veers a point or half a point; now the tide flows slowly or rapidly; it may be favourable or contrary—you will know if you use log and soundings. You must study the colour of the water, whether deep or shallow, and the appearance of the mountains, whether far or near, but the force of the water rising and falling must always be carefully observed. You must be frugal of sleep. If you make an error of a tiny fraction, you may lose a thousand *li*, and then it will be too late for regret.

If the wind rises from east, west, south or north, there may be a change of a whole compass-point. Those who observe this must immediately take the proper steps. If it is a question of raising a sail, this must be set in accordance with the compass bearing, changing according to necessity, sometimes more, sometimes less, and then when the wind becomes favourable, you work back to the original course.

As for difficulties of direction, there are passages which you can look for, and men whom you can ask. Mountains and islands are not always recognisable, and when you are far from land, when the ocean waters blend with the sky, then you must base your proper course upon the method of Chou Kung. The compass needle and the books of sailing directions will then be the guide. Even if you meet with high winds and waves, and come near to reefs and shoals, they can all be avoided by making good use of the compass and the sailing directions.

A point of some interest is that the MS. contains a liturgical form for use in the ship's chapel or before its shrine at the beginning of a voyage, in which the mariner's compass is prominent. A litany incorporates as saints and sages the names of a number of Taoist geomancers, both legendary and real,[a] thus giving further proof, if indeed it were necessary, that the mariner's compass derived from the geomantic compass.

Actual examples of Chinese maritime compasses are not very rare. Probably the oldest type includes those which are like flat bronze plates not more than 6 in. in diameter, having a bowl-shaped depression at the centre for the water on which the needle is floated (Fig. 333). Two of these are illustrated by Wang Chen-To.[b] A particularly elaborate one may be seen in the Palace Museum at Peking, dating from the late Yuan or early Ming period; it is marked not only with the 24 compass-points and the eight *kua*, but also with a concentric circle giving the five elements, three occupying each segment.[c] Much more massively constructed, but probably much later, is the wooden floating-compass bowl[d] shown in Figs. 334 and 335. In this specimen the regular twelve cyclical characters (see below, p. 297) are used, each main division of 30° being divided into two parts by the word *chung*,[1] and each of these 15° intervals

---

[a] P. 5b. These include Chhing Wu Tzu (the Blue Raven Master), a Han geomantic expert, as also Pai Ho Tzu (the White Crane Master); then Yang Chiu-Phin[2] (perhaps another name for Yang Yün-Sung); Wangtzu Chhiao[3] a character from the −6th century appropriated by the Taoists; Li Shun-Fêng the +7th-century mathematician and astronomer; and Chhen Thuan, +10th-century Taoist.

[b] (5), pl. vii.

[c] Noted during my visit in the summer of 1958.

[d] From the collection of Mr Armand A. Mick, on loan to the Old Ashmolean Museum at Oxford. Grateful thanks are due to the kindness of Mr Mick and Dr C. H. Josten for permission to study and figure this bowl.

¹ 中　　　² 楊救貧　　　³ 王子喬

PLATE CXVI

Fig. 333. Mariner's floating-compass of bronze, Ming period, diam.
c. 3¼ in. (from Wang Chen-To, *5*). South is at 'half-past ten o'clock',
and a fiducial line on the bottom of the water-compartment seems to
make allowance for a declination slightly west of north. The 24
compass-points in the outer band, the 8 *kua* in the inner.

PLATE CXVII

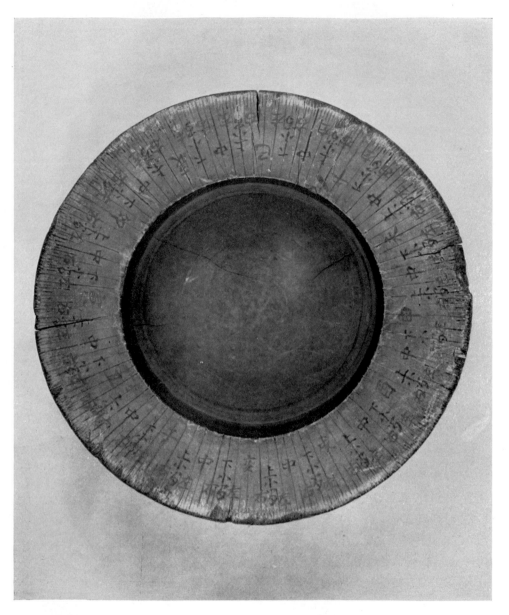

Fig. 334. Mariner's floating-compass of painted wood, Ming period (perhaps Japanese), diam. *c.* 6¼ in. (Collection of Mr A. A. Mick and Old Ashmolean Museum, Oxford). South is at 'one o'clock'. There are only 12 compass-points, but each 30° division so formed is divided into two parts by the mark *chung*, and each of these again bisected by the marks *shang* and *hsia* denoting the 'little intervals' (*hsiao chien*). The finest divisions, marked *tso*, left, and *yu*, right, are thus of 3¾°. The red and black colours of the graduation markings can be easily distinguished in the photograph.

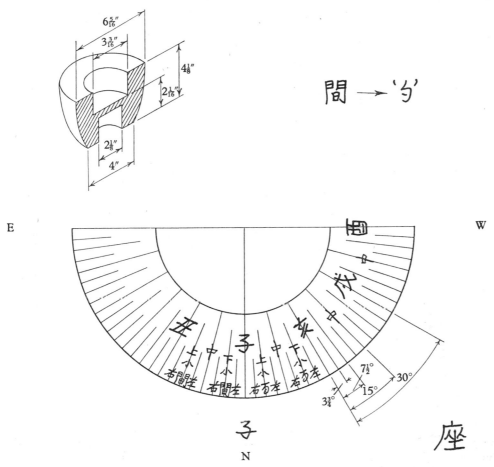

Fig. 335. Diagrams to elucidate the mariner's compass of painted wood. Above, left, the dimensions of a perspective cross-section; right, the abbreviation of the character *chien* used by the instrument-maker in all but two places; below, the graduations; to the right the lubber's point, *tso*, 'seat', carved in the side of the bowl underneath the north point.

bisected again by lines at $7\frac{1}{2}°$ marked 'upper' and 'lower, little intervals' (*shang, hsia, hsiao chien*[1]) respectively. The finest divisions, marked *tso* (left) and *yu* (right), are thus of $3\frac{3}{4}°$. On the side of the bowl, underneath the north point, is carved the character *tso*,[2] indicating some kind of lubber's point.[a] The circumstances shortly to be described would indicate a date in the first half of the +16th century, especially as the writing of the graduation inscriptions has a Japanese flavour. Here we may recall the bronze planispheric bowl with its two depressions for dry-mounted needles

[a] This was a mark on the compass indicating the ship's axis which came into use in the West in the early part of the +16th century, when night sailing increased, and the bow of the ship could often not be seen by the helmsman. Cf. Waters (7) and the discussion immediately below on the compass-card.

[1] 上 下 小 間   [2] 座

already described and illustrated.[a] This was probably of the +17th century. Mariner's compasses of the late Ming and Chhing periods with dry-mounted needles (Fig. 337) are always much simpler than those used by the geomancers, rarely containing more than one ring of direction-points.[b]

It would be needless to follow further developments of a purely nautical character. It may suffice to say that illustrations of Chinese ship compasses may be found in many Chhing books, such as the *Liu-Chhiu Kuo Chih Lüeh*[1] (Account of the Liu-Chhiu Islands) by Chou Huang[2] (Fig. 336), together with his admirable pictures of ships (cf. Sect. 29 below). Hsiung Ming-Yü[3] too gave an account of the compass in his *Ko Chih Tshao*[4] (Scientific Sketches) of +1620 (cf. Hummel, 6).

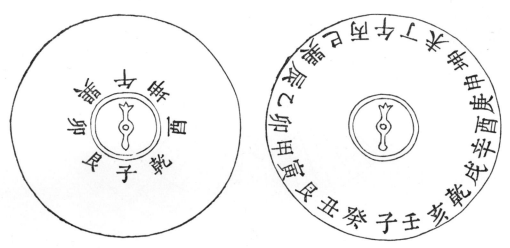

Fig. 336. Illustrations of mariner's dry-pivoted compasses from the *Liu-Chhiu Kuo Chih Lüeh* (Thu hui sect., p. 35*a*), +1757. On the left, the simplest form with eight directions; on the right the normal form with 24 compass-points. South is at the top in both.

(i) *The mariner's compass and the compass-card*

A few words remain to be said about the suspension of the Chinese mariner's compass, and the compass-card. If a magnet-needle is suspended, by whatever means, within a circular box carrying upon its rim the compass-points, the helmsman who wishes to direct his ship in a given direction has to keep the bow and stern in line with an invisible axis across the face of the compass. As time went on it was found convenient to affix to the pivoted magnet itself a light card whereon was painted the 'wind-rose' with its compass-points. The whole was then enclosed in a circular box which carried nothing but a fiducial line in the ship's longitudinal axis. The helmsman was thus

---

[a] Vol. 3, p. 279 and Fig. 108.

[b] And often only the most important 12 rather than the 24. Cf. Hommel (1), p. 336, fig. 507, and many illustrations of examples collected by Wang Chen-To (1), whose study of the subject is the most elaborate yet made.

[1] 琉球國志略   [2] 周煌   [3] 熊明遇   [4] 格致草

PLATE CXVIII

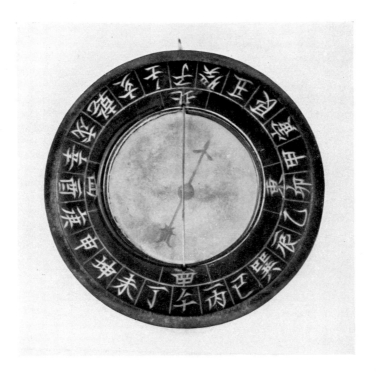

Fig. 337. Mariner's dry-pivoted compass, early Chhing period, diam. *c.* 3½ in. (Museum of the History of Science, Florence). South is at the bottom. The 24 compass-points in the outer band, the four cardinal directions in the inner.

PLATE CXIX

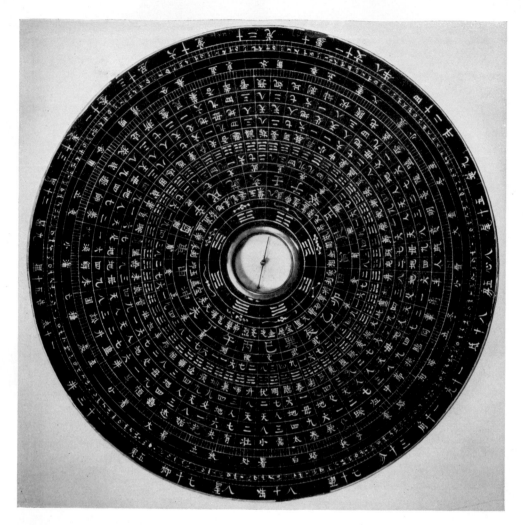

Fig. 338. Geomantic compass with dry-pivoted needle; markings in red and gold on black lacquer. South is at 'seven o'clock'. The 8 *kua* are in the first circular band from the centre, and the 24 compass-points in the fourth. The outermost circle contains the lunar mansions (*hsiu*) with their equatorial extensions in degrees (cf. Vol. 3, pp. 233 ff.). The remaining circles are all of divinatory character (see the description of Fig. 340). Bought by the author in Shanghai, 1946. Diam. 7½ in.

enabled to attend only to the compass itself, and to keep on course with much greater accuracy.[a] This was the invention which has been connected with the name of Flavio Gioja, and though the personage may be unhistorical, there is little doubt that the improvement originated somewhere in south Italy, probably at Amalfi, shortly after +1300 (Sylvanus Thompson, 1). De Saussure (35) was quite right in suspecting that the Chinese did not have the compass-card until it was introduced to their waters by Europeans in the +16th century, for Wang Chen-To (5) has now been able to find texts which show this. In his *Chhing Wu Hsü Yen*[1] (Introduction to the Blue Raven Canon), Li Yü-Hêng[2] wrote, about +1570:

The needle floating on water and giving the north and south directions, is ordinarily called the Wet Compass (*shui lo ching*[3]). In the Chia-Chhing reign-period (+1522 to +1566) there were attacks of Japanese pirates (on the coast), so from that time onwards Japanese methods began to be used. Thus the needle was placed in the compass box, and a paper was stuck on to it carrying all the directions, so that no matter what direction is taken the Tzu Wu signs are always situated at the north and south. This is called the Dry Compass (*han lo ching*[4]).[b]

So also in another book, *Thui Phêng Wu Yü*[5] (Talks on Awaking at Sea), the same author says:[c]

Recently the people of Chiangsu, Chekiang, Fukien and Kuangtung have all suffered from the attacks of Japanese pirates. Japanese ships always use at the stern the dry-pivoted compass to fix their course; and our people, having captured this, imitated it so that the method became common in Chiangsu. The needle is prepared by the use of the lodestone or heating,[d] but after a time its property weakens[e] and then it is no longer serviceable. It is not so precise as the water compass.[f]

Other books of the late Ming speak of the *phan sui chen*[6] which is obviously the compass-card. Then Wang Ta-Hai, writing about +1790 in his *Hai Tao I Chih Chai Lüeh* (Brief Selection of Lost Records of the Isles of the Sea),[g] said:

The Dutch ships did not use the (simple) needle in their compasses, but a shuttle-shaped plate of iron sharpened at both ends and wide in the middle; at the centre there was a small

---

[a] Du Halde (vol. 2, pp. 280ff.) related the observations of some Jesuits who travelled from Siam to China in a Chinese ship in 1687. They noted that the compass-box was marked with 24 points and that a silk thread was stretched across as a fiducial line from north to south. In sailing, either the desired rhumb was placed parallel to the ship's axis and the needle kept under the string, or the string was placed parallel, and the needle allowed to point to a rhumb the opposite of that desired, e.g. N.E. instead of N.W. We have just seen an example of this latter practice in the form of a lubber's point on a late Japanese floating-compass.

[b] Tr. auct.   [c] Ch. 7, tr. auct.

[d] Perhaps another indication that the use of thermo-remanence was still current.

[e] Presumably because good steel was not used.

[f] The dissatisfaction here expressed may have been due to the poorness of the suspension, a difficulty which had not arisen in the flotation method. It is surprising that it never occurred to the Chinese, so far as we know, to use the 'Cardan' suspension, or gimbals, for the support of the compass, since, as will be shown below (Sect. 27d), this device is far older in China than in Europe.

[g] Already mentioned above, p. 235. Here ch. 5, tr. auct. adjuv. Anon. (37).

[1] 青鳥緒言   [2] 李豫亨   [3] 水羅經   [4] 旱羅經   [5] 推篷寤語
[6] 盤隨針

cavity (*wa*[1]) having a pointed pin underneath to support it. It turned like an umbrella, and on the surface (of the card) there were Dutch characters written for sixteen directions, E., W., S., N.; S.E., N.E., S.W., N.W.; S.S.E., E.S.E., N.N.E., E.N.E., W.S.W., E.S.W., N.N.W. and W.N.W. This formed a complete round. When Chinese sailors wanted to set a course, they turned the characters of the compass to accommodate it to the direction of the vessel ('north' being always in line with the bows); but Western mariners navigating towards any quarter turned the vessel so that it pointed in a direction shown on the compass(-card). The principle is one and the same, only the instruments were of different construction.

From all this and other evidence it seems fairly certain that the general use of the dry-pivoted compass and the compass-card was transmitted by European, probably Dutch or Portuguese, ships to Japan in the latter part of the +16th century, after which it was gradually adopted by the Chinese.[a]

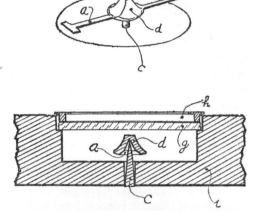

When Barrow visited China in Wang Ta-Hai's time (+1797) he was impressed by the way in which Chinese compass-needles were pivoted.[b] He found them extremely thin, no longer than an inch, and mounted in a very sensitive way. The needle's centre was attached to a small inverted conical copper bowl which rested on a fine steel pivot, both well polished.[c] The bowl being much larger

Fig. 339. Sketches to illustrate the Chinese method of suspending compass-needles towards the end of the +18th century (after Wang Chen-To, 5). *a*, flat needle; *c*, sharply-pointed steel pivot; *d*, inverted conical copper bowl; *g*, glass cover; *h*, copper ring; *i*, ivory plate.

than the pivot, the needle tended to retain its position no matter how the compass was moved (Fig. 339). Moreover, the fact that the centre of gravity was well below the point of suspension meant that it was quite sufficient to overcome the dip or inclination of the needle downwards.[d] Europeans had avoided this by counter-balancing the opposite end of the needle, but this was not very satisfactory because of the variation of dip in different parts of the world.

To recapitulate, therefore, it is clear that the first dry-pivoted compass, namely that described in the Sung *Shih Lin Kuang Chi* (p. 255 above), was Chinese. Presumably it travelled to Europe together with the floating-needle or water-compass during the

---

[a] Wang Chen-To (5) instances, among other examples of adoption from Western ships at this time, the sand-filled hour-glass. See, for instance, the *Liu-Chhiu Kuo Chih Lüeh* (+1757), Thu hui sect. p. 34*b*. We shall come back to this subject in the discussion of mechanical clockwork, Sect. 27*j* below, and in Sect. 29. He also notes the fact that the small portable sundials combined with compasses, which were introduced and made by the Jesuits, all had dry pivots (cf. Vol. 3, pp. 310ff., 312 above).

[b] Staunton (2), vol. 1, pp. 411, 441; Klaproth (1), p. 97.

[c] After this time, the pivot suspension became universal on all geomantic compasses, as in my own example (Fig. 338).

[d] On this, see below, p. 314.

[1] 凹

+ 12th century, but while in the West the latter was soon superseded by the former, in China the floating type continued as the most prevalent, both for geomancy and for navigation, until the middle of the + 16th century.[a] The Italian compass-card then spread over East Asia as an accompaniment of the revived dry pivot. Specimens of nautical and other water-compasses have (as we have seen) been found and studied by Wang Chen-To (5), see Fig. 333. They often comprise a small separate 'teapot'[b] which held the water to be poured into the central space.[c] Ships carried lodestones for remagnetisation[d] (*yang chen*[1]), and sometimes special kinds of water were prescribed, as in the *Shun Fêng Hsiang Sung*.[e] There were superstitions about the exact manner in which the needle floated, and ceremonial libations to be used at the time of preparing the compass.[f]

Most landsmen would be inclined to put down the persistence of the floating-compass in Chinese ships to the celebrated conservatism of that people. But with the usual irony of history modern navigation came back to well-damped floating magnets in the end.[g] Though the light-card Thomson (Kelvin) compass was adopted in the British Navy in 1889, it lasted only until 1906, when it was replaced by various forms of liquid compass. This was more than a century after the first experiments in which a refuge had been sought in liquid-filled globes from the unendurable oscillation and vibration of dry-pivoted magnets.

We may take our leave of the practical sailors by quoting the ninth chapter of the *Tung Hsi Yang Khao* (Studies on the Oceans East and West) by Chang Hsieh (+ 1618).[h]

When (the ship) goes out through the entrance to the sea, the spray from the breaking waves flies up to heaven; the white horses seem to leap up to the Milky Way itself—no longer can one follow the banks of the coast, no longer can one recognise villages and count distances stage by stage. The captain and the experienced sailors (*chhang nien san lao*[2])[i] ply the oars and raise the sails, then cut through the flood of waves relying only on the south-pointing needle to lead and direct them. Sometimes they use only one (compass-bearing), sometimes

---

a So Friar Martín de Rada saw it, about + 1575, 'a very sensitive little tongue of steel' floating 'in a little saucer full of sea-water' (cf. Boxer (1), p. 295). As late as + 1643 Georges Fournier could write (p. 526, and 1667 ed. p. 400): 'Les boussoles de la Chine, dont ils usent encore de présent dans leurs Ioncos, ne sont qu'un médiocre vaisseau plein d'eau, sur laquelle ils font flotter un petit triangle de fil de fer touché d'aymant, soutenu d'un peu de liège, et il est très-croyable que l'usage en est fort ancien chez eux.' It is interesting to know that the floating-needle compass had at this time by no means disappeared from Europe. In + 1661 J. B. Riccioli wrote (p. 473): 'Esto in Mari Balthico et Oceano Germanico multorum Versorium nempe triangulum ex filo ferreo ope trium frustulorum suberis innatet aquae vasculo inclusae, censeaturque; hic usus apud illos valde antiquus.'

b *Chen hu*,[3] called fancifully by the sailors *huan shui shen chün*[4] (Water-Change Spirit) or *huan shui thung lang*[5] (Water-Change Serving-Boy).

c *Chin chung*,[6] also called *chan*,[7] the cup. *Shui Chan Shen Chê*[8] was its tutelary deity, to whom the sailors prayed.

d Li Yü-Hêng in the *Thui Phêng Wu Yü*. Cf. Bromehead (5) for occidental practice.

e P. 7a.   f P. 6a.

g The story is briefly told by May (2).   h Ch. 9, p. 1a, tr. auct. adjuv. Mills (4).

i This term goes back to the Thang (+8th century) as the celebrated poet, Tu Fu, used it (cf. *Nan Hao Shih Hua*, p. 18b).

---

¹ 養針   ² 長年三老   ³ 針壺   ⁴ 換水神君   ⁵ 換水童郎

⁶ 金鍾   ⁷ 盞   ⁸ 水盞神者

the needle must point to the space between two—they rely on the guidance which it provides, and the ship forges ahead. . . . Again, they sink a line to take soundings, calculating the depth of water at such and such a place in so many fathoms. . . . The superintendent of the compass needle is the mate, the 'fire-commander' (Huo Chhang[1]);[a] though the waves may be boisterous and the way long, all obey the movement of the (south-)pointer.

A certain air about this, reminiscent of the Greek Anthology,[b] makes one feel again that the Chinese have too often been underestimated as a maritime people.

Our next topic must be the Chinese discovery of magnetic declination, which will involve an explanation of the fully developed geomantic compass. Before passing to this, however, there is one curious point which must be taken up.

(ii) *The direction-finder on the imperial lake*

In what has gone before it has been said repeatedly that the south-pointing carriage and the magnetic compass have nothing whatever in common. While this is in general true, there are two places where the subjects touch: one concerns the existence of a 'south-pointing boat', and the other the existence of a south-pointing carriage which was clearly stated to have no mechanism inside it, but to be directed by a concealed man.

Taking the former first, the story of a south-pointing boat appears, mentioned in an offhand way, in the important account of the history of the south-pointing carriage in the *Sung Shu*.[c] 'Under the Chin dynasty',[d] it says, 'there was also a south-pointing ship.'[e] The *Yuan Chien Lei Han* encyclopaedia adds[f] from another source[g] that it used to sail about on the Ling Chih Chhih[2] lake in the palace gardens.[h] Of course this may have been simply a popular legend based on that of the south-pointing carriage. But if the story had some real basis then one might be inclined to see in it a very early application of the magnetic compass. For such rough experiments by palace magicians on the smooth waters of a lake, even the lodestone spoon would have done, and it is curious that the date corresponds rather well with that which we have been led to suggest for the first beginnings of the floating magnetised needle. An alternative would be that the boat was a paddle-boat, and that someone attempted to apply to its two wheels the same kind of device as had already been applied to the carriage.[i] Probably no one will ever find out.

---

[a] This at first sight strange title seems to have arisen from the fact that in the +12th century fire was allowed only on the leading ship of a convoy (the 'kitchen-ship'), and this also carried the chief pilots and navigators (see *Chhing Hsü Tsa Chu Pu Chhüeh*,[3] p. 5*a*, etc.).

[b] Compare for instance IX, 546.   [c] Ch. 18, p. 5*b*. Cf. Sect. 27*e* below.

[d] +265 to +420.   [e] Giles (5), vol. I, p. 112.

[f] Ch. 386, p. 3*a*.

[g] The *Chin Kung Ko Chi*[4] (Record of the Palaces of the Chin).

[h] As Edkins (12) noted. Hirth (3, 8) also, but he was too confused between the carriage and the compass to see the real significance of it.

[i] It will later be shown (Sect. 27*i*) that the use of paddle-wheels on boats goes back a long way in Chinese technological history.

[1] 火長   [2] 靈芝池   [3] 清虛雜著補闕   [4] 晉宮閣記

The other statement occurs in the biography of Tsu Chhung-Chih, the eminent
+5th-century mathematician and engineer.[a]

When Wu of the Sung subdued Kuan-chung, he obtained the south-pointing carriage of
Yao Hsing,[1] but it had only the shell, and no machinery inside. Whenever it moved, a man
was stationed inside to turn it.[b]

If this is to be taken as something more than a mere deception, it must imply that the
man inside turned the visible pointer to correspond with the south according to some
information which he had with him, and this could only have been a form of the
compass. Here again, if the carriage was used only on ceremonial occasions under
conditions where it moved on a good flat surface, a lodestone spoon on a *shih* could
perhaps have been used, yet the date would allow of a floating 'fish' or even a needle
being available.

(6) MAGNETIC DECLINATION

In the quotations already given from Shen Kua and Khou Tsung-Shih, we noticed
that not only the directivity of the magnetic needle was spoken of, but also its declination, i.e. the fact that it did not point directly north and south. There are many other
passages in Chinese literature which refer to this property, and which, moreover,
antedate the time at which even directivity itself was first known to the occident.
In order to appreciate this, we must start by examining the Chinese geomantic
compass in its fully developed form, for medieval knowledge of the declination is in
fact embedded in the traditional layout of this instrument.

(i) *Description of the geomantic compass*

When one first contemplates a Chinese compass in its most elaborate form (Figs. 340
and 341), one is bewildered by its complexity,[c] and many Chinese scholars as well
as most Western sinologists have been only too willing to dismiss it as a monument
of superstitious web-spinning. Geomancy, said de Groot, 'fully shows the dense
cloud of ignorance which hovers over the Chinese people; it exhibits in all its naked-
ness the low condition of their native culture, illustrating the fact that natural philosophy in that part of the globe is a huge mountain of learning without a single trace
of true knowledge in it'.[d] We have already seen enough to appreciate the frank

[a] *Nan Chhi Shu*, ch. 52, p. 20*b*; *Nan Shih*, ch. 72, p. 11*b*; tr. Moule (7).
[b] Yao Hsing was emperor of the Later Chhin dynasty, *r.* +393 onwards, *d.* +416. His south-
pointing carriage had been made by Linghu Shêng; cf. Sect. 27*e* below.
[c] Certainly scholars in 17th-century Europe must have been quite baffled. Just as Robert Boyle,
Robert Hooke and Samuel Pepys puzzled over Chinese calendars (cf. Vol. 3, p. 391), so at Oxford
John Selden possessed a 'Pixis Sortilega Sinensium', a Chinese geomantic compass, dating from about
+1600. Selden (*d.* +1654) was a lawyer, historian, antiquary, orientalist, and one of the encouragers
of Thomas Hyde. His 'Pixis' is now in the Old Ashmolean Museum; what he made of it we shall
never know.
[d] (2), vol. 3, p. 938.
[1] 姚興

Legends to Figs. 340 and 341

Translation of inscription painted by the maker on the back of the Chinese geomantic compass presented by Mrs John Couch Adams, in the Whipple Museum of the History of Science, Cambridge. Made under the supervision of Wang Yang-Chhi (courtesy name, Yuan-Shan) of the Yün-I Studio at Haiyang near Hsin-an (in Kuangtung province). [Not dated, but of the Chhing, probably late +18th century.]

IDENTIFICATION TABLE, FROM WITHIN OUTWARDS

	1	The Heaven Pool (a name derived from the medieval floating-compass, and from the ancient diviner's board), with the South-Pointing Needle.
1st circle	2	The Eight *Kua* (Trigrams of the *I Ching*, the 'Book of Changes') arranged in the *Hsien Thien* (Prior to Heaven) order [cf. Vol. 2, p. 313].
2nd circle	3	The Twelve Terrestrial *Chih* (cyclical characters) [cf. Vol. 1, p. 79; Vol. 3, pp. 396, 398].
3rd circle	4	Nine Stars (i.e. fate-categories, in 24 divisions) concerned with the Orientation of Dwellings.
4th circle	5	(24) Constellations controlling success in Civil-Service Careers (including the imperial examinations).
5th circle	6	Earth Plate (i.e. graduations controlling the Inner Region of the Compass Disc; with 20 cyclical characters and 4 *kua* as the standard Azimuthal Direction Signs) arranged in the *Chêng Chen* (lit. 'Correct Needle') positions (i.e. according to the astronomical north-south), with pure Yin and pure Yang (influences shown, by means of red and black colour. This arrangement is associated with the name of Chhiu Yen-Han, +8th century).
6th circle	7	The (24) *Chieh Chhi* (fortnightly divisions) of the Four Seasons (the tropic year) [cf. Vol. 3, p. 405].
7th circle	8	Terrestrial Record (or Cycle) of the Mountain-Penetrating Tiger (60 combinations of the cyclical characters, i.e. the 12 *chih* and the 10 *kan*, in 72 divisions—concerned with underground watercourses, veins, and foundations) [cf. Vol. 3, pp. 396, 397].
8th circle	9	The Nine Halls of the *Tun Chia* (method of divination; 72 divisions containing the numbers from 1 to 9, which also appear in the nine cells (Nine Halls) of the *Lo Shu* magic square [cf. Vol. 3, p. 57]. Twelve divisions are left empty as in the previous row. This method was popular in the +5th and +6th centuries).
9th circle	10	The *Fên Chin* of the Inner Part of the Compass (48 combinations of the cyclical characters, distributed among the Five Elements in 120 divisions) [cf. Vol. 2, pp. 243 ff.].
10th circle	11	Orderly Arrangement, equally dividing the Dragon (Influence; 60 combinations of the cyclical characters in 60 divisions).
11th circle	12	The Five Elements (in 60 divisions) for the *Nei Yin* (method of divination; this dates from the −1st century).
12th circle	13	Man Plate (i.e. graduations controlling the Middle Region of the Compass Disc; with 20 cyclical characters and 4 *kua* as the standard Azimuthal Direction Signs, as in row no. 6) arranged in the *Chung Chen* (lit. 'Middle Needle') positions (i.e. all points shifted $7\frac{1}{2}°$ W. of N. as introduced in the +12th century, traditionally by) Lai Kung (Lai Wên-Chün) (to take account of the westward declination observed at that time).
13th circle	14	Mr Tshai's Earth-Penetrating Dragon (method; attributed to Tshai Shen-Yü, +10th century; 60 combinations of the cyclical characters).
14th circle	15	Threes and Sevens riding the *Chhi* (60 divisions containing 24 azimuthal compass-points corresponding with those in row no. 6; and 36 divisions containing odd numbers).
15th circle	16	The New Degrees adopted by the Present (i.e. the Chhing) Dynasty (360 divisions, the numbering starting afresh at the beginning of each division of the following row, no. 17. The division of the circle into 360° was a Jesuit innovation, the Chinese having formerly divided it into $365\frac{1}{4}°$. At the same time the old Chinese division of the degree (*tu*) into 100 minutes (*fên*), and the minute into 100 seconds (*miao*), was replaced, unfortunately, by the usual Western sexagesimal system [cf. Vol. 3, p. 374]).
16th circle	17	The (twenty-eight) Lunar Mansions (*Hsiu*; constellations dividing the Equator) with the number of degrees of equatorial extension occupied by each. (N.B. These two rows are placed with the Middle Region of the Compass Disc because it agrees with the westward declination still observed in China in the +18th century, but as modern introductions they are separated from it by a double line.)

(*continued on p. 295*)

PLATE CXX

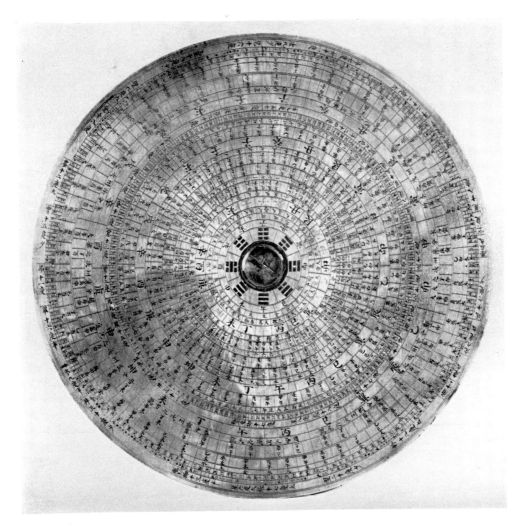

Fig. 340. The Chinese geomantic compass in full dress: an example presented by the late Mrs John Couch Adams to the Cambridge Astronomical Observatory, and now in the Whipple Museum. South at the bottom. The standard 24 compass-points appear in the fifth circle from the centre. They then appear again in the 12th circle staggered $7\frac{1}{2}°$ W. of N., and in the 17th circle staggered $7\frac{1}{2}°$ E. of N.; these, it is believed, are the permanent traces of western and eastern declinations observed at different times during the Middle Ages and later (see text). This specimen bears upon its under-surface an explanatory table which is translated and elucidated overleaf. Diam. 12·6 in.

PLATE CXXI

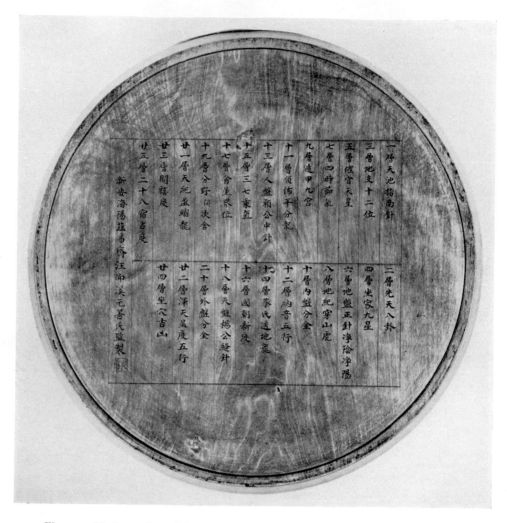

Fig. 341. Under-surface of the John Couch Adams compass with its inscription and
explanatory table, translated on pp. 294 and 295.

absurdity of this kind of judgment, and we shall now see that although geomancy itself was of course always a pseudo-science, it was, nevertheless, the true mother of our knowledge of terrestrial magnetism, just as astrology was of astronomy and alchemy of chemistry.

The compass (*lo ching*,[1,2] or *lo phan*[3] or *chen phan*[4])[a] of the geomancers' art (*fêng shui*,[5] *khan yü*[6]) is composed of some 18–24 concentric circles (*tshêng*[7]) surrounding the small balanced needle at the centre.[b] Let us look first at the first and fifth circles starting from the centre (Fig. 340). They are composed of the eight *kua* (trigrams), depicted in the first, and (four of them) written in the fifth. With the aid of Table 13 (Vol. 2, p. 313) it will immediately be seen that there is a discrepancy between them, for in the *kua* as depicted *Chhien* (the first of the *kua*, signifying bright male creativity) occupies the south, while *Khun* (the second, signifying dark female receptivity) occupies the north; whereas in the *kua* as written *Chhien* is in the north-west and *Khun* in the south-west.

---

[a] Wang Chen-To (5), p. 106, lists some 25 variant names. Chao Fang[8] of the Yuan was one of the earliest geomancers to use the afterwards widespread term *lo ching*.

[b] There is a good deal of local variation in the design of these compasses; from Anhui examples with 33 and 40 concentric circles are known (Wang Chen-To (5), p. 121).

[1] 羅經   [2] 羅鏡   [3] 羅盤   [4] 針盤   [5] 風水   [6] 堪輿
[7] 層   [8] 趙汸

Legends to Figs. 340 and 341 *(continued)*

17th circle	18	Heaven Plate (i.e. graduations controlling the Outer Region of the Compass Disc; with 20 cyclical characters and 4 *kua* as the standard Azimuthal Direction Signs, as in rows no. 6 and 13) arranged in the *Fêng Chen* (lit. 'Seam Needle') positions (i.e. all points shifted $7\frac{1}{2}°$ E. of N., as introduced in the $+$9th century, traditionally by) Yang Kung (Yang Yün-Sung) (to take account of the eastward declination observed at that time).
18th circle	19	The *Fên Yeh* (12 positions, i.e. the astrological controls of the different provinces, the names of which are given; a system popular in the $+$7th century [cf. Vol. 3, p. 545]; and the (12) *Tzhu* positions (i.e. the 12 Jupiter-stations, their astronomical, not astrological, names being used for the years; they were also employed for the months of a single year and the double-hours of a single day [cf. Vol. 3, p. 403]. This circle is further evidence for a Thang date for this part of the compass).
19th circle	20	The *Fên Chin* of the Outer Part of the Compass (identical with row no. 10, but shifted eastwards. Another double line follows).
20th circle	21	Celestial Record (or Cycle) of the Fully Coiled Dragon (60 combinations of the cyclical characters in 60 unequal divisions; corresponds to row no. 8—concerned with hill ridges).
21st circle	22	The Five Elements (distributed among 61 unequal divisions) corresponding to the (Equatorial Extension) Degrees of the Constellations of the Celestial Sphere.
22nd circle	23	(Equatorial Extensions of the *Hsiu*, the Lunar Mansions, in) Degrees (Chinese), as determined in the Khai-Hsi reign-period ($+$1206).
23rd circle	24	Fortunate and Unfortunate Positions for the Siting of Tombs (indicated by red and black symbols; corresponding with row no. 4).
24th circle	25	(Equatorial Extensions of the) twenty-eight *Hsiu*, the Lunar Mansions, in (Chinese) Degrees for Divination (for divination because the $365\frac{1}{4}°$ system was that used in medieval times when the divination rules were fixed).

For further explanations see pp. 297, 299, 305, 307, and 312.

This difference need not detain us long since it does not affect the knowledge of declination or any other matter of scientific importance, yet it is interesting archaeologically, because the north-west and south-west positions were those on the ancestor of the compass, the *shih*.[a] As is well known, there were in Han times two alternative azimuthal arrangements of the eight trigrams. Both of these are in the *I Ching*, and

	N.				N.	
	Khun				Khan	
	Kên	Chen		*Chhien*	Kên	
W. Khan		Li E.	W. Tui			Chen E.
	Sun	Tui		*Khun*	Sun	
	Chhien				Li	
	S.				S.	
	The 'Fu-Hsi' system			The 'Wên Wang' system		

it is extremely difficult to be precise as to their antiquity; presumably they are somewhat pre-Han.[b] The allegedly more ancient system, named after the legendary sage Fu-Hsi,[1] with *Chhien* in the south, was the so-called Hsien Thien[2] ('prior to Heaven') system, and corresponded with the Ho Thu[3] diagram (cf. Vol. 3, pp. 56 ff.).[c] The allegedly later system, named after the Chou High King Wên Wang,[4] with *Chhien* in the north-west, was the so-called Hou Thien[5] ('posterior to Heaven') system, and corresponded with the Lo Shu[6] magic square (cf. Vol. 3, pp. 56 ff.).[d] Since, however, the *shih* followed the latter system, it was probably really the earlier of the two. The difference was only in symbolism.[e] This impression as to date is strengthened by a conversation which has come down to us recorded in the *San Kuo Chih*.[f] Kuan Lo[7] (+209 to +256) was a famous geomancer; we shall shortly meet with him again.

Kuan Lo, talking to his friend Liu Fên,[8] said: 'I really do not understand why our ancient sages placed *Chhien* at the north-west and *Khun* at the south-west. After all, *Chhien* and *Khun* are the symbols of Heaven and Earth, the supreme things. The former represents kingship and paternity, and covers all the myriad things in the world, just as the latter bears and carries them. How could these two be reduced to the position of the other six *kua*?... How could they occupy side positions?

[a] And on the computing device of the *Shu Shu Chi I* (cf. p. 260 above) which evidently derived from the *shih*.

[b] It is thought, however, that the *Shuo Kua*, in which they occur, is not earlier than the Han diviners Chiao Kan[9] and Ching Fang[10] (−1st century). Moreover, the diagrams derived from it and the attributions to Fu-Hsi and Wên Wang probably date only from Shao Yung[11] (+11th century) or Chhen Thuan[12] (+10th). Cf. Vol. 2, p. 341.

[c] R. Wilhelm (2), vol. 1, p. 200; Baynes tr. vol. 1, p. 285; *I Ching*, Shuo Kua, ch. 2, §3.

[d] R. Wilhelm (2), vol. 1, p. 202; Baynes tr. vol. 1, p. 288; *I Ching*, Shuo Kua, ch. 2, §5. On both these cf. Granet (5), p. 186.

[e] The Fu-Hsi system allocated the *kua* cosmically; the Wên Wang system did so according to the march of the seasons of the year, starting at the south and going round clockwise.

[f] *Wei Shu*, ch. 29, p. 23 a, b, commentary. Tr. auct.

[1] 伏羲   [2] 先天   [3] 河圖   [4] 文王   [5] 後天   [6] 洛書
[7] 管輅   [8] 劉邠   [9] 焦贛   [10] 京房   [11] 邵雍   [12] 陳摶

The interest of this lies in the fact that in the +3rd century, just about the time when the lodestone spoon was giving place to the needle, it was felt that the old *shih* arrangement was unsatisfactory, and that a change should be made. Hu Wei, in the +18th century, considering this passage,[a] believed that the change could be traced back to Wei Po-Yang the alchemist, in the Later Han. In any case, the compass in its final form, though adopting the Fu-Hsi Hsien Thien system for its depicted *kua*, retained the Wên Wang or Hou Thien system embedded in its round of azimuthal written compass-points.[b]

These points, 24 in number, are seen in the fifth circle (Fig. 340). The accompanying table (p. 298) shows that they divide the circle into segments of 15° each. The whole duodenary cycle of *chih*[1] characters is represented, but the denary cycle of *kan*[2] characters has dropped out two, i.e. precisely those which were represented twice over on the *shih* (Fig. 326 above); *wu*[3] and *chi*,[4] symbolising Earth. This omission[c] left four places, which were filled by the four most important *kua*, *Chhien*, *Khun*, *Sun* and *Kên*, as shown in Table 51 and in Fig. 340. The remaining four of the eight trigrams do not appear on the round of written azimuth points.

Compasses used by navigators reduced all this to its simplest form, using only the 24 points, or even reducing them to 12 or 8. Since in China the geomantic compass long preceded the navigational one, and since the azimuthal role of the abstract denary and duodenary cyclical characters was so old there, the 'wind-rose' of Chinese mariners probably developed in rather a different way from that of the West. In Greece the points of solstitial sunrise and sunset on the horizon seem to have generated by the −4th century northern and southern 'companion' directions neighbouring due east and west by about 27°. The addition of similar companions by analogy to north and south gave the classical twelvefold system of points which were associated with a series of standard winds.[d] On the other hand the ancient Mediterranean pilots also used a round of eight points and winds obtained simply by halving the four basic quadrants.[e] So obvious a plan certainly developed (as far as this) in China too, but the former one was rather foreign in spirit, for Chinese starcraft, giving from the beginning a great deal of attention to the equator and the circumpolar stars, paid relatively little to ecliptic risings and settings.[f] Yet in practice it approximated more closely to the deeply sexagesimal Chinese system than the rounds of

---

[a] *I Thu Ming Pien*, ch. 8, p. 10a. Cf. *Tshan Thung Chhi*, ch. 1, p. 1a.

[b] Thus we see at the outset its syncretistic character, still more evident in what follows.

[c] The process is clearly described in Yang Yün-Sung's *Chhing Nang Ao Chih*[5] (Mysterious Principles of the Blue Bag, i.e. the Universe), cf. *TSCC, I shu tien*, ch. 665, *hui khao* 15, p. 6a. Shiba (1) suggested very plausibly that the omission was essentially practical, since *wu* and *chi* would so readily be confused with *hsü* and *ssu*. If we knew the date at which the change took place, we might fix better the time when the sailors took over from the geomancers. But the fact emphasised by Wada (1), p. 152, namely that *wu* and *chi* symbolised the element Earth, may also have had its importance, for Earth was associated with the Centre, and its symbols were accordingly hardly suitable for peripheral azimuth points.

[d] This is associated with the name of the −3rd-century Egyptian admiral, Timosthenes of Rhodes, and his book of sailing directions for the Mediterranean. Cf. Vol. 3, p. 532.

[e] Taylor (8), pp. 6ff., 14ff., 53ff.   [f] Cf. Vol. 3, pp. 229ff. But also p. 306.

¹ 支   ² 干   ³ 戊   ⁴ 己   ⁵ 青靈奧旨

16 and 32 points used later on in Europe, possibly deriving from Etruscan augury but certainly obtained by simple duplation from the fundamental four.

Beyond this point most exponents of the Chinese compass have not proceeded.

Table 51. *Azimuth compass-points*

Modern compass-points			Chinese names[a]		
°					
0	N.		Tzu		
15		N. 15° E.		*Kuei*	
22·5	N.N.E.				Tzu-Chhou
30		N. 30° E.		Chhou	
45	N.E.		KÊN		
60		N. 60° E.		Yin	
67·5	E.N.E.				Yin-Mao
75		N. 75° E.		*Chia*	
90	E.		Mao		
105		S. 75° E.		*I*	
112·5	E.S.E.				Mao-Chhen
120		S. 60° E.		Chhen	
135	S.E.		SUN		
150		S. 30° E.		Ssu	
157·5	S.S.E.				Ssu-Wu
165		S. 15° E.		*Ping*	
180	S.		Wu		
195		S. 15° W.		*Ting*	
202·5	S.S.W.				Wu-Wei
210		S. 30° W.		Wei	
225	S.W.		KHUN		
240		S. 60° W.		Shen	
247·5	W.S.W.				Shen-Yu
255		S. 75° W.		*Kêng*	
270	W.		Yu		
285		N. 75° W.		*Hsin*	
292·5	W.N.W.				Yu-Hsü
300		N. 60° W.		Hsü	
315	N.W.		CHHIEN		
330		N. 30° W.		Hai	
337·5	N.N.W.				Hai-Tzu
345		N. 15° W.		*Jen*	

Kua are shown in capitals; denary characters in italics; duodenary in roman. The Chinese characters for all these will be found in Vol. 1, p. 79, Vol. 2, p. 313, and Vol. 3, p. 396, as well as in Figs. 338 and 340.

[a] The double terms shown in the third column, first given by Klaproth (1), p. 102, may be found in geographical works, but the navigators hardly used them. For S.S.W. their instruction was 'sail between Ting and Wei', not 'towards Wu-Wei'.

(ii) *The three circles of Master Chhiu, Master Yang and Master Lai*

If now the fifth circle is compared with the twelfth and the seventeenth, it will at once be seen that the two outer ones, while repeating the 24 points, are (as we shall say) 'staggered'. Thus the south point is moved $7\frac{1}{2}°$ eastwards in the twelfth circle, and $7\frac{1}{2}°$ westwards in the seventeenth. This was noticed by Klaproth (1), Eitel (2), de Groot (2) and others, who also gave the obvious significance of all the subsidiary circles (cf. pp. 294 ff.), but they could not explain it, nor did Wylie (11) in his important paper to which we shall shortly refer. It was Edkins (12) alone whose understanding of geomantic literature enabled him to give the right explanation in 1877, but his paper has remained little known.[a] It may be summarised by saying that the two main subsidiary circles were introduced in the Thang and Sung respectively to allow for the declinations of the magnetic needle which were then observed, and that the geomantic compass thus still contains these old observations in fossil form.

According to tradition, the compass-points were stabilised in their present system at least by the time of Chhiu Yen-Han,[1] a famous geomancer whose approximate *floruit* was $+713$ to $+741$; hence a contemporary of I-Hsing.[b] The technical term for these points, assimilated to the astronomical north–south line, was Chêng Chen[2] (lit. 'correct needle'). Rather over a century later, about $+880$, another great geomancer, Yang Yün-Sung,[3] in order to take account of an eastern declination which was then observed, introduced a second, staggered, circle of points, i.e. those $7\frac{1}{2}°$ westwards of the former. These were (and are) technically known as Fêng Chen[4] (lit. 'seam needle').[c] But by the $+12$th century the declination had moved over to the westwards, so a third geomancer, Lai Wên-Chün,[5] introduced a third, staggered, circle, to allow for this, each point being $7\frac{1}{2}°$ eastwards of the astronomical north–south line.[d] There can be no doubt that both these added circles were attempts to make an average correction for the declination. Lai's new circle was termed Chung Chen[6] (lit. 'central needle'); we think because it was inserted on the compass between the two original circles (cf. Fig. 340).

For the assistance of the reader in what follows it may be worth while to tabulate these correlations (p. 300). It is difficult to validate this presentation from Ming writers, because the idea of gradual change of declination with time became consciously held only in the Chhing. One of them, however, Wu Thien-Hung,[7] is worth listening to.

---

[a] Though attention was ineffectually drawn to it by Schück (4).

[b] Chhiu Yen-Han's most famous book was the *Thien Chi Su Shu*[8] (which one would be tempted to translate 'Pure Book of Celestial Mechanics', if this were not to invite misunderstanding). He also wrote a *Nei Chuan Thien Huang Ao Chi Chen Shih Shu*[9] (Atlas-Tortoise Geomancy).

[c] His *Chhing Nang Ao Chih* (*TSCC, I shu tien*, ch. 665, *hui khao* 15 ff.) mentions the 24 directions ('mountains') but not, so far as we can see, the Fêng Chen. On Yang Yün-Sung cf. pp. 241, 282 above.

[d] His best-known book is the *Tshui Kuan Phien*[10] (On Official Promotion), i.e. how to get it by geomancy.

[1] 丘延翰　[2] 正針　[3] 楊筠松　[4] 縫針　[5] 賴文俊
[6] 中針　[7] 吳天洪　[8] 天機素書　[9] 內傳天皇鼇極鎮世書
[10] 催官篇

In his *Lo Ching Chih Nan Pho Wu Chi*[1] (A South-Pointer to Disperse the Fog about the Geomantic Compass), probably of the +16th century, he wrote:

Master Chhiu got (his knowledge from) Thai I Lao Jen[2] (the Old Man of the Great Unity), and (established) the Chêng Chen, (but) there is also a 'Heaven-measurement' (*thien chi*[3]) and an 'Earth-record' (*ti chi*[4]). The Fên Chin divisions are arranged in three (concentric circles), so that although for the earth one follows the Chêng Chen (system), as everyone knows; in the north (Tzu) (the needle) (once) declined to the N.E., and in the south (Wu) it declined to the S.W. Therefore Master Yang added the Fêng Chen (system). But in the 'Heaven-measurement', the needle in the north (Tzu) declined to the N.W., and in the south (Wu) it declined to the S.E. Therefore Master Lai added the Chung Chen (system).[a]

	Century		
Chhiu Yen-Han	+8th	Chêng Chen 正	astronomical north-south
Yang Yün-Sung	+9th	Fêng Chen 縫	$7\frac{1}{2}°$ east of N., west of S.
Lai Wên-Chün	+12th	Chung Chen 中	$7\frac{1}{2}°$ west of N., east of S.

This is about as clear as could be expected, and the general course of events is confirmed by other Ming writers on geomancy, notably Wu Wang-Kang[5] in his *Lo Ching Chieh*[6] (Analysis of the Magnetic Compass), and Kan Shih-Wang[7] in his *Lo Ching Pi Chhiao*[8] (Confidential Intelligence about the Magnetic Compass).[b]

That the magnetic declination varied in different places was an idea fully accepted in the astronomical chapters of the *Ming Shih*. But it is not clearly stated there that it changed with time. Fan I-Pin,[9] however, who wrote his *Lo Ching Ching I Chieh*[10] (Analysis of the Essential Features of the Magnetic Compass) some time between +1736 and +1795, stated clearly that there had been changes in the declination. He wrote:

Moreover, many people reason that as the needle belongs to (the element) Metal, Metal is afraid of Fire which corresponds to the south,[c] and this would be why the metal deviates towards its 'mother-position' more than 3°. It is also said that in accordance with the *kua*

---

[a] Tr. auct.

[b] The Ming and Chhing literature on geomancy is, however, extremely confused, and seemingly became more and more so as time went on. Quite different statements are to be found in Liu Kung-Chung's[11] *Khan Yü Phi Miu Chhuan Chen*[12] (A Brushing Away of Mistakes and Establishment of Right Theory in Geomancy), and in Hu Shen-An's[13] *Lo Ching Chieh Ting*[14] (Definitive Analysis of the Magnetic Compass)—both of the +18th century. Liu thought that the Fêng Chen was established according to the astronomical meridian by the gnomon shadow (*nieh ying*[15]). Fan I-Pin[9] affirmed, on the contrary, that the Chêng Chen was the astronomical north–south, but doubted the tradition of the introduction of the Fêng Chen and Chung Chen by Yang and Lai respectively. Details of this literature may be found in Wang Chen-To (*4, 5*). But he himself seems to think that there was some difference between the gnomon-shadow meridian and the pole-star meridian, and that Fêng Chen and Chung Chen corresponded to these two—an idea which we find incomprehensible. He does, however, agree that one of the extra circles was added to take account of an observed declination.

[c] Fire takes precedence of Metal in the Mutual Conquest Order, cf. Vol. 2, p. 257.

[1] 羅經指南潑霧集   [2] 太乙老人   [3] 天紀   [4] 地記   [5] 吳望崗
[6] 羅經解   [7] 甘時望   [8] 羅經秘竅   [9] 范宜賓   [10] 羅經精一解
[11] 劉公中   [12] 堪輿闢謬傳眞   [13] 胡愼庵   [14] 羅經解定
[15] 臬影

of Fu-Hsi, the influences of which are constantly in motion, the 'Yang head' inclines to the left, and the 'Yin head' to the right. Others again say that (in) the south there is an ascending (influence) following the Yang and pulling to the left, while (in) the north there is a falling (influence) following the Yin and pulling to the right. Yet others, that (in) the Hsien Thien ('prior to Heaven' system) the *kua Tui* (serenity) and Metal are in Ssu;[1] therefore the needle inclines to the left. Then there is the notion that since Fire contains Earth,[a] the true south point (Wu) of heaven is in the west, and therefore the needle head had to incline to the west, to follow its 'mother-position'.

There are many theories of this sort and all are vain. The important thing to know is that nowadays the direction of the needle, between Hsü and Wei, shows in fact what the true north–south was in the days of Thang Yü (the legendary emperor). So also the position of the sun was in Chou times in 2° of (the *hsiu*) Nü, but between Yuan and Ming it fell to 3° in Chi.[b] Ordinary people do not realise that even Heaven has its (slow) motions and changes. They insist that the Wei direction is something fixed and unalterable.[c]

Presumably Fan I-Pin is here saying that the change in declination of the magnetic needle is a natural phenomenon of saecular change, analogous to the precessional movement of the celestial pole. He may of course have thought that the magnetic meridian alone remained constant while the celestial coordinates changed in a kind of trepidation.

(iii) *Early observations of declination*

Let us now see how a group of texts on the declination fits in with these traditionally accepted facts. The earliest of them are unfortunately difficult to date. First, there is the observation attributed to the Buddhist astronomer, I-Hsing, which would have been made at some time not far from +720. This was reported by Wylie (11)[d] as follows:

On comparing the needle with the north pole, he found that the former pointed between the constellations Hsü and Wei. The pole was just in 6° of Hsü, from which the needle declined to the right (east) 2° 95'.

This does not quite sound like a contemporary account,[e] and in any case Wylie omitted to give any reference.[f] The text has been diligently sought for by sinologists ever since Wylie first gave it (1859), but so far without success. Hashimoto (1)

---

[a] Fire takes precedence of Earth in the Mutual Production Order, cf. Vol. 2, p. 257.

[b] On the recognition of equinoctial precession cf. Vol. 3, pp. 247, 251.   [c] Tr. auct.

[d] Repr. Wylie (5), Sci. sect., p. 156.

[e] Though Wylie placed all of it in quotation-marks.

[f] An *I-Hsing Ti Li Ching*[2] (I-Hsing's Geomantic Manual) is twice mentioned in the Sung bibliography (*Sung Shih*, ch. 206, pp. 20b, 23a; cf. *TSCC*, *I shu tien*, ch. 680, *tsa lu*, p. 2a), but it might be hard to find the text of it. Apparently he also completed another work on the same subject, for the Sung bibliography also lists a *Li Shun-Fêng I-Hsing Chhan-Shih Tsang Lü Pi Mi Ching*[3] (Confidential Manual on the Principles of Tomb Siting, by Li Shun-Fêng and the Zen Master I-Hsing), ch. 206, p. 23b. Though Li died two years before I-Hsing was born, the latter was in many ways his disciple and successor. For instance, Li's value for the obliquity of the ecliptic was accepted as standard by I-Hsing in his meridian arc computations (Beer *et al.* 1).

[1] 巳   [2] 一行地理經   [3] 李淳風一行禪師葬律秘密經

indicated passages in the astronomical chapters of the *Thang Shu* which he thought might have misled Wylie, and similar suggestions have recently been put forward by Wang Chen-To,[a] but they are not very convincing. The astronomical chapters do indeed say[b] that the pole has a different position with relation to the *hsiu* Hsü and Wei from that which it had anciently, because they are talking about proper motion;[c] but there is no mention of needles, nor any declination of 2° 95′. One can only await the results of further enquiry, but it is at least interesting that two other texts, which are certainly old, though difficult to date, also speak of an eastern declination, thus supporting Wylie's interpretation of his I-Hsing reference.[d]

The first passage is preserved in encyclopaedias from the *Kuan shih Ti Li Chih Mêng*[1] (Master Kuan's Geomantic Instructor). The book is attributed to Kuan Lo of the +3rd century, but this cannot be credited, and the work is more likely (from internal evidence) to be of the late Thang time.[e] In it we find the following:[f]

The lodestone follows a maternal principle. The needle is struck out from the iron (originally a stone)[g] and the nature of mother and son is that each influences the other, and they communicate together. The nature of the needle is to return to its original completeness. As its body is very light and straight, it must indicate straight lines. It responds to the *chhi* by orientation (*chao*[2]), being central to the earth and deviating in various directions. (To the south) it points to the Hsüan-Yuan[3] constellation, hence to the *hsiu* Hsing,[4] (and therefore to) the *hsiu* Hsü[5] (in the north), along the axis Ting–Kuei.[6] The yearly differences (*sui chha*[7]) follow the ecliptic, and all such phenomena can be understood.[h]

---

[a] (4), p. 206. He remarks how difficult it is for Western scholars not to fall into error when studying Chinese history and history of science, but Wylie was a very accomplished sinologist, and I fear that the puzzle remains.

[b] *Chiu Thang Shu*, ch. 35, p. 4*b*; *Hsin Thang Shu*, ch. 31, pp. 3*b*, 4*a*.

[c] Cf. Vol. 3, pp. 270ff.

[d] The interpretation itself was indeed questioned by de Saussure (35), pp. 26ff. who maintained that I-Hsing's words imply a western declination, but we have been forced with reluctance to reject his arguments, for reasons which will be evident if careful thought is given to them. His view that the declination was 4°, the figure above being the biographer's mistake, is however plausible.

[e] A *Kuan shih Chih Mêng* is in the extension of the *Sung Shih* bibliography (*Pu*, p. 22*a*), but it is hard to be sure whether lost books with titles such as this were not rather concerned with the philosopher Kuan Tzu,[8] and therefore dealt with politics and economics rather than with geomancy. Such was certainly the *Kuan shih Chih Lüeh*[9] mentioned in the *Hsin Thang Shu* bibliography (ch. 59, p. 8*b*), by Tu Yu;[10] it is there classed with Fa Chia writings. As the *Kuan shih Ti Li Chih Mêng* seems not to mention Fêng Chen, it dates presumably from before Yang Yün-Sung's time, but must be later than the time when the compass-points assumed their traditional positions. It is interesting that the *Sui Shu* bibliography (ch. 34, p. 25*a*) attributes to Kuan Lo a *Shih-erh Ling Chhi Pu Ching*[11] (Manual of Divination by the Twelve Magic Chessmen), cf. below, p. 316. Of course Kuan Lo himself knew the attractive power of the lodestone (cf. Phei Sung-Chih's[12] commentary on the *San Kuo Chih* (*Wei Shu*), ch. 29, p. 24*b*).

[f] *TSCC, I shu tien*, ch. 655, *hui khao* 5, p. 18*b*; tr. auct.

[g] Here the theory of the transformation of ores and metals in the earth will be remembered; cf. Vol. 3, pp. 637ff.

[h] What the writer is suggesting in the last sentence is that the declination is a phenomenon like the extra quarter day in the sidereal year.

---

[1] 管氏地理指蒙   [2] 召   [3] 軒轅   [4] 星   [5] 虛   [6] 丁癸
[7] 歲差   [8] 管子   [9] 管氏指略   [10] 杜佑   [11] 十二靈棊卜經
[12] 裴松之

This is a perfectly definite statement that the declination was some 15° E. For the date of the passage the middle of the +9th century would be a good guess.

Secondly, a passage in the *Chiu Thien Hsüan Nü Chhing Nang Hai Chio Ching* discusses Chêng Chen and Fêng Chen, implying thereby an eastern declination. If it was really Yang Yün-Sung who introduced the first of the subsidiary staggered circles (Fêng Chen), then the passage would be of about +900.[a]

Unfortunately, it is very difficult to date some of these obscure books which are quoted or reproduced in the geomantic sections of encyclopaedias. This would require a special investigation not so far attempted. The *Chiu Thien Hsüan Nü Chhing Nang Hai Chio Ching*[1] has a title which almost defies translation, but one might say: 'The Nine-Heaven Mysterious-Girl[b] Blue-Bag[c] Sea Angle Manual.' Note the reference to the sea. It must be at least Sung, because it had a preface by Chang Shih-Yuan[2] of that period.[d] A diagram from it entitled 'The Directions and *Chhi* of the Floating Needle'[e] is preserved in the *Thu Shu Chi Chhêng* encyclopaedia (see Fig. 342). Beside it, we read:[f]

In the daytime Hsüan Nü determined the directions by the rising and setting of the sun.[g] In the night she determined the directions by the divisions of the *hsiu*.[h] It was Chhih-Yu[3] (the ancient rebel)[i] who invented the 'south-pointer'.[j] Later on, the details of the compass-bearings were fixed, using the 'ten celestial stems' (*thien kan*[4]) for the directions (*fang so*[5]), and the 'twelve horary characters' (*ti chih*[6]) for the *chhi* of the directions (*fang chhi*[7]).[k]

[a] The attribution of this book to Kuo Pho of the Chin, claimed in the Ming preface by Wu Yuan-Chio,[8] is of course not acceptable, though it may well be that some parts are quite old. The *Sung Shih* bibliography (ch. 206, p. 23b) gives the author of the *Hai Chio Ching* as Chhih Sung Tzu,[9] the 'Red Pine Master'. This is a hoary pseudonym; the *Yün Chi Chhi Chhien* (ch. 108, p. 1a), for example, places the bearer of it in legendary antiquity. But it was also the Taoist name of Huang Chhu-Phing,[10] allotted by Wieger (3, 6) to the Chin, and by most standard biographical dictionaries to the Later Han. Under his alternative name of Thai Hsü Chen Jen[11] he appears to be responsible for four books in the *Tao Tsang* (TT 108, 610, 901 and 1357), one of which has to do with alchemy. He may be worth watching in connection with the origin of the compass. On the other hand the *Shansi Thung Chih*[12] and other sources say that Chhiu Yen-Han (fl. +713) received the *Hai Chio Ching* from a mysterious immortal, which may mean that he wrote it, or a first version of it (*TSCC, I shu tien*, ch. 679, *lieh chuan*, p. 5a). The passage we quote would then have been a later addition. Of course Yang Yün-Sung may have written it himself. Another possibility is Liu Chhien[13] who, according to the *Chiangsi Thung Chih*,[14] ch. 106, p. 28a, wrote a *Nang Ching*. His grandfather, Liu Chiang-Tung,[15] had been a pupil of Yang Yün-Sung.

[b] One wonders what relation this feminine character could have had with her namesake whom we have met in another connection (cf. Vol. 2, p. 147 above). Her 'biography' is in *Yün Chi Chhi Chhien*, ch. 114, pp. 15bff.

[c] A favourite Taoist term for the universe (cf. pp. 256, 297, 299 above, and Vol. 2, p. 360).

[d] We have not been able to find out much about this scholar, but he was probably a Chiangsi geomancer.

[e] The fact that it was a floating needle pleads to some extent for an early date of the text.

[f] *I shu tien*, ch. 651, *hui khao* 1, p. 16a; tr. auct.

[g] Cf. p. 297 above on the origins of the Greek wind-rose.

[h] Equatorial constellations, it will be remembered, transferred to the horizon as azimuthal bearings.

[i] Cf. Vol. 2, p. 115 above.

[j] The reference is to the legends about the south-pointing carriage (cf. Sect. 27e below).

[k] Cf. the description just given of the compass-points and their origins.

[1] 九天玄女青囊海角經   [2] 張士元   [3] 蚩尤   [4] 天干

[5] 方所   [6] 地支   [7] 方氣   [8] 吳元爵   [9] 赤松子   [10] 黃初平

[11] 太虛眞人   [12] 山西通誌   [13] 劉謙   [14] 江西通志   [15] 劉江東

Afterwards a copper plate was made with exactly 24 azimuth points (chosen from) the denary cyclical series (which had been associated with the) 'heaven-plate' (*thien phan*[1]) (of the Han diviner's board, *shih*[2]), and the duodenary cyclical series (which had been associated with the) 'earth-plate' (*ti phan*[3]) (of the Han diviner's board, *shih*[2]). The former are called *li hsiang na shui*,[4] and the latter *ko lung shou sha*.[5] Nowadays geomancers use the Chêng Chen[6] (astronomical north and south points) and the 'heaven-plate' denary azimuth points to 'find out where the dragon is' (*ko lung*[5]); and they use the Fêng Chen[7] (magnetic north and south points) to perform other divinations (*li chan*[8]). Naturally the round (plate) follows (i.e. symbolises) heaven, and the square (plate) follows earth.[a]

Fig. 342. 'The Directions and Chhi of the Floating Needle', a diagram from the *Chiu Thien Hsüan Nü Chhing Nang Hai Chio Ching* (probably early +10th century), preserved in the *Thu Shu Chi Chhêng* encyclopaedia. South is at the top. The 12 primary compass-points are in the second ring from the centre, the 12 secondary ones in the first; both are combined in the fourth. But between the second and the fourth there is a ring in which the points are shifted 7½° E. of N.; this is the Fêng Chen system based upon an eastern declination of the needle. Here we see by itself the first of the accretions to the geomantic compass which were introduced to allow for observed declinations.

Though at first sight it may seem surprising that so elaborate a description of the geomantic compass could be of the Sung period, there is nothing in it which fails to fit with the general pattern of development which we are gradually uncovering. It is probable that the fixation of the azimuth compass-points from a selection of denary and duodenary characters, and from the *kua*, took place at least as early as the Han. But here in this passage we meet for the first time with one of the technical terms for declination directions. Similar statements are encountered in other geomantic books which seem to be of Sung date, such as the *Shen Pao Ching*[9] (Spiritual Precious Canon)[b] of Hsieh Ho-Chhing.[10]

[a] Note that nothing is said about the Chung Chen.

[b] *TSCC, I shu tien*, ch. 667, *hui khao* 17, p. 10*a, b*. If it is significant that Hsieh is quoted first among these geomantic writers, he may have flourished earlier than Wang Chi (see opposite), in which case his references to the needle, dating from the +10th century, would be the earliest we have. He was a contemporary of the geomancer Liu Yuan-Tsê,[11] who might be a help in dating both of them.

[1] 天盤	[2] 式	[3] 地盤	[4] 立向納水	[5] 格龍收沙	[6] 正針
[7] 縫針	[8] 立占	[9] 神寶經	[10] 謝和卿	[11] 劉淵則	

One argument which strengthens the relatively early datings of all these references to the magnetic needle is the occurrence of the term 'suspended needle' in the 'tadpole passage' of the *Ku Chin Chu*, as quoted in the *Thai-Phing Yü Lan* encyclopaedia of +983 (cf. p. 274 above).

What is probably the earliest datable reference to magnetic declination occurs as early as the first certain reference to a floating magnet. Wang Chi[1] (Chao-Chhing[2, 3]), the founder of the Fukien school of geomancers, left a poem on the magnet needle (*Chen Fa Shih*[4]) some lines of which are preserved in encyclopaedias. The relevant verse runs:[a]

Between (the *hsiu*) Hsü[5] and Wei[6] points clearly the needle's path (*chen lu*[7])[b]
(But) to the south there is (the *hsiu*) Chang[8], which 'rides upon all three' (*shang san chhêng*[9]);
(The *kua*) Khan[10] and Li[11] stand due north and south, though people cannot recognise (their subtleties),
And if there is the slightest mistake there will be no correct predictions.

Since it can be established[c] that Wang Chi's *floruit* was in the neighbourhood of +1030 to +1050, these lines are in fact the oldest distinct statement of the use of the needle which we possess. It will be remembered that Tsêng Kung-Liang's instrument (+1027 or before) was a floating 'fish'. The direction referred to in the first line is clearly the astronomical north–south (Chêng Chen), but it can be seen by looking at a geomantic compass (e.g. 24th circle, Fig. 340) that the southern *hsiu* Chang is so broad that it can include all the three 'souths', i.e. those for the two declinations, formerly eastern and now western (Fêng Chen and Chung Chen) as well as Chêng

---

[a] *TSCC, I shu tien*, ch. 655, *hui khao* 5, p. 18*b*; tr. auct.
[b] Note that this was the term later used for nautical compass bearings.
[c] Wang Chi's works, *Hsin Ching*[12] and *Wên Ta Yü Lu*,[13] were published after his death by his pupil Yeh Shu-Liang,[14] and a postface to them was written by Fan Shun-Jen[15] (G 534) whose dates, +1026 to +1101, are known (*Sung Shih*, ch. 314, p. 29*b*). According to the *Chhu-chou Fu Chih*[16] (quoted in *TSCC, I shu tien*, ch. 679, *lieh chuan*, p. 10*b*), Wang Chi predicted geomantically the success of the unborn sons of three Fukienese families who consulted him. Since the dates of their careers are known, the date of Wang Chi can be approximately fixed. The three men were Kuan Shih-Jen[17] (+1044 to +1109; see *Sung Shih*, ch. 20, p. 9*a*; ch. 351, p. 20*b*), Ho Chih-Chung[18] (+1042 to +1116; see *Sung Shih*, ch. 351, pp. 10*b*, 11*a*), and Chang Shang-Ying[19] (+1042 to +1121; see *Sung Shih*, ch. 351, p. 6*a*). Wang Chi's grandfather, Wang Chhu-No,[20] was an astronomer and mathematician who had been involved in a famous controversy about calendrical matters with Wang Pho[21] in the Later Chou dynasty (*Sung Shih*, ch. 21 ff., ch. 461, p. 3*a, b*). The grandfather must, therefore, have flourished between +950 and +970. Wang Chhu-No's son, Wang Hsi-Yuan,[22] died in +1018 at the age of 58 (*Sung Shih*, ch. 461, p. 4*a*). From all these facts, the year +990 would be probable for the birth of Wang Chi. Wang Chen-To (*4*, p. 116; *5*, p. 210, etc.) does not seem to have realised that Wang Chi and Wang Chao-Chhing were one and the same person. Elsewhere (*5*, p. 121) he has an argument that the *Chen Fa Shih* cannot be earlier than the last decade of the +12th century, but we have not been convinced by it. Since Wang Chi was little known outside geomantic circles, it seems very unlikely that anyone writing about +1200 would have attributed a poem to him. He was not one of the three who left their mark on the geomantic compass, nor even one of their famous pupils. The two forms of the character in Wang Chao-Chhing's name are due to the tabu system; it was necessary to avoid certain characters at certain times for court reasons (cf. Li Yeh and Li Chih, in Sect. 19 above).

[1] 王伋	[2] 鑒卿	[3] 趙卿	[4] 針法詩	[5] 虛	[6] 危
[7] 針路	[8] 張	[9] 上三乘	[10] 坎	[11] 離	[12] 心經
[13] 問答語錄	[14] 葉叔亮	[15] 范純仁	[16] 處州府志	[17] 管師仁	
[18] 何執中	[19] 張商英	[20] 王處訥	[21] 王朴	[22] 王熙元	

Chen.[a] This must surely be the explanation of the expression 'it rides upon all three'.[b] The reference to the *kua Khan* and *Li* would simply be that they represent north and south only in a general way, and that there was some tendency in nature for the magnetic axis to diverge slightly from the astronomical meridian; this was the subtlety which ordinary people could not understand.

After +1050 all the descriptions agree that the declination was to the west. We are familiar with the statement of Shen Kua (p. 249 above) that in +1086 the needle pointed slightly east of south; and with the observation of Khou Tsung-Shih (p. 251 above) in +1115 that it pointed to the sign Ping (165°). The next record is one which Hashimoto discovered—an entry in the *Thung Hua Lu*[1] (Mutual Discussions), written by Tsêng San-I[2] in the neighbourhood of +1189.[c]

> The North–South Needle.
>
> As for the use of the 'Earth-Disc' (*ti lo*[3])[d] (the compass), there is the Chêng Chen (astronomical north–south) Tzu–Wu system, but there is also a system which uses a Fêng Chen (seam needle)[e] axis between the true north–south and the Ping–Jen axis.
>
> Since heaven and earth are correctly placed (on the astronomical north–south meridian) one ought to use the Tzu–Wu line, but since there is what is now called the 'sloping-off of the land south of the river', it is difficult to use the Tzu–Wu, Chêng Chen, system; and the Ping–Jen axis checks up better.
>
> Anciently, men measured the sun-shadow at Loyang, considering that place to be the centre of the earth. But others later thought that Yang-chhêng, a *hsien* away from the capital (was more central).[f] As the ground 'sloped away' somewhat, the measurements (at Loyang) could therefore hardly be used. So the principle of the difference is quite understandable.[g]

The declination was thus something less than 15° W., but the interest of the passage lies in its search for a theory to account for the declination. As we saw, Khou Tsung-Shih had appealed to five-element theory, to the mutual relations of Fire and Metal. Here is a more sophisticated idea, namely that there must be some central place or meridian on the earth's surface where the declination would be found to be zero.[h] This can only be called an inspired guess, for in fact there is a line of null declination, which moves slowly across the earth's surface.

[a] Hsü and Wei only include two each.

[b] Of course the middle (Chung Chen) circle was not (according to tradition) incorporated in the geomantic compass until the century after Wang Chi. In those days it would be natural to tread warily in such a matter.

[c] Embodied in *Shuo Fu*, ch. 23 (though not in all editions). Tr. auct.

[d] *Lo* really means a conch or screw-like shape, here spiral, loosely applied to the concentric rings of the compass-plate.

[e] Note that Tsêng San-I refers to the new axis by the old name of Fêng Chen, which afterwards meant only the Ting–Kuei line of eastern declination. Evidently in his time the technical term Chung Chen for the western declination had not come into general use.

[f] Cf. above, concerning the giant gnomon at Yang-chhêng, Vol. 3, pp. 296 ff.; and above, pp. 46 ff.

[g] This was the passage which Hashimoto (1) did not dare to translate himself, and which Mitchell (2) considered so obscure as to be meaningless.

[h] Cf. the +16th-century European doctrine of the 'true meridian' (Taylor (8), pp. 173, 183, 186).

[1] 同話錄   [2] 曾三異   [3] 地螺

Tsêng San-I was thus theorising about the declination before Europeans knew even of the polarity. A little later there is quite a long argument about it, involving the introduction of a new element, namely the second subsidiary staggered circle, the series of Chung Chen points, traditionally due to Lai Wên-Chün about +1150. If we dip into the first chapter of the *Chhu I Shuo Tsuan*[1] (Discussions on the Dispersal of Doubts), written by Chhu Hua-Ku[2] (Chhu Yung) about +1230, we find the following:[a]

On Choosing Needles.[b]

[Nowadays the geomancers make use of both the Chung Chen and the Chêng Chen.][c]

The theories of the Yin–Yang school are very prevalent, and it is not easy to discuss lightly their value and truth. They should be tested by facts and rationally analysed.

The first thing in geomancy is to distinguish the 'twenty-four mountains'[d] (the 24 compass-points), and to know the Chêng (the astronomical north-south meridian line). Then there are the 120 Fên Chin[3] divisions.[e] If we use the Ping–Wu, or Chung Chen, system, there will be a difference of $2\frac{1}{2}$ of these divisions, to the south-west. If we use the Tzu–Wu, or Chêng Chen, system, there will be a difference of $2\frac{1}{2}$ of these divisions, to the south-east. (Where there is so much difference in systems of needle positions, predictions of) fortune and misfortune should be very clearly distinguishable.

When my father was selecting lucky ground, the geomancers differed among themselves. One thought the Ping–Wu direction, in accordance with the Chung Chen system, was the right one to take, but the other preferred the Tzu–Wu (meridian) line, according to the Chêng Chen system. Both insisted on the correctness of their knowledge, handed down by their respective teachers. As there are now no sages left in the world, no one could tell us which was the right choice to make. Actually, there are good reasons for both views.

Those who believe in the Ping–Wu direction, or Chung Chen system, say that the ancient *Hu Shou*[4] books explain it quite clearly—'The direction from Tzu (N.) to Ping (S. 15° E.) is south-east, and therefore indicates the Yang; the direction from Wu (S.) to Jen (N. 15° W.) is north-west, and therefore indicates the Yin. The Jen–Tzu–Ping–Wu area is the centre of heaven and earth.' And they argue that although the needle points to the south, its original nature inclines to the north (so that it tends to veer back there).[f] This theory has some foundation.

(Those who take this view) also say that the twelve cyclical characters (*chih*[5]) have the Tzu–Wu axis as their correct centre, and that the 64 *kua* are arranged among the 24 compass-points. But Ping is side by side with Wu, and shares its position (on the Chêng Chen circle). The point between them is at the centre of the round of 12 *chih*, i.e. at Wu (on the Chung Chen circle). This idea is also reasonable.

---

[a] Ch. 1, p. 2*b*, tr. auct. This passage has not previously been noticed. The version in the *Pai Hai* collection is abbreviated and unintelligible; that in the *Pai Chhuan Hsüeh Hai* seems to be complete, and has here been used.

[b] I.e. systems of magnet-needle bearings for divination.

[c] This sentence is prefixed, in small characters, to the *Pai Hai* version only.

[d] Cf. Yang Yün-Sung's *Erh-shih-ssu Shan Hsiang Chüeh*[6] (Oral Instruction on the 24 Mountain Directions), in *TSCC, I shu tien*, ch. 654, *hui khao* 4, p. 10*b*.

[e] These are the circles 9 and 19 in Fig. 340. They consist of combinations of the cyclical signs, distributed among the five elements.

[f] The idea at the back of this is probably mineralogical, connected with the transformation of ores and metals in the earth, see p. 302 above.

[1] 祛疑說纂　　[2] 儲華谷　　[3] 分金　　[4] 狐首　　[5] 支　　[6] 二十四山向訣

But those who choose the Tzu–Wu line and Chêng Chen system say that Fu-Hsi determined the eight directions by means of the 8 *kua*; *Li* and *Khan* fixing the true north and south. Ping and Ting flank *Li* (in the south); Jen and Kuei flank *Khan* (in the north). Thus the eight directions are divided into 24 positions; the south getting Ping, Wu and Ting; the north getting Jen, Tzu and Kuei. The Tzu–Wu (meridian) axis is indeed central. This argument also is reasonable and cannot possibly be neglected.

They go on to argue that the position of the sun at the Ping (double-) hour[a] is at Ping, and at noon (the Wu (double-) hour) it is at Wu. What is now called the Ping (double-) hour was formerly two fixed positions ahead.

Thus I made a record of these arguments for the benefit of those who like to busy themselves with such matters. Those who use the Chung Chen system often had successful results; this is because the divination method was based exactly on the original positions.

To this may be added the fact that when in +1280 Chhêng Chhi quoted the words of the *Pên Tshao Yen I* about the magnet in his *San Liu Hsüan Tsa Chih*,[b] he added only a remark about the 'Tzu–Wu Ping–Jen chih li'[1]—the principles of the Chêng Chen and Chung Chen systems—showing that the latter was well known in his time also.[c] He said that they were used by the geomancers (Yin-Yang chia[2]).

The main point of this is that in the time of Chhu Yung and Chhêng Chhi, the declination was 15° W. or less; the fact that they talk about a Ping–Wu position suggests that it was about half that. Chhu's theory is retrograde compared with that of Tsêng San-I, falling back as it does upon the *kua*, and offering obscure symbolic explanations. We are, however, introduced to a current of geomantic literature not previously met with, namely the 'Fox-Head Books' (*Hu Shou*).[d]

All this was still two centuries before the declination was first observed in Europe. In special studies, Winter (4, 5) concluded that it was quite unknown to Pierre de Maricourt and his contemporaries (+1270). On the other hand, Mitchell (2), after careful examination of the tradition[e] that Columbus discovered it on his voyage of +1492, decided that this must also be rejected.[f] He and Chapman agree, however, with Hellmann and Wolkenhauer, that from a date not later than +1450 German makers of portable sundials which embodied compasses by which to set the noon line were including a special mark on the dials to which the needle should point, thus showing empirical knowledge of the declination. This is found also on Nuremberg

---

[a] Ordinarily there is no such thing as the Ping double-hour, for the hours were designated by *chih* and not by *kan*, so we are not sure of the meaning of this argument. See Needham, Wang & Price (1), pp. 199 ff.

[b] Referred to above, p. 252.   [c] Wang Chen-To (4), p. 194.

[d] Chêng Ssu-Hsiao (d. +1332) says, in his *So-Nan Wên Chi*, p. 16a, that only one *Hu Shou Ching*[3] still existed in his time, and he thought it was by Kuo Pho of the Chin. Whether it has any connection with *Hu Tzu*[4] books, two of which are mentioned in the Thang bibliography, or with the *Hu Kang Tzu*[5] books, one of which occurs in the Sui and three in the Sung bibliographies, we have not been able to find out. But the fact that the *Hu Shou* tradition seems to have been connected with the western declination would suggest that in fact it was not older than the beginning of the +11th century. It is much to be wished that some scholar would make a serious attempt to bring order and light into the maze of geomantic literature.

[e] Bertelli (2).   [f] Taylor (8), p. 172, concurs.

[1] 子午丙壬之理   [2] 陰陽家   [3] 狐首經   [4] 狐子   [5] 狐剛子

maps just before +1500. Lange puts forward an argument in favour of Geoffrey Chaucer (+1380), but the more likely date is the early +15th century.

Chinese books of about +1580, such as Hsü Chih-Mo's[1] *Lo Ching Chien I Thu Chieh*[2] (Illustrated Easy Explanation of the Magnetic Compass), offer other theories, suggesting, for example, that there is declination because the pole-star is not quite polar. Fang I-Chih in his *Wu Li Hsiao Shih* (+1664) mentioned[a] that in Europe (Ta-Chhin) at that time the declination was eastern, which was quite correct (Edkins, 12). He wondered whether this might be due to the intervening massif of the Himalayas (Khun-Lun mountains).

It may be convenient to summarise the information contained in this Section in the form of a table (Table 52). The general upshot is that the first knowledge of declination in China dates back to the +9th century at least. Eastern before the Sung, it left its traces in the traditional form of the geomantic compass as the outer subsidiary staggered circle, the Fêng Chen; then during the Sung about +1000 it became western, and gave rise to the inner staggered circle, the Chung Chen. De Saussure[b] was quite right in his statement that the Chinese knew the magnetic declination long before Europeans knew the polarity itself.[c] And once again de Groot was hopelessly at sea when he wrote 'there is not the slightest indication that the Chinese possess any knowledge of the declination of the compass, or that they are able to make a distinction between the magnetic north and the exact north'.[d]

It is good to know that we now have before us the possibility of verification of the old Chinese figures and traditions. At an earlier stage, in connection with the *Wu Ching Tsung Yao* quotation, mention was made of the phenomenon of thermo-remanence. A bar of iron or steel allowed to cool slowly from the Curie point (*c.* 700°) in the earth's magnetic field acquires the properties of a magnet without any contact with a lodestone. But this is also true, at a much more feeble level of intensity, of most sedimentary rocks, lava flows and natural clays, which contain small amounts of iron oxides.[e] Measurements with delicate methods now actively proceeding,[f] therefore, will be able to give detailed information not only about the properties of the earth's magnetic field at the time when the rocks and clays were being formed and deposited, but also as it manifested itself at the time of baking of tiles, bricks, kiln walls and pottery. These developments are thus of great interest to archaeologists as well as to geologists, for if once a systematised body of knowledge could be built up of the earth's

---

[a] Ch. 8, p. 18*b*.   [b] (35), p. 31.

[c] Mitchell (2) says that according to Lecomte (+1699), the Chinese knew nothing of declination, but on checking his text (Eng. ed. p. 229) I find that his words can be interpreted to mean that they did know it when Lecomte explained sufficiently clearly what he was referring to. In any case, even if Lecomte was as unfortunate as de Groot, it means very little in the light of the other evidence of this Section.

[d] (2), vol. 3, p. 974. I would repeat that the 'anthropological' method of de Groot would be exactly paralleled by the efforts of a Chinese scholar to find out what Englishmen knew of nuclear physics by interrogating fishwives and traditional morris dancers.

[e] Cf. the introductory article by Runcorn (1).

[f] See Cook & Belshé (1); Aitken (1).

[1] 徐之鏌   [2] 羅經簡易圖解

Table 52. *Details on Chinese compass observations including magnetic declination, mainly before +1500*

Date or approximate date	Author	Name of book	Probable place of observation	Lat. N. ° '	Long. E. ° '	Details	Declination °
c. +720	I-Hsing	Unknown (Wylie, 11)	Chhang-an (Sian)	34 16	108 57	Between Hsü and Wei 'to the right'	3–4 E.
		This reference is very doubtful, see p. 301.					
(c. +730)	Chhiu Yen-Han.	The 24 azimuth points stabilised.	Chêng Chen				
Mid +9th	Unknown	Kuan shih Ti Li Chih Mêng	Probably Sian	34 16	108 57	Ting-Kuei axis	c. 15 E.
(c. +880)	Yang Yün-Sung adds new divisions each one 7½° to the left of Chhiu's, for eastern declination, and speaks of Fêng Chen in his *Chhing Nang Ao Chih*						
c. +900	Unknown	Chiu Thien Hsüan Nü Chhing Nang Hai Chio Ching	Probably Sian	34 16	108 57	Speaks of Fêng Chen	c. 7½ E.
c. +1030	Wang Chi	In comm. on *Kuan Shih*…	Probably Khaifêng	34 52	114 38		Slightly W.
c. +1086	Shen Kua	Mêng Chhi Pi Than	Khaifêng	34 52	114 38	'Slightly E.' ('of south')	5–10 W.
+1115	Khou Tsung-Shih	Pên Tshao Yen I	Khaifêng	34 52	114 38	To Ping	c. 15 W.
(c. +1150)	Lai Wên-Chün adds new divisions each one 7½° to the right of Chhiu's, for western declination, and introduces the term Chung Chen						
c. +1174	Tsêng San-I	Thung Hua Lu	Hangchow	30 17	120 10	near Ping-Jen axis	5–10 W.
(+1190)	first knowledge of magnetic polarity in Europe						
c. +1230	Chhu Hua-Ku	Chhu I Shuo Tsuan	Probably Hangchow	30 17	120 10	Ping-Wu direction and Chung Chen	7½ W.
c. +1280	Chhêng Chhi	San Liu Hsüan Tsa Chih	Probably Hangchow	30 17	120 10	Tzu-Wu and Ping-Jen axes	7½ W.
(c. +1450)	first knowledge of magnetic declination in Europe						
c. +1580	Hsü Chih-Mo	Lo Ching Chieh	Probably Peking	39 54	116 28		c. 7½ W.
c. +1625	J. A. Schall and Hsü Kuang-Chhi	(Wylie, 11)	Peking	39 54	116 28		5½–7½ W.
c. +1680	Mei Wên-Ting	Khuei Jih Chi Yao [a]					
+1690	de Fontaney		Nanking	32 4	118 47		3 W.
			Suchow	31 23	120 25		2½ W.
			Canton	23 8	111 16		2½ W.
+1708	Régis & Jartoux	(Wylie, 11) (Gaubil (1), p. 209)[b]	Shanhaikuan	40 2	119 37		2 W.
			Chiayükuan	39 49	98 32		3 W.
			Canton	23 8	111 16		null
+1817	Wylie (11)		Canton	23 8	111 16		null
+1829	Wylie (11)[c]		Peking	39 54	116 28		1½ W.

[a] This reference is given by Wylie (11) from a tractate which he did not quote in Chinese characters, giving only the title in his peculiar romanisation system. We conjecture it to be *Khuei Jih Chi Yao*.[1] He says it was a 'small work on the sundial'. Unfortunately, it is not listed in the *Wu-An Li Suan Shu Mu*[2] where, however, we find an *Jih Kuei*[3] *Pei Khao* (p. 29a), a *Khuei Jih Chhi*[4] (p. 30a), and a *Khuei Jih Chhien Shuo*[5] (p. 30b). Perhaps Wylie was quoting one of these from memory. We have not been able to gain any further light from Li Nien's detailed biographical bibliography of Mei Wên-Ting, (21), vol. 3, pp. 544ff., and must leave the matter to be cleared up by others to whom the works of the great Chinese seventeenth-century mathematician and astronomer are more accessible.

[b] It is not generally known that the Khang-Hsi emperor himself wrote on the declination of the compass. This was in his *Khang-Hsi Chi Hsia Ko Wu Pien*[6] (Scientific Observations made in Leisure Hours), finished about +1710, and abstracted in French by Cibot (7) in +1779. The emperor said that the declination at Peking had been 3° W. in +1683, and had fallen to 2½° W. at the time when he wrote; in some provinces an eastern declination was still observable. Further observations, also in the neighbourhood of Peking, varying up to 4½° W., were reported by Amiot (5) in +1780 and +1782.

[c] On the saecular variation in China in subsequent times see de Moidrey & Lu (1).

field properties (declination, dip, etc.) at different places throughout historical time, a method of great value for dating any objects showing thermo-remanence would become available. Evidently it will also in time be possible to trace the saecular variation of declination in China, and thus to test the general picture sketched in this Section.

The variation which we have deduced in China may be compared with the results of a series of direct observations made in London from about +1580 onwards.[a] The declination passed from east to west about +1660 and reached its western maximum in 1820, just as we have been led to assume that a passage from east to west occurred in China about +1000. Of the two extrapolations on the left of Fig. 343 the upper one derives from observations by Cook & Belshé of +15th-century kiln and hearth wall thermo-remanence in England, and is thus more suitable than the lower one which

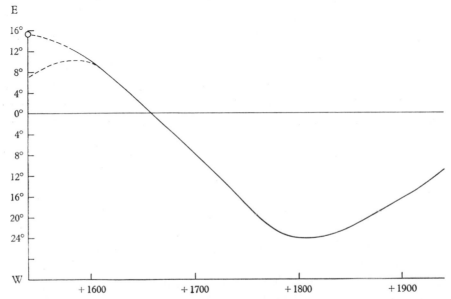

Fig. 343. Plot of the saecular variation of magnetic declination at London from the beginning of the +16th century onwards.

is based on observations made at Rome from +1540 onwards.[b] Here we are plotting the declination against time, but it may be more useful to plot declination against dip, as in the famous diagram of L. A. Bauer, which for the period and region concerned gives almost a complete circle.

Data for the fired structures of the Romano-British period so far available suffer equally from scatter of readings and vagueness of dating,[c] so that it is not yet possible to plot them on a graph similar to that in Fig. 343. However, in the relation of declina-

---

[a] See, e.g. Chapman & Bartels (1), vol. 1, p. 130.

[b] Kiln walls of the +12th century in England gave values of 6° and 7° E., as if the declination was then moving towards its easterly maximum. Thellier (1, 2) and Thellier & Thellier have done much to establish similar thermo-remanent records for France and North Africa. Cf. Aitken (2).

[c] Between the +1st and +4th centuries the declination varies from 13° W. to ½° E.

tion to dip a clear trend was perceptible, the latter being greater the more western the declination. Another study, having regard only to the inclination, has been made on Chinese Yüeh ware, the pieces of which may be assumed to have been baked in an upright position, and the results compared with the thermo-remanence figures of kilns and lavas in Japan.[a] A dating of $+1080 \pm 120$ resulted—quite consistent with the stylistic estimates. Similarly, high-fired stoneware from Shao-hsing, attributed to the -2nd and -3rd centuries, could be confirmed as to period by this method. But the whole approach is as yet in its infancy, and great developments may be expected of it.

In the course of time the saecular variation of magnetic declination at particular points on the earth's surface will receive explanation in terms of general geophysical theory. Ever since the first recognition of the magnetic pole by Mercator in $+1546$,[b] knowledge of these changes has been steadily accumulating[c] and towards the end of the last century was summarised in the classical charts of van Bemmelen. We now realise that they depend upon the existence of centres of magnetic perturbation which slowly travel back and forth across the surface of the earth. But the full meaning of these remains to be discovered.

It now only remains to clear up the explanation of Fig. 340 by explaining the nature of some of the remaining circles on the geomantic compass which have not so far been mentioned. The outermost circle (no. 24)[d] always shows the equatorial *hsiu* divisions (cf. Vol. 3, pp. 233 ff.); inside it the number of (Chinese) degrees of each *hsiu* are given (no. 22),[e] and geomantic denotations as to the favourable or unfavourable nature of some of these degrees (no. 23). The 6th circle gives the 24 fortnightly periods (cf. Vol. 3, p. 405 above), and the 4th gives 24 constellation and star names. Analysis of the rest would take us too far from the scientific significance of the instrument.[f]

(iv) *Traces in urban orientations*

One last word may be devoted to the question of the orientation of city-walls and the like, always a subject full of uncertainty.[g] Gaubil (11) drew attention in $+1763$ to the fact that the walls of Peking, built about $+1410$, were $2\frac{1}{2}°$ out of the meridian to the west. In recent years, Mr Brian Harland, when working in Shantan[1] city, Kansu, one of the towns on the Old Silk Road, about 1945, noticed that the street-plan seemed to show two different alignments. His plan, part of which is reproduced here (Fig. 344), demonstrates clearly the persistence of two old city alignments differing by about $11°$, one due north–south and one east.[h] It seems quite possible that such differences

[a] Aitken (1); Watanabe (1, 1, 2).

[b] Cf. the translation of his letter by Harradon (2).

[c] Cf. the translations of $+16$th-century discussions by Chapman & Harradon (2), and the summary in Taylor (8), pp. 172 ff., 181 ff. The first clear recognition of the saecular variation was due to Henry Gellibrand in $+1635$. In the winter of $+1699$ Edmund Halley made a classic voyage to study the declination, and drew up a world map on Mercator's projection marked with isogonic lines.

[d] Here also the 16th.   [e] Here also the 15th, in modern degrees.

[f] See the caption to Figs. 340 and 341.

[g] Cf. p. 249 above.

[h] This would suggest that in the Thang, or at some time before the Sung, the street grid was oriented to the needle's eastern magnetic declination.

[1] 山丹

PLATE CXXII

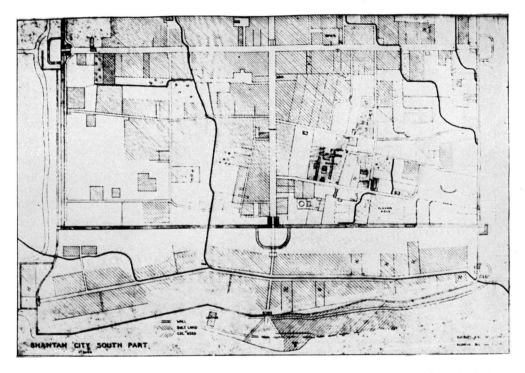

Fig. 344. Plan of the southern part of the city of Shantan, Kansu, made by Mr Brian Harland about 1945. Within the walls there are traces of two different alignments, one due north and south, the other following a line about 11° E. of N. Such differences, which are also to be found in the plans of other Chinese cities, may originate from the divergent advice of geomancers after the discovery of magnetic declination.

are the remains of medieval discussions between partisans of Chêng Chen, Fêng Chen and Chung Chen.

Extensive material is available in old Chinese city-plans for a special study of this subject. The southern part of the city of Nanking has a true north–south axis, but the walls, which most of the streets follow, are inclined 13° E.[a] In Chhêngtu, Szechuan, the street grid also follows the walls, which have an inclination as much as 25° E., but the central palace is oriented to the meridian.[b] On the other hand at Khaifêng, though the walls are inclined 11½° E., the street grid follows the Sung palace which is oriented due north and south;[c] in this case, therefore, one could deduce that the interior of the city must have been rebuilt in Sung times. Again, the Chin Tartar capital at Huining has, as one would expect, a westwards inclination of 5½°.[d] At Chhang-an (Sian), the Thang capital, the main north–south street is some 2° E. in relation to the walls, and other streets a little more.[e] Such eastward orientations conform with what has been said above about the eastward declination of the needle in Thang times.

At first sight it would appear to be a *reductio ad absurdum* of such suggestions when we find from Shih Chang-Ju (2) that the tombs at Anyang, dating from about − 1300, are all oriented from 5° to 12° E. of our present true north. But for such very ancient sites precessional change becomes a factor. And indeed a glance at Fig. 94 in Vol. 3 will show that this difference is not far from what we should expect if the Shang people had taken care to site their tombs in accordance with the astronomical north of their time.

(7) Magnetic Variation and Inclination

In spite of the ancient legends about the attraction of ships with iron nails to certain islands which contained masses of lodestone, the discovery of real local variations was not a pre-Renaissance one in Europe.[f] To William Barlowe (+ 1597) is owing the realisation that ships' magnets were disturbed by iron in the vessel itself.[g] Chinese sailors, however, seem to have been acquainted with local variation from the + 15th century onwards. Fei Hsin remarked, in his *Hsing Chha Shêng Lan* of + 1436 (cf. Vol. 3, p. 558 above):[h]

There is a seamen's saying 'To the north we are afraid of the Seven Islands; to the south we fear the Khun-Lun (Pulo Condor Island)'. At these places the needle may err (*chen mi*[1]), and if that happens, or the steering is inaccurate, both men and ships will be lost.

Then, in the (*Thai-Wan*) *Tan-shui Thing Chih*[2] (Local History and Topography of Tan-shui, Formosa), a work of late Chhing date (*c.* 1871), we hear of the *fan ching shih*,[3] the rocks which bewilder the compass.

---

[a] Herrmann (1), p. 57.   [b] Gutkind (1), fig. 42.   [c] Herrmann (1), p. 48.
[d] Herrmann (1), p. 47.   [e] Herrmann (1), p. 21.
[f] Famous cases now known are Skye in the Hebrides and Kursk in the Soviet Union.
[g] Cf. May (1).
[h] Quoted in *Tung Hsi Yang Khao*, ch. 9, p. 4*b*, tr. auct.

[1] 針迷   [2] 臺灣淡水廳志   [3] 反經石

The Fan-Ching-Shih are near Hsi-Yün Yen (West Cloud Cliff) on Kuan-Yin mountain. There are two large stones, one of them saddle-shaped. If a compass is placed on them, its needle, instead of pointing north and south, turns and points east and west. Hence the name of the rocks. Another rock is near the Mushroom-Orchid Village (Chih-lan Pao) on Shih-Ko mountain, but here the needle turns to point west and east.[a]

How far back this observation went we cannot ascertain, but it was closely similar to the classical experiments of the Portuguese commander João de Castro on the island of Chaul off the west coast of India in +1539.[b]

So far as it has been possible to ascertain, the vertical component of the earth's magnetic field (magnetic dip or inclination) was never discovered in China. According to Mitchell (3) the first observation of it was due to Hartmann in +1544,[c] and the first correct estimate of it to Robert Norman in +1581. We have just seen something of the help which it may give in archaeological dating.

(8) THE MAGNET, DIVINATION, AND CHESS

The whole story of magnetism in China so far has been extraordinary enough, but there is still a little more to be said. As an addendum to it we propose to explore some dark by-paths in the fields of divination-technique from which games such as chess derive, for there are indications that they alone contain the clue to the first use of a material earthly 'south-pointer'. We must keep such an exploration as brief as possible, avoiding the temptation to stray far afield into the realms of anthropology and archaeology.

The essence of the problem is—how did the *shao* get on the *shih*? Why should anyone have thought of making a model of the Great Bear at all, and of placing it on a board? That it should be made of lodestone, shaped from a bar of magnetite into the form of a dipper, is a thing less difficult to understand. But if it can be shown that there were many ancient systems of divination which involved the use of 'pieces' similar to chess-men or other game component units, and that these 'men' often represented the celestial bodies, then the whole process of thought begins to reveal itself.

Many remarkable scholars have devoted years of work to the history of chess (such as van der Linde and H. J. R. Murray). The generally accepted view resulting from their studies is that chess, the war-game as we know it today, developed first in +7th-century India, whence it radiated to Persia, to the Muslim world, and ultimately Europe.[d] But its antecedents have so far been very mysterious. Strictly

[a] Tr. auct.

[b] The descriptions of these are in his own route books (1, 2, 3), translated by Harradon & Ferraz (2) and summarised by Taylor (8), pp. 183 ff. From +1538 onwards de Castro studied systematically the variations in declination between Portugal and India, using the 'shadow instrument' of Nunes (cf. Harradon & Ferraz, 1), and these are the oldest systematic observations now extant. The boulders of Chaul, which turned his needle right round by 180°, were probably of basaltic lava like the Deccan trap-rocks. De Castro satisfied himself that the rock was not lodestone, but the effect was almost certainly due to the thermo-remanence we have already discussed, with a field too weak to attract iron filings. In connection with the Formosa description it is interesting that in 1879 Knipping described a clear case of magnetic deviation (local variation) at the summit of the Futarasan mountains in Japan.

[c] See the translation of his letter by Harradon (1).   [d] As also to China and Japan.

speaking, 'chess-men' should be defined as the sets of pieces of the two, three, or four contending sides in the war-game; but for the present purpose we shall find it convenient to define them (for lack of any better term), as any collection, set, side or team, of small symbolic models, which may represent anything, not only component parts of an army, but animals, or (significantly) celestial bodies such as the sun and moon, planets, stars and zodiacal houses. Though China is the only civilisation where a close connection between the magnet and 'chess'[a] can be shown to exist, the connection between chess and astronomical-astrological symbolism is widespread in all civilisations.

In following the argument to be presented in the next few pages, it will be best to have before us a summary of it at the outset. What I propose is along these lines. (1) The diviner's board (shih[1]) was either used with, or related to, a technique which consisted in throwing a set of 'chess'-men on to a board and noting where they came to rest, these pieces being identified with various celestial bodies. The adjustment of the heaven-plate on the ground-plate would be made first. (2) The stages in the discovery of the lodestone's directive power would have involved (a) the throwing of the pieces, (b) the decision to make some of them of lodestone because of its obviously magic attractive power, (c) the decision to model the Great Bear in magnetite in the shape of a spoon, (d) the observation that it took up an oriented direction. Putting the matter in its shortest way, the lodestone spoon was a 'chess'-man of one of the divinatory forms of proto-chess. The Chinese language makes no distinction between these forms and the war-game of chess itself, both being expressed by the character chhi.[2,3] The game or divination-procedure is also called hsi.[4]

(i) *The fighting chess-men of Luan Ta*

The first steps in the process of discovery were being taken in the −2nd century. The text which originally attracted my attention as of great interest, and which led to the present train of thought, has come down to us in five versions, and is associated with the name of Luan Ta,[5] one of the magicians of the Han emperor Wu Ti. There is a good deal of information about him in the *Shih Chi*:[b]

> That spring (−113) (Ting I[6]), the marquis of Lo-Chhêng, presented a memorial recommending Luan Ta to the emperor. Luan Ta had been one of the eunuchs of the Prince of Chiao-Tung, and had had the same teachers as the Perfected-Learning General (Shao Ong,[c] another imperial magician). It was this fact which had made him Magician and Pharmacist-Royal to the Prince of Chiao-Tung....

---

[a] Although the word 'chess' is not properly applicable to anything except the specific war-game, whether the pieces have a cosmological or purely military significance, inverted commas will usually be dropped from the word henceforward, and the reader is asked to bear in mind the wider sense in which it is being used, i.e. the many forms of divination involving the throwing or moving of symbolic pieces. We may sometimes speak of 'proto-chess'.

[b] Ch. 28, p. 27b, tr. auct. adjuv. Chavannes (1), vol. 3, p. 479. Identical with ch. 12, p. 9b, where the Thang commentator Chang Shou-Chieh makes Kao Yu quote the *Huai Nan Wan Pi Shu*.

[c] Cf. Vol. 1, p. 108 above.

[1] 式   [2] 棊   [3] 碁   [4] 戲   [5] 欒大   [6] 丁義

The emperor was now regretting that he had put to death the Perfected-Learning General, and that the fullness of his arts had not been experienced, so he welcomed Luan Ta warmly. Luan Ta was tall and a brilliant talker, fertile in techniques, and daring in promises, never hesitating. He said to the emperor 'Your subject has often been overseas and seen An Chhi, Hsienmên (Kao) and other great magicians, but as I was an ordinary commoner they despised me and did not take me into their confidence.... My master maintained that yellow gold can be produced (artificially), that the breach in the Yellow River can be closed, that the herb of immortality can be found, and that the *hsien* can be made to appear. But all your subjects are afraid that they will meet with the same fate as the Perfected-Learning General, so none of them dare to open their mouths. How then should I dare to speak to you of my arts?'[a]...At the end of the interview, the emperor asked Luan Ta to demonstrate one of his lesser arts by making chess-men (*chhi*[1]) fight automatically, and indeed they did mutually hit against each other (*hsiang chhu chi*[2]).

The *Chhien Han Shu's* version[b] of this feat is identically worded. So far the lodestone has not been mentioned, but it prominently appears in the other three versions, all from the *Huai Nan Wan Pi Shu*,[c] which, if genuinely connected with Liu An, would be contemporary (c. − 120). The excerpt preserved in *Thai-Phing Yü Lan*, ch. 736 is almost identical with that which the Thang commentator Ssuma Chên[3] added to the *Shih Chi* text. It runs:[d]

The lodestone lifts (animates) (*thi*[4]) chess-men.
The blood of a cock is rubbed up with needle-iron (filings) and pounded to mix. (Then when) lodestone chess-men are set up on the board, they will move of themselves and bump against each other (*hsiang thou*[5]).[e]

The excerpt preserved in *Thai-Phing Yü Lan*, ch. 988 is a little different.[f]

Take the blood of a cock and mix it with iron (filings) from the grinding of needles, pounding it with lodestone powder. In the day-time, put the paste on to the heads of chess-men (*chhi*[6]) and let it dry in the sun. Then put them on the board and they will constantly bounce against and repel (*chü*[7]) one another.

Several things are noteworthy in the above. Exactly how it was that the magnetised chess-men were 'animated' is not clear; they may have been lodestone balls with iron underneath the board, or some of them may have been lodestone while others were iron. Powdered magnetite would have little attractive power. The connection with needles is interesting, and suggests that the demonstration of polarity using needles may really have been older than we thought (p. 278 above).[g] But in any case the important thing is

---

[a] Note the emphasis on alchemical arts, cf. pp. 269, 277 above. The emperor alleged that Shao Ong had died accidentally by eating horse liver, and raised Luan Ta to high honours, but he in his turn was executed in − 112.

[b] Ch. 25A, p. 24a.   [c] Already mentioned, p. 232 above.

[d] P. 8a, tr. auct.

[e] Note that the word used here is the same as that which Wang Chhung uses for 'rotating' the lodestone spoon. Ssuma Chên has *ti chi*,[8] 'repel and hit'.   [f] P. 3b, tr. auct.

[g] In the passage from the *Shih Chi* above, we followed Chavannes in translating *hsiao fang*[9] as 'lesser arts', but it is possible that it might refer to the square ground-plate of the diviner's board.

[1] 棊   [2] 相觸擊   [3] 司馬貞   [4] 提   [5] 相投   [6] 碁
[7] 拒   [8] 抵擊   [9] 小方

the association of the magnet with the men or pieces used in divinatory proto-chess. This is what gives significance to the several mentions of chess in connection with magnetism which we have noticed above (pp. 233, 268 and 302), as well as those yet to be mentioned.

Luan Ta's magnetised chess-men were not forgotten, as the following examples show. Ko Hung referred to them[a] as an example of a technique proved to be effective, and elsewhere[b] speaks of

playing with the three (sets of) chess-men, in order to foretell the success or failure of military enterprises.[c]

It will be remembered that Ko Hung spent the latter part of his life in alchemical pursuits in the Lo-fou Shan mountains north of Canton; we are not, therefore, surprised to read in the *Lo-Fou Shan Chih*[1]:[d]

Under the Shih-Lou Fêng peak there is a stone as smooth as a mirror, and on it there used to be 18 chess-men, some black, some white. They moved to and fro and pushed each other about, yet if you tried to pick them up you could not do so. This was called 'spirit chess' (*hsien i*[2]).[e]

Again, Jen Kuang[3] gives a term for chess-men in +1126, *pai yao hsüan shih*,[4] the 'white-jade magnetite objects' (cf. p. 232 above, and Table 53).[f] And there are later references to wooden horses and paper men being made to dance by the use of magnets.[g]

Some food for thought is provided by occidental legends which parallel this material. In the Far West, St Augustine (+354 to +430), a long time after Luan Ta, was much struck by seeing pieces of iron move about on a silver plate under the influence of a lodestone manipulated below.[h] An episcopal colleague, Severus, saw this done at the house of Bathanarius ('some-time a Count of Africa') and later demonstrated it to Augustine.[i] But a flotation method also enters in, for when about +325 Julius Valerius made that Latin translation of Pseudo-Callisthenes[j] which (as the *Res Gestae Alexandri Macedonis*) became one of the great sources of the medieval

[a] *Pao Phu Tzu* (*Nei Phien*), ch. 3, p. 5*b* (Feifel (1), p. 201).

[b] *Pao Phu Tzu* (*Nei Phien*), ch. 3, p. 20*a* (Feifel (1), p. 195).

[c] The meaning of this will be understood later; see p. 321.

[d] By Thao Ching-I[5] (+1716) but based on the early history of the region by Kuo Chih-Mei[6] (before +1053).

[e] Ch. 1, p. 27*b*, tr. auct. I myself have often seen stone chess-boards in grottoes of Taoist temples, such as the beautiful San Chhing Ko, near Kunming. Cf. Vol. 2, Fig. 40.

[f] *Shu Hsü Chih Nan*[7] (Literary South-Pointer), ch. 9, p. 13*a*.

[g] E.g. in the Sung, the encyclopaedia *Shih Lin Kuang Chi* (Guide through the Forest of Affairs), ch. 10; in the Chhing, the *Wu Li Hsiao Shih* (Small Encyclopaedia of the Principles of Things), entry under lodestone, ch. 12, pp. 2*a*ff.

[h] *De Civitate Dei*, XXI, 4; cf. Jenkins (1).

[i] Nothing ever seems to die in these affairs. At Christmas 1959 I was delighted at being asked to participate in a game of magnetised footballers, the figures being moved from beneath the board in the age-old way.

[j] This core of the Alexander Romance was, it is thought, composed at Alexandria in the −2nd century, but the later embroiderings and accretions were voluminous.

[1] 羅浮山志　　　[2] 仙弈　　　[3] 任廣　　　[4] 白瑤玄石　　　[5] 陶敬益
[6] 郭之美　　　[7] 書叙指南

'Alexander Romance',[a] he incorporated a very relevant passage. According to this corpus of legend Alexander's father was the last of the Egyptian kings, Nectanebus the magician. It was his custom to make divination about the outcome of naval battles by floating model ships of wax with model crews in a basin of water. They began to move, says the text, seeming to be alive, but if the omens were good they would sink when he passed his ebony wand around the basin, uttering invocations to the gods of the upper and lower regions.[b] Though the legend does not mention the presence of lodestone and iron, it surely must have been invented, as Taylor says,[c] by someone who had seen or practised such 'magic'. The story is particularly interesting because it forms a 'pre-historic' background for the development of the floating magnetic needle of which our first description is Chinese of the early +11th century. Evidence already presented (p. 277) indicates that this could have originated in China at any time after the +4th. One might almost venture to predict that somewhere in the vast mass of Chinese legendary material there will be found some parallel account of Luan Ta's chess-men turned marines and gone to sea. Nothing suggests that the European stories were not quite independent. But there is equally nothing to indicate that the knowledge behind them led to the invention of the floating-compass.

(ii) *Chess and astronomical symbolism*

Let us now approach the matter from another angle. A great wealth of information exists about chess in medieval Europe[d] and it is likely, on philological grounds, that the earliest knowledge of it by Europeans was early in the +10th century. But there is no specific mention of it before the +11th; the first date having now been pushed back from the famous letter of Peter Damiani (+1061) to certain wills and bequests on the Pyrenean frontier about +1010.[e] The entry of the game to Europe was thus almost certainly through Spain from the Muslim world, where it had long been well-known.[f] In the +10th century Firdausī in Persia had played it,[g] and al-Mas'ūdī had written, or attempted to write, its history.[h] It seems quite certain that the Arabs obtained it from India, where the earliest references to it (as *chaturaṅga*) occur early in the +7th century. There it had developed from an earlier game which had also used a checker-board (*ashṭāpada*) and which had probably been a race-game in which dice were used.[i] Most of the authorities have considered that Chinese chess (using the word in its most precise sense) was derived from India, but their grounds for this are very weak; its 'Indian ancestry', said Murray,[j] rests upon 'the identity of certain

---

[a] For the general background of this see Cary (1).
[b] In medieval pictures illustrating the Alexander Romance Nectanebus is generally represented with his wand and basin as well as a sighting-tube for observing the stars (cf. Vol. 3, p. 333).
[c] (8), pp. 93 ff.
[d] v. d. Linde (1, 2); v. d. Lasa (1); H. J. R. Murray (1).
[e] Murray (1), p. 402; v. d. Linde (2), p. 54; Garner (1).
[f] The Latin name *scacus*, or *ludus scacorum*, is a direct derivative of Persian *shah*, king; and *mate* comes from Persian *māt*.
[g] Murray (1), p. 155.   [h] Murray (1), p. 209.
[i] Murray (1), pp. 33, 42; v. d. Linde (1), p. 34.
[j] (1), p. 119.

essential features of the two games, and partly upon what is known of the indebtedness of China to India in religion, culture, and above all, in games'. Board-games as such are of course found everywhere, going back to at least the −11th century in Egypt, and there has been much speculation as to how some of them were played in Greek and Roman times.[a]

The oldest Chinese name for a chess-like game played on a board is i,[1] to which there are two references in Mencius (−4th century),[b] but there is no sure information as to what it was, or how played.[c] In the +1st century, Pan Ku said[d] that this game was what people in the south call $chhi$,[2] so it is quite probable that it was similar to, if not identical with, the game known as $wei\ chhi$[3] from the San Kuo period onwards. This is also a war-game, but played with some 150 pieces a side, moving along the lines of a rectangular grid of 19 units, thus giving 361 places in all.[e] Its object is, not to capture the opposing pieces, but to surround them and occupy as many as possible of the available cross-points. This was not the only kind of $chhi$[2] other than chess proper, as we shall see.

Chess proper, i.e. the game with the sets of pieces differing among themselves in value and move, was known in China as $hsiang\ chhi$,[4] and became common during the Thang period.[f] The name has usually been interpreted as meaning 'elephant-chess', and there were indeed generally four elephants among the pieces,[g] corresponding broadly to the bishops,[h] but it can equally well be taken as meaning 'image' ($hsiang$[5]), 'model' or 'figure' chess, so-called in order to distinguish it from other earlier games in which all the pieces were identical. Among the numerous Sung mentions of true chess, that in Ouyang Hsiu's[6] $Kuei\ Thien\ Lu$[7] (On Returning Home), of +1067, may be adduced. But the earliest description of the game recognisably identical with chess proper occurs in the $Yu\ Kuai\ Lu$[8] written by Niu Sêng-Ju[9] at the end of the +8th century—he has a story[i] of a man who dreamed that he was present at a ceremonial

[a] Cf. Ridgeway (1); Austin (1); H. J. R. Murray (2).

[b] *Mêng Tzu*, IV (2), xxx, 2 (Legge (3), p. 213); VI (1), lx, 3 (Legge (3), p. 286).

[c] Another ancient reference is in the *Mu Thien Tzu Chuan*, ch. 5 (tr. Chêng Tê-Khun (2), p. 137), but its words are not very explicit.

[d] In his essay on chess, *TPYL*, ch. 753, p. 5a; *TSCC*, *I shu tien*, ch. 799, *i wên*, p. 1b.

[e] Descriptions by H. A. Giles (6); Volpicelli (1); Cheshire (1). There are many quotations about it in *TPYL*, ch. 753.

[f] Many have given descriptions of true chess, as it is now played in China, e.g. Holtz (1); Hollingworth (1); Schlegel (4); Himly (7); v. Möllendorff (1); Holt (1); Volpicelli (2); Wilkinson (1); Gruber (1); Slobodchikov (1); and Tu Chung-Ming (1). All the papers are of the last century except the three last-named. The account of Irwin in +1793 is interesting because he actually played it with his friend 'Tinqua' (Phan Chen-Kuan[10]) in China. Phan wrote for him a memorandum in which the invention of the game was attributed to Han Hsin (the −2nd-century general), but though this was said to be in a quotation from 'Chinese annals', it has never been possible to identify it (Murray (1), p. 123).

[g] Not always, however, for some sets of pieces have another character, $hsiang$,[11] meaning diviner, instead of elephant. This is often regarded as a mistake, but in fact it may have been the earlier form, and the change to 'elephant' may have been made under later Indian influence. On military diviners see Sects. 18 and 30. The word $hsiang$[11] may also be interpreted as counsellor.

[h] Thus the medieval term for chess bishop was 'Alfil'. The elephant as castle in Indian chess sets is a modern development.

[i] Translation given in Murray (1), p. 123.

[1] 弈   [2] 棊   [3] 圍棊   [4] 象棊   [5] 像   [6] 歐陽修

[7] 歸田錄   [8] 幽怪錄   [9] 牛僧孺   [10] 潘珍官   [11] 相

battle, the moves being those of chess, and who afterwards found a chess set in an old tomb on the other side of the wall. This event is referred to +762. But it has a background which, though known at second or third hand to historians of chess, has hardly yet been appreciated at its full value.

The *Tan Chhien Tsung Lu* (Red Lead Record), compiled by Yang Shen shortly before +1554, summarises this as follows:[a]

Tradition handed down says that image-chess (*hsiang chhi*[1]) was invented by the Emperor Wu[2] of the (Northern) Chou dynasty (+561 to +578).[b] According to the *Hou Chou Shu*, it was in the 4th year of the Thien-Ho reign-period (+569) that the emperor finished writing his *Hsiang Ching*[3] (Image-Chess Manual). He assembled all his officials in a palace hall and gave lectures to them about it.[c]

The bibliography of the *Sui Shu* records this *Hsiang Ching* in one chapter, as written by Chou Wu Ti, with commentaries by Wang Pao,[4] Wang Yü,[5] and Ho Tho.[6]

There was also a *Hsiang Ching Fa Thi I*[7] (The Substance and Main Idea of the Image-Chess Manual) (written by someone else).

Then the story-tellers say that in the *Hsiang Ching* (it was stated that) images of the sun, moon, stars, and constellations were used. From this it is to be supposed that the playing-board had (divinatory technical terms such as) *ping-chi*,[8] *ku-hsü*,[9] and *chhung-pho*[10] marked on it. It was not the same as our modern chess (*hsiang hsi*[11]), where chariots, horses, etc. are in play. Had it been like our modern chess, even ordinary people or children could have understood it without much difficulty. Yet it was necessary to have scholarly commentaries on it, and lectures to the hundred officials.

Confirmatory evidence of these traditions comes from another Ming scholar, Wang Shih-Chên[12] in his *Yen-Chou Ssu Pu*[13] (Talks at Yenchow on the Four Branches of Literature)[d] and indeed they are often referred to. Already in the Sung, Kao Chhêng had said:[e]

Image-chess (*hsiang hsi*[11]) was invented by Chou Wu Ti. It had as chess-men (*chhi*[14]) the sun, moon, stars and constellations, and was quite different from modern *hsiang chhi*. Probably it was the modern chess which was referred to by Niu Sêng-Ju of the Thang in his *Yu Kuai Lu*.

Now although the emperor's manual has disappeared, we are fortunate to have the preface[f] which Wang Pao[4] wrote for it.[g] This runs:

The first (great significance) of image-chess (*hsiang hsi*[11]) is astrological, for (among the pieces are represented) heaven, the sun, the moon and the stars. The second concerns the

---

a Ch. 8, p. 14*b*, tr. auct. adjuv. Himly. The passage is also to be found in *TSCC, I shu tien*, ch. 801, *tsa lu*, p. 4*a*.

b This was Yüwên Yung,[15] a rather successful ruler, but opposed to Buddhism, and (more curiously) to Taoism.   c (*Hou*) *Chou Shu*, ch. 5, p. 15*a*; *Pei Shih*, ch. 10, p. 7*a*.

d Quoted *TSCC, I shu tien*, ch. 801, *i chhi pu, tsa lu*, p. 3*b*.

e *Shih Wu Chi Yuan*, ch. 48, p. 30*a*; cit. *TSCC, I shu tien*, ch. 801, *tsa lu*, p. 4*b*; tr. auct.

f Preserved in *TPYL*, ch. 755, p. 7*a*; tr. auct.

g His biography is in (*Hou*) *Chou Shu*, ch. 41; *Pei Shih*, ch. 83.

¹ 象棊   ² 武   ³ 象經   ⁴ 王褒   ⁵ 王裕   ⁶ 何妥
⁷ 象經發題義   ⁸ 兵機   ⁹ 孤虛   ¹⁰ 衝破   ¹¹ 象戲
¹² 王世貞   ¹³ 弇州四部   ¹⁴ 棊   ¹⁵ 宇文邕

earth, for (among the pieces are represented) earth, water, fire, wood and metal. The third concerns the Yin and the Yang; if we start from an even number it signifies Yang and Heaven; if we start from an odd one it signifies Yin and Earth. The fourth concerns the seasons; the colour of the east is green,[a] and the other three directions have each their proper colour. The fifth concerns the following of permutations and combinations, according to the changes of position of the heavenly bodies and the five elements. The sixth concerns the musical tones, following the dispersion of the *chhi*. The Tzu position (among the compass-points) takes the cyclical sign *wei*, the Wu position takes *chhou*, and so on. The seventh concerns the 8 *kua*, fixing their position; *Chen* takes *Tui*, *Li* takes *Khan*, and so on.[b] The eighth concerns loyalty and filial piety. . . . The ninth concerns ruler and minister. . . . The tenth concerns peace and war. . . . The eleventh concerns rites and ceremonies. . . . The twelfth concerns the recognition of virtue and the punishment of vice (i.e. promotions and demotions, etc.). . . .

Here again is a definite statement that not only the heavenly bodies, but also the five elements were represented among the set of pieces. It also looks as if the position of the pieces at the outset differed according to the position of the celestial bodies and the situation of the cyclical characters at the time when play was begun. The latter part of the preface seems to refer to the kind of question which this complicated divination technique was required to answer. A parallel source is the essay on *Hsiang Hsi*[1] (the Image-Chess Game) by Yü Hsin,[2] a +6th-century cavalry general.[c] Using obscure hidden language, he speaks of a board (*chü*[3]) made round (*yuan*[4]) according to *Chhien*[5] (the main Yang *kua*; Heaven), and another made square (*fang*[6]) according to *Khun*[7] (the main Yin *kua*; Earth). This is valuable information because it links up the emperor's inventions with the ancient diviner's board (*shih*[8]). He then speaks of the model pieces carrying ivory tablets (*chin hu*[9]) like officials, and placed about according to sidereal reference-points. The boards had diagrams on them (*wên chih hua*[10]). It will be agreed that Himly (2, 3, 4) had abundant justification for his remark that the more one investigates the origins of chess in Asia, the more intimate its connections with astrology and astronomy appear to be.

Before alluding to a few outstanding items in the mass of other evidence which associates chess with astrology and cosmic speculation, let us pause for a moment to try and reconstruct what the emperor Wu had in mind. The ultimate object of this account, be it remembered, is to explain how a lodestone model of the Northern Dipper found itself on the diviner's board. But that was in the +1st century; in the +6th the diviners were much more sophisticated. Our immediate object must be to explain how a cosmic–astrological technique used for divination could have turned into a war-game used for recreation. The answer is not far to seek; the image-chess of the emperor Wu was nothing but a mimicry of the eternal contest between the two great forces in the universe, Yang and Yin. It was desired to determine the general balance between Yin and Yang in the existing cosmic situation, and if the model

---

[a] Cf. Vol. 2, pp. 238, 262, 263. [b] Cf. p. 296 above.
[c] Preserved in *TSCC*, *I shu tien*, ch. 799, *i wên* 1, p. 6a.

[1] 象戲 [2] 庾信 [3] 局 [4] 圓 [5] 乾 [6] 方 [7] 坤
[8] 式 [9] 搢笏 [10] 文之畫

pieces were well chosen, their moves properly adjusted, and the board oriented and arranged according to the concrete circumstances, the players, being themselves part of that situation,[a] could not fail to proceed to a valid and informative decision. The idea of the stars fighting against one another was quite old in Chinese astrology. One has only to open the first of the astronomical chapters in the *Hou Han Shu* to find it said that in the time of Wang Mang 'large stars and small stars were struggling against one another in the palaces of the heavens.'[b] The word *tou*[1] (combat) occurs frequently in this sense.

> What chariots, what horses, against us shall bide
> While the stars in their courses do fight on our side?[c]

Superstition of course the image-chess was, yet it must have seemed at the time a brilliant device, evoking respect somewhat analogous to that accorded to the elaborate computing-machines of today.

After all, though essentially the good, for the Chinese mind, was the perfect balance between Yin and Yang, it had always been realised, for example in medical circles discussing the aetiology of disease, that Yin and Yang did not always balance. Image-chess was a way of detecting the extent of the unbalance at the time in question. As for the disposition of the sides, the details will presumably never be known, but it is easy to picture that the 28 *hsiu* (equatorial constellations) were the pawns,[d] while the two kings would be sun and moon, and the eight planets (including counter-Jupiter, Rahu and Ketu) would be divided between the sides.[e] Cannon and chariots (our knights and rooks) may well have been comets. And the remaining places may have been taken up by the five elements (perhaps represented on both sides), with sundry bright stars such as Canopus or Algol. The 'river' dividing the Chinese chess-board across the middle still retains its original name of the Milky Way (*thien ho*[2]).

That this interpretation is on the right track is confirmed by no less an authority than Pan Ku the historian (+ 1st century), who understood the astrological significance, not of *hsiang chhi*[3] (image-chess) which had not been invented in his time, but of a game or technique which was probably *wei chhi*.[4] In his essay on 'chess' he says:[f]

Northerners call *chhi* by the name of *i*.[5] It has a deep significance. The board (*chü*[6]) has to be square, for it signifies the earth, and its right angles signify uprightness. The pieces (of the two sides) are yellow and black; this difference signifies the Yin and the Yang—scattered in groups all over the board, they represent the heavenly bodies.

---

a No doubt there was some accompaniment of abstinence and liturgy.

b *Hou Han Shu*, ch. 20, p. 2*b*; *ta hsing yü hsiao hsing tou yu kung chung*.[7]

c From Rudyard Kipling's 'Astrologer's Song' in *Rewards and Fairies*. Cf. Judges v., 20.

d There had been, as early as the Han, an astrological 'game' in which the names of all the *hsiu* had been written on sets of bamboo slips. Some of these have survived (Schindler (4), p. 222).

e Since each of the planets and pseudo-planets was associated with one or other of the five elements, the sides were no doubt set up in accordance with the position of the elements in the Yin–Yang field (cf. Vol. 2, p. 461). Yin and Yang themselves were sufficiently represented by the moon and the sun.

f *TPYL*, ch. 753, p. 5*a*; *TSCC*, *I shu tien*, ch. 799, *i wên* 1, p. 1*b*; tr. auct.

¹ 鬭   ² 天河   ³ 象棊   ⁴ 圍棊   ⁵ 弈   ⁶ 局
⁷ 大星與小星鬭于宮中

These significances being manifest it is up to the players to make the moves, and this is connected with kingship. Following what the rules permit, both opponents are subject to them—this is the rigour of the Tao.

He could hardly be more explicit.[a]

Moreover, one can quote the converse. For the *Chin Shu* says:[b]

The heavens are round in shape like an open umbrella, while the earth is square like a chess-board (*chhi-chü*[1]).

To such an extent was the analogy fixed in the Chinese mind.

It would be tempting to pursue this further, but I will only add that evidence supporting the division of the *hsiu* into two teams of fourteen each according to Yin and Yang may be found in Sung Taoist books such as the *Wu Chen Phien Chih Chih Hsiang Shuo San Chhêng Pi Yao*[2] (Precise Explanation of the Difficult Essentials of the 'Poetical Essay on the Understanding of the Truth', according to the Three Scriptures),[c]

Fig. 345. Bronze chess-men (British Museum, after Murray, 1). On the right, two pawns; on the left a cannon (equivalent to the knight).

attributed to Ong Pao-Kuang,[3] where tables are given separating all kinds of natural objects into Yin and Yang things, and the *hsiu* appear in them. A very large number of astrological 'pieces', some of which look like coins or medallions but also resemble the discoidal true chess-men figured by Murray (Fig. 345),[d] were illustrated by Li Tso-Hsien,[4] in his classical study of Chinese coinage (*Ku Chhuan Hui*[5]) (cf. Figs. 346–8).[e] There are numerous tokens representing the Great Bear (significant because of the round heaven-plate of the *shih*),[f] some of which have references to the spirits of its seven stars (five men and two women);[g] other pieces show what may be other constellations.[h] Larger discs have radiating arrangements reminiscent of the geomantic compass,[i] and of the non-Chinese 'star-chess' which will shortly be referred

[a] There may have been astrological significance in the 361 places of the *wei chhi* board if they represented the days in the year (Culin (1), p. 870).

[b] Ch. 11, p. 2*a*.   [c] *TT* 140.
[d] (1), p. 126.   [e] Pt. IV (Chên).
[f] Ch. 2, p. 12*b*; ch. 4, pp. 3*a*, 3*b*, 4*b*, 12*a*, 13*b*; ch. 5, pp. 1*a*, 2*b*, 3*b*.
[g] Ch. 5, p. 13*a*; ch. 6, p. 10*a*.   [h] Ch. 7, p. 13*a*.
[i] Ch. 7, p. 11*a*, *b*; ch. 8, p. 7*a*.

[1] 碁局   [2] 悟眞篇直指祥說三乘祕要   [3] 翁葆光   [4] 李佐賢
[5] 古泉匯

Fig. 346   Fig. 347

Fig. 348

Fig. 346. Tokens resembling the pieces probably used in +6th-century 'star-chess'. This one depicts the Great Bear. Above, an inscription recalls the five male and two female spirits of its stars; the latter are represented to left and right, the former are on the reverse. Below there is a sword, and then the tortoise and serpent, symbolic of the northern palace of the heavens (cf. Vol. 3, p. 242). From *Ku Chhuan Hui* (Chên sect.), ch. 6, p. 10*a, b*.

Fig. 347. 'Star-chess' token representing the planet Mercury (Chhen hsing, cf. Vol. 3, p. 398). The imbricated lozenges on each side are of uncertain meaning, but the constellation with its attendant spirits on the reverse is probably the four-star Chih fa, which, like the planet itself, governs judges, punishments and executions (*Hsing Ching*, p. 3*a*; *Chin Shu*, ch. 11, p. 9*a*, ch. 12, p. 2*a, b*). From *Ku Chhuan Hui* (Chên sect.), ch. 8, p. 4*a*.

Fig. 348. 'Star-chess' token representing the *chih* cyclical character Wu, sign of the south among the compass-points and of the noon double-hour of the day. It is accompanied by its symbolic animal the horse. From *Ku Chhuan Hui* (Chên sect.), ch. 8, p. 5*a*.

to. Some of these give the azimuth compass-points.[a] Others show pictures of star-spirits,[b] or give the 8 *kua*.[c] It seems extremely likely that though these may have been used in some dynasties as distributed temple tokens, they may also have been connected with the pieces used in the emperor's image-chess.[d] In any case, the important thing to notice is that in China, and in China alone, on account of the dominance of the Yin–Yang theory of the macrocosm, could a divination technique or 'pre-game' have been devised which was both astrological and yet had a sufficient combat element to enable it to be vulgarised into a purely military symbolism.

There is no need to commit ourselves to any definite conclusion as to when and where the 'militarisation' of astrological image-chess took place; it may well have

[a] Ch. 7, numerous examples.   [b] Ch. 8, p. 5*a*.
[c] Ch. 9, pp. 1*a*, 1*b*, 2*a*ff.
[d] Another connection lies in the fact that down to modern times the moves and pieces in *hsiang chhi* were employed in the procedures of glyphomantic soothsayers (see Sect. 14*a*). Before 1928 there used to be many of them outside the Fu Tzu Miao at Nanking and our collaborator Dr Lu Gwei-Djen often watched them at work.

been in India in the following century. The appearance of elephants may indeed have been a misunderstanding, since *hsiang* can mean both 'image' (of a celestial body) and 'elephant'. It may even have been a substituted homophone for another word meaning 'diviner'. But if our general conclusions so far about the origin of true chess are right, we might expect to find widespread traces of astronomical symbolism clinging to it throughout later centuries. All historians of chess have agreed that this was in fact the case, though none of them has explained why. We must give a few examples.

In his *Murūj al-Dhabab*, al-Mas'ūdī, writing about +950, attributed the invention of chess to an Indian king, Balhit, saying:[a]

He also made of this game a kind of allegory of the celestial bodies, such as the seven planets and the twelve zodiacal signs, and allotted each piece to a star. The chess-board became a school of government and defence; it was *consulted*[b] in time of war, when it was necessary to have recourse to stratagems, and study the more or less rapid movement of troops. The Indians give a mysterious meaning to the houses (mansions, squares)[c] of the chess-board, and establish a relation between the First Cause which soars above the spheres, and on which everything depends, and the sum of the squares of the houses....

Here one can see the old Asian ideas subject to modification on coming within the sphere of Aristotelian learning. Al-Mas'ūdī is talking about something that must have been very like true chess played on a square board, the lineal descendant (I would suggest) of the square ground-plate of the *shih*. But what is extremely interesting is that there were several forms of chess played on discoidal boards with radial divisions, as if the round heaven-plate of the *shih* also tenaciously lived on. Al-Mas'ūdī and al-Amulī describe[d] a form of this in which the board was called *al-falakīya* (the celestial). It seems to have been particularly popular in the Eastern Roman Empire, and is often known as Byzantine star-chess.[e] In due course this found its way to Spain, where it was known as 'Los Escaques' and is described in the MS. *Libro del Acedrex* of Alfonso X of Castile,[f]—'los Escaques que se juega por astronomia'.[g] The board consisted of seven concentric rings, divided radially into twelve parts.[h] Some wind of this got through to later western Europe, where, for example, a +13th-century Latin poem gives the astrological symbolism of each chess-piece.[i] And as late as the +16th century the old tree was still budding, for Fulke in +1571 produced a new *Uranomachia, seu Astrologorum Ludus*.[j]

[a] Tr. de Meynard & de Courteille; v. d. Linde (1); eng. Murray (1), p. 210.

[b] Italics mine.   [c] Cf. the *hsiu* as pawns.

[d] Murray (1), p. 343. There were seven pieces of different colours to represent the five planets and the two luminaries.

[e] See v. d. Linde (2), p. 251; Weil (1). The Arabs knew it as *al-Rūmīya* (Byzantine) and *al-muddawara* (circular). Murray (1), p. 342, describes it from surviving MSS. Weil and Lemoine show the connection between this discoidal chess-board and the *zā'irjat al-'ālam* or circular table of the universe (cf. *maṇḍala*), each sector of which carried letters corresponding to stars and numbers; there was also a table of fates, and the connection between the two was made by the finger-game, in divination.

[f] The monarch whom we have so often met in Vol. 3 (Sect. 20), and whom we shall meet again below (Sect. 27*j*) in connection with the history of mechanical clocks.

[g] Murray (1), p. 349; v. d. Linde (2), p. 254.

[h] The Alfonsine MS. also describes a four-handed chess of the four seasons, which seems an extremely Chinese idea (*Acedrex de los quatro tiempos*).

[i] v. d. Linde (2), p. 68.   [j] v. d. Linde (1), vol. 2, p. 374; Murray (1), p. 351.

Related to these disc-like chess-boards were the astrological dicing boards of which the most famous example is perhaps the Bianchini Table, discovered on the Aventine in +1705.[a] It is supposed to date from the +2nd century. In a series of concentric circles it shows, from the centre outwards, the twelve zodiacal animals, the twelve signs twice over (according to the fixed and moving ecliptics), the thirty-six decan-gods, and lastly thirty-six faces of the seven planet-gods repeated in septizonium order.

(iii) *Divination by throwing*

This provides the transition to an operation of the diviners not yet discussed, namely the throwing of objects on to a board, or the tossing of them on it or off it. In astronomical image-chess or in true chess, as in other board-games, there are moves made by the pieces of two or more sides paralleling tactics and strategy. But in forms perhaps more primitive, the pieces were actually thrown on to the board, conclusions being drawn from where they came to rest. The pieces thus approximated to dice, and the procedure had no intrinsic combat element. Hence the interest of the fact that Chinese literature contains a number of references to 'spirit-chess'. In the *Tao Tsang* there are two books with the title *Ling Chhi Ching*[1] (Spirit-Chess Manual), one[b] attributed to Tungfang Shuo[2] and possibly of Later Han or San Kuo date; the other[c] a development of it and bearing the name of Yen Yu-Ming[3] of the Sung. Significantly, the term for the board used in these works is the *shih*.[4] Their prefaces are bibliographical and the bulk of the text consists of interpretations of combinations (e.g. 'one upper, four middle, three lower'), for there were four pieces marked *shang*,[5] four marked *chung*[6] and four *hsia*.[7] The board, of jujube wood, was round, like the heaven-plate. More information comes from the Sung encyclopaedist, Kao Chhêng:[d]

The *I Yuan* (Garden of Strange Things)[e] says: 'Divination using the twelve chess-men (*chhi pu*[8]) started with Chang Liang[9] (d. −187), who got it from Huang Shih Kung[10] (the Old Gentleman of the Yellow Stone—the Sage of Miao-tai-tzu in Shensi).'

Our present-day (Sung) method is to divide the twelve 'chess' or divination pieces into three classes, upper, middle, and lower; these are then thrown (on to the board) and by the result obtained the decision is given as to good or evil fortune.

The introduction of Thang Shih-Yuan[11] to the *Chhi Ching*[12] says 'We do not know when *ling chhi*[1] was invented—some say it was by Tungfang Shuo[2] in Han Wu Ti's time, who used it for divination and always proved right; others say that Chang Liang[9] learned it from Huang Shih Kung;[10] others again refer it to the Prince of Huai-Nan who was taught it by a guest but kept it secret afterwards.'

In other words the method was so old that no one knew whence it originated.

---

[a] Described by Boll, Bezold & Gundel (1 b), p. 60, pl. XVIII; Gundel (2), pl. 16; Boll (1), p. 303; and Eisler (1), pp. 82, 112, 267, pl. VIb.
[b] *TT* 285.   [c] *TT* 1029.
[d] *Shih Wu Chi Yuan*, ch. 39, p. 39 b; tr. auct.
[e] By Liu Ching-Shu, probably of the +5th century.

[1] 靈棊經   [2] 東方朔   [3] 顏幼明   [4] 式   [5] 上   [6] 中
[7] 下   [8] 棊卜   [9] 張良   [10] 黃石公   [11] 唐事遠   [12] 棊經

Another game or divination-technique which seems to have been given off by the *shih*[1] was 'crossbow-bullet chess' (*tan chhi*[2]).[a] This also had its *Tan Chhi Ching*[3] or manual. Whatever it was, it seems to have originated in the Han, for essays on it still survive by Later Han and San Kuo people, such as Tshai Yung,[4] Tshao Phei[5] (Emperor Wên of (San Kuo) Wei), Ting I,[6] and Hsiahou Tun,[7] a general, who said that the stone pieces grouped themselves in various ways like the stars in heaven. In the +3rd century, Hantan Chhun,[8] in his book (*I Ching*[9]) on arts,[b] gave a short description of *tan chhi*, and from this as well as from other later descriptions one gets the impression that the procedure involved both throwing the pieces on to the board, and combat moves following this chance placing. The twelve pieces, red and black, seem to have symbolised the twelve animals of the 'zodiacal' animal-cycle,[c] and each player started with six. Significantly, Wang Pao[10] of Northern Chou, the expert on image-chess, wrote an essay on this form also. By the Thang the number of pieces had increased to twenty-four (according to the *Yu-Yang Tsa Tsu*), and from an essay of Lu Yü[11] we find that 'the shape of the board is square below like the earth and round above like the heavens'.[d] The pieces 'fly up when the board is quickly knocked, and scatter to different positions'. Sometimes the two sides represented two orders of society, commoners and officials (*chien*[12] and *kuei*[13]), or prognosticated about this or that alternative social fate, as in the Thang essays of Liu Tsung-Yuan[14] and Wei Ying-Wu.[15] Several sources associate the game with the Taoists, and some repeat a story that it was invented by Tungfang Shuo to induce Han Wu Ti to stop playing the active ball-games of which he was fond.

Yet another game or divination-procedure not to be forgotten which has a close connection with astronomy was that of *liu-po*[16] ('the Six Learned Scholars') which, as we saw above,[e] was played with twelve *chhi*[17] pieces on a board almost identical with the plate of the Han sundials. Fig. 349 shows a group of tomb-figures engaged in it. As Yang Lien-Shêng (1, 2) has shown, this can be traced back without difficulty to the −3rd century.[f] The moves of the pieces were determined by the throwing of six sticks (*chu*[18]), and they were divided into two 'sides', each piece being marked with one of the four animals symbolising the four directions of space.[g] There seems to have been a central belt of water, like the Milky Way in later systems, and when a piece arrived there it was promoted to be a 'leading piece' with greater powers. The relation of all these systems to one another remains obscure, however.

[a] There is a good deal about it in *TSCC, I shu tien*, ch. 801 (*tan chhi pu*), *hui khao* 2, *i wên* 1, pp. 1aff.
[b] Preserved in *YHSF*, ch. 78, p. 69a.   [c] Cf. Vol. 3, p. 405.
[d] The remark of the ancient cosmologist that the heavens were 'as round as a crossbow bullet' (cf. Vol. 3, p. 217 above) may be worth remembering in this connection.
[e] Cf. Vol. 3, p. 305.
[f] Particularly interesting are certain stories (*Shih Chi*, ch. 3, p. 9b; *Han Fei Tzu*, ch. 11, p 6b) that earthly kings played *liu-po* against deities and sometimes won.
[g] Cf. Vol. 3, p. 242.

[1] 式	[2] 彈棊	[3] 彈棊經	[4] 蔡邕	[5] 曹丕	[6] 丁廙
[7] 夏侯惇	[8] 邯鄲淳	[9] 藝經	[10] 王褒	[11] 盧諶	[12] 賤
[13] 貴	[14] 柳宗元	[15] 韋應物	[16] 六博	[17] 棊	[18] 箸

(iv) *Comparative physiology of games*

It would be impossible here to embark on a history of all the games and divination-techniques on which the Chinese encyclopaedias have such a mass of information. It is obvious that the throwing of things lent itself to divination as well as to games from the earliest periods. One of the oldest of such procedures was the 'pitch-pot' game, the throwing of arrows into a pot (*thou-hu*[1]); a game the history of which has been analysed by Montell (1) and by Rudolph (3).[a] The *locus classicus* for this is a long passage in the *Li Chi*, where indeed a whole chapter is devoted to it.[b] This is evidence for early Han and possibly late Chou times. Better evidence for Chou is the mention in the *Tso Chuan*,[c] under date −529, where the game was recreational. The *Shih Chi* has it[d] in connection with Shunyu Khun,[2] the −4th-century philosopher,[e] and the *Hou Han Shu*[f] in connection with a +1st-century general, Tshai Tsun.[3] In the Chin dynasty there was a special work on it, the *Thou Hu Pien*[4] (Changes and Chances of the Pitchpot)[g] by Yü Than,[5] and later many, such as the *Thou Hu Hsin Ko*[6] of Ssuma Kuang[7] of the +11th century. That it was used for divination appears from numerous quotations in the *Thu Shu Chi Chhêng* encyclopaedia.[h] Rudolph (3) illustrates several pictures of the game from Han tomb carvings (Fig. 350).

Some social anthropologist will produce some day a fully integrated and connected evolutionary story, quite biological in character, showing how all these games and divination-techniques were genetically connected. It would only need markings or numbers on the arrows to have an object which by compression would become a cubical dice, and this again by extension or unfolding would give rise to dominoes on the one hand and playing-cards on the other. Cubical dice (*chhu-phu* or *yü-phu*[8]) are ancient, examples having been found in Egypt and India, and from Graeco-Roman times; it is generally supposed that they reached China from India, and this we may provisionally accept.[i] But it is also now rather well established that dominoes and playing-cards were originally Chinese developments from dice.[j] There are indications, says Carter, that the transition from dice to cards (leaf-dice, sheet-dice; *yeh tzu*,[9]

---

[a] Be it remembered that the very name of this game includes the same character, *thou*, which appears in the *locus classicus* in the *Lun Hêng* for the 'throwing' of the lodestone spoon on the *shih*'s ground-plate (above, p. 262).

[b] Ch. 40 (tr. Legge (7), vol. 2, pp. 397ff.).

[c] Duke Chao, 12th year (Couvreur (1), vol. 3, p. 195).

[d] Ch. 126, p. 3*a*.   [e] Vol. 2, pp. 234ff.

[f] Ch. 50, p. 10*b*.   [g] Preserved in *YHSF*, ch. 78, pp. 72*a*ff.

[h] *I shu tien*, ch. 747.

[i] The transmission must have occurred quite early. Waley (8), p. 140, has suggested that the prominence of the number six in the *Book of Changes* was derived from the six sides of a cubical dice. There are Chinese mentions in +406 (Goodrich (1), p. 92) and +501 (Carter (1), 1st ed. p. 139, 2nd ed. p. 183). But Dubs (2), vol. 1, p. 292, refers dice to the Early Han time (−2nd century).

[j] Hummel (15); Wilkinson (2); Carter (1), 1st ed. p. 140, 2nd ed. p. 184. Culin (2) has described modern Chinese games with dice and dominoes. Other influences in the development not of course to be excluded are those of the drawing of lots by long and short sticks (cf. Vol. 2, pp. 305, 347), and of paper money.

---

[1] 投壺   [2] 淳于髠   [3] 蔡遵   [4] 投壺變   [5] 虞潭

[6] 投壺新格   [7] 司馬光   [8] 摴蒲   [9] 葉子

PLATE CXXIII

Fig. 349. A group of Han tomb-figures engaged in the game of *liu-po* (from Yang Lien-Shêng, 2). British Museum. Cf. Vol. 3, p. 305 and Fig. 130.

Fig. 350. Han scholars playing at *thou-hu* (from Rudolph, 3). A Later Han relief of the +2nd century recovered from a tomb near Nanyang in Honan province. In this we see again a small ladle something like the lodestone spoon, standing free upon a small table, perhaps upon the earth-plate of a *shih*. The relation between divination, games of chance, and the origin of the magnetic compass is discussed in the text.

phai[1])[a] took place at about the same time as the transition from manuscript rolls to paged books. These cards, at their first appearance towards the close of the Thang, must have been among the earliest examples of block printing (cf. Sect. 32 below). After the beginning of the Sung, their evolution forked into two directions, one leading to playing-cards as we know them, the other to dominoes (*ya phai*[2] or *ku phai*[3]), from which again in its turn the famous game of *ma chhiao*[4] ('Mah Jongg') derived.[b] But while the story of playing-cards in China includes firm dates as early as +969, when one of the Liao emperors had card games at court,[c] the earliest reference in Europe is +1377 in Germany.[d] It is strange that there are no references to them in Arabic literature, for the most obvious route of transmission would have been through the Islamic world, and many early European sources say that they came 'from the Saracens'. Still, the dates would readily permit a direct transmission through merchant contacts of the Mongol period more or less contemporary with Marco Polo. The whole question is important with regard to the origins of block printing, for the Chinese cards had long been printed when Europeans came first in contact with them, and it seems that from +1400 some of the European cards were also printed. The earliest European dated religious prints, the Virgin and Child of +1418 and the St Christopher of +1423, coincide closely with St Bernardino's famous sermon against card-playing. Carter concludes that playing-cards have a very important place in the transmission of the art of printing to Europe.[e]

The case of dominoes is quite similar. Western encyclopaedists say that this game was not known in Europe till the +18th century, and that it was invented in Italy. But all Chinese manuals on dominoes, of which there are many, point back to the year +1120, just before the move of the Sung capital to Hangchow, when a set of thirty-two pieces, totalling 227 pips, was presented to the emperor. The *Shuo Fu* collection contains several books on dominoes, one with a preface dated +1368. The chief work on them is the *Phai Thung Fu Yü*[5] of +1639, described by Hummel (15).

There remains one important piece of evidence to fit in to the picture. Culin (1) has given us a remarkable monograph on chess and playing-cards and their related games and divination-techniques in all civilisations. Much of this work is concerned with the divination-techniques and gambling games widespread among the North American Indian tribes, where the pieces are tossed on to, or in, certain special baskets.[f] When I first read this, my attention was caught by the striking fact that in many cases the pieces used were not simple, like draughts-men, or counters, nor numbered, like dice, but complex, like chess-men. The Chippewa Indians, for example,

---

[a] Carter (1), 1st ed. p. 243.

[b] See Culin (3).

[c] *Liao Shih*, ch. 7, p. 5a; *TSCC, I shu tien*, ch. 807, p. 8a. Schlegel (4), p. 20, maintained that the first of all card-games was fully developed in the course of the +9th century.

[d] Full details in Carter (1), 1st ed. p. 141, 2nd ed. p. 185.

[e] How characteristic it was of a scholarly civilisation that both metal money and bone or ivory dice should all in the end turn into paper.

[f] Analogous to the *shih*. Cf. Weltfish (1).

[1] 牌   [2] 牙牌   [3] 骨牌   [4] 麻雀   [5] 牌統孚玉

had a set consisting of two human figures (rulers), two amphibious monsters, one or two war-clubs, one or two fish, four plain counters ('pawns'), and three ducks.[a] The Central Eskimo have a set of various kinds of animals,[b] and that of the Assiniboin of the Upper Missouri includes large and small crows' claws, and various fruit-stones.[c] Pieces themselves might be marked with the four cosmic directions.[d] This throws a good deal of light on the Chinese situation, encouraging the view we have already formed that sets of pieces existed symbolising the celestial bodies, and that divination took place by noting their fall on a prepared board. In this way the model Northern Dipper placed on the *shih* becomes quite comprehensible, and finds an appropriate context in divination-techniques in other parts of the world.

In order to summarise some of the points which have been made about the genetic origin of games and divination-techniques, a provisional chart is offered (Table 53). Pending the appearance of a comprehensive work on the subject, the details are put forward with all reserve, but the chart may well raise other useful suggestions for the early history of science.

(9) GENERAL SUMMARY

Looking back over the course of our argument, we see in broad survey a long and slow developmental period in China, followed by sudden appearance and later more rapid advance in the West. Some transmission from east to west our careful chronological titration compels us to recognise. But since the crucial couple of centuries before Alexander Neckam have so far afforded no trace or clue from intermediary regions such as the Arabic, Persian or Indian culture-areas, the possibility arises that this Chinese transmission occurred not in the maritime context at all, but by some over-land route through the hands of astronomers and surveyors who were primarily interested in establishing the meridian of their place.[e] Certainly Petrus Peregrinus devotes loving description to two azimuth dial instruments with alidades and inserted compasses (floating in one, dry-suspended on a pivoted spindle in the other). The determination of the meridian on land was of course important not only for carto-graphy but also for such operations as the proper adjustment of sundials, and Europeans at this time still had no horologes more accurate than these. It is certainly a striking fact that as late as the +17th century the needles used in the compasses of surveyors and astronomers were all made so as to indicate the south (in contradistinction to the north-pointing sailor's needles), exactly as all the Chinese needles had done for perhaps as much as a millennium previously.[f] If this conception found favour we might have to envisage an overland westward transmission of the 'astronomer's compass', followed by a western application of it by mariners independently paralleling the earlier application by the sea-captains of China. What we know of the level of culture of the Russians and their Central Asian or Siberian neighbours during the two or three centuries preceding the Mongol invasions, however, might at first sight hardly

[a] Culin (1), p. 694.   [b] Culin (1), p. 717.   [c] Culin (1), p. 750.
[d] Culin (1), p. 701.
[e] So Lynn White (5), p. 524.   [f] This was shown by Taylor (6).

Table 53. *Chart to show the genetic relationships of games and divination-techniques in relation to the development of the magnetic compass* Numbers indicate page-references to the catalogue of Culin (1). Chinese examples underlined

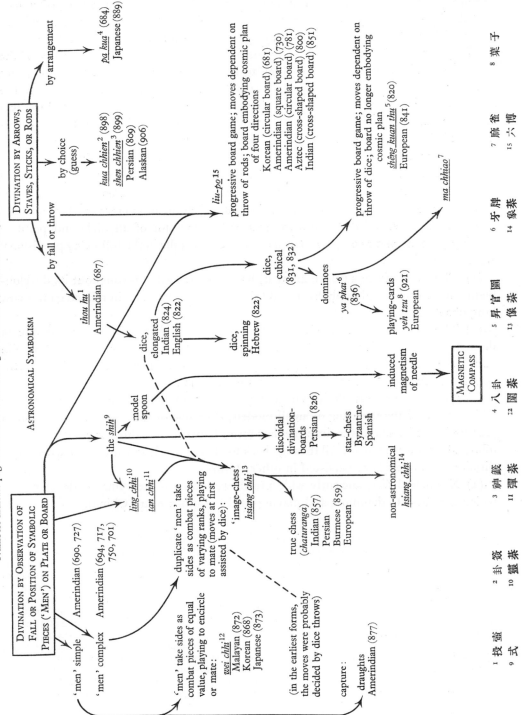

1 投壼　2 卦　3 神籤　4 八卦　5 昇官圖　6 牙牌　7 麻雀　8 葉子

9 式　10 靈棊　11 彈棊　12 圍棊　13 像棊　14 象棊　15 六博

encourage us to believe in such a route for a scientific discovery. It is true, of course, that the feats of the needle and the stone may well have been thought of in a magical–technological rather than a scientific way by those who would have carried them across the length of the heartland. Evidently there is room for much further study of the possibilities of transmission from China to Europe through the Steppe peoples and the Russians, avoiding the Islamic, Byzantine and Indian culture-areas;[a] in the meantime many may well prefer to believe that in fact the mariner's compass was what was handed on, and that texts as yet unknown in Arabic or more easterly sources will yet inform us about how the sailors of the Indian Ocean did it.

The subsection just preceding may have seemed somewhat anomalous in a Section devoted to the history of physics in China. The modern physicist who may have read it will have found himself straying in fields which at first sight could have no conceivable connection with that science as it is today. Yet the problem was a fundamental one, namely to elucidate the first origins of the ancestor of all dial- and pointer-readings, the magnetic compass. Let us sum up the results in the following provisional way.

(1) The game of chess (as we know it) has been associated throughout its development with astronomical symbolism, and this was even more overt in related games now long obsolete.

(2) The battle element of chess seems to have developed from a technique of divination in which it was desired to ascertain the balance of ever-contending Yin and Yang forces in the universe (+6th-century China, whence it passed to +7th-century India, generating there the recreational game).

(3) This 'image-chess' derived in its turn from a number of divination-techniques which involved the throwing of small models, symbolic of the celestial bodies, on to prepared boards. There were intermediate forms between pure throwing and placement followed by combat moves. All these go back to Han or pre-Han China (−3rd century). Similar techniques have persisted in other cultures.

(4) Numbered dice, anciently widespread, were on a related line of development which gave rise to dominoes and playing-cards (+9th-century China).

---

[a] Here the Qarā-Khiṭāi or Western Liao kingdom presents interesting possibilities. It will be remembered (cf. Vol. 3, pp. 118, 457) that persistent tradition in China has ascribed to this country the transmission of scientific and technical knowledge to the West. The 'Black Chhi-tan' was a succession-state of the Liao dynasty established in +1124 by exiles who followed Yehlü Ta-Shih as Gurkhan to Western Sinkiang after the liquidation of that empire by the Jurchen (Chin) Tartars. It lasted under a succession of rulers until the gathering might of the Mongols swept it away in +1211. Centred on Kashgar, it included Samarqand in the west and Turfan in the east; Chinese was its official language, and Chinese literature had prestige value as the 'Latin' of the East in its culture. In religious matters the Qarā-Khiṭāi were very tolerant, and Christianity flourished side by side with Shamanism, Buddhism and Islam (cf. Wittfogel, Fêng Chia-Shêng et al. (1) pp. 670ff.); indeed we have already noted (Vol. 1, p. 133) that the Western legend of Prester John originated precisely from the existence of this State. Assuredly it had mercantile and cultural contacts with the Russian princedoms of Novgorod, Vladimir and Kiev. In fact, its dates, spanning as they do the +12th century, are exactly right for the spread of knowledge of the magnetic compass to the West. Yet we shall find, disturbingly, in Sect. 29, that another maritime invention, that of the stern-post rudder, also reached Europe from China towards the close of the +12th century—and the transmission of something like this by a non-maritime route would be still harder to believe.

(5) The most significant of the ancient boards was the *shih* (used from the late Warring States period onwards), a double-decked cosmical diagram having a square earth-plate surmounted by a rotatable discoidal heaven-plate, both being marked with cyclical and astronomical signs (compass–points, *hsiu*, etc.) as well as *kua* and technical terms used only in divination. 'Pieces' or symbolic models were employed with this in a variety of different ways. In the round heaven-plate of the *shih* we may recognise the lineal ancestor of all compass-dials.

(6) Among the symbolic models used was one representing the Great Bear (the Northern Dipper)—so important in Chinese polar–equatorial astronomy—carved into the shape of a spoon. This replaced the picture of the Great Bear carved on the heaven-plate of the diviner's board.

(7) The model spoon was probably first of wood, stone or pottery, but in the + 1st century (and possibly already in the − 2nd century), the unique properties of magnetite suggested in China the use of this substance. Since polarity would establish itself along the main axis of a bar of the mineral, whether or not it was removed from the rock in a north–south direction (i.e. in the earth's magnetic field), the 'south-pointing spoon' was discovered. Some examples, of course, must have pointed north, as indeed the texts indicate.

(8) During later centuries the frictional drag of the lodestone spoon on its base-plate was avoided by inserting the piece of lodestone in a piece of wood with pointed ends, which could be floated, or balanced upon an upward-projecting pin. Such methods were used as late as the + 13th century.

(9) Some time between the + 1st and the + 6th century it was found in China that the directive property of the lodestone could be transferred to (induced in) the small pieces of iron which the lodestone attracted. These also could be made to float upon water by suitable devices. The earliest description still extant of such water-compasses, from which all subsequent forms must derive, is of the early + 11th century.

(10) Some time before the +11th century it was discovered in China that magnet-isation could be carried out not only by rubbing the pieces of iron on the lodestone, but by cooling them (quenching) from red heat, through the Curie point, while held in a north–south direction (the earth's magnetic field).

(11) Probably by the + 7th or + 8th century the needle was replacing the lodestone, and pieces of iron of other shapes, on account of the much greater precision with which its readings could be taken.

(12) Although the first clear and accurately datable descriptions of the magnetic compass, with needle, antedate European knowledge of it only by one or two centuries (Shen Kua, Wang Chi, Hsieh Ho-Chhing), it is probable that the Chinese use of the compass-needle is some three or four centuries older.

(13) By the late Thang period (+ 8th or + 9th century) the declination, as well as the polarity, of the magnet, had been discovered, antedating European knowledge of the declination by some six centuries. The Chinese were theorising about the declination before Europe knew even of the polarity (end of the + 12th century).

(14) The successive declinations, first eastern and then western, were embodied in

the design of the Chinese geomantic compass as concentric circles which have persisted until our own time.

(15) The compass was undoubtedly employed in China for geomancy a long time before it was used to assist navigation.

(16) The first clear and accurately datable description of the use of the compass for navigation on Chinese ships antedates the first knowledge of this technique in Europe by just under a century, but there are indications that it was used for this purpose in China somewhat earlier.

(17) Chinese sailors remained faithful to the floating-compass for many centuries. Although the dry pivoted compass had been described early in the +12th century, it did not become common on Chinese vessels until it was reintroduced from the West in the +16th century by the Dutch and Portuguese by way of Japan. Associated then with it was the compass-card (the wind-rose attached to the magnet) which had probably been an Italian invention at the beginning of the +14th century.

(18) The ancestor of all dial- and pointer-readings, the greatest single factor in the voyages of discovery, and the oldest instrument of magnetic–electrical science, may thus perhaps be said to have begun as a proto-'chess'-man used in a divination-technique.

(19) Magnetical science, unlike Euclidean geometry and Ptolemaic planetary astronomy, was an essential component of nascent modern science the antecedents of which were not primarily Greek (cf. pp. 60, 236 above). All the preparation for Pierre de Maricourt, and hence for the ideas of Gilbert and Kepler on the cosmic role of magnetism, had been Chinese; and they in their turn, with their belief that gravity must be something like magnetic influence, were an important part of the preparation for Newton. The field physics of still later times, firmly established in Clerk Maxwell's classical equations and more congruent with organic thought than Greek atomic materialism, can again be traced back to the same root. Much is owing to the faithful and magnificent experimenters of medieval China.

BIBLIOGRAPHIES

A CHINESE BOOKS BEFORE +1800

B CHINESE AND JAPANESE BOOKS AND JOURNAL ARTICLES SINCE +1800

C BOOKS AND JOURNAL ARTICLES IN WESTERN LANGUAGES

In Bibliographies A and B there are two modifications of the Roman alphabetical sequence: transliterated *Chh-* comes after all other entries under *Ch-*, and transliterated *Hs-* comes after all other entries under *H-*. Thus *Chhen* comes after *Chung* and *Hsi* comes after *Huai*. This system applies only to the first words of the titles. Moreover, where *Chh-* and *Hs-* occur in words used in Bibliography C, i.e. in a Western language context, the normal sequence of the Roman alphabet is observed.

When obsolete or unusual romanisations of Chinese words occur in entries in Bibliography C, they are followed, wherever possible, by the romanisations adopted as standard in the present work. If inserted in the title, these are enclosed in square brackets; if they follow it, in round brackets. When Chinese words or phrases occur romanised according to the Wade–Giles system or related systems, they are assimilated to the system here adopted without indication of any change. Additional notes are added in round brackets. The reference numbers do not necessarily begin with (1), nor are they necessarily consecutive, because only those references required for this volume of the series are given.

ABBREVIATIONS

See also p. xix.

A	*Archeion*
AA	*Artibus Asiae*
AAA	*Archaeologia*
AAN	*American Anthropologist*
A/AIHS	*Archives Internationales d'Histoire des Sciences* (contin. of *Archeion*)
AC	*l'Antiquité Classique*
ACLS	American Council of Learned Societies
ACP	*Annales de Chimie et Physique*
ACSS	*Annual of the China Society of Singapore*
AE	*Ancient Egypt*
AEST	*Annales de l'Est* (Fac. des Lettres, Univ. Nancy)
AFGR	*Atti della Fondazione Georgio Ronchi* (Arcetri)
AGMNT	See *QSGNM*
AGNL	*Archiv f. d. gesamte Naturlehre*
AGNT	See *QSGNM*
AH	*Asian Horizon*
AHAW/PH	*Abhandlungen d. Heidelberger Akademie d. Wissenschaften* (Phil.-Hist. Klasse)
AHES/AESC	*Annales d'Hist., Econ. et Sociale; Economies, Sociétés, Civilisations*
AHMM	*Annalen d. Hydrogr. u. maritimen Meteorologie*
AHOR	*Antiquarian Horology*
AHR	*American Historical Review*
AIPG	*Annales de l'Inst. de Physique du Globe* (Paris)
AJ	*Asiatic Journal and Monthly Register for British and Foreign India, China and Australia*
AKG	*Archiv f. Kulturgeschichte*
AKML	*Abhandlungen f. d. Kunde des Morgenlandes*
AM	*Asia Major*
AMG	*Annales du Musée Guimet*
AMM	*American Mathematical Monthly*
AMT	*Archives du Musée Teyler* (Haarlem)
AMW	*Archiv f. Musikwissenschaft*
AMY	*Archaeometry* (Oxford)
AN	*Anthropos*
ANP	*Annalen d. Physik*
ANS	*Annals of Science*
ANTJ	*Antiquaries Journal*
APPCM	*Archiv f. Physiol., Pathol., Chem. u. Mikroskopie*
AQ	*Antiquity*

AQC	*Antique Collector*
ARC	Agricultural Research Council
ARLC/DO	*Annual Reports of the Librarian of Congress* (Division of Orientalia)
ARSI	*Annual Reports of the Smithsonian Institution*
ARUSNM	*Annual Reports of the U.S. National Museum*
ASAE	*Annales du Service des Antiquités de l'Égypte*
AS/BIE	*Bulletin of the Institute of Ethnology, Academia Sinica* (Thaiwan)
AS/BIHP	*Kuo-Li Chung-Yang* (now *Chung-Kuo Kho-Hsüeh*) *Yen-Chiu Yuan, Li-Shih Yü-Yen Yen-Chiu So Chi-Khan* (*Bulletin of the Institute of History and Philology, Academia Sinica*)
AS/CJA	*Chung-Kuo Khao-Ku Hsüeh Pao* (*Chinese Journal of Archaeology, Academia Sinica*)
ASI	*Actualités scientifiques et industrielles*
ASPN	*Archives des Sciences physiques et naturelles*
ASSB	*Annales de la Société scientifique de Bruxelles*
ASURG	*Annals of Surgery*
AX	*Ambix*
BA	*Baessler Archiv* (*Beiträge z. Völkerkunde herausgeg. a. d. Mitteln d. Baessler Instituts, Berlin*)
BBSSMF	*Bollettino di Bibliografia e di Storia delle Scienze Matematiche e fisiche* (Boncompagni's)
BCS	*Chung-Kuo Wên-Hua Yen-Chiu Hui Khan* (*Bulletin of Chinese Studies*, Chhêngtu)
BDPG	*Berichte d. deutsch. physikal. Gesellschaft*
BEFEO	*Bulletin de l'École française de l'Extrême Orient* (Hanoi)
BGTI	*Beitr. z. Gesch. d. Technik u. Industrie* (changed to *Technik Geschichte BGTI/TG* in 1933)
BIHM	*Bulletin of the (Johns Hopkins) Institute of the History of Medicine*
BJPS	*British Journal for the Philosophy of Science*

BLSOAS	Bulletin of the London School of Oriental and African Studies	ENG	Engineering
BMFEA	Bulletin of the Museum of Far Eastern Antiquities (Stockholm)	ES	Encyclopaedia Sinica (ed. Couling)
		ETH	Ethnos
BMJ	British Medical Journal	FF	Forschungen und Fortschritte
BNI	Bijdragen tot de taal-, land- en volken-kunde v. Nederlandsch-Indië	FJHC	Fu-Jen Hsüeh-Chih (Journal of Fu-Jen University, Peking)
		FL	Folklore
BNYAM	Bull. New York Acad. of Med.	FLS	Folklore Studies (Peiping)
BQR	Bodleian (Library) Quarterly Record (Oxford)	FMNHP/AS	Field Museum of Natural History (Chicago) Publications, Anthropological Series
BRMQ	Brooklyn Museum Quarterly		
BSG	Bulletin de la Société de Géographie (contin. as La Géographie)	G	Geimon (Art Journal)
		GGM	Geographical Magazine
BSGI	Boll. Soc. Geogr. Ital.	GLB	Glastechnische Berichte (Zeitschr. f. Glaskunde)
BTG	Blätter f. Technikgeschichte (Vienna)		
		GS	Gakugei Shirin
BUA	Bulletin de l'Université de l'Aurore (Shanghai)	GSJ	Galpin Society Journal
		GTIG	Geschichtsblätter f. Technik, Industrie u. Gewerbe
BUM	Burlington Magazine		
BUSNM	Bulletin of the U.S. National Museum	GW	Geographische Wochenschrift
		HCCC	Huang Chhing Ching Chieh
CAM	Communications de l'Académie de Marine (Brussels)	HCUKY	Hua-chung Ta-Hsüeh Kuo-Hsüeh Yen-chiu Lun-Wên Chuan-Khan (Huachung Univ. Sinological Research Reports)
CAMR	Cambridge Review		
CEN	Centaurus		
CHER	Chhing-Hua (University) Engineering Reports	HGB	Hansische Geschichtsblätter
		HH	Han Hiue (Han Hsüeh); Bulletin du Centre d'Études Sinologiques de Pékin
CHESJ	Chhing-Hua Engineering Journal (Kung-Chhêng Hsüeh Hui Hui Khan)		
		HJAS	Harvard Journal of Asiatic Studies
CHJ/T	Chhing-hua (Tsing-hua) Journal of Chinese Studies (New Series pub. Thaiwan)	HKM	Hua Kuo Yüeh Khan (Szechuan Magazine)
		HMSO	Her Majesty's Stationery Office (London)
CIB	China Institute Journal		
CJ	China Journal of Science and Arts		
CM	Cambridge Magazine (1912–1922)	ILN	Illustrated London News
CMAG	China Magazine (New York)	IM	Imago Mundi: Yearbook of Early Cartography
CMIS	Chinese Miscellany		
CMJ	China Medical Journal	IN	Indian [Amerindian] Notes (Mus. of the Amer. Indian, New York)
CNRS	Centre Nationale de la Recherche Scientifique		
CPH	Contemporary Physics	IPR	Institute of Pacific Relations
CPICT	China Pictorial	IRAQ	Iraq (British Sch. Archaeol. Iraq)
CQ	Classical Quarterly	ISIS	Isis
CR	China Review (Hong Kong and Shanghai)	JA	Journal asiatique
		JAOPS	Journal of the American Optical Society
CRAS	Comptes Rendus de l'Académie des Sciences (Paris)	JAOS	Journal of the American Oriental Society
CREC	China Reconstructs		
CRR	Chinese Recorder	JCE	Journal of Chemical Education
CZOM	Centralzeitung f. Optik u. Mechanik u. verw. Berufszweige	JDZWT	Japanisch-deutsche Zeitschrift f. Wissenschaft u. Technik
		JEA	Journal of Egyptian Archaeology
DI	Der Islam	JEB	Journal of Experimental Biology
DNAT	Die Natur (Halle a/d Saale)	JEFDS	Journal of the English Folk-Dance and Song Society
EB	Encyclopaedia Britannica	JEZ	Journal of Experimental Zoology
EHR	Economic History Review	JFP	Jahrbuch f. Photographie
END	Endeavour		

JFSUT	*Journal of the Faculty of Science Univ. Tokyo*	*MDGNVO*	*Mitteilungen d. deutsch. Gesellschaft f. Natur- u. Volkskunde Ostasiens*
JHI	*Journal of the History of Ideas*	*MGGM*	*Mitteilungen d. Gesellsch. f. Geschichte d. Medizin*
JHMAS	*Journal of the History of Medicine and Allied Sciences*	*MGMNW*	*Mitteilungen z. Geschichte d. Medizin u. Naturwissenschaft*
JHS	*Journal of Hellenic Studies*	*MGSC*	*Ti Chih Chuan Pao* (*Memoirs*
JIN	*Journal of the Institute of Navigation*		*of the Chinese Geological Survey*)
JOP	*Journal of Physiology*	*MHJ*	*Middlesex Hospital Journal*
JOSP	*Journal of Oriental Studies* (Poona, India)	*MMA*	*Mineralogical Magazine*
JPH	*Journal de Physique*	*MMI*	*Mariner's Mirror*
JPOS	*Journal of the Peking Oriental Society*	*MO*	*Monist*
		MRASP	*Mémoires de l'Acad. royale des Sciences* (Paris)
JRAS/B	*Journal of the (Royal) Asiatic Society of Bengal*	*MRDTB*	*Memoirs of the Research Department of the Tōyō Bunko* (Tokyo)
JRAS/KB	*Journal (or Transactions) of the Korea Branch of the Royal Asiatic Society*	*MS*	*Monumenta Serica*
		MSLP	*Mémoires de la Société de Linguistique de Paris*
JRAS/M	*Journal of the Malayan Branch of the Royal Asiatic Society*	*MSOS*	*Mitteilungen d. Seminars f. orientalische Sprachen* (Berlin)
JRAS/NCB	*Journal (or Transactions) of the North China Branch of the Royal Asiatic Society*	*MUJ*	*Museum Journal* (Philadelphia)
		MUQ	*Musical Quarterly*
JRSA	*Journal of the Royal Society of Arts*	*N*	*Nature*
JS	*Journal des Savants*	*NAW*	*Nieuw Archief voor Wiskunde*
JWAS	*Journal of the Washington Academy of Science*	*NCH*	North China Herald Publishing House
JWCBRS	*Journal of the West China Border Research Society*	*NCH*	*North China Herald*
		NCR	*New China Review*
JWCI	*Journal of the Warburg and Courtauld Institutes*	*NKKZ*	*Nihon Kagaku Koten Zensho*
		NSN	*New Statesman and Nation* (London)
K	*Kagaku* (*Science*)	*NYSOAYB*	*New York State Optometric Association Year Book*
KDVS/AKM	*Kgl. Danske Videnskabernes Selskab* (Archaeol.-Kunsthist. Medd.)		
		O	*Observatory*
KHHP	*Kho Hsüeh Hua Pao* (*Science Illustrated*)	*OAZ*	*Ostasiatische Zeitung*
		OC	*Open Court*
KHTP	*Kho-Hsüeh Thung Pao* (*Scientific Correspondent*)	*OL*	*Old Lore; Miscellany of Orkney, Shetland, Caithness and Sutherland*
KK	*Kokka* (*Art History*)		
		OR	*Oriens*
LNC	*La Nouvelle Clio* (Brussels)	*ORA*	*Oriental Art*
LSYC	*Li Shih Yen Chiu* (Peking) (*Journ. Historical Research*)	*ORE*	*Oriens Extremus*
		OSIS	*Osiris*
MAAA	*Memoirs of the American Anthropological Association*	*PA*	*Pacific Affairs*
		PAKJS	*Pakistan Journal of Science*
MAI/LTR	*Mémoires de Litt. tirés des Registres de l'Acad. des Inscr. et Belles-Lettres* (Paris)	*PASP*	*Publications of the Astronomical Society of the Pacific*
		PBA	*Proceedings of the British Academy*
MARSL	*Memorias del Academia Real d. Sciencias de Lisboa*	*PC*	*People's China*
		PEW	*Philosophy East and West* (University of Hawaii)
MAS/B	*Memoirs of the Asiatic Society of Bengal*	*PHR*	*Philosophical Review*
MCHSAMUC	*Mémoires concernant l'Histoire, les Sciences, les Arts, les Mœurs et les Usages, des Chinois, par les Missionnaires de Pékin*, Paris, 1776–1814	*PHY*	*Physis* (Florence)
		PHYR	*Physical Review*
		PINO	*Pubblicazioni dell'Istituto Nazionale di Ottica* (Arcetri)
		PMG	*Philosophical Magazine*

PN	*Philosophia Naturalis*	SPCK	Society for the Promotion of Christian Knowledge
PNHB	*Peking Natural History Bulletin*		
PPHS	*Proceedings of the Prehistoric Society*	SPMSE	*Sitzungsber. d. physik.-med. Soc. Erlangen*
PRO	*Proteus*	SPR	*Science Progress*
PRS	*Proceedings of the Royal Society*	SS	*Science and Society* (New York)
PRSA	*Proceedings of the Royal Society* (Ser. A)	SSA	*Scripta Serica, Bulletin bibliographique* (Centre francochinois d'Études sinologiques, Peking)
PRSB	*Proceedings of the Royal Society* (Ser. B)		
PRSG	*Publicaciones de la Real Sociedad Geográfica* (Spain)	SSE	*Hua-Hsi Ta-Hsüeh Wên Shih Chi-Khan (Studia Serica; West China Union University Literary and Historical Journal)*
PTRS	*Philosophical Transactions of the Royal Society*		
PTRSA	*Philosophical Transactions of the Royal Society* (Series A)	SSIP/P	*Shanghai Science Institute Publications* (Physics Series)
		STMF	*Svensk. Tidskr. för Musikforskning*
QJGS	*Quarterly Journal of the Geological Society of London*	SWAW/PH	*Sitzungsber. d. (österreichischen) Akad. Wiss. Wien (Vienna)* (Phil.-hist. Klasse)
QSGNM	*Quellen u. Studien z. Geschichte d. Naturwiss. u. d. Medizin* (contin. of *Archiv f. Gesch. d. Math., d. Naturwiss. u. d. Technik* (AGMNT), formerly *Archiv f. d. Gesch. d. Naturwiss. u. d. Technik* (AGNT))	TAPS	*Transactions of the American Philosophical Society*
		TAS/J	*Transactions of the Asiatic Society of Japan*
		TFTC	*Tung Fang Tsa Chih (Eastern Miscellany)*
R	*Research*	TG/K	*Tōhō Gakuhō, Kyōto (Kyoto Journal of Oriental Studies)*
RA	*Revue archéologique*		
RAA/AMG	*Revue des Arts asiatiques (Annales du Musée Guimet)*	TG/T	*Tōhō Gakuhō, Tōkyō (Tokyo Journal of Oriental Studies)*
RGI	*Rivista Geografica Italiana*	TGAS	*Transactions of the Glasgow Archaeological Society*
RHS	*Revue d'Histoire des Sciences*		
RSO	*Rivista di Studi Orientali*	TH	*Thien Hsia Monthly* (Shanghai)
RSPT	*Revue des Sciences Philosophiques et Théologiques*	TKP/WS	*Ta Kung Pao (Wên Shih Chou Khan) (Lit. Supplement)*
		TM	*Terrestrial Magnetism and Atmospheric Electricity* (continued as *Journal of Geophysical Research*)
S	*Sinologica*		
SA	*Sinica* (originally *Chinesische Blätter f. Wissenschaft u. Kunst*)	TNS	*Transactions of the Newcomen Society*
SAM	*Scientific American*	TOCS	*Transactions of the Oriental Ceramic Society*
SBE	*Sacred Books of the East* Series		
SBIMG	*Sammelbände d. Internationalen Musik-Gesellschaft*	TOPS	*Transactions of the Optical Society*
SC	*Science*	TP	*T'oung Pao (Archives concernant l'Histoire, les Langues, la Géographie, l'Ethnographie et les Arts de l'Asie Orientale,* Leiden)
SCI	*Scientia*		
SCIS	*Sciences* (Paris)		
SCSR	*School Science Review*		
SGK	*Shinagaku Kenkyū (Journ. Sinol. Studies)*	TRIA	*Transactions of the Royal Irish Academy*
SIO	*The Student and Intellectual Observer of Science, Literature and Art*	TRSE	*Transactions of the Royal Society of Edinburgh*
SKCS	*Ssu Khu Chhüan Shu*	TS	*Tōhō Shūkyō (Journal of East Asian Religions)*
SM	*Scientific Monthly* (formerly *Popular Science Monthly*)		
		TSFFA	*Technical Studies in the Field of the Fine Arts*
SOS	*Semitic and Oriental Studies* (Berkeley)	TSGT	*Transactions of the Society of Glass Technology*
SP	*Speculum*		

TSSC	*Transactions of the Science Society of China*	*YAHS*	*Yenching Shih-Hsüeh Nien Pao (Yenching Annual of Historical Studies) or Yenching Historical Annual*
TTCY	*Tao Tsang Chi Yao*		
TYG	*Tōyō Gakuhō (Reports of the Oriental Society of Tokyo)*	*YCCC*	*Yün Chi Chhi Chhien*
		YCHP	*Yenching Hsüeh Pao (Yenching University Journal of Chinese Studies)*
UMN	*Unterrichtsblätter f. Math. u. Naturwiss.*		
VA	*Vistas in Astronomy*	*ZAC*	*Zeitschrift f. angewandte Chemie*
VAG	*Vierteljahrsschrift d. astronomischen Gesellschaft*	*ZDMG*	*Zeitschrift d. deutsch. morgenländischen Gesellschaft*
VBGE	*Verhandlungen d. Berliner Gesellschaft f. Ethnologie*	*ZGEB*	*Zeitschrift d. Gesellschaft f. Erdkunde (Berlin)*
VDPG	*Verhandlungen d. deutsch. Physikal. Gesellschaft*	*ZHWK*	*Zeitschrift f. historische Wappenkunde (contin. as Zeitschrift f. historische Wappen- und Kostumkunde)*
VS	*Variétés Sinologiques Series*		
WWTK	*Wên Wu Tshan Khao Tzu Liao (Reference Materials for History and Archaeology)*	*ZMNWU*	*Zeitschrift f. Math. u. Naturwiss. Unterricht*
		ZP	*Zeitschrift f. Physik*

A. CHINESE BOOKS BEFORE +1800

Each entry gives particulars in the following order:

(a) title, alphabetically arranged, with characters;

(b) alternative title, if any;

(c) translation of title;

(d) cross-reference to closely related book, if any;

(e) dynasty;

(f) date as accurate as possible;

(g) name of author or editor, with characters;

(h) title of other book, if the text of the work now exists only incorporated therein; or, in special cases, references to sinological studies of it;

(i) references to translations, if any, given by the name of the translator in Bibliography C;

(j) notice of any index or concordance to the book if such a work exists;

(k) reference to the number of the book in the *Tao Tsang* catalogue of Wieger (6), if applicable;

(l) reference to the number of the book in the *San Tsang* (Tripiṭaka) catalogues of Nanjio (1) and Takakusu & Watanabe, if applicable.

Words which assist in the translation of titles are added in round brackets.

Alternative titles or explanatory additions to the titles are added in square brackets.

It will be remembered (p. 335 above) that in Chinese indexes words beginning *Chh-* are all listed together after *Ch-*, and *Hs-* after *H-*, but that this applies to initial words of titles only.

Where there are any differences between the entries in these bibliographies and those in vol. 1, the information here given is to be taken as more correct.

References to the editions used in the present work, and to the *tshung-shu* collections in which books are available, will be given in a later volume.

ABBREVIATIONS

C/Han	Former Han.
E/Wei	Eastern Wei.
H/Han	Later Han.
H/Shu	Later Shu (Wu Tai).
H/Thang	Later Thang (Wu Tai).
J/Chin	Jurchen Chin.
L/Sung	Liu Sung.
N/Chou	Northern Chou.
N/Chhi	Northern Chhi.
N/Sung	Northern Sung (before the removal of the capital to Hangchow).
N/Wei	Northern Wei.
S/Chhi	Southern Chhi.
S/Sung	Southern Sung (after the removal of the capital to Hangchow).
W/Wei	Western Wei.

Akhak Kwebŏm 樂學軌範.
    Standard Patterns in Musicology (encyclopaedia).
    Korea (Chosŏn), +1493; later editions +1610 and +1655.
    Sŏng Hyŏn *et al.* 成俔 based on the work of Pak Yŏn *et al.* 朴堧.

Chan Kuo Tshê 戰國策.
    Records of the Warring States.
    Chhin.
    Writer unknown.

Chang Tzu Chhüan Shu 張子全書.
    Complete Works of Master Chang (Tsai) (d. +1077) with commentary by Chu Hsi.
    Sung (ed. Chhing), *editio princeps* +1719.
    Chang Tsai.
    Ed. Chu Shih 朱軾 & Tuan Chih-Hsi 段志熙.

Chang Yen Kung Chi 張燕公集.
    Collected Writings of Chang Yüeh.
    Thang, *c.* +730.
    Chang Yüeh 張說.

Chen-La Fêng Thu Chi 眞臘風土記.
    Description of Cambodia.
    Yuan, +1297.
    Chou Ta-Kuan 周達觀.

Chêng Lei Pên Tshao 證類本草.
    [*Chhung Hsiu Chêng-Ho Ching Shih Chêng Lei Pei Yung Pên Tshao.*]
    Reorganised Pharmacopoeia.
    N/Sung, +1108, enlarged +1116: re-edited in J/Chin, +1204, and definitively re-published in Yuan, +1249; re-printed many times afterwards, e.g. in Ming, +1468.
    Original compiler: Thang Shen-Wei 唐慎微.
    Cf. Hummel (13), Lung Po-Chien (1).

Chêng Mêng 正蒙.
    Right Teaching for Youth [or, Intellectual Discipline for Beginners].
    Sung, *c.* +1060.
    Chang Tsai 張載.

Chi Chiu Phien (or Chang) 急就篇 (章).
    Dictionary for Urgent Use.
    C/Han, −48 to −32.
    Shih Yu 史游. With +13th-century commentary by Wang Ying-Lin 王應麟.

Chi Ku Lu 集古錄.
    Collection of Ancient Inscriptions.
    Sung, *c.* +1050.
    Ouyang Hsiu 歐陽修.

Chi Ku Lu Pa Wei 集古錄跋尾.
    Postscript to the *Collection of Ancient Inscriptions.*
    Sung, *c.* +1060.
    Ouyang Hsiu 歐陽修.

Chi Lu Hui Pien 紀錄彙編.
    Classified Records of a Decennium.

Chi Lu Hui Pien (*cont.*).
  Ming, *c.* +1530.
  Shen Chieh-Fu 沈節甫 .

Chi Sêng Fang 濟生方 .
  Life-Saving Prescriptions.
  Sung, +1267.
  Yen Yung-Ho 嚴用和 .

Chia-Shen Tsa Chi 甲申雜記 .
  Miscellany of the Chia-Shen Year.
  Sung, +1104, pr. +1163. (Deals with
    events from +1023 to +1104.)
  The second of the three parts of the *Chhing
    Hsü Tsa Chu*, q.v.
  Wang Kung 王鞏 .

Chia-Yu Tsa Chih 嘉祐雜志 .
  Miscellaneous Records of the Chia-Yu
    Reign-period.
  Sung, +1062.
  Chiang Lin-Chi 江鄰幾 .

Chiangsi Thung Chih 江西通志 .
  Provincial Historical Geography of Chiangsi.
  Chhing, +1732.
  Ed. Hsieh Min *et al.* 謝旻 .

Chieh Ku Lu 羯鼓錄 .
  On the History and Use of Drums.
  Thang, +848.
  Nan Cho 南卓 .

Chih Lin Hsin Shu 志林新書 .
  New Book of Miscellaneous Records.
  Chin, +4th century.
  Yü Hsi 虞喜 .

Chin Hou Lüeh Chi 晉後略記 .
  Brief Records set down after the (Western)
    Chin (dynasty).
  Chin, after +317.
  Hsün Cho 荀綽 .

Chin Lou Tzu 金樓子 .
  Book of the Golden Hall Master.
  Liang, *c.* +550.
  Hsiao I 蕭繹 .
    (Liang Yuan Ti 梁元帝 .)

Chin Shih Tzu Chang 金師子章 .
  Essay on the Golden Lion.
  Thang, +704.
  Fa-Tsang 法藏 .
  TW/1880.

Chin Shu 晉書 .
  History of the Chin Dynasty [+265 to
    +419].
  Thang, +635.
  Fang Hsüan-Ling 房玄齡 .
  A few chs. tr. Pfizmaier (54–57); the astro-
    nomical chs. tr. Ho Ping-Yü (1). For
    translations of passages see the index of
    Frankel (1).

Ching Ching Ling Chhih
  See Chêng Fu-Kuang (1).

Chiu Chang Suan Shu 九章算術 .
  Nine Chapters on the Mathematical Art.

H/Han, +1st century (containing much
  material from C/Han and perhaps Chhin).
  Writer unknown.

Chiu Ku Khao 九穀考 .
  A Study of the Nine (Cereal) Grains.
  Chhing, *c.* +1790.
  Chhêng Yao-Thien 程瑤田 .
  (In *HCCC*, ch. 551.)

Chiu Thang Shu 舊唐書 .
  Old History of the Thang Dynasty [+618
    to +906].
  Wu Tai, +945.
  Liu Hsü 劉昫 .
  For translations of passages see the index of
    Frankel (1).

*Chiu Thien Hsüan Nü Chhing Nang Hai Chio
  Ching* 九天玄女青囊海角經 .
  The Nine-Heaven Mysterious Girl Blue-
    Bag Sea Angle Manual [geomantic].
  Thang.
  Writer unknown.

Chiu Wu Tai Shih 舊五代史 .
  Old History of the Five Dynasties [+907 to
    +959].
  Sung, +974.
  Hsüeh Chü-Chêng 薛居正 .
  For translations of passages see the index of
    Frankel (1).

Cho Kêng Lu 輟耕錄 .
  [sometimes *Nan Tshun Cho Kêng Lu*.]
  Talks (at South Village) while the Plough is
    Resting.
  Yuan, +1366.
  Thao Tsung-I 陶宗儀 .

Chou Li 周禮 .
  Record of the Institutions (lit. Rites) of
    (the) Chou (Dynasty) [descriptions of all
    government official posts and their duties].
  C/Han, perhaps containing some material
    from late Chou.
  Compilers unknown.
  Tr. E. Biot (1).

Chou Li Chêng I 周禮正義 .
  Amended Text of the *Record of the Institu-
    tions* (lit. *Rites*) *of* (the) *Chou* (*Dynasty*)
    with Discussions (including the H/Han
    commentary of Chêng Hsüan 鄭玄).
  C/Han, perhaps containing some material
    from late Chou.
  Compilers unknown.
  Ed. Sun I-Jang (1899) 孫詒讓 .

Chou Li I I Chü Yao 周禮疑義舉要 .
  Discussion of the Most Important Doubtful
    Matters in the *Record of the Institutions*
    (lit. *Rites*) *of* (the) *Chou* (*Dynasty*).
  Chhing, +1791.
  Chiang Yung. 江永 .

Chou Pei Suan Ching 周髀算經 .
  The Arithmetical Classic of the Gnomon
    and the Circular Paths (of Heaven).
  Chou, Chhin and Han. Text stabilised
    about the −1st century, but including

Chou Pei Suan Ching (*cont.*)
    parts which must be of the late Warring
    States period (*c.* −4th century) and some
    even pre-Confucian (−6th century).
    Writers unknown.
Chou Shu 周書.
    History of the (Northern) Chou Dynasty
    [+557 to +581].
    Thang, +625.
    Linghu Tê-Fên 令狐德棻.
    For translations of passages see the index of
    Frankel (1).
Chu Fan Chih 諸蕃志.
    Records of Foreign Peoples.
    Sung, *c.* +1225. (This is Pelliot's dating;
    Hirth & Rockhill favoured between
    +1242 and +1258.)
    Chao Ju-Kua 趙汝适.
    Tr. Hirth & Rockhill (1).
Chu Tzu Chhüan Shu 朱子全書.
    Collected Works of Master Chu (Hsi).
    Sung (ed. Ming), *editio princeps* +1713.
    Chu Hsi 朱熹.
    Ed. Li Kuang-Ti 李光地 (Chhing).
    Partial trs. Bruce (1); le Gall (1).
Chuang Tzu 莊子.
    [= *Nan Hua Chen Ching*.]
    The Book of Master Chuang.
    Chou, *c.* −290.
    Chuang Chou 莊周.
    Tr. Legge (5); Fêng Yu-Lan (5); Lin Yü-
    Thang (1).
    Yin-Tê Index no. (suppl.) 20.
Chuang Tzu Pu Chêng
    The Text of *Chuang Tzu*, Annotated and
    Corrected.
    See Liu Wên-Tien (1).
Chung Hua Ku Chin Chu 中華古今注.
    Commentary on Things Old and New in
    China.
    Wu Tai (H/Thang), +923 to +936.
    Ma Kao 馬縞.
    See des Rotours (1), p. xcix.
Chung Lü Wei 鐘律緯.
    Apocryphal Treatise on Bells and Pipes.
    Liang, *c.* +540.
    Hsiao Yen 蕭衍.
      (Liang Wu Ti 梁武帝.)
    (*YHSF*, ch. 31.)
Chung Yung 中庸.
    Doctrine of the Mean.
    Chou (enlarged in Chhin and Han), −4th
    century, with additions of −3rd.
    Trad. attrib. Khung Chi 孔伋 (Khung
    Tzu-Ssu 孔子思).
    Tr. Legge (2); Lyall & Ching Chien-Chün
    (1); Hughes (2).
Chhang-Chhun Chen Jen Hsi Yu Chi 長春眞人
    西遊記.
    The Western Journey of the Taoist (Chhiu)
    Chhang-Chhun.
    Yuan, +1228.

    Li Chih-Chhang 李志常.
Chhi Ching 棊經.
    Chess Manual.
    Sung.
    Chang I 張擬.
Chhi Ching 棊經.
    Chess Manual.
    Sung.
    Yen Thien-Chang 晏天章.
Chhi Hsiu Lei Kao 七修類稿.
    Seven Compilations of Classified
    Manuscripts.
    Ming, *c.* +1530.
    Lang Ying 郎瑛.
Chhi Min Yao Shu 齊民要術.
    Important Arts for the People's Welfare
    [lit. Equality].
    N/Wei (and E/Wei or W/Wei), between
    +533 and +544.
    Chia Ssu-Hsieh 賈思勰.
    See des Rotours (1), p. c; Shih Shêng-Han (1).
Chhi Tung Yeh Yü 齊東野語.
    Rustic Talks in Eastern Chhi.
    Sung, *c.* +1290.
    Chou Mi 周密.
Chhien Han Shu 前漢書.
    History of the Former Han Dynasty
    [−206 to +24].
    H/Han (begun about +65), *c.* +100.
    Pan Ku 班固, and (after his death in +92)
    his sister Pan Chao 班昭.
    Partial trs. Dubs (2), Pfizmaier (32–34,
    37–51), Wylie (2, 3, 10), Swann (1).
    Yin-Tê Index, no. 36.
Chhin-Ting Hsü Wên Hsien Thung Khao 欽定
    續文獻通考.
    Imperially Commissioned Continuation of
    the *Comprehensive Study of* (*the History
    of*) *Civilisation* (cf. *Wên Hsien Thung
    Khao* and *Hsü Wên Hsien Thung Khao*).
    Chhing, ordered +1747, pr. +1772
    (+1784).
    Ed. Chhi Shao-Nan 齊召南, Hsi Huang
    嵇璜 *et al.*
    This parallels, but does not replace, Wang
    Chhi's *Hsü Wên Hsien Thung Khao*.
Chhin-Ting Ku Chin Thu Shu Chi Chhêng 欽定
    古今圖書集成.
    See *Thu Shu Chi Chhêng*.
Chhing Hsü Tsa Chu 清虛雜著.
    The 'Pure Emptiness' Miscellaneous
    Record. [Comprising *Wên Chien Chin
    Lu*, *Chia-Shen Tsa Chi*, and *Sui Shou Tsa
    Lu*, q.v.]
    Sung, collected posthumously +1163.
    Wang Kung 王鞏.
Chhing Hsü Tsa Chu Pu Chhüeh 清虛雜著補闕.
    Additions to the '*Pure Emptiness*' Miscel-
    laneous Record.
    Sung, *c.* +1163.
    Wang Tshung-Chin 王從謹.

Chhing I Lu 清異錄.
Records of the Unworldly and the Strange.
Wu Tai, c. +950.
Thao Ku 陶穀.

Chhing Nang Ao Chih 青囊奧旨.
Mysterious Principles of the Blue Bag (i.e. the Universe) [geomantic].
Thang, c. +880.
Attrib. Yang Yün-Sung 楊筠松.

Chhing Wu Hsü Yen 青烏緒言.
Introduction to the *Blue Raven Manual* [geomancy and the use of the mariner's compass].
Ming, c. +1570.
Li Yü-Hêng 李豫亨.

Chhou Jen Chuan 疇人傳.
Biographies of Mathematicians and Astronomers.
Chhing, +1799.
Juan Yuan 阮元.
With continuations by Lo Shih-Lin 羅士琳, Chu Kho-Pao 諸可寶 and Huang Chung-Chün 黃鍾駿.
(In *HCCC*, chs. 159 ff.)

Chhu Hsüeh Chi 初學記.
Entry into Learning [encyclopaedia].
Thang, +700.
Hsü Chien 徐堅.

Chhü I Shuo Tsuan 祛疑說纂.
Discussions on the Dispersal of Doubts.
Sung, c. +1230.
Chhu Yung 儲泳.

Chhu Tzhu 楚辭.
Elegies of Chhu (State) [or, Songs of the South].
Chou, c. −300 (with Han additions).
Chhü Yuan 屈原 (& Chia I 賈誼, Yen Chi 嚴忌, Sung Yü 宋玉, Huainan Hsiao-Shan 淮南小山 *et al.*).
Partial tr. Waley (23); tr. Hawkes (1).

Chhu Tzhu Pu Chu 楚辭補註.
Supplementary Annotations to the *Elegies of Chhu*.
Sung, c. +1140.
Ed. Hung Hsing-Tsu 洪興祖.

Chhü Wei Chiu Wên 曲洧舊聞.
Talks about Bygone Things beside the Winding Wei (River in Honan).
Sung, c. +1130.
Chu Pien 朱弁.

Chhun Chhiu Fan Lu 春秋繁露.
String of Pearls on the *Spring and Autumn Annals*.
C/Han, c. −135.
Tung Chung-Shu 董仲舒.
See Wu Khang (1).
Partial trs. Wieger (2); Hughes (1); d'Hormon (ed.).
Chung-Fa Index, no. 4.

Chhun Chhiu Wei Khao I Yu 春秋緯考異郵.
Apocryphal Treatise on the *Spring and Autumn Annals*; Investigation of the Strange and Extreme Penetration (of the Mutual Influences of Things).
C/Han, −1st century.
Writer unknown.
Comm. Sung Chün 宋均.

Chhun Chu Chi Wên 春渚紀聞.
Record of Things Heard at Spring Island.
Sung, c. +1095.
Ho Wei 何薳.

Chhung Hsiu Chêng-Ho Ching Shih Chêng Lei Pei Yung Pên Tshao 重修政和經史證類備用本草.
The Official Practical Reclassified Pharmacopoeia of the Chêng-Ho reign-period (+1116), re-edited.
See *Chêng Lei Pên Tshao*.

Erekiteru Yakusetsu エレキテル譯説.
A Translated Discourse on Electricity.
Japan, late +18th century.
Hashimoto Donsai 橋本曇齋.
In *NKKZ*, vol. 6; cf.
Hashimoto Donsai (1).

Erh-shih-ssu Shan Hsiang Chüeh 二十四山向訣.
Oral Instructions on the 24 Mountain Directions (Compass-Points) [geomantic].
Thang, c. +880.
Yang Yün-Sung 楊筠松.

Erh Ya 爾雅.
Literary Expositor [dictionary].
Chou material, stabilised in Chhin or C/Han.
Compiler unknown.
Enlarged and commented on c. +300 by Kuo Pho 郭璞.
Yin-Tê Index no. (suppl.) 18.

Fang-Chou Tsa Yen 方洲雜言.
Reminiscences of (Chang) Fang-Chou.
Ming, c. +1452.
Chang Ning 張寧.

Fêng Su Thung I 風俗通義.
The Meaning of Popular Traditions and Customs.
H/Han, +175.
Ying Shao 應劭.
Chung-Fa Index, no. 3.

Hai Chio Ching
See *Chiu Thien Hsüan Nü Chhing Nang Hai Chio Ching*.

Hai Nei Shih Chou Chi 海內十洲記.
Record of the Ten Sea Islands [or, of the Ten Continents in the World Ocean].
Ascr. Han; prob. +4th or +5th century.
Attrib. Tungfang Shuo 東方朔.

Hai Tao Chen Ching 海道針經.
Seaways Compass Manual.
Yuan or Ming, +14th century.
Writer unknown.

Hai Tao Ching 海道經.
  Manual of Sailing Directions.
  Yuan, +14th century.
  Compilers unknown.
Hai Tao I Chih Chai Lüeh 海島逸誌摘略.
  Brief Selection of Lost Records of the Isles
    of the Sea [or, a Desultory Account of the
    Malayan Archipelago].
  Chhing, between +1783 and +1790,
    preface +1791.
  Wang Ta-Hai 王大海.
  Tr. Anon. (37).
Hai Yen Hsien Thu Ching 海鹽縣圖經.
  Illustrated Historical Geography of Sea-
    Salt City.
  Ming, c. +1528.
  Phêng Tsung-Mêng 彭宗孟.
Han Fei Tzu 韓非子.
  The Book of Master Han Fei.
  Chou, early −3rd century.
  Han Fei 韓非.
  Partial tr. Liao Wên-Kuei (1).
Han Lung Ching 撼龍經.
  Manual of the Moving Dragon [geomantic].
  Thang, c. +880.
  Yang Yün-Sung 楊筠松.
Han Shih Wai Chuan 韓詩外傳.
  Moral Discourses illustrating the Han Text
    of the *Book of Odes*.
  C/Han, c. −135.
  Han Ying 韓嬰.
Han Wei Tshung-Shu 漢魏叢書.
  Collection of Books of the Han and Wei
    Dynasties [first only 38, later increased
    to 96].
  Ming, +1592.
  Ed. Thu Lung 屠隆.
Han Yü Thung 函宇通.
  General Survey of the Universe [incl. *Ko
    Chih Tshao* on astronomy and cosmo-
    logy, and *Ti Wei* on geography, q.v.].
  Chhing, +1648.
  Hsiung Ming-Yü 熊明遇 & Hsiung Jen-
    Lin 熊人霖.
Hang Hai Chen Ching 航海針經.
  Sailors' Compass Manual.
  Yuan or Ming, +14th century.
  Writer unknown.
Hou Chou Shu 後周書.
  See *Chou Shu*.
Hou Han Shu 後漢書.
  History of the Later Han Dynasty [+25 to
    +220].
  L/Sung, +450.
  Fan Yeh 范曄.
  The monograph chapters by Ssuma Piao
    司馬彪.
  A few chs. tr. Chavannes (6, 16), Pfizmaier
    (52, 53).
  Yin-Tê Index, no. 41.
Hou Shan Than Tshung 後山談叢.
  Collected Discussions at Hou-Shan.

  Sung, +11th century.
  Chhen Shih-Tao 陳師道.
Hua Man Chi 畫墁集.
  Painted Walls.
  Sung, c. +1110.
  Chang Shun-Min 張舜民.
Hua Shu 化書.
  Book of the Transformations (in Nature).
  H/Thang, c. +940.
  Attrib. Than Chhiao 譚峭.
  TT/1032.
Hua Yang Kuo Chih 華陽國志
  Record of the Country South of Mount
    Hua [historical geography of Szechuan
    down to +138].
  Chin, +347.
  Chhang Chhü 常璩.
Huai Nan Hung Lieh Chieh 淮南鴻烈解.
  See *Huai Nan Tzu*.
Huai Nan Tzu 淮南子.
  [= *Huai Nan Hung Lieh Chieh*.]
  The Book of (the Prince of) Huai-Nan
    [compendium of natural philosophy].
  C/Han, c. −120.
  Written by the group of scholars gathered by
    Lui An (prince of Huai-Nan) 劉安.
  Partial trs. Morgan (1); Erkes (1);
    Hughes (1); Chatley (1); Wieger (2).
  Chung-Fa Index, no. 5.
  TT/1170.
Huai Nan (Wang) Wan Pi Shu 淮南(王)萬
    畢術.
  [Prob. = *Chen-Chung Hung-Pao
    Yuan-Pi Shu* and variants.]
  The Ten Thousand Infallible Arts of (the
    Prince of) Huai-Nan [Taoist alchemical
    and technical recipes].
  C/Han, −2nd century.
  No longer a separate book but fragments
    contained in *TPYL*, ch. 736 and else-
    where.
  Reconstituted texts by Yeh Tê-Hui in
    Kuan Ku Thang So Chu Shu, and Sun
    Fêng-I in *Wên Ching Thang Tshung-Shu*.
  Attr. Liu An 劉安.
  See Kaltenmark (2), p. 32.
  It is probable that the terms *Chen-Chung*
    枕中 Confidential Pillow-Book; *Hung-
    Pao* 鴻寶 Infinite Treasure; *Wan-Pi*
    萬畢 Ten Thousand Infallible; and
    Yuan-Pi 苑祕 Garden of Secrets; were
    originally titles of parts of a *Huai-Nan
    Wang Shu* 淮南王書 (Writings of the
    Prince of Huai-Nan) forming the Chung
    Phien 中篇 (and perhaps also the Wai
    Shu 外書) of which the present *Huai
    Nan Tzu* book (q.v.) was the Nei Shu
    內書.
Huang Chhing Ching Chieh 皇清經解.
  Collection of (more than 180) Monographs
    on Classical Subjects written during the
    Chhing Dynasty.

Huang Chhing Ching Chieh (*cont.*)
See Yen Chieh (*1*) (ed.).

Huang Ti Nei Ching, Ling Shu 黃帝內經靈樞.
The Yellow Emperor's Manual of Internal Medicine; The Spiritual Pivot (or Gate, or Driving-shaft, or Motive Power) [medical physiology and anatomy].
Probably C/Han, *c.* — 1st century.
Writers unknown.
Edited +762 by Wang Ping 王冰.
Analysis by Huang Wên (*1*).
Tr. Chamfrault & Ung Kang-Sam (*1*).

Huang Ti Nei Ching, Su Wên 黃帝內經素問.
The Yellow Emperor's Manual of Internal Medicine; The Plain Questions (and Answers) [clinical medicine]. (Cf. *Pu Chu Huang Ti Nei Ching, Su Wên.*)
Chou, remodelled in Chhin and Han, *c.* — 2nd century.
Writers unknown.
Partial trs. Hübotter (*1*), chs. 4, 5, 10, 11, 21; Veith (*1*); complete, Chamfrault & Ung Kang-Sam (*1*).
See Wang & Wu (*1*), pp. 28 ff.; Huang Wên (*1*).

Huang Ti Nei Ching Su Wên Chi Chu 黃帝內經素問集註.
The *Yellow Emperor's Manual of Internal Medicine; The Plain Questions (and Answers)*; with Commentaries.
Chhing, +1679.
Chang Chih-Tshung 張志聰.
(The Ming commentary of Ma Shih 馬蒔 is given in *TSCC, I shu tien,* chs. 21–66.)

Huang Ti Su Wên Ling Shu Ching 黃帝素問靈樞經.
See *Huang Ti Nei Ching, Ling Shu.*

Huang Ti Su Wên Nei Ching 黃帝素問內經.
See *Huang Ti Nei Ching, Su Wên.*

Huang-Yu Hsin Yo Thu Chi 皇祐新樂圖記.
New Illustrated Record of Musical Matters of the Huang-Yu reign-period.
Sung, *c.* +1050.
Juan I 阮逸.

Hun Thien Hsiang Shuo (or *Chu*) 渾天象説 (注).
Discourse on Uranographic Models.
San Kuo, *c.* +260.
Wang Fan 王蕃.
(*CSHK* (San Kuo sect.), ch. 72, pp. 1 *a* ff.)

Hsi Chhi Tshung Hua 西溪叢話.
(*SKCS* has *Yü*) (語).
Western Pool Collected Remarks.
Sung, +11th century.
Yao Khuan 姚寬.

Hsi Ching Tsa Chi 西京雜記.
Miscellaneous Records of the Western Capital.
Liang or Chhen, mid +6th century.

Attrib. to Liu Hsin 劉歆 (C/Han) and to Ko Hung 葛洪 (Chin), but probably Wu Chün 吳均.

Hsi-Yang Chhao Kung Tien Lu 西洋朝貢典錄.
Record of the Tributary Countries of the Western Oceans [relative to the voyages of Chêng Ho].
Ming, +1520.
Huang Shêng-Tsêng 黃省曾.
Ed. by Sun Yün-Chia 孫允伽 & Chao Khai-Mei 趙開美.
Tr. Mayers (3).

Hsi-Yang Chi 西洋記.
See *San-Pao Thai-Chien Hsia Hsi-Yang Chi Thung Su Yen I.*

Hsi Yu Chi.
See *Chhang-Chhun Chen Jen Hsi Yu Chi.*

Hsia Jih Chi 暇日記.
Records of Leisure Hours.
Sung, *c.* +1100.
Liu Chhi 劉跂.

Hsiang Shan Yeh Lu 湘山野錄.
Rustic Notes from Hsiang-shan.
Sung, *c.* +1060.
Wên-Jung 文瑩.

Hsiang Yen Lu 湘烟錄.
The Smoke of Hunan Hearths [Hunanese matters].
Ming.
Min Yuan-Ching 閔元京.

Hsiao Ching 孝經.
Filial Piety Classic.
Chhin and C/Han.
Attrib. Tsêng Shen (pupil of Confucius) 曾參.
Tr. de Rosny (2); Legge (1).

Hsiao Wei Yuan Shen Chhi 孝緯援神契.
Apocryphal Treatise on the *Filial Piety Classic*; Cantraps for Salvation by the Spirits.
C/Han, — 1st century.
Writer unknown.

Hsin Ching 心經.
Canon of the Core (of the Earth) [geomantic].
Sung, *c.* +1060.
Wang Chi 王伋.

Hsin Hsiu Pên Tshao 新修本草.
Newly Reorganised Pharmacopoeia.
Thang, +659.
Su Ching (ed.) 蘇敬 and a commission of 21 collaborators under the direction first of Chhangsun Wu-Chi 長孫無忌 then of Li Chi 李勣.
This pharmacopoeia was sometimes afterwards commonly but inaccurately known as the *Thang Pên Tshao.* It was lost in China but five chapters have been preserved as they were copied by a Japanese monk in +731 and survived in Japan.

Hsin Thang Shu 新唐書.
　　New History of the Thang Dynasty [+618 to +906].
　　Sung, +1061.
　　Ouyang Hsiu 歐陽修 & Sung Chhi 宋祁.
　　Partial trs. des Rotours (1, 2); Pfizmaier (66–74). For translations of passages see the index of Frankel (1).
　　Yin-Tê Index, no. 16.
Hsing Chha Shêng Lan 星槎勝覽.
　　Triumphant Visions of the Starry Raft [accounts of the voyages of Chêng Ho, whose ship, as carrying an ambassador, is thus styled].
　　Ming, +1436.
　　Fei Hsin 費信.
Hsing Li Ta Chhüan (Shu) 性理大全(書).
　　Collected Works of (120) Philosophers of the Hsing-Li (Neo-Confucian) School [*Hsing* = human nature; *Li* = the principle of organisation in all Nature].
　　Ming, +1415.
　　Ed. Hu Kuang *et al.* 胡廣.
Hsü Han Shu 續漢書.
　　Addenda to the Han History.
　　San Kuo, +3rd century.
　　Hsieh Chhêng 謝承.
Hsü Po Wu Chih 續博物志.
　　Supplement to the *Record of the Investigation of Things* (cf. *Po Wu Chih*).
　　Sung, mid +12th century.
　　Li Shih 李石.
Hsü Shih Shih 續事始.
　　Supplement to the *Beginnings of All Affairs* (cf. *Shih Shih*).
　　H/Shu, c. +960.
　　Ma Chien 馬鑑.
Hsü Wên Hsien Thung Khao 續文獻通考.
　　Continuation of the *Comprehensive Study of (the History of) Civilisation* (cf. *Wên Hsien Thung Khao* and *Chhin-Ting Hsü Wên Hsien Thung Khao*).
　　Ming, finished +1586, pr. +1603.
　　Ed. Wang Chhi 王圻.
　　This covers the Liao, J/Chin, Yuan and Ming dynasties, adding some new material for the end of S/Sung from +1224 onwards.
Hsüan Chieh Lu 玄解錄.
　　See *Hsüan Chieh Lu* 懸解錄.
Hsüan Chieh Lu 懸解錄.
　　A Record of Explanations [alchemical].
　　Thang, anonymous preface of +855.
　　Writer unknown.
　　TT/921, YCCC, ch. 64.
Hsüan-Ho Fêng Shih Kao-Li Thu Ching 宣和奉使高麗圖經.
　　Illustrated Record of an Embassy to Korea in the Hsüan-Ho reign-period.
　　Sung, +1124 (+1167).
　　Hsü Ching 徐兢.
Hsüan Kuai Lu 玄怪錄.
　　See *Yu Kuai Lu*.

Hsüeh Ku Phien 學古編.
　　On our Knowledge of Ancient Objects [seal inscriptions].
　　Yuan, +1307.
　　Wuchhiu Yen 吾邱衍.
Hsün Tzu 荀子.
　　The Book of Master Hsün.
　　Chou, c. −240.
　　Hsün Chhing 荀卿.
　　Tr. Dubs (7).

I Chao Liao Tsa Chi 猗覺寮雜記.
　　Miscellaneous Records from the I-Chao Cottage.
　　Sung, c. +1200.
　　Chu I 朱翌.
I Chhieh Ching Yin I 一切經音義.
　　Dictionary of Sounds and Meanings of Words in the *Vinaya* [part of the Buddhist Tripiṭaka].
　　Thang, c. +649, enlarged c. +730.
　　Hsüan-Ying 玄應.
　　Enlarged by Hui-Lin 慧琳.
　　N/1605; TW/2178.
I Ching 易經.
　　The Classic of Changes [or, Book of Changes].
　　Chou with C/Han additions.
　　Compilers unknown.
　　See Li Ching-Chhih (1, 2); Wu Shih-Chhang (1).
　　Tr. R. Wilhelm (2); Legge (9); de Harlez (1).
　　Yin-Tê Index, no. (suppl.) 10.
I Li 儀禮.
　　The Personal Conduct Ritual [one of the 'Three Rituals', with the *Li Chi* and the *Chou Li*].
　　Chhin and Han, based on Chou material, some of which may be as old as Confucius.
　　Edited in C/Han by Kaothang Sêng 高堂生.
　　Tr. Steele (1).
I Lin 意林.
　　Forest of Ideas [philosophical encyclopaedia].
　　Thang.
　　Ma Tsung 馬總.
　　TT/1244.
I Thu Ming Pien 易圖明辨.
　　Clarification of the Diagrams in the (*Book of*) *Changes* [historical analysis].
　　Chhing, +1706.
　　Hu Wei 胡渭.
I Yuan 異苑.
　　Garden of Strange Things.
　　L/Sung, c. +460.
　　Liu Ching-Shu 劉敬叔.

Jen-Chai Chih Chih Fang (Lun) 仁齋直指方(論).
　　(Yang) Jen-Chai's Basic Principles of (Paediatric) Prescribing.

Jen-Chai Chih Chih Fang (Lun) (cont.)
    Sung, +1264.
    Yang Shih-Ying　楊士瀛.

Jih Chih Lu　日知錄.
    Daily Additions to Knowledge.
    Chhing, +1673.
    Ku Yen-Wu　顧炎武.

Jih Hua (Chu Chia) Pên Tshao　日華 (諸家) 本草.
    Master Jih-Hua's Pharmacopoeia (of All the Schools).
    Sung, *c.* +970.
    Ta Ming (Jih-Hua Tzu)　大明 (日華子).

Jih Kuei Pei Khao　日晷備考.
    A Study on the Construction of Sundials.
    Chhing, *c.* +1690.
    Mei Wên-Ting　梅文鼎.

Ju Thang Chhiu Fa Hsün Li Hsing Chi.
    See *Nittō Guhō Junrei Gyōki.*

Kan Chhüan Fu　甘泉賦.
    Rhapsodic Ode on the Sweetwater Springs.
    C/Han, *c.* −10.
    Yang Hsiung　揚雄.

Kan Ying Ching　感應經.
    On Stimulus and Response (the Resonance of Phenomena in Nature).
    Thang, *c.* +640.
    Li Shun-Fêng　李淳風.
    See Ho & Needham (2).

Kan Ying Lei Tshung Chih　感應類從志.
    Record of the Mutual Resonances of Things according to their Categories.
    Chin, *c.* +295.
    Chang Hua　張華.
    See Ho & Needham (2).

Kao-Li Thu Ching.
    See *Hsüan-Ho Fêng Shih Kao-Li Thu Ching.*

Khai-Yuan Chan Ching　開元占經.
    The Khai-Yuan reign-period Treatise on Astrology (and Astronomy).
    Thang, +729.
    (Some parts, such as the *Chiu Chih* (*Navagrāha*) calendar, had been written as early as +718.)
    Chhüthan Hsi-Ta　瞿曇悉達.

Khan Yü Phi Miu Chhuan Chen　堪輿闢謬傳眞.
    A Brushing Away of Mistakes and Establishment of Right Theory in Geomancy.
    Chhing, end +18th century.
    Liu Kung-Chung　劉公中.

Khang-Hsi Chi Hsia Ko Wu Pien　康熙幾暇格物編.
    Observations in the Natural Sciences, made in Our Leisure Hours, by the Khang-Hsi Emperor [geology, mineralogy, zoology, botany and agriculture].
    Chhing, *c.* +1711.
    Aihsin-chüehlo Hsüan-Yeh (emperor of the Chhing)　愛新覺羅玄燁.

Cf. *BEFEO*, 1903, **3**, 747.

Khang-Hsi Tzu Tien　康熙字典.
    Imperial Dictionary of the Khang-Hsi reign-period.
    Chhing, +1716.
    Ed. Chang Yü-Shu　張玉書.

Khao Kung Chi　考工記.
    The Artificers' Record [a section of the *Chou Li*, q.v.].
    Chou and Han, perhaps originally an official document of Chhi State, incorporated *c.* −140.
    Compiler unknown.
    Tr. E. Biot (1).
    Cf. Kuo Mo-Jo (1); Yang Lien-Shêng (7).

Khao Kung Chi Thu　考工記圖.
    Illustrations for the *Artificers' Record* (of the *Chou Li*) (with a critical archaeological analysis).
    Chhing, +1746.
    Tai Chen　戴震.
    (In *HCCC*, chs. 563, 564; reprinted Shanghai, 1955.)
    See Kondō (1).

Kho Tso Chui Yü　客座贅語.
    My Boring Discourses to my Guests [memorabilia of Nanking].
    Ming, *c.* +1628.
    Ku Chhi-Yuan　顧起元.

Khuei Jih Chhi　揆日器.
    Apparatus for Determining the Sun's Position.
    Chhing, +1675.
    Mei Wên-Ting　梅文鼎.

Khuei Jih Chhien Shuo　揆日淺說.
    Elementary Account of the Sun's Motion.
    Chhing, *c.* +1695.
    Mei Wên-Ting　梅文鼎.

Khuei Jih Chi Yao　揆日紀要.
    Essentials of the Sundial [includes data on magnetic declination].
    Chhing, *c.* +1680.
    Mei Wên-Ting　梅文鼎.

Khung Tzu Chia Yü　孔子家語.
    Table Talk of Confucius.
    H/Han or more probably San Kuo, early +3rd century (but compiled from earlier sources).
    Ed. Wang Su　王肅.
    Partial trs. Kramers (1); A. B. Hutchinson (1); de Harlez (2).

Ko Chih Tshao　格致草.
    Scientific Sketches [astronomy and cosmology; part of *Han Yü Thung*, q.v.].
    Ming, +1620, pr. +1648.
    Hsiung Ming-Yü　熊明遇.

Ko Ku Yao Lun　格古要論.
    Handbook of Archaeology.
    Ming, +1387, enlarged and reissued +1459.
    Tshao Chao　曹昭.

Ko Wu Tshu Than　格物麤談.

Ko Wu Tshu Than (*cont.*)
　　Simple Discourses on the Investigation of
　　　Things.
　　Sung, *c.* +1080.
　　Su Tung-Pho　蘇東坡.
Ku Chhüan Hui.
　　See Li Tso-Hsien (*1*).
Ku Chin Chu　古今註.
　　Commentary on Things Old and New.
　　Chin, *c.* +300.
　　Tshui Pao　崔豹.
　　See des Rotours (*1*), p. xcviii.
Ku Chin Lü Li Khao　古今律曆考.
　　Investigation of the (Chinese) Calendars,
　　　New and Old.
　　Ming, *c.* +1600.
　　Hsing Yün-Lu　邢雲路.
Ku Chin Yo Lu　古今樂錄.
　　Records of Acoustic and Musical Matters,
　　　Old and New.
　　Chhen, *c.* +580.
　　Chih-Chiang　智匠.
Ku Wei Shu　古微書.
　　Old Mysterious Books [a collection of the
　　　apocryphal Chhan-Wei treatises].
　　Date uncertain, in part C/Han.
　　Ed. Sun Chio　孫瑴 (Ming).
Kuan shih Ti Li Chih Mêng　管氏地理指蒙.
　　Master Kuan's Geomantic Instructor.
　　Ascr. San Kuo, +3rd century; prob.
　　　Thang, +8th century.
　　Attrib. Kuan Lo　管輅.
Kuan Tzu　管子.
　　The Book of Master Kuan.
　　Chou and C/Han. Perhaps mainly com-
　　　piled in the Chi-Hsia Academy (late
　　　−4th century) in part from older
　　　materials.
　　Attrib. Kuan Chung　管仲.
　　Partial trs. Haloun (2, 5); Than Po-Fu *et al.*
　　　(*1*).
Kuan Yin Tzu　關尹子.
　　[= *Wên Shih Chen Ching.*]
　　The Book of Master Kuan Yin.
　　Thang, +742 (may be later Thang or Wu
　　　Tai). A work with this title existed in
　　　the Han, but the text is lost.
　　Prob. Thien Thung-Hsiu　田同秀.
Kuang-Chhuan Shu Po　廣川書跋.
　　The Kuang-Chhuan Bibliographical Notes.
　　Sung, *c.* +1125.
　　Tung Yu　董逌.
Kuang Chih　廣志.
　　Extensive Records of Remarkable Things.
　　Chin, +4th.
　　Kuo I-Kung　郭義恭.
　　(*YHSF*, ch. 74.)
Kuang Lun.
　　See Chang Fu-Hsi (*1*).
Kuang Ya　廣雅.
　　Enlargement of the *Erh Ya*; *Literary
　　　Expositor* [dictionary].

San Kuo (Wei), +230.
　　Chang I　張揖.
Kuei-Hsin Tsa Chih　癸辛雜識.
　　Miscellaneous Information from Kuei-
　　　Hsin Street (in Hangchow).
　　Sung, late +13th century, perhaps not
　　　finished before +1308.
　　Chou Mi　周密.
　　See des Rotours (*1*), p. cxii; H. Franke (*14*).
Kuei Ku Tzu　鬼谷子.
　　Book of the Devil Valley Master.
　　Chou, −4th century (perhaps partly Han
　　　or later).
　　Writer unknown; possibly Su Chhin　蘇秦
　　　or some other member of the School of
　　　Politicians (Tsung-Hêng Chia).
Kuei Thien Lu　歸田錄.
　　On Returning Home.
　　Sung, +1067.
　　Ouyang Hsiu　歐陽修.
Kungsun Lung Tzu　公孫龍子.
　　The Book of Master Kungsun Lung (cf.
　　　Shou Pai Lun).
　　Chou, −4th century.
　　Kungsun Lung　公孫龍.
　　Tr. Ku Pao-Ku (*1*); Perleberg (*1*); Mei Yi-
　　　Pao (*3*).
Kuo Yü　國語.
　　Discourses on the (ancient feudal) States.
　　Late Chou, Chhin and C/Han, containing
　　　early material from ancient written
　　　records.
　　Writers unknown.

Lao Hsüeh An Pi Chi　老學庵筆記.
　　Notes from the Hall of Learned Old Age.
　　Sung, *c.* +1190.
　　Lu Yu　陸游.
Lei Kung Phao Chih　雷公炮製.
　　(Handbook based on the) *Venerable Master
　　　Lei's* (*Treatise on*) *the Preparation* (*of
　　　Drugs*).
　　L/Sung, *c.* +470.
　　Lei Hsiao　雷斅.
　　Ed. Chang Kuang-Tou　張光斗 (Chhing,
　　　1871).
Lei Kung Phao Chih Lun　雷公炮炙論.
　　The Venerable Master Lei's Treatise on
　　　the Decoction and Preparation (of Drugs).
　　L/Sung, *c.* +470.
　　Lei Hsiao　雷斅.
　　(Preserved only in quotations in *Chêng Lei
　　　Pên Tshao* and elsewhere, and reconstituted
　　　by Chang Chi (1932); see Lung Po-
　　　Chien (*1*), p. 116.)
Lei Kung Phao Chih Yao Hsing (*Fu*) *Chieh*
　　雷公炮製藥性(賦)解.
　　(Essays and) Studies on the *Venerable
　　　Master Lei's* (*Treatise on*) *the Natures of
　　　Drugs and their Preparation*.
　　First four chapters J/Chin, *c.* +1220.
　　Li Kao　李杲.

Lei Kung Phao Chih Yao Hsing (Fu) Chieh (cont.)
Last six chapters Chhing, *c.* +1650.
Li Chung-Tzu 李中梓.
(Contains many quotations from earlier Lei
Kung books, +5th century onwards.)

Lêng Chai Yeh Hua 冷齋夜話.
Night Talks in a Cool Library.
Sung, *c.* +1110.
Hui-Hung 惠洪.

Li Chi 禮記.
[= *Hsiao Tai Li Chi.*]
Record of Rites [compiled by Tai the
Younger].
(Cf. *Ta Tai Li Chi.*)
Ascr. C/Han, *c.* −70 to −50, but really
H/Han, between +80 and +105, though
the earliest pieces included may date
from the time of the *Analects* (*c.* −465
to −450).
Attrib. ed. Tai Shêng 戴聖.
Actual ed. Tshao Pao 曹褒.
Trs. Legge (7); Couvreur (3); R. Wilhelm
(6).
Yin-Tê Index, no. 27.

Li Chi Chu Su 禮記注疏.
Record of Rites, with assembled
Commentaries.
Text C/Han, commentaries of all periods.
Ed. Juan Yuan (1816) 阮元.

Li Fa Hsin Shu 曆法新書.
New Treatise on Calendar Science.
Ming, *c.* +1590.
Yuan Huang 袁黃.

Li Sao 離騷.
Elegy on Encountering Sorrow [ode].
Chou (Chhu), *c.* −295.
Chhü Yuan 屈原.
Tr. Hawkes (1).

Li Sao Tshao Mu Su 離騷草木疏.
On the Trees and Plants mentioned in the
Elegy on Encountering Sorrow.
Sung, +1197.
Wu Jen-Chieh 吳仁傑.

Li Shu 曆書.
Calendrical Opus.
Ming, *c.* +1601.
Chu Tsai-Yü (prince of the Ming)
朱載堉.

Li Suan Shu Mu.
See *Wu-An Li Suan Shu Mu.*

Li Wei Tou Wei I 禮緯斗威儀.
Apocryphal Treatise on the *Record of Rites*;
System of the Majesty of the Ladle [the
Great Bear].
C/Han, −1st century or later.
Writer unknown.

Liang Chhi Man Chih 梁溪漫志.
Bridge Pool Essays.
Sung, +1192.
Fei Kun 費衮.

Liang Shu 梁書.

History of the Liang Dynasty [+502 to
+556].
Thang, +629.
Yao Chha 姚察 and his son Yao Ssu-
Lien 姚思廉.
For translations of passages see the index of
Frankel (1).

Liang Ssu Kung Chi 梁四公記.
Tales of the Four Lords of Liang.
Thang, *c.* +695.
Chang Yüeh 張說.

Liao Shih 遼史.
History of the Liao (Chhi-tan) Dynasty
[+916 to +1125].
Yuan, *c.* +1350.
Tho-Tho (Toktaga) 脫脫 & Ouyang
Hsüan 歐陽玄.
Partial tr. Wittfogel, Fêng Chia-Shêng *et al.*
Yin-Tê Index, no. 35.

Lieh Tzu 列子.
[= *Chhung Hsü Chen Ching.*]
The Book of Master Lieh.
Chou and C/Han — 5th to −1st centuries.
Ancient fragments of miscellaneous origin
finally cemented together with much new
material about +380.
Attrib. Lieh Yü-Khou 列禦寇.
Tr. R. Wilhelm (4); L. Giles (4); Wieger
(7).
TT/663.

Ling Chhi Ching 靈棋經.
Spirit-Chess Manual.
Prob. Chin.
Attrib. Tungfang Shuo 東方朔.
TT/285.

Ling Chhi Ching 靈棋經.
Spirit-Chess Manual.
Sung.
Yen Yu-Ming 顏幼明.
TT/1029.

Ling Hsien 靈憲.
The Spiritual Constitution (or Mysterious
Organisation) of the Universe [cosmo-
logical and astronomical].
H/Han, +118.
Chang Hêng 張衡.
(In *YHSF*, ch. 76.)

Ling Shu Ching.
See *Huang Ti Nei Ching, Ling Shu.*

Liu Chhing Jih Cha 留青日札.
Diary on Bamboo Tablets.
Ming, +1579.
Thien I-Hêng 田藝衡.

Liu-Chhiu Kuo Chih Lüeh 琉球國志略.
Account of the Liu-Chhiu Islands.
Chhing, +1757.
Chou Huang 周煌.

Liu Tzu 劉子.
The Book of Master Liu.
N/Chhi, *c.* +550.
Prob. Liu Chou 劉晝.
TT/1018.

Lo Ching Chieh 羅經解.
　Analysis of the Magnetic Compass.
　Ming.
　Wu Wang-Kang 吳望崗.
Lo Ching Chieh Ting 羅經解定.
　Definitive Analysis of the Magnetic
　　Compass.
　Chhing, between +1660 and +1720.
　Hu Shen-An 胡慎庵.
Lo Ching Chien I Thu Chieh 羅經簡易圖解.
　Illustrated Easy Explanation of the Magnetic
　　Compass.
　Ming, *c.* +1580.
　Hsü Chih-Mo 徐之鏌.
Lo Ching Chih Nan Pho Wu Chi 羅經指南潑
　霧集.
　A South-Pointer to Disperse the Fog about
　　the Geomantic Compass.
　Ming.
　Wu Thien-Hung 吳天洪.
Lo Ching Ching I Chieh 羅經精一解.
　Analysis of the Essential Features of the
　　Magnetic Compass.
　Chhing, between +1736 and +1795.
　Fan I-Pin 范宜賓.
Lo Ching Pi Chhiao 羅經秘竅.
　Confidential Intelligence about the
　　Magnetic Compass.
　Ming.
　Kan Shih-Wang 甘時望.
Lo Ching Ting Mên Chen 羅經頂門鍼.
　Portal of Highest Knowledge on the
　　Magnetic Compass.
　Ming, *c.* +1580.
　Hsü Chih-Mo 徐之鏌.
Lo-Fou Shan Chih 羅浮山志.
　History and Topography of the Lo-Fou
　　Mountains (north of Canton).
　Chhing, +1716 (but based on older
　　histories).
　Thao Ching-I 陶敬益.
Lü Hsüeh Hsin Shuo 律學新說.
　A New Account of the Science of the
　　Pitch-Pipes.
　Ming, +1584.
　Chu Tsai-Yü (prince of the Ming)
　　朱載堉.
Lü Li Chih 律曆志.
　Memoir on the Calendar.
　H/Han, +178.
　Liu Hung 劉洪 & Tshai Yung 蔡邕.
　(In *Hou Han Shu*, ch. 13.)
Lü Li Yuan Yuan 律曆淵源.
　Ocean of Calendrical and Acoustic
　　Calculations (compiled by Imperial
　　Order) [includes *Li Hsiang Khao Chhêng,
　　Shu Li Ching Yün, Lü Lü Chêng I*, q.v.].
　Chhing, +1723; printing probably not
　　finished before +1730.
　Ed. Mei Ku-Chhêng 梅穀成 & Ho
　　Kuo-Tsung 何國宗.

Cf. Hummel (2), p. 285; Wylie (1),
　pp. 96 ff.
Lü Li Yung Thung 律曆融通.
　The Pitch-Pipes and their Calendrical
　　Concordances [one of the books embodied
　　in the Calendrical Opus, *Li Shu*, q.v.].
　Ming, +1581.
　Chu Tsai-Yü (prince of the Ming) 朱載
　　堉.
Lü Lü Chêng I 律呂正義.
　Collected Basic Principles of Music (com-
　　piled by Imperial Order) [part of *Lü Li
　　Yuan Yuan*, q.v.].
　Chhing, +1713 (+1723).
　Ed. Mei Ku-Chhêng 梅穀成 & Ho Kuo-
　　Tsung 何國宗.
　Cf. Hummel (2), p. 285.
Lü Lü Ching I 律呂精義.
　The Essential Meaning of the Standard
　　Pitch-Pipes [the first part of the Pitch-
　　Pipe Opus, *Lü Shu*, q.v.].
　Ming, +1596.
　Chu Tsai-Yü (prince of the Ming)
　　朱載堉.
Lü Lü Hsin Lun 律呂新論.
　New Discourse on Acoustics and Music.
　Chhing, *c.* +1740.
　Chiang Yung 江永.
Lü Lü Hsin Shu 律呂新書.
　New Treatise on Acoustics and Music.
　Sung, *c.* +1180.
　Tshai Yuan-Ting 蔡元定.
Lü Lü Lun 律呂論.
　A Discourse on the Pitch-Pipes.
　Ming, *c.* +1520.
　Wang Thing-Hsiang 王廷相.
Lu Shih 路史.
　The Peripatetic History [a collection of
　　fabulous and legendary material put
　　together in the style of the dynastic
　　histories, but containing much curious
　　information on techniques].
　Sung.
　Lo Pi 羅泌.
Lü Shih Chhun Chhiu 呂氏春秋.
　Master Lü's Spring and Autumn Annals
　　[compendium of natural philosophy].
　Chou (Chhin), −239.
　Written by the group of scholars gathered
　　by Lü Pu-Wei 呂不韋.
　Tr. R. Wilhelm (3).
　Chung-Fa Index, no. 2.
Lü Shu 律書.
　The Pitch-Pipe Opus [in two parts, of
　　which the first is the *Lü Lü Ching I*, q.v.].
　Ming, +1596.
　Chu Tsai-Yü (prince of the Ming)
　　朱載堉.
Lun Hêng 論衡.
　Discourses Weighed in the Balance.
　H/Han, +82 or +83.
　Wang Chhung 王充.

Lun Hêng (*cont.*)
Tr. Forke (4); cf. Leslie (3).
Chung-Fa Index, no. 1.

Lun Thien 論天.
Discourse on the Heavens.
San Kuo or Chin, *c.* +274.
Liu Chih 劉智.
(In *CSHK* (Chin sect.), ch. 39, pp. 5*a* ff.)

Lun Yü 論語.
Conversations and Discourses (of Confucius)
[perhaps Discussed Sayings, Normative
Sayings, or Selected Sayings]; Analects.
Chou (Lu), *c.* −465 to −450.
Compiled by disciples of Confucius (chs.
16, 17, 18 and 20 are later interpolations).
Tr. Legge (2); Lyall (2); Waley (5); Ku
Hung-Ming (1).
Yin-Tê Index no. (suppl.) 16.

Lung Hu Huan Tan Chüeh 龍虎還丹訣.
Explanation of the Dragon-and-Tiger
Cyclically Transformed Elixir.
Prob. Sung.
Chin Ling Tzu 金陵子.
TT/902.

Mei Tzu Hsin Lun 梅子新論.
New Discourse of Master Mei.
Chin.
Mei shih 梅氏.

Mêng Chhi Pi Than 夢溪筆談.
Dream Pool Essays.
Sung, +1086; last supplement dated +1091.
Shen Kua 沈括.
Ed. Hu Tao-Ching (1); cf. Holzman (1).

Mêng Liang Lu 夢粱錄.
The Past seems a Dream [description of the
capital, Hangchow].
Sung, +1275.
Wu Tzu-Mu 吳自牧.

Mêng Tzu 孟子.
The Book of Master Mêng (Mencius).
Chou, *c.* −290.
Mêng Kho 孟軻.
Tr. Legge (3); Lyall (1).
Yin-Tê Index, no. (suppl.) 17.

Miao Chieh Lu 妙解錄.
See *Yen Mên Kung Miao Chieh Lu.*

Ming I Pieh Lu 名醫別錄.
Informal Records of Famous Physicians.
Liang, *c.* +510.
Thao Hung-Ching 陶宏景.
(Lost as a separate work but constantly
quoted in the pharmaceutical compendia.
Reconstitution by Huang Yü, in *Chhen
Hsiu-Yuan hsien-sêng I Shu.*)

Ming Shih 明史.
History of the Ming Dynasty [+1368 to
+1643].
Chhing, begun +1646, completed +1736,
first pr. +1739.
Chang Thing-Yü 張廷玉 *et al.*

Mo Ching 墨經.
See *Mo Tzu.*

Mo Chuang Man Lu 墨莊漫錄.
Recollections from the Literary Cottage.
Sung, *c.* +1131.
Chang Pang-Chi 張邦基.

Mo Tzu (incl. *Mo Ching*) 墨子.
The Book of Master Mo.
Chou, −4th century.
Mo Ti (and disciples) 墨翟.
Tr. Mei Yi-Pao (1); Forke (3).
Yin-Tê Index, no. (suppl.) 21.
TT/1162.

Mu Thien Tzu Chuan 穆天子傳.
Account of the Travels of the Emperor Mu.
Chou, before −245. (Found in the tomb of
An Li Wang, a prince of the Wei State,
r. −276 to −245; in +281.)
Writer unknown.
Tr. Eitel (1); Chêng Tê-Khun (2).

Nan Chhi Shu 南齊書.
History of the Southern Chhi Dynasty
[+479 to +501].
Liang, +520.
Hsiao Tzu-Hsien 蕭子顯.
For translations of passages see the index of
Frankel (1).

Nan Chou I Wu Chih 南州異物志.
Strange Things of the South.
Chin, +3rd or +4th century.
Wan Chen 萬震.

Nan Hao Shih Hua 南濠詩話.
Essays of the Retired Scholar dwelling by
the Southern Moat [literary criticism].
Ming, +1513.
Tu Mu 都穆.

Nan Shih 南史.
History of the Southern Dynasties [Nan Pei
Chhao period, +420 to +589].
Thang, *c.* +670.
Li Yen-Shou 李延壽.
For translations of passages see the index of
Frankel (1).

Nan Tshun Cho Kêng Lu 南村輟耕錄.
See *Cho Kêng Lu.*

Nei Ching.
See *Huang Ti Nei Ching, Su Wên.*

Nei Chuan Thien Huang Ao Chi Chen Shih Shu
內傳天皇鼇極鎮世書.
Atlas-Tortoise Geomantic Treatise.
Thang, *c.* +730.
Chhiu Yen-Han 邱延翰.

Nêng Kai Chai Man Lu 能改齋漫錄.
Miscellaneous Records of the Nêng-Kai
Studio.
Sung, mid +12th century.
Wu Tshêng 吳曾.

Nittō Guhō Junrei Gyōki 入唐求法巡禮行
記.
Record of a Pilgrimage to China in search
of the (Buddhist) Law.

Nittō Guhō Junrei Gyōki (*cont.*)
Thang, +838 to +847.
Ennin 圓仁.

Pai Chhuan Hsüeh Hai 百川學海.
The Hundred Rivers Sea of Learning [a
collection of separate books; the first
tshung-shu].
Sung, late +12th or early +13th century.
Compiled and edited by Tso Kuei 左圭.

Pai Hai 稗海.
The Sea of Wild Weeds [a *tshung-shu*
collection of 74 books].
Ming.
Compiled and edited by Shang Chün
商濬.

Pai Hu Thung Tê Lun 白虎通德論.
Comprehensive Discussions at the White
Tiger Lodge.
H/Han, *c.* +80.
Pan Ku 班固.
Tr. Tsêng Chu-Sên (1).

Pai Shih Hui Pien 稗史彙編.
Informal History [a collection of classified
quotations].
Ming, *c.* +1590.
Wang Chhi 王圻.

Pai Shih Tao-Jen Shih Chi Ko Chhü 白石道
人詩集歌曲.
Collected Poems and Songs of the White-
Stone Taoist.
Sung, *c.* +1210.
Chiang Khuei 姜夔.

Pao Phu Tzu 抱樸(朴)子.
Book of the Preservation-of-Solidarity
Master.
Chin, early +4th century.
Ko Hung 葛洪.
Partial trs. Feifel (1, 2); Wu & Davis (2);
etc.
TT/1171–1173.

Pei Chhi Shu 北齊書.
History of the Northern Chhi Dynasty
[+550 to +577].
Thang, +640.
Li Tê-Lin 李德林, and his son Li Pai-
Yao 李百藥.
A few chs. tr. Pfizmaier (60).
For translations of passages see the index of
Frankel (1).

Pei Chou Shu 北周書.
See *Chou Shu*.

Pei Mêng So Yen 北夢瑣言.
Dreams of the North and Trifling Talk.
Wu Tai, *c.* +930.
Sun Kuang-Hsien 孫光憲.

Pei Shih 北史.
History of the Northern Dynasties (Nan Pei
Chhao period, +386 to +581].
Thang, *c.* +670.
Li Yen-Shou 李延壽.

For translations of passages see the index of
Frankel (1).

Pên Tshao Kang Mu 本草綱目.
The Great Pharmacopoeia.
Ming, +1596.
Li Shih-Chen 李時珍.
Paraphrased and abridged tr. Read &
collaborators (1–7) and Read & Pak (1)
with indexes.

Pên Tshao Shih I 本草拾遺.
Omissions from Previous Pharmacopoeias.
Thang, *c.* +725.
Chhen Tshang-Chhi 陳藏器.

Pên Tshao Thu Ching 本草圖經.
The Illustrated Pharmacopoeia.
Sung, *c.* +1070 (presented +1062).
Su Sung 蘇頌.
Now contained only as quotations in the
Thu Ching Yen I Pên Tshao (*TT*/761)
and later pharmacopoeias.

Pên Tshao Yen I 本草衍義.
The Meaning of the Pharmacopoeia
Elucidated.
Sung, +1116.
Khou Tsung-Shih 寇宗奭.
Partly contained also in the *Thu Ching Yen
I Pên Tshao* (*TT*/761) and as quotations
in later pharmacopoeias.

Phei Wên Yün Fu 佩文韻府.
Encyclopaedia of Phrases and Allusions
arranged according to Rhyme.
Chhing, +1711.
Ed. Chang Yü-Shu 張玉書 *et al.*

Phi-Pha Fu 琵琶賦.
Rhapsodic Ode on the Phi-Pha Lute.
San Kuo or Chin, *c.* +265.
Fu Hsüan 傅玄.

Phi Ya 埤雅.
A Heap of Elegances.
Sung, +1096.
Lu Tien 陸佃.

Phing-chou Kho Than 萍州可談.
Phingchow Table-Talk.
Sung, +1119 (referring to +1086 onwards).
Chu Yü 朱彧.

Pien Chêng Lun 辨正論.
Discourse on Proper Distinctions.
Thang, *c.* +630.
Fa-Lin 法琳.

Po Wu Chi 博物記.
Notes on the Investigation of Things.
H/Han, *c.* +190.
Thang Mêng (*b*) 唐蒙.

Po Wu Chih 博物志.
Record of the Investigation of Things (cf.
Hsü Po Wu Chih).
Chin, *c.* +290 (begun about +270).
Chang Hua 張華.

Pu Chu Huang Ti Nei Ching Su Wên 補注黃
帝內經素問.

Pu Chu Huang Ti Nei Ching Su Wên (cont.)
The *Yellow Emperor's Manual of Internal Medicine*; *The Plain Questions (and Answers)*; with Commentaries.
Thang, +762.
Wang Ping　王冰.
Re-edited in Sung, c. +1050.
Lin I　林億 *et al.*

Pu-Li Kho Than　歩里客談.
Discussions with Guests at Pu-li.
Sung, c. +1110.
Chhen Chhang-Fang　陳長方.

San Chhin Chi　三秦記.
Record of the Three Princedoms of Chhin [into which that State was divided after the Chhin and before the Han].
Chin.
Writer unknown.

San Chü Hsin Hua　山居新話.
Conversations on Recent Events in the Mountain Retreat.
Yuan, +1360.
Yang Yü　楊瑀.
Tr. H. Franke (2).

San Fu Chiu Shih　三輔舊事.
Stories of the Three Districts in the Capital (Chhang-an, Sian).
Between Chin and Thang.
Writer unknown.

San Fu Huang Thu　三輔皇圖.
Illustrated Description of the Three Districts in the Capital (Chhang-an, Sian).
Chin, late +3rd century, or perhaps H/Han.
Attrib. Miao Chhang-Yen　苗昌言.

San Kuo Chih　三國志.
History of the Three Kingdoms [+220 to +280].
Chin, c. +290.
Chhen Shou　陳壽.
Yin-Tê Index, no. 33.
For translations of passages see the index of Frankel (1).

San Liu Hsüan Tsa Chih　三柳軒雜識.
Three Willows Miscellany.
Sung, c. +1280.
Chhêng Chhi　程棨.

San-Pao Thai-Chien Hsia Hsi-Yang Chi Thung Su Yen I　三寶太監下西洋記通俗演義.
Popular Instructive Story of the Voyages and Traffics of the Three-Jewel Eunuch (Admiral, Chêng Ho), in the Western Oceans [novel].
Ming, +1597.
Lo Mou-Têng　羅懋登.

San Tshai Thu Hui　三才圖會.
Universal Encyclopaedia.
Ming, +1609.
Wang Chhi　王圻.

Shan Hai Ching　山海經.
Classic of the Mountains and Rivers.

Chou and C/Han.
Writers unknown.
Partial tr. de Rosny (1).
Chung-Fa Index, no. 9.

Shan Thang Ssu Khao　山堂肆考.
Books seen in the Mountain Hall Library.
Ming, +1595.
Phêng Ta-I　彭大翼.

Shang Lin Fu　上林賦.
Rhapsodic Ode on the Imperial Hunting Park.
C/Han, c. −130.
Ssuma Hsiang-Ju　司馬相如.

Shang Shu Wei Khao Ling Yao　尚書緯考靈曜.
Apocryphal Treatise on the *Shang Shu* Section of the *Historical Classic*; Investigation of the Mysterious Brightnesses.
C/Han, −1st century.
Writer unknown.
(Now contained in *Ku Wei Shu*, chs. 1 and 2.)

Shansi Thung Chih　山西通志.
Provincial Historical Geography of Shansi.
Chhing, +1733.
Ed. Lo Shih-Lin　羅石麟 *et al.*

Shen Hsien Chuan　神仙傳.
Lives of the Divine Hsien. (Cf. *Lieh Hsien Chuan* and *Hsü Shen Hsien Chuan*.)
Chin, early +4th century.
Attrib. Ko Hung　葛洪.

Shen Nung Pên Tshao Ching　神農本草經.
Pharmacopoeia of the Heavenly Husbandman.
H/Han, c. +2nd century.
Writers unknown.
(Lost as a separate work but formed the basis of all subsequent pharmaceutical compendia, in which it is constantly quoted. Reconstituted and annotated by many scholars; see Lung Po-Chien (1), pp. 2 ff., 12 ff. Best reconstruction by Mori Tateyuki, 1845.)

Shen Tzu　慎子.
The Book of Master Shen.
Date unknown, probably between +2nd and +8th centuries.
Attrib. Shen Tao (Chou philosopher)　慎到.

Shêng Hui Hsüan Fang　聖惠選方.
Selected Imperial Solicitude Prescriptions.
Sung, +1046.
Ho Hsi-Ying　何希影.
(Now lost but the contents was a selection from *Thai-Phing Shêng Hui Fang*, q.v.)

Shêng Shou Wan Nien Li　聖壽萬年曆.
The Imperial Longevity Permanent Calendar.
(One of the books in the Calendrical Opus, *Li Shu*, q.v.)
Ming, +1595.

Shêng Shou Wan Nien Li (*cont.*)
Chu Tsai-Yü (prince of the Ming) 朱載
堉.

Shih Chi 史記.
Historical Records [or perhaps better:
Memoirs of the Historiographer (-Royal);
down to −99].
C/Han, *c.* −90 [first pr. *c.* +1000].
Ssuma Chhien 司馬遷, and his father
Ssuma Than 司馬談.
Partial trs. Chavannes (1); Pfizmaier (13–
36); Hirth (2); Wu Khang (1); Swann (1),
etc.
Yin-Tê Index, no. 40.

Shih Ching 詩經.
Book of Odes [ancient folksongs].
Chou, −9th to −5th centuries.
Writers and compilers unknown.
Tr. Legge (8); Waley (1); Karlgren (14).

Shih I Chi 拾遺記.
Memoirs on Neglected Matters.
Chin, *c.* +370.
Wang Chia 王嘉.
Cf. Eichhorn (5).

Shih Lin Kuang Chi 事林廣記.
Guide through the Forest of Affairs
[encyclopaedia].
Sung, between +1100 and +1250; first pr.
+1325.
Chhen Yuan-Ching 陳元靚.
(A Ming edition of +1478 is in the Cam-
bridge University Library.)

Shih Ming 釋名.
Explanation of Names [dictionary].
H/Han, *c.* +100.
Liu Hsi 劉熙.

Shih Shuo Hsin Yü 世說新語.
New Discourses on the Talk of the Times
[notes of minor incidents from Han to
Chin]. Cf. *Hsü Shih Shuo.*
L/Sung, +5th century.
Liu I-Chhing 劉義慶.
Commentary by Liu Hsün 劉峻 (Liang).

Shih Tzu 尸子.
The Book of Master Shih.
Ascr. Chou, −4th century; probably +3rd
or +4th century.
Attrib. Shih Chiao 尸佼.

Shih Wei Fan Li Shu 詩緯汜歷樞.
Apocryphal Treatise on the [*Book of*] Odes;
the Pivot of the Infinite Calendar.
C/Han or H/Han.
Writer unknown.

Shih Wu Chi Yuan 事物紀原.
Records of the Origins of Affairs and
Things.
Sung, *c.* +1085.
Kao Chhêng 高承.

Shih Wu Kan Chu 事物紺珠.
Valuable Observations on Political Affairs,
etc.

Ming.
Huang I-Chêng 黃一正.

Shih Yu Than Chi 師友談記.
Records of Discussions with my Teachers
and Friends.
Sung, end +11th century.
Li Chih 李廌.

Shou Pai Lun 守白論.
A Treatise in Defence of (the Doctrine of)
Whiteness (and Hardness).
Alternative title for *Kungsun Lung Tzu*
(q.v.).

Shu Ching 書經.
Historical Classic [or, Book of Documents].
The 29 'Chin Wên' chapters mainly Chou
(a few pieces possibly Shang); the 21 'Ku
Wên' chapters a 'forgery' by Mei Tsê
梅賾, *c.* +320, using fragments of
genuine antiquity. Of the former, 13 are
considered to go back to the −10th
century, 10 to the −8th, and 6 not before
the −5th. Some scholars accept only 16
or 17 as pre-Confucian.
Writers unknown.
See Wu Shih-Chhang (1); Creel (4).
Tr. Medhurst (1); Legge (1, 10); Karlgren
(12).

Shu Chü Tzu 叔苴子.
Book of the Hemp-seed Master.
Ming, +15th or +16th century.
Chuang Yuan-Chhen 莊元臣.

Shu Hsü Chih Nan 書敘指南.
The Literary South-Pointer [guide to style
in letter-writing, and to technical terms].
Sung, +1126.
Jen Kuang 任廣.

Shu Pên Tshao 蜀本草.
Szechuan Pharmacopoeia.
Wu Tai (Shu), +925 to +950.
Ed. Han Pao-Shêng 韓保昇.

Shu Shu Chi I 數術記遺.
Memoir on some Traditions of Mathe-
matical Art.
H/Han, +190 (?).
Hsü Yo 徐岳.

Shui Ching 水經.
The Waterways Classic [geographical
account of rivers and canals].
Ascr. C/Han, prob. San Kuo.
Attrib. Sang Chhin 桑欽.

Shui Ching Chu 水經注.
Commentary on the *Waterways Classic*
[geographical account greatly extended].
N/Wei, late +5th or early +6th century.
Li Tao-Yuan 酈道元.

Shun Fêng Hsiang Sung 順風相送.
Fair Winds for Escort [pilot's handbook].
Ming, *c.* +1430.
Author unknown.
(MS., Bodleian Library, Laud Or. no. 145.)

Shuo Fu 說郛.
Florilegium of (Unofficial) Literature.

Shuo Fu (cont.).
Yuan, *c.* +1368.
Thao Tsung-I 陶宗儀.
See Ching Phei-Yuan (1).

Shuo Wên.
See *Shuo Wên Chieh Tzu*.

Shuo Wên Chieh Tzu 說文解字.
Analytical Dictionary of Characters.
H/Han, +121.
Hsü Shen 許慎.

Shuo Yuan 說苑.
Garden of Discourses.
Han, *c.* −20.
Liu Hsiang 劉向.

So-Nan Wên Chi 所南文集.
Collected Writings of (Chêng) So-Nan
(Chêng Ssu-Hsiao).
Yuan, *c.* +1340.
Chêng Ssu-Hsiao 鄭思肖.

Su Shu 素書.
Book of Pure Counsels.
Ascr. Chhin or C/Han.
Attrib. Huang Shih Kung 黃石公
(for Ku Chieh-Kang's opinion on its
authenticity see the Fa-hsing edition,
Chungking, 1945).

Suan Hsüeh Hsin Shuo 算學新說.
A New Account of the Science of Calcula-
tion (in Acoustics and Music).
Ming, +1603.
Chu Tsai-Yü (prince of the Ming) 朱
載堉.

Sui Shou Tsa Lu 隨手雜錄.
Random Reminiscences.
Sung, *c.* +1067, pr. +1163. (Deals with
Wu Tai and Sung memorabilia down to
+1067.) [The third of the three parts of
the *Chhing Hsü Tsa Chu*, q.v.]
Wang Kung 王鞏.

Sui Shu 隋書.
History of the Sui Dynasty [+581 to
+617].
Thang, +636 (annals and biographies);
+656 (monographs and bibliography).
Wei Chêng 魏徵 *et al.*
Partial trs. Pfizmaier (61–65); Balazs (7, 8);
Ware (1).
For translations of passages see the index of
Frankel (1).

Sun Tzu Ping Fa 孫子兵法.
Master Sun's Art of War.
Chou (Chhi), *c.* −345.
Attrib. Sun Wu 孫武, more probably by
Sun Pin 孫臏.

Sung Kao Sêng Chuan 宋高僧傳.
Sung Compilation of Biographies of
Eminent (Buddhist) Monks. (Cf. *Kao
Sêng Chuan* and *Hsü Kao Sêng Chuan*.)
Sung, +988.
Tsan-Ning 贊寧.
TW/2061.

Sung Shih 宋史.
History of the Sung Dynasty [+960 to
+1279].
Yuan, *c.* +1345.
Tho-Tho (Toktaga) 脫脫 & Ouyang
Hsüan 歐陽玄.
Yin-Tê Index, no. 34.

Sung Shu 宋書.
History of the (Liu) Sung Dynasty [+420
to +478].
S/Chhi, +500.
Shen Yo 沈約.
A few chs. tr. Pfizmaier (58).
For translations of passages see the index of
Frankel (1).

Sung Ssu Tzu Chhao Shih 宋四子抄釋.
Selections from the Writings of the Four
Sung (Neo-Confucian) Philosophers
[excl. Chu Hsi].
Sung (ed. Ming, +1536).
Ed. Lü Jan 呂柟.

Ta-Kuan Ching Shih Chêng Lei Pên Tshao
大觀經史證類本草.
Ta-Kuan reign-period Reorganised
Pharmacopoeia.
See *Chêng Lei Pên Tshao*.

Ta Thang Hsi Yü Chi 大唐西域記.
Record of (a Pilgrimage to) the Western
Countries in the time of the Thang.
Thang, +646.
Hsüan-Chuang 玄奘.
Text by Pien-Chi 辯機.
Tr. Julien (1); Beal (2).

Ta Yuan Hai Yün Chi 大元海運記.
Records of Maritime Transportation of the
Yuan Dynasty (orginally part of the *Yuan
Ching Shih Ta Tien*).
Yuan, before +1331.
Compiler unknown.
Ed. Hu Ching (Chhing) 胡敬 (清).

Tai Tsui Phien 代醉編.
On Substitutes for getting Drunk.
Ming.
Chang Ting-Ssu 張鼎思.

Tan Chhi Ching 彈棊經.
Manual of Crossbow-Bullet Chess.
Chin, *c.* +400.
Hsü Kuang 徐廣.

Tan Chhien Tsung Lu 丹鉛總錄.
Red Lead Record.
Ming, +1542.
Yang Shen 楊慎.

Tao Tê Ching 道德經.
Canon of the Tao and its Virtue.
Chou, before −300.
Attrib. Li Erh (Lao Tzu) 李耳 (老子).
Tr. Waley (4); Chhu Ta-Kao (2); Lin Yü-
Thang (1); Wieger (7); Duyvendak (18);
and very many others.

Tao Tsang 道藏.
The Taoist Patrology [containing 1464 Taoist works].
All periods, but first collected in the Thang about +730, then again about +870 and definitively in +1019. First printed in the Sung (+1111 to +1117). Also printed in J/Chin (+1186 to +1191), Yuan (+1244), and Ming (+1445, +1598 and +1607).
Writers numerous.
Indexed by Wieger (6), on which see Pelliot's review; and Ong Tu-Chien (Yin-Tê Index, no. 25).

Thai-Chhing Shih Pi Chi 太清石壁記.
The Records in the Rock Chamber; a Thai-Chhing Scripture.
Liang, early +6th, but includes earlier work as old as the late +3rd century, attributed to Su Yuan-Ming.
Edited by Chhu Tsê hsien-sêng 楚澤先生.
Original writer, Su Yuan-Ming 蘇元明.
TT/874.

Thai Chi Chen-Jen Tsa Tan Yao Fang 太極眞人雜丹藥方.
Tractate of the Supreme-Pole Adept on Miscellaneous Elixir Recipes (with illustrations of alchemical apparatus).
Date unknown, but perhaps Sung on account of the philosophical significance of the pseudonym.
Writer unknown.
TT/939.

Thai-Phing Kuang Chi 太平廣記.
Miscellaneous Records collected in the Thai-Phing reign-period.
Sung, +981.
Ed. Li Fang 李昉.

Thai-Phing Shêng Hui Fang 太平聖惠方.
Prescriptions Collected by Imperial Solicitude in the Thai-Phing reign-period.
Sung, +992.
Ed. Wang Huai-Yin 王懷隱.

Thai-Phing Yü Lan 太平御覽.
Thai-Phing reign-period Imperial Encyclopaedia [lit. the Emperor's Daily Readings].
Sung, +983.
Ed. Li Fang 李昉.
Some chs. tr. Pfizmaier (84–106).
Yin-Tê Index, no. 23.

(Thai-Wan) Tan-shui Thing Chih (臺灣)淡水廳志.
See Chhen Phei-Kuei (1).

Thang Chhüeh Shih 唐闕史.
Forgotten Stories of the Thang Dynasty.
Wu Tai, +10th century.
Kao Yen-Hsiu 高彥休.

Thang Hui Yao 唐會要.

History of the Administrative Statutes of the Thang Dynasty.
Sung, +961.
Wang Phu 王溥.

Thang Liu Tien 唐六典.
Institutions of the Thang Dynasty.
Thang, +738 or +739.
Ed. Li Lin-Fu 李林甫.

Thang Pên Tshao 唐本草.
Pharmacopoeia of the Thang Dynasty.
[= *Hsin Hsiu Pên Tshao*, q.v.]

Thang Yin Pi Shih 棠陰比事.
Parallel Cases from under the Pear-Tree [comparable legal cases solved by eminent judges].
Sung, +1211, pr. +1222, +1234.
Kuei Wan-Jung 桂萬榮.
Tr. von Gulik (6).

Thang Yü Lin 唐語林.
Miscellanea of the Thang Dynasty.
Sung, collected c. +1107.
Wang Tang 王讜.

Thieh Wei Shan Tshung Than 鐵圍山叢談.
Collected Conversations at Iron-Fence Mountain.
Sung, c. +1115.
Tshai Thao 蔡條.

Thien Chi Su Shu 天機素書.
Pure Book of Celestial Mechanics [geomancy].
Thang, c. +730.
Chhiu Yen-Han 邱延翰.

Thien Hsiang Lou Ou Tê 天香樓偶得.
Occasional Discoveries at the Heavenly Fragrance Pavilion.
Chhing, before +1740.
Yü Chao-Lung 虞兆㴖.

Thien Kung Khai Wu 天工開物.
The Exploitation of the Works of Nature.
Ming, +1637.
Sung Ying-Hsing 宋應星.

Thou Hu Hsin Ko 投壺新格.
New Examination of the Pitchpot Game.
Sung, +11th century.
Ssuma Kuang 司馬光.

Thou Hu Pien 投壺變.
Changes and Chances of the Pitchpot Game.
Chin, +4th century.
Yü Than 虞潭.

Thu Ching Pên Tshao 圖經本草.
See *Pên Tshao Thu Ching*.
The name belonged originally to a work prepared in the Thang (c. +658) which by the 11th century had become lost. Su Sung's *Pên Tshao Thu Ching* was prepared as a replacement. The name *Thu Ching Pên Tshao* has often been applied to Su Sung's work, but wrongly.

Thu Ching Yen I Pên Tshao 圖經衍義本草.
The Illustrated and Elucidated Pharmacopoeia.

Thu Ching Yen I Pên Tshao (cont.)
Largely a conflation of the *Pên Tshao Yen I*
and the *Pên Tshao Thu Ching* but with
many additional quotations.
Sung, c. +1120.
Khou Tsung-Shih 寇宗奭.
TT/761.

Thu Shu Chi Chhêng 圖書集成.
Imperial Encyclopaedia.
Chhing, +1726.
Ed. Chhen Mêng-Lei 陳夢雷 *et al.*
Index by L. Giles (2).

Thui Phêng Wu Yü 推蓬寤語.
Talks on Awaking at Sea.
Ming, c. +1570.
Li Yü-Hêng 李豫亨.

Thung Chien Chhien Pien 通鑑前編.
History of Ancient China (down to the
point at which the *Comprehensive Mirror
of History* begins).
J/Chin, c. +1275.
Chin Li-Hsiang 金履祥.

Thung Chih 通志.
Historical Collections.
Sung, c. +1150.
Chêng Chhiao 鄭樵.

Thung Chih Lüeh 通志略.
Compendium of Information [part of
Thung Chih, q.v.].

Thung Hua Lu 同話錄.
Mutual Discussions.
Sung, c. +1189.
Tsêng San-I 曾三異.

Thung Thien Hsiao 通天曉.
Book of General Information [including
techniques, etc.].
Chhing, end +18th century, pr. 1816, 1837,
1856 at Canton.
Wang Hsiang-Thang 王纕堂.

Ti Li Cho Yü Fu 地理琢玉斧.
Precious Tools of Geomancy.
Ming, c. +1570.
Hsü Chih-Mo 徐之鏌.
Transmitted and re-edited +1716 by
Chang Chiu-I 張九義 *et al.*

Ti Li Wu Chüeh 地理五訣.
Five Transmitted Teachings in Geomancy.
Chhing, +1786.
Chao Chiu-Fêng 趙九峯.

Ti Wang Shih Chi 帝王世紀.
Stories of the Ancient Monarchs.
San Kuo or Chin, c. +270.
Huangfu Mi 皇甫謐.

Ti Wei 地緯.
Outlines of Geography [part of *Han Yü
Thung*, q.v.].
Ming, +1624, pr. +1638 and +1648.
Hsiung Jen-Lin 熊人霖.

Tshan Thung Chhi 參同契.
The Kinship of the Three; or, The
Accordance (of the *Book of Changes*) with

the Phenomena of Composite Things
[alchemy].
H/Han, +142.
Wei Po-Yang 魏伯陽.
Comm. by Yin Chhang-Sêng 陰長生.
Tr. Wu & Davis (1).
TT/990.

Tshan Thung Chhi Fa Hui 參同契發揮.
Elucidations of the *Kinship of the Three*
[alchemy].
Yuan, +1284.
Yü Yen 俞琰.
TT/996.

Tshan Thung Chhi Fên Chang Chu Chieh 參同
契分章註解.
The *Kinship of the Three* divided into
Chapters, with Commentary and Analysis.
Yuan, c. +1330.
Chhen Chih-Hsü 陳致虛 (Shang Yang
Tzu) 上陽子.
TTCY, *pên* 93.

Tshan Thung Chhi Khao I 參同契考異.
A Study of the *Kinship of the Three*.
Sung, +1197.
Chu Hsi 朱熹 (originally using pseudonym
Tsou Hsin 鄒訢).
TT/992.

Tshao Man Ku Yo Phu 操縵古樂譜.
Melodies for Harmonious Ancient Music.
Ming, +1606.
Chu Tsai-Yü 朱載堉.

Tshê Fu Yuan Kuei 册府元龜.
Collection of Material on the Lives of
Emperors and Ministers.
Sung, +1013.
Ed. Wang Chhin-Jo 王欽若 & Yang I
楊億.

Tshui Kuan Phien 催官篇.
On Official Promotion [i.e. on how to get it
by geomancy].
Sung, c. +1150.
Lai Wên-Chün 賴文俊.

Tso Chuan 左傳.
Master Tsochhiu's Enlargement of the
Chhun Chhiu (*Spring and Autumn Annals*)
[dealing with the period −722 to −453].
Late Chou, compiled between −430 and
−250, but with additions and changes by
Confucian scholars of the Chhin and
Han, especially Liu Hsin. Greatest of the
three commentaries on the *Chhun Chhiu*,
the others being the *Kungyang Chuan*
and the *Kuliang Chuan*, but unlike them,
probably originally itself an independent
book of history.
Attrib. Tsochhiu Ming 左邱明.
See Karlgren (8); Maspero (1); Chhi Ssu-
Ho (1); Wu Khang (1); Wu Shih-
Chhang (1); Eberhard, Müller &
Henseling.

Tso Chuan (cont.)
Tr. Couvreur (1); Legge (11); Pfizmaier
(1–12).
Index by Fraser & Lockhart (1).

Tu Hsing Tsa Chih 獨醒雜志.
Miscellaneous Records of the Lone Watcher.
Sung, +1176.
Tsêng Min-Hsing 曾敏行.

Tung Ching Fu 東京賦.
Ode on the Eastern Capital (Loyang).
H/Han, +107.
Chang Hêng 張衡.

Tung Ching Mêng Hua Lu 東京夢華錄.
Dreams of the Glories of the Eastern
Capital (Khaifêng).
S/Sung, +1148 (referring to the two decades
which ended with the fall of the capital
of N/Sung in +1126 and the completion
of the move to Hangchow in +1135),
first pr. +1187.
Mêng Yuan-Lao 孟元老.

Tung Hsi Yang Khao 東西洋考.
Studies on the Oceans East and West.
Ming, +1618.
Chang Hsieh 張燮.

Tung Thien Chhing Lu [*Chi*] 洞天清錄 [集].
Clarifications of Strange Things [Taoist].
Sung, c. +1240.
Chao Hsi-Ku 趙希鵠.

Wamyō Ruijushō 和名類聚抄.
General Encyclopaedic Dictionary.
Japan, +934.
Minamoto no Shitagau 源順.

Wan Pi Shu 萬畢書.
See *Huai Nan Wan Pi Shu*.

Wang Chung Wên Kung Chi 王忠文公集.
Collected Writings of Wang I.
Ming, c. +1375.
Wang I 王禕.

Wei Lüeh 魏略.
Memorable Things of the Wei Kingdom
(San Kuo).
San Kuo (Wei) or Chin, c. +264.
Yü Huan 魚豢.

Wei Shu 魏書.
History of the (Northern) Wei Dynasty
[+386 to +550, including the Eastern
Wei successor State].
N/Chhi, +554, revised +572.
Wei Shou 魏收.
See Ware (3).
One ch. tr. Ware (1, 4).
For translations of passages, see the index of
Frankel (1).

Wên Chien Chin Lu 聞見近錄.
New Records of Things Heard and Seen.
Sung, wr. +1085 to +1104, pr. +1163.
(Deals with events from +954 to +1085.)

[First of the three parts of the *Chhing Hsü
Tsa Chu*, q.v.]
Wang Kung 王鞏.

Wên Hsien Thung Khao 文獻通考.
Comprehensive Study of (the History of)
Civilisation.
Sung, begun c. +1254, finished c. +1280, but
not published until +1319.
Ma Tuan-Lin 馬端臨.
A few chs. tr. Julien (2); St Denys (1).

Wên Hsin Tiao Lung 文心雕龍.
On the Carving of the Dragon of Litera-
ture; or, The Anatomy of the Literary
Mind [the earliest book on literary
criticism].
Liang, +6th century.
Liu Hsieh 劉勰.

Wên Hsüan 文選.
General Anthology of Prose and Verse.
Liang, +530.
Ed. Hsiao Thung (prince of the Liang)
蕭統.
Tr. von Zach (6).

Wên Shih Chen Ching 文始眞經.
True Classic of the Original Word (of Lao
Chün, third person of the Taoist Trinity)
(= *Kuan Yin Tzu*, q.v.).

Wu-An Li Suan Shu Mu 勿菴曆算書目.
Bibliography of Mei Wên-Ting's (Wu-An's)
Mathematical and Astronomical Writings.
Chhing, +1702.
Mei Wên-Ting 梅文鼎.

Wu Chen Phien 悟眞篇.
Poetical Essay on the Understanding of the
Truth [alchemy, both spiritual and
practical].
Sung, +1075.
Chang Po-Tuan 張伯端.
Tr. Davis & Chao Yün-Tshung (7).
TT/138.

*Wu Chen Phien Chih Chih Hsiang Shuo San
Chhêng Pi Yao* 悟眞篇直指祥說三
乘祕要.
Precise Explanation of the Difficult
Essentials of the *Poetical Essay on the
Understanding of the Truth* according to
the Three Scriptures.
Sung.
Attrib. Ong Pao-Kuang 翁葆光.
TT/140.

Wu Ching Tsung Yao 武經總要.
Collection of the most important Military
Techniques (compiled by Imperial
Order).
Sung, +1040 (+1044).
Ed. Tsêng Kung-Liang 曾公亮.

Wu Lei Hsiang Kan Chih 物類相感志.
On the Mutual Responses of Things
according to their Categories.
Sung, c. +1080.
Su Tung-Pho 蘇東坡.

Wu Li Hsiao Shih 物理小識.
　Small Encyclopaedia of the Principles of
　　Things.
　Chhing, +1664.
　Fang I-Chih 方以智.
　Cf. Hou Wai-Lu (3, 4).

Wu Lin Chiu Shih 武林舊事.
　Institutions and Customs of the Old Capital
　　(Hangchow).
　Sung, c. +1270 (but referring to events
　　from about +1165 onwards).
　Chou Mi 周密.

Wu Lu 吳錄.
　Record of the Kingdom of Wu.
　San Kuo, +3rd century.
　Chang Pho 張勃.

Wu Lu Ti Li Chih 吳錄地理志.
　[= *Wu Lu*, q.v.]

Wu Tsa Tsu 五雜俎.
　Five Assorted Offering-Trays [miscellaneous
　　memorabilia in five sections].
　Ming, c. +1600.
　Hsieh Tsai-Hang 謝在杭.

Wu Tu Fu 吳都賦.
　Rhapsodic Ode on the Capital of Wu
　　(Kingdom).
　San Kuo, c. +260.
　Tso Ssu 左思.

Yang-chhêng Ku Chhao.
　See Chhou Chhih-Shih (1).

Yeh Chung Chi 鄴中記.
　Record of Affairs at the Capital of the
　　Later Chao Dynasty.
　Chin.
　Lu Hui 陸翽.

Yeh Kho Tshung Shu 野客叢書.
　Collected Notes of the Rustic Guest.
　Sung, +1201.
　Wang Mou 王楙.

Yen-ching Sui Shih Chi.
　See Tun Li-Chhen (1).

Yen-Chou Ssu Pu 弇州四部.
　Talks at Yenchow on the Four Branches of
　　Literature.
　Ming.
　Wang Shih-Chên 王世貞.

Yen Fan Lu 演繁露.
　Extension of the *String of Pearls on the
　　Spring and Autumn Annals* [on the mean-
　　ing of many Thang and Sung expressions].
　Sung, +1180.
　Chhêng Ta-Chhang 程大昌.
　See des Rotours (1), p. cix.

Yen Mên Kung Miao Chieh Lu 鴈門公妙解錄.
　The Venerable Yen Mên's Explanations of
　　the Mysteries [alchemy and elixir
　　poisoning].
　Thang, probably in the neighbourhood of
　　+855 since the text is substantially

identical with the *Hsüan Chieh Lu* (q.v.) of
　this date.
　Yen Mên 鴈門.
　TT/937.

Yen shih Chia Hsün 顏氏家訓.
　Mr Yen's Advice to his Family.
　Sui, c. +590.
　Yen Chih-Thui 顏之推.

Yen Thieh Lun 鹽鐵論.
　Discourses on Salt and Iron [record of the
　　debate of −81 on State control of
　　commerce and industry].
　C/Han, c. −80.
　Huan Khuan 桓寬.
　Partial tr. Gale (1); Gale, Boodberg & Lin.

Yin Yang Erh Chai Chhüan Shu 陰陽二宅全
　書.
　Complete Treatise on Siting in relation to
　　the Two Geodic Currents [geomancy].
　Chhing, +1744.
　Yao Shan-Chhi 姚瞻旂.

Yo Chi 樂記.
　Record of Ritual Music and Dance.
　Late Chou.
　Writer unknown.
　(Incorporated into *Li Chi*, chs. 18 and 19;
　　and into *Shih Chi*, ch. 24. Also in
　　YHSF, ch. 30.)

Yo Chi 樂記.
　Music Record.
　Han, −1st century.
　Liu Hsiang 劉向.
　(*YHSF*, ch. 30.)

Yo Ching 樂經.
　Music Classic.
　Chou.
　Writer unknown.
　(Only a fragment left in *YHSF*, ch. 30.)

Yo Fu Tsa Lu 樂府雜錄.
　Miscellaneous Notes on the Bureau of
　　Music.
　Thang or Wu Tai, +10th century.
　Tuan An-Chieh 段安節.

Yo Lü Chhüan Shu 樂律全書.
　Collected Works on Music and the Pitch-
　　Pipes. [Contains *Lü Hsüeh Hsin Shuo*,
　　Lü Lü Ching I, and *Suan Hsüeh Hsin
　　Shuo*, q.v.]
　Ming, c. +1620.
　Chu Tsai-Yü (prince of the Ming)
　　朱載堉.

Yo Lü I 樂律義.
　The Basic Idea of the Acoustic Pipes.
　N/Chou, c. +570.
　Shen Chung 沈重.
　(In *YHSF*, ch. 31.)

Yo Shu 樂書.
　Book of Acoustics and Music.
　N/Wei, c. +525 or E/Wei, c. +540.
　Hsintu Fang 信都芳.
　(In *YHSF*, ch. 31, pp. 19a ff.)

Yo Shu 樂書.
Treatise on Acoustics and Music.
Sung, +1101.
Chhen Yang 陳暘.

Yo Shu Chu Thu Fa 樂書註圖法.
Commentary and Illustrations for the *Book of Acoustics and Music*.
N/Wei, *c.* +525 or E/Wei, *c.* +540.
Hsintu Fang 信都芳.
(Partially preserved in *Yo Shu Yao Lu*, ch. 6, pp. 18*a* ff.)

Yo Shu Yao Lu 樂書要錄.
Record of the Essentials in the Books on Music (and Acoustics).
Thang, *c.* +670.
Wu Huang Hou 武皇后 (the Empress Wu, later known as Wu Tsê Thien), probably written while emperor Kao Tsung was still reigning.
Incomplete, and only preserved because copied by Kibi no Makibi 吉備眞備 between +716 and +735.

Yü-Chih Lü Lü Chêng I 御製律呂正義.
See *Lü Lü Chêng I*.

Yu Huan Chi Wên 游宦紀聞.
Things Seen and Heard on my Official Travels.
Sung, *c.* +1230.
Chang Shih-Nan 張世南.

Yu Kuai Lu 幽怪錄.
Record of Things Dark and Strange [or, Mysteries and Monsters].
Thang, end +8th century.
Niu Sêng-Ju 牛僧孺 or Wang Yün 王惲.

Yu Li Tzu 郁離子.
The Book of Master Yu Li.
Yuan, *c.* +1360.
Liu Chi 劉基.

Yü Lieh Fu 羽獵賦.
The Imperial Hunt [ode].
H/Han, +124.
Chang Hêng 張衡.

Yü Tung Hsü Lu 餘冬序錄.
Late Winter Talks.
Ming.
Ho Mêng-Chhun 何孟春.

Yu-Yang Tsa Tsu 酉陽雜俎.
Miscellany of the Yu-yang Mountain (Cave) [in S.E. Szechuan].
Thang, +863.
Tuan Chhêng-Shih 段成式.
See des Rotours (1), p. civ.

Yuan Chien Lei Han 淵鑑類函.
The Deep Mirror of Classified Knowledge [literary encyclopaedia; a conflation of Thang encyclopaedias].
Chhing, +1710.
Ed. Chang Ying 張英 *et al.*

Yuan Ching Shuo 遠鏡說.
The Far-Seeing Optick Glass [account of the telescope].
Ming, +1626.
Thang Jo-Wang (J. A. Schall von Bell) 湯若望.

Yuan Hai Yün Chih 元海運志.
A Sketch of Maritime Transportation during the Yuan Period.
Yuan or Ming, late +14th century.
Wei Su 危素.

Yuan-Ho Chün Hsien Thu Chih 元和郡縣圖志.
Yuan-Ho reign-period General Geography.
Thang, +814.
Li Chi-Fu 李吉甫.

Yuan Shih 元史.
History of the Yuan (Mongol) Dynasty [+1206 to +1367].
Ming, *c.* +1370.
Sung Lien 宋濂 *et al.*
Yin-Tê Index, no. 35.

Yüeh Ling 月令.
Monthly Ordinances (of the Chou Dynasty).
Chou, between −7th and −3rd centuries.
Writers unknown.
Incorporated in the *Hsiao Tai Li Chi* and the *Lü Shih Chhun Chhiu*, q.v.
Tr. Legge (7); R. Wilhelm (3).

Yüeh Ling Chang Chü 月令章句.
Commentary on the *Monthly Ordinances*.
H/Han, *c.* +175.
Tshai Yung 蔡邕.

Yün Chi 韻集.
Rhyme Dictionary.
Chin, +4th century.
Lü Ching 呂靜.

Yün Chi Chhi Chhien 雲笈七籤.
The Seven Bamboo Tablets of the Cloudy Satchel [an important collection of Taoist material made by the editor of the first definitive form of the *Tao Tsang* (+1019), and including much material which is not in the Patrology as we now have it].
Sung, *c.* +1025.
Chang Chün-Fang 張君房.
TT/1020.

Yün Lin Shih Phu 雲林石譜.
Cloud Forest Lapidary.
Sung, +1133.
Tu Wan 杜綰.

Yung Chhêng Shih Hua 榕城詩話.
Plantain City (Fuchow) Essays [literary criticism].
Chhing, +1732.
Hang Shih-Chün 杭世駿.

B. CHINESE AND JAPANESE BOOKS AND JOURNAL ARTICLES SINCE +1800

Anon. (7).
Shantung I-nan Han Hua Hsiang Shih Mo
山東沂南漢畫像石墓.
On the [recently excavated] Han Tomb with
Sculptured Reliefs at I-nan in Shantung.
WWTK, 1954, no. 8 (no. 48), 35.

Anon. (10).
Tunhuang Pi Hua Chi 敦煌壁畫集.
Album of Coloured Reproductions of the
fresco-paintings at the Tunhuang cave-
temples.
Peking, 1957.

Anon. (11).
Chhangsha Fa Chüeh Pao-Kao 長沙發掘
報告.
Report on the Excavations (of Tombs of
the Chhu State, of the Warring States
period; and of the Han Dynasties) at
Chhangsha.
Acad. Sinica Archaeol. Inst., Kho-Hsüeh,
Peking, 1957.

Anon. (17).
Shou-hsien Tshai Hou Mo Chhu Thu I Wu
壽縣蔡侯墓出土遺物.
Objects Excavated from the Tomb of the
Duke of Tshai at Shou-hsien.
Acad. Sinica Archaeol. Inst., Peking, 1956.

Aoji Rinsō (1) 青地林宗.
Kikai Kanran 氣海觀瀾.
A Survey of the Ocean of Pneuma
[astronomy and meteorological physics].
Japan, 1825; enlarged in 1851.
In *NKKZ*, vol. 6.

Chan Chien-Fêng (1) 詹劍峰.
Mo Chia ti Hsing-Shih Lo-Chi 墨家的形
式邏輯.
The Formal Logic of the Mohists.
Hupei Jen-Min, Wuhan, 1956, 1957.

Chang Chhi-Yün (1) (ed.) 張其昀.
*Hsü Hsia-Kho hsien-sêng Shih Shih San Pei
Chou Nien Chi Nien Khan* 徐霞客先
生逝世三百週年紀念刊.
Essays in Commemoration of the 300th
Anniversary of the Death of Hsü Hsia-
Kho (+1586 to +1641) [11 contributors].
Chekiang Univ. Res. Inst. Hist. and
Geogr. Publications, no. 4: Tsunyi, 1942.

Chang Fu-Hsi (1) 張福僖.
Kuang Lun 光論.
Discourse on Optics.
c. 1840.

Chang Hung-Chao (1) 章鴻釗.
Shih Ya 石雅.
Lapidarium Sinicum; a Study of the Rocks,
Fossils and Minerals as known in Chinese
Literature.
Chinese Geol. Survey, Peiping: 1st ed.
1921, 2nd ed. 1927.
MGSC (ser. B), no. 2. (With Engl. summary.)
Crit. P. Demiéville, *BEFEO*, 1924, **24**, 276.

Chang Li-Chhien (1) 張禮千.
'*Tung Hsi Yang Khao*' *chung chih Chen Lu*
東西洋考中之針路.
The Compass-Bearings in *Studies on the
Oceans, East and West*.
TFTC, 1945, **41** (no. 1), 49.

Chang Thai-Yen (1) 張太炎.
Chih Nan Chen Khao 指南鍼考.
A Study of the South-Pointing Needle.
HKM, **1** (no. 5).

Chang Yin-Lin (2) 張蔭麟.
Chung-Kuo Li-Shih shang chih '*Chhi Chhi*'
chi chhi Tso-chê 中國歷史上之「奇器」
及其作者.
Scientific Inventions and Inventors in
Chinese History.
YCHP, 1928, **1** (no. 3), 359.

Chêng Fu-Kuang (1) 鄭復光.
Ching Ching Ling Chhih 鏡鏡詅癡.
Treatise on Optics by an Untalented
Scholar.
Contains as appendix *Huo Lun Chhuan Thu
Shuo* 火輪船圖說 On Steam Paddle-
boat Machinery, with Illustrations.
c. 1835.
Pref. 1846, pr. 1847.

Chin Ching-Chen (1) 今井楳.
Chung-Kuo Tzhu Chen Shih Lüeh 中國磁
針史略.
Materials on the History of the Lodestone
and Magnet-Needle in China.
SSIP/P, 1942, **12**, 147.

[Chhang Shu-Hung] (1) (ed.) 常書鴻.
Tunhuang Mo-kao-khu 敦煌莫高窟.
(The Cave-Temples at) Mo-kao-khu
[Chhien-fo-tung] near Tunhuang.
Kansu Jen-min, Lanchow, 1957.

Chhen Phei-Kuei (1) 陳培桂 *et al.*
(*Thaiwan*) *Tan-shui Thing Chih* (臺灣)淡
水廳志.
Local Gazetteer of Tan-shui Thing in
Formosa.
1871.

Chhen Wei-Chhi 陳維祺, with Yeh Yao-Yuan 葉燿元, Sun Pin-I 孫斌翼 *et al.* (*1*).
Chung-Hsi Suan-Hsüeh Ta Chhêng 中西算學大成.
Compendium of Chinese and Western Mathematics [and Physics].
1889.

Chhen Wên-Thao (*1*) 陳文濤.
Hsien Chhin Tzu-Jan Hsüeh Kai Lun 先秦自然學概論.
History of Science in China during the Chou and Chhin periods.
Com. Press, Shanghai, 1934.

Chhêng Su-Lo (*1*) 程溯洛.
Chung-Kuo Ku-Tai Chih Nan Chen ti Fa-Ming chi chhi yü Hang Hai ti Kuan-Hsi 中國古代指南針的發明及其與航海的關係.
The Discovery of the Magnetic Compass in Ancient China and its Relation to Navigation.
Essay in Li Kuang-Pi & Chhien Chün-Hua (q.v.), p. 21.
Peking, 1955.

Chhêng Yao-Thien (*2*) 程瑤田.
Khao Kung Chhuang Wu Hsiao Chi 考工創物小記.
Brief Notes on the Specifications (for the Manufacture of Objects) in the *Artificers' Record* (of the *Chou Li*).
Peking, *c.* 1805.
In *HCCC*, chs. 536–539.

Chhêng Yao-Thien (*3*) 程瑤田.
Chiu Ku Khao 九穀考.
A Study of the Nine Grains (monograph on the history of cereal agriculture).
Peking, *c.* 1805.
In *HCCC*, ch. 548.

Chhi Ssu-Ho (*1*) 齊思和.
Huang Ti chih Chih Chhi Ku Shih 黃帝之制器故事.
Stories of the Inventions of Huang Ti [and his Ministers].
YAHS, 1934, **2** (no. 1), 21.

Chhien Chün-Thao (*1*) 錢君匋.
Chung-Kuo Yin-Yo Shih Tshan-Khao Thu Phien 中國音樂史參考圖片.
Album of Photographs to illustrate the History of Chinese Music and Musical Instruments.
Ed. Central School of Music Research Institute, New Music Pub. Co. Shanghai, 1954.

Chhien Lin-Chao (*1*) 錢臨照.
Shih 'Mo Ching' chung Kuang-hsüeh Li-hsüeh chu Thiao 釋墨經中光學力學諸條.
Expositions of the Optics and Mechanics in the Mohist Canons.
In *Studies presented to Mr Li Shih-Tsêng on his 60th Birthday* 李石曾先生六十歲紀念論文集.
Nat. Peiping Academy, Kunming, 1940.

Chhien Lin-Chao (*2*) 錢臨照.
Yang Sui 陽燧.
On Burning-Mirrors (description of three Sung specimens).
WWTK, 1958 (no. 7), 28.

Chhien Wên-Hsüan (*1*) 錢文選.
Chhien shih so Tshang Khan-Yü Shu Thi Yao 錢氏所藏堪輿書提要.
Descriptive Catalogue of the Geomantic Books collected by Mr Chhien.
Peking.
Cit. Wang Chen-To (*5*), p. 121.

Fêng Yün-Phêng 馮雲鵬 & Fêng Yün-Yuan 馮雲鵷.
Chin Shih So 金石索.
Collection of Carvings, Reliefs and Inscriptions.
(This was the first modern publication of the Han tomb-shrine reliefs.)
1821.

Harada Yoshito 原田淑人 & Tazawa Kingo 田澤金吾 (*1*).
Rakurō Gokan-en Ō Ku no Fumbo 樂浪五官掾王盱の墳墓.
Lo-lang; a Report on the Excavation of Wang Hsü's Tomb in Lo-lang Province (an ancient Chinese Colony in Korea).
Tokyo Univ., Tokyo, 1930.

Hashimoto Donsai (*1*) 橋本曇齋.
Erekiteru Kyūrigen エレキテル究理原.
A Study of the Basic Principles of Electricity.
Cf. *Erekiteru Yakusetsu*.
Japan, 1811.
In *NKKZ*, vol. 6.

Hashimoto Masukichi (*3*) 橋本增吉.
Shinan Sha Kō 指南車考.
An Investigation of the South-Pointing Carriage.
TYG, 1918, **8**, 249, 325; 1924, **14**, 412; 1925, **15**, 219.

Hayashi Kenzō (*1*) 林謙三.
Sui Thang Yen Yo Tiao Yen-Chiu 隋唐燕樂調研究.
Researches on the Music of the Sui and Thang Periods [and the Foreign Influences thereon].
Tr. Kuo Mo-Jo 郭沫若.
Com. Press Shanghai, 1936; 2nd ed. 1955.

Hiraoka Teikichi (*1*) 平岡禎吉.
Ki no Shisō Seiritsu tsuite ni 「氣」の思想成立について.
On the Development of the Concept of *chhi* (*pneuma*).
SGK, 1955, **13**, 34.

Hou Wai-Lu (*3*) 侯外廬.
Fang I-Chih—Chung-Kuo ti Pai Kho Chhüan Shu Phai Ta Chê-Hsüeh Chia 方以智——中國的百科全書派大哲學家.

Hou Wai-Lu (*cont.*)
 Fang I-Chih—China's Great Encyclopaedist
   Philosopher.
 LSYC, 1957 (no. 6), 1; 1957 (no. 7), 1.
Hou Wai-Lu (*4*) 侯外廬 .
 *Shih-liu Shih-Chi Chung-Kuo ti Chin-Pu ti
   Chê-Hsüeh Ssu-Chhao Kai-Shu* 十六世
 紀中國的進步的哲學思潮概述 .
 Progressive Philosophical Thinking in
   +16th-century China.
 LSYC, 1959 (no. 10), 39.
Hou Wai-Lu 侯外廬 , Chao Chi-Pin 趙紀彬 ,
   Tu Kuo-Hsiang 杜國庠 & Chhiu Han-
   Sêng 邱漢生 (*1*).
 Chung-Kuo Ssu-Hsiang Thung Shih 中國
 思想通史 .
 General History of Chinese Thought.
 5 vols.
 Jen-min, Peking, 1957.
Hu Tao-Ching (*1*) 胡道靜 .
 '*Mêng Chhi Pi Than*' *Chiao Chêng* 夢溪
 筆談校證 .
 Complete Annotated and Collated Edition
   of the *Dream Pool Essays* (of Shen Kua,
   +1086).
 2 vols.
 Shanghai Pub. Co., Shanghai, 1956.
 Crit. rev. Nguyen Tran-Huan, *RHS*, 1957,
   10, 182.
Hsiang Ta (*2*) 向達 .
 Shuo Shih 説式 .
 On the Diviner's Board (ancestor of the
   magnetic compass).
 TKP/WS, 1947, no. 30.
Hsü Chia-Chen (*1*) 徐家珍 .
 Fêng Chêng Hsiao Chi 風箏小記 .
 A Note on Aeolian Whistles (attached to
   kites).
 WWTK, 1959 (no. 2), 27.
Hsü Chung-Shu (*6*) 徐中舒 .
 Ching-Thien Chih Tu Than Yuan 井田制
 度探源 .
 A Study of the Origin of the 'Well-Field'
   (Land) System.
 BCS, 1944, **4** (no. 1), 121.
 Tr. Sun Zen E-Tu & de Francis (1), p. 3.

Jung Kêng (*1*) 容庚 .
 Han Wu Liang Tzhu Hua Hsiang Khao Shih
 漢武梁祠畫像考釋 .
 Investigations on the Carved Reliefs of the
   Wu Liang Tomb-shrines of the [Later]
   Han Dynasty.
 2 vols.
 Yenching Univ. Archaeol. Soc.
   Peking, 1936.

Kamata Ryūkō (*1*) 鎌田柳泓 .
 Rigaku Hiketsu 理學秘訣 .
 Elements (lit. Mysteries) of Physics.
 Japan, 1815.
 In *NKKZ*, vol. 6.

Koizumi Akio (*1*) 小泉顯夫 .
 Rakurō Saikyōzuka 樂浪彩篋冢 .
 The Tomb of the Painted Basket, [and two
   other tombs] of Lo-lang.
 Koseki Chōsa Hōkoku (Archaeol. Res. Rep.)
   no. 1.
 Soc. Stud. Korean Antiq., Seoul, 1934.
 (With Engl. summary.)
Kondō Mitsuo (*1*) 近藤光男 .
 Tai Shin no Kōkōkizu ni tsuite 戴震の考
 工記圖について .
 On Tai Chen and his *Khao Kung Chi Thu*.
 TG/T, 1955, **11**, 1; abstr. *RBS*, 1955, **1**,
   no. 452.
Kuroda Genji (*1*) 黑田源次 .
 Ki 氣 .
 On *chhi* (*pneuma*) [psychological and medico-
   biological conceptions].
 TS, 1953 (no. 4–5), 1; 1955 (no. 7), 16.
Kuwaki Ayao (*1*) 桑木彧雄 .
 *Jishaku oyobi Kohaku ni Kansuru Tōyō
   Kagaku Zasshi* 磁石及琥珀に關す
 る東洋科學雜史 .
 Magnet and Amber in Oriental Science.
 SS, 1935, 169.

Li Hui (*1*) 李卉 .
 *Thaiwan chi Tung-Ya Ko Ti-Thu Chu Min-
   Tsu-ti Khou-Chhin chih Pi-Chiao Yen-
   Chiu* 臺灣及東亞各地土著民族的
 口琴之比較研究 .
 A Comparative Study of the 'Jew's Harp'
   among the Aborigines of Formosa and
   East Asia.
 AS/BIE 1956, **1**, 85.
Li Nien (*21*) 李儼 .
 Chung Suan Shih Lun Tshung (second series)
 中算史論叢 .
 Collected Essays on the History of Chinese
   Mathematics—vol. 1, 1954; vol. 2, 1954;
   vol. 3, 1955; vol. 4, 1955; vol. 5, 1955.
 Kho-hsüeh, Peking.
Li Shu-Hua (*1*) 李書華 .
 Chih Nan Chen ti Chhi-Yuan 指南針的
 起源 .
 The Origins of the Magnetic Compass.
 Ta-Lu Tsa Chih Pub. Office, Thaipei,
   Thaiwan, 1954.
Li Shun-I (*1*) 李純一 .
 Chung-Kuo Ku-Tai Yin-Yo Shih Kao 中
 國古代音樂史稿 .
 Sketch of a History of Ancient Chinese
   Music.
 Yin-yo, Peking, 1958.
Li Shun-I (*2*) 李純一 .
 Kuan-yü Yin Chung ti Yen-Chiu 關於殷
 鐘的研究 .
 A Study of Shang-Yin Bells.
 AS/CJA, 1957 (no. 3), 41.
Li Tso-Hsien (*1*) 李佐賢 .
 Ku Chhüan Hui 古泉匯 .

Li Tso-Hsien (*cont.*)
Treatise on (Chinese) Numismatics.
1859.

Liu Fu (2) 劉復.
Tshung Wu Yin Liu Lü Shuo tao San-pai-liu-shih Lü 從五音六律說到三百六十律.
The Development of the Chinese Musical Scale from the 5 Sounds and the 6 Tones to 360 Tones.
FJHC, 1930, **2**, 1.

Liu Fu (3) 劉復.
Shih-erh Têng Lü ti Fa-Ming-chê, Chu Tsai-Yü 十二等律的發明者朱載堉.
Chu Tsai-Yü, Inventor of the Chromatic Scale of Equal Temperament.
In *Studies presented to Mr Tshai Yuan-Phei on his 65th Birthday* 慶祝蔡元培先生六十五歲論文集.
2 vols.
Acad. Sin., Peiping, 1933, 1935, p. 279.
Abstr. *CIB*, 1940, **4**, 78.

Liu Hsien-Chou (1) 劉仙洲.
Chung-Kuo Chi-Hsieh Kung-Chhêng Shih-Liao 中國機械工程史料.
Materials for the History of Engineering in China.
CHESJ, 1935, **3**, and **4** (no. 2), 27. Reprinted Chhinghua Univ. Press, Peiping, 1935. With supplement.
CHER, 1948, **3**, 135.

Liu Ming-Shu (2) 劉名恕.
Chêng Ho Hang Hai Shih Chi chih Tsai Than 鄭和航海事蹟之再探.
Further Investigations on the Sea Voyages of Chêng Ho.
BCS, 1943, **3**, 131.
Crit. ref. by A. Rygalov, *HH*, 1949, **2**, 425.

Liu Wên-Tien (1) 劉文典.
Chuang Tzu Pu Chêng 莊子補正.
Emended Text of *Chuang Tzu*.
Com. Press, Shanghai, 1947.

Lo Chen-Yü (2) 羅振玉.
Ku Ching Thu Lu 古鏡圖錄.
Illustrated Discussion of Ancient Mirrors.
1916.

Luan Tiao-Fu (1) 樂調甫.
Mo Tzu Yen-Chiu Lun Wên Chi 墨子研究論文集.
Collected Essays on Mohist Researches.
Jen-min, Peking, 1957.

Lung Po-Chien (1) 龍伯堅.
Hsien Tshun Pên Tshao Shu Lu 現存本草書錄.
Bibliographical Study of Extant Pharmacopoeias (from all periods).
Jen-min Wei-sêng, Peking, 1957.

Ma Hêng (1) 馬衡.
'*Sui Shu (Lü Li Chih)*' *Shih-wu Têng Chhih* 隋書律曆志十五等尺.
The Fifteen different Classes of Measures as given in the Memoir on Acoustics and Calendar (by Li Shun-Fêng) in the *History of the Sui Dynasty*.
Pr. pr. Peiping, 1932, with translation by J. C. Ferguson (see Ma Hêng, 1).

Ma Kuo-Han (1) (ed.) 馬國翰.
Yü Han Shan Fang Chi I Shu 玉函山房輯佚書.
The Jade-Box Mountain Studio Collection of (Reconstituted) Lost Books.
1853.

Ōhashi Totsuan (1) 大橋訥菴.
Hekija Shōgen 闢邪小言.
False Science Exposed (lit. Insignificant Words Exposing Error).
c. 1854.
In *Meiji Bunka Zenshū*, vol. 15.
Tokyo, 1929.

Phan Chieh-Tzu (1) 潘絜茲.
Tunhuang Mo-kao-khu I-Shu 敦煌莫高窟藝術.
The Art of the Cave-Temples at Mo-kao-khu [Chhien-fo-tung] near Tunhuang.
Jen-min, Shanghui, 1957.

Sekino Tadashi 關野貞, Yatsui Sai-ichi 谷井濟一, Kuriyama Shun-ichi 栗山俊一, Oha Tsunekichi 小場恒吉, Ogawa Keikichi 小川敬吉 & Nomori Takeshi 野守健 (1).
Rakurō-gun Jidai no Iseki 樂浪郡時代の遺蹟.
Archaeological Researches on the Ancient Lo-lang District (Korea); 1 vol. text, 1 vol. plates.
Sp. Rep. Serv. Antiq. Govt. Gen., Chosen, 1925, no. 4, and 1927, no. 8.

Shan Chhing-Lin (1) 單慶麟.
Thungchow Hsin-Chhu-Thu Fo-Ting-Tsun Shêng Tho-Lo-Ni Chhuang chih Yen-Chiu 通州新出土佛頂尊勝陀羅尼幢之研究.
A Study of a Stone Stele (of the +11th century) depicting the Victorious Dhāraṇī-Buddha recently recovered at Thungchow.
AS/CJA, 1957 (no. 4), 107.

Shan Shih-Yuan (1) 單士元.
Kung Têng 宮燈.
On Ornamental Lanterns.
WWTK, 1959 (no. 2), 22.

Shiba Kentarō (1) 柴謙太郎.
Jūkara no Ichi ni tsuite Gojin no Kenkai 重迦羅の位置に就いて吾人の見解.
Our View about the Location of Chung-chia-lo (in the East Indies) [contains also discussion of the medieval Chinese compass-points].
TYG, 1914, **4**, 99.

Shih Chang-Ju (2) 石璋如.
Honan Anyang Hou-Kang ti Yin Mo 河南
安陽後岡的殷墓.
Burials of the Yin (Shang) Dynasty at Hou-
kang, Anyang.
AS/BIHP, 1948, 13, 21.

Than Chieh-Fu (1) 譚戒甫.
Mo Ching I Chieh 墨經易解.
Analysis of the Mohist Canon.
Com. Press (for Wuhan University),
Shanghai, 1935.
Thang Lan (1) (ed.) 唐蘭.
Wu Shêng Chhu Thu Chhung-Yao Wên-Wu
Chan-Lan Thu Lu 五省出土重要文
物展覽圖錄.
Album of the Exhibition of Important
Archaeological Objects excavated in Five
Provinces (Shensi, Chiangsu, Jehol,
Anhui and Shansi).
Wên-wu, Peking, 1958.
Thang Po-Huang (1) 唐擘黃.
'Yang-Sui Chhü Huo yü Fang-Chu chhü
Shui' 「陽燧取火與方諸取水」.
On the Statement that 'The Burning
Mirror attracts Fire and the Dew Mirror
attracts Water'.
AS/BIHP, 1935, 5 (no. 2), 271.
Tomioka Kenzō (1) 富岡謙藏.
Kokyō no Kenkyū 古鏡の研究.
Researches on Ancient Mirrors.
Tokyo, 1918.
Tsêng Chao-Yü 曾昭燏, Chiang Pao-Kêng
蔣寶庚 & Li Chung-I (1) 黎忠義.
I-nan Ku Hua Hsiang Shih Mo Fa-Chüeh
Pao-Kao 沂南古畫像石墓發掘報
告.
Report on the Excavation of an Ancient
[Han] Tomb with Sculptured Reliefs at
I-nan [in Shantung] (c. +193).
Nanking Museum, Shantung Provincial
Dept. of Antiquities, and Ministry of
Culture, Shanghai, 1956.
Tun Li-Chhen (1) 敦禮臣.
Yenching Sui Shih Chi 燕京歲時記.
Annual Customs and Festivals of Peking.
Peking, 1900.
Tr. Bodde (12).

Wang Chen-To (2) 王振鐸.
Ssu-Nan Chih-Nan Chen yü Lo-Ching Phan
(Shang) 司南指南針與羅經盤(上).
Discovery and Application of Magnetic
Phenomena in China, I (The Lodestone
Spoon of the Han).
AS/CJA, 1948, 3, 119.
Wang Chen-To (4) 王振鐸.
Ssu-Nan Chih-Nan Chen yü Lo-Ching Phan
(Chung) 司南指南針與羅經盤(中).
Discovery and Application of Magnetic
Phenomena in China, II (The 'Fish'

Compass, the Needle Compass, and Early
Work on Declination).
AS/CJA, 1950, 4, 185.
Wang Chen-To (5) 王振鐸.
Ssu-Nan Chih-Nan Chen yü Lo-Ching Phan
(Hsia) 司南指南針與羅經盤(下).
Discovery and Application of Magnetic
Phenomena in China, III (Origin and
Development of the Chinese Compass
Dial).
AS/CJA, 1951, 5 (n.s. 1), 101.
Wang Chin-Kuang (1) 王錦光.
Tsu Kuo Ku-Tai tsai Kuang-Hsüeh shang ti
Chhêng-Chiu 祖國古代在光學上的
成就.
Ancient Chinese Achievements in Physical
Optics.
KHHP, 1955 (no. 5), 178.
Wang Hsien-Chhien (1) (ed.) 王先謙.
Huang Chhing Ching Chieh Hsü Pien 皇清
經解續編.
Continuation of the Collection of Monographs
on Classical Subjects written during the
Chhing Dynasty.
1888.
See Yen Chieh (1).
Wang Hsien-Chhien (2) 王先謙.
Chuang Tzu Chi Chieh 莊子集解.
Collected Commentaries on the Chuang Tzu
book.
1909.
Wang Kuang-Hsi (1) 王光祈.
Chungkuo Yin-Yo Shih 中國音樂史.
History of Chinese Music.
2 vols.
Chung-hua, Shanghai, 1934.
Repr. Yin-yo, Peking, 1957.
Watanabe Naotsune (1) 渡邊直經.
Chijiki Nendai Gaku 地磁氣年代學.
Geomagneto-chronology; Dating by the
Direction of Magnetism of Baked Earth.
(Thermo-remanent Magnetism in
Japanese Pottery, Kiln Walls and Lava
Flows).
K, 1958, 28 (no. 1), 24.
Wei Chü-Hsien (1) 衛聚賢.
Chung-Kuo Khao-Ku-Hsüeh Shih 中國考
古學史.
History of Archaeology in China.
Com. Press, Shanghai, 1937.
Wu Chhêng-Lo (2) 吳承洛.
Chung-Kuo Tu Liang Hêng Shih 中國度
量衡史.
History of Chinese Metrology [weights and
measures].
Com. Press, Shanghai, 1937; 2nd ed.
Shanghai, 1957.
Wu Jen-Ching 吳仁敬 & Hsin An-Chhao
辛安潮 (1).
Chung-Kuo Thao Tzhu Shih 中國陶瓷史.
History of Chinese Pottery and Porcelain.
Com. Press, Shanghai, 1936.

Wu Nan-Hsün (*1*) 吳南薰.
　　Chung-Kuo Wu-Li-Hsüeh Shih 中國物
　　理學史.
　　A History of Physics in China (preliminary
　　draft, based on courses of lectures).
　　Pr. pr. Dept. of Physics, Wuhan Univ.,
　　Wuhan, 1954.
Wu Yü-Chiang (*1*) (ed.) 吳毓江.
　　Mo Tzu Chiao Chu 墨子校注.
　　The Collected Commentaries on the *Mo
　　Tzu* book (including the Mohist Canon).
　　Tu-li, Chungking, 1944.

Yabuuchi Kiyoshi (*18*) 藪內清.
　　Jūni Rikkan ni tsuite 十二律管について.
　　On the Twelve Standard Pitch-Pipes.
　　TG/K, 1939, **10**, 280.
Yang Khuan (*3*) 楊寬.
　　Chan Kuo Shih 戰國史.
　　History of the Warring States Period.
　　Jen-min, Shanghai, 1955, 1956.
Yang Khuan (*4*) 楊寬.
　　Chung-Kuo Li-Tai Chhih Tu Khao 中國
　　歷代尺度考.
　　A Study of the Chinese Foot-Measure
　　through the Ages.
　　Com. Press, Shanghai, 1938; revised and
　　amplified ed. 1955.
Yang Tsung-Jung (*1*) 楊宗榮.
　　Chan Kuo Hui Hua Tzu-Liao 戰國繪畫
　　資料.
　　Materials for the Study of the Graphic Art
　　of the Warring States Period.
　　Ku-tien I-shu, Peking, 1957.
Yang Yin-Liu (*1*) 楊蔭瀏.
　　Phing Chün Lü Suan Chieh Mu Lu 平均
　　律算解目錄.
　　The History of the Search for Equal
　　Temperament (in Chinese Acoustics and
　　Music).
　　YCHP, 1937, **21**, 1.
Yang Yin-Liu 楊蔭瀏 & Yin Fa-Lu
　　陰法魯 (*1*).
　　*Sung Chiang Pai-Shih Chhuang-Tso Ko-
　　Chhü Yen-Chiu* 宋姜白石創作歌曲
　　研究.
　　Studies on the Songs and Tunes of Chiang
　　Khuei of the Sung Period.
　　Peking, 1957.

Yeh Yao-Yuan (*1*) 葉耀元.
　　Chung Hsüeh 重學.
　　Mechanics and Dynamics.
　　Chapters in *Chung-Hsi Suan-Hsüeh Ta
　　Chhêng* 中西算學大成.
　　Complete Textbook of Chinese and Western
　　Mathematics (and Physics).
　　Thung-wên, Shanghai, 1889.
Yen Chieh (*1*) (ed.) 嚴杰.
　　Huang Chhing Ching Chieh 皇清經解.
　　Collection of [more than 180] Monographs
　　on Classical Subjects written during the
　　Chhing Dynasty.
　　1829; 2nd ed. Kêng Shen Pu Khan, 1860.
　　Cf. Wang Hsien-Chien (*1*).
Yen Kho-Chün (*1*) (ed.) 嚴可均.
　　*Chhüan Shang-Ku San-Tai Chhin Han San-
　　Kuo Liu Chhao Wên* 全上古三代秦
　　漢三國六朝文.
　　Complete Collection of Prose Literature
　　(including Fragments) from Remote
　　Antiquity through the Chhin and Han
　　Dynasties, the Three Kingdoms and the
　　Six Dynasties.
　　Finished 1836; published 1887–93.
Yen Tun-Chieh (*14*) 嚴敦傑.
　　*Chung-Kuo Ku-Tai Shu-Hsüeh ti Chhêng-
　　Chiu* 中國古代數學的成就.
　　Contributions of Ancient and Medieval
　　Chinese Mathematics.
　　Chung-hua (Chhüan Kuo Kho-Hsüeh Chi-
　　Shu Phu-Chi Hsieh-Hui), Peking, 1956.
Yen Tun-Chieh (*15*) 嚴敦傑.
　　Po Liu Jen Shih Phan 跋六壬式盤.
　　A Note on the Liu-Jen Diviner's Board.
　　WWTK, 1958 (no. 7), 20.
Yin Fa-Lu (*1*) 陰法魯.
　　Hsien Han Yo Lü Chhu Than 先漢樂律
　　初探.
　　A Preliminary Investigation of Han and
　　pre-Han Music and Acoustics.
　　Tali, 1944.
Yin Fa-Lu (*2*) 陰法魯.
　　*Thang Sung Ta Chhü chih Lai-Yuan chi
　　chhi Tsu-Chih* 唐宋大曲之來源及其
　　組織.
　　Origin and Structure of the 'Extended
　　Melody' [orchestral compositions] of the
　　Thang and Sung periods.
　　HCUKY, 1945, **1** (no. 4), 104.

ADDENDUM TO BIBLIOGRAPHY B

Umehara Sueji, Oba Tsunekichi & Kayamoto
　　Kamejirō (*1*) 梅原末治 小場恒吉
　　榧本龜次耶.
　　Rakurō 樂浪王光墓.
　　The Tomb of Wang Kuang at Lolang
　　(Korea).

Soc. for the Study of Korean Antiquities.
　　Chōsen Koseki Kenkyū-Kwai 朝鮮古蹟
　　研究會.
　　Detailed Reports of Archaeological Research,
　　vol. 2.
　　Seoul (Keijo), 1935; 2 vols.

C. BOOKS AND JOURNAL ARTICLES IN
WESTERN LANGUAGES

VAN AALST, J. A. (1). *Chinese Music.* Chinese Imp. Maritime Customs Reports, II, Special Series no. 6. Shanghai, 1884. (Repr. Vetch, Peiping, 1933.)

ADAM, N. K. (1). *Physics and Chemistry of Surfaces.* Oxford, 1930.

AITKEN, M. J. (1). 'Magnetic Dating' (remanent magnetism in Chinese porcelain). *AMY,* 1958, 1 (no. 1), 16; also 1960, 3, 41.

ALLEN, M. R. (1). 'Early Chinese Lamps.' *ORA,* 1950, 2, 133.

AMIOT, J. J.-M. (1). 'Mémoire sur la Musique des Chinois tant anciens que modernes.' *MCHSAMUC,* 1780 (written in 1776), 6, 1.

AMIOT, J. J.-M. (5). Observations of Magnetic Declination. *MCHSAMUC,* 1780, 9, 2; 1782, 10, 142.

ANDERSON, E. W. (1). 'The Development of the Organ.' *TNS,* 1928, 8, 1.

ANDRADE, E. N. DA C. (1). 'Robert Hooke' (Wilkins Lecture). *PRSA,* 1950, 201, 439. (The quotation concerning fossils is taken from the advance notice, December 1949.) Also *N,* 1953, 171, 365.

ANDRADE, E. N. DA C. (2). 'The Early History of the Permanent Magnet.' *END,* 1958, 17, 22.

ANNANDALE, N., MEERWARTH, G. H. & GRAVES, H. G. (1). 'Weighing Apparatus from the Southern Shan States' (with appendices on the bismar in Russia and on the elementary mechanics of balances and steelyards). *MAS/B,* 1917, 5, 195.

ANON. (12). 'Without Mirrors' (scientific 'levitation'; magnetic suspension of metal objects in air and the melting of them in that position by high-frequency currents). *SAM,* 1952, 187 (no. 1), 37.

ANON. (37) (tr.). 'The Chinaman Abroad; or, a Desultory Account of the Malayan Archipelago, particularly of Java, by Ong-Tae-Hae' [Wang Ta-Hai's *Hai Tao I Chih Chai Lüeh* of +1791]. *CMIS,* 1849 (no. 2), 1.

D'ANVILLE, J. B. B. (1). *Mémoire sur la Chine.* Chez l'auteur, Paris, 1776.

D'ANVILLE, J. B. B. (2). 'Mémoire sur le *li,* Mésure itinéraire des Chinois.' *MAI/LTR,* 1761, 28, 487.

APEL, W. (ed.) (1). *The Harvard Dictionary of Music.* Harvard Univ. Press, Cambridge, Mass., 1946.

ARAGO, D. F. (1). Presentation of a Chinese 'Magic Mirror' to the Académie des Sciences. *CRAS,* 1844, 19, 234.

ARBERRY, A. J. (2) (tr.). *The Ring of the Dove* [the *Ṭauq al-Ḥamāma* of Abū Muḥammad 'Ali ibn Ḥazm al-Andalusī]. Luzac, London, 1953.

ARCHIBALD, R. C. (2). 'Mathematics and Music'. *AMM,* 1924, 31, 1.

ARDSHEAL (1). 'Weighing the Elephant.' *EAM,* 1903, 2, 357; also *Actes du 14ème Congrès International des Orientalistes,* 1903, 357.

ATKINSON, R. W. (1). 'Japanese Magic Mirrors.' *N,* 1877, 16, 62.

AUSTIN, R. G. (1). 'Greek Board-Games.' *AQ,* 1940, 14, 257.

AYRTON, W. E. (1). 'The Mirror of Japan and its Magic Quality.' 1879, 25. (Unidentifiable reprint.)

AYRTON, W. E. & PERRY, J. (1). 'The Magic Mirror of Japan, I.' *PRS,* 1878, 28, 127. Fr. tr. (with illustrations), *ACP,* 1880 (5e sér.), 20, 110.

AYRTON, W. E. & PERRY, J. (2). 'On the Expansion produced by Amalgamation.' *PMG,* 1886 (5th ser.), 22, 327.

BAILEY, K. C. (1). *The Elder Pliny's Chapters on Chemical Subjects.* 2 vols. Arnold, London, 1929 and 1932.

BAKER, I. (1). 'The Story of Amber.' *AQC,* 1951, 22 (no. 1), 25.

BALMER, H. (1). *Beiträge zur Geschichte der Erkenntnis des Erdmagnetismus.* Aarau, 1956.

BARLOWE, WM. (1). *The Navigator's supply, conteining many things of principall importance belonging to Navigation, with the description and use of diverse Instruments framed chiefly for that purpose; but serving also for sundry other of Cosmography in generall.* Bishop, Newbery & Barker, London, 1597.

BARRINGTON, D. (1). 'An Historical Disquisition on the Game of Chess.' *AAA,* 1789, 9, 16.

BAXTER, W. (1) (tr.). 'Pleasure not Attainable according to Epicurus.' In *The Works of Plutarch.* Ed. Morgan. London, 1694.

BAYON, H. P. (1). 'William Harvey, Physician and Biologist; his Precursors, Opponents and Successors.' *ANS,* 1938, 3, 59, 83, 435; 1939, 4, 65, 329.

BECK, H. C. & SELIGMAN, C. G. (1). 'Barium in Ancient Glass.' *N*, 1934, **133**, 982. (*Proc. 1st Internat. Congr. Prehist. and Protohist. Sci.* London, 1932.)

BECKMANN, J. (1). *A History of Inventions, Discoveries and Origins.* 1st German ed., 5 vols. 1786 to 1805. 4th ed., 2 vols. tr. by W. Johnston, Bohn, London, 1846. Enlarged ed., 2 vols. Bell & Daldy, London, 1872. Bibl. in John Ferguson (2).

BEER, A., HO PING-YÜ, LU GWEI-DJEN, NEEDHAM, JOSEPH, PULLEYBLANK, E. G. & THOMPSON, G. I. (1). 'An 8th-century Meridian Line; I-Hsing's Chain of Gnomons and the Pre-History of the Metric System.' *VA*, 1960.

BELAIEV, N. T. (5). 'The Bismar in Ancient India.' *AE*, 1933, 76.

VAN BEMMELEN, W. (1). *De Isogonen in de XVIde en XVIIde Eeuw.* Inaug. Diss. Leiden. Utrecht, 1893. The charts of magnetic variation (+1540 to +1700) were republished in S. Günther (2).

VAN BEMMELEN, W. (2). 'Die Abweichung der Magnetnadel; Beobachtungen, Säcular-variation, Wert- und Isogonen-systeme bis zur Mitte des XVIIIten Jahrhundert.' *Observations Magn. & Meteorol. Batavia Observatory* Suppl. 1899, **21**, 109.

BENTON, W. A. (1). On Roman, Scandinavian and Chinese Steelyards (contribution to discussion of Chatley, 2). *TNS*, 1942, **22**, 135.

BERNARD-MAÎTRE, H. (1). *Matteo Ricci's Scientific Contribution to China*, tr. by E. T. C. Werner. Vetch, Peiping, 1935. Orig. pub. as *L'Apport Scientifique du Père Matthieu Ricci à la Chine*, Hsienhsien, Tientsin, 1935 (rev. Chang Yü-Chê, *TH*, 1936, **3**, 538).

BERNARD-MAÎTRE, H. (15). 'Les Sources Mongoles et Chinoises de l'Atlas Martini, (1655).' *MS*, 1947, **12**, 127.

BERRIMAN, A. W. (1). *Historical Metrology.* Dent, London, 1953 (rev. A. W. Richeson, *ISIS*, 1954, **45**, 111).

BERRY, A. (1). *A Short History of Astronomy.* Murray, London, 1898.

BERSON, G. H. (1). 'On the Japanese Magic Mirror.' *GS*, 1880, **7**, 276.

BERTELLI, T. (1). 'Sopra Pietro Peregrino di Maricourt e la sua Epistola *De Magnete*.' *BBSSMF*, 1868, **1**, 1, 65, 319; *RGI*, 1891, **1**, 335; 1902, **9**, 281, 353, 409; 1903, **10**, 1, 105, 314; 1904, **11**, 433; *BSGI*, 1903, **3**, 178. Crit. Hazard, D. L. *TM*, 1903, **8**, 179.

BERTELLI, T. (2). 'The Discovery of Magnetic Declination made by Christopher Columbus.' *Proc. Internat. Meteorol. Congr. Chicago*, 1893. Weather Bureau Bull. no. 11. Wash. D.C. 1894–6, 486. *La Declinazione Magnetica e la sua Variazione nello Spazio scoperte da Christoforo Colombo.* Rome, 1892.

BERTIN, A. (1). 'Étude sur les Miroirs Magiques.' *ACP*, 1881 (5ᵉ sér.), **22**, 472.

BERTIN, A. & DUBOSCQ, J. (1). 'Production Artificielle des Miroirs Magiques.' *ACP*, 1880 (5ᵉ sér.), **20**, 143.

BIOT, E. (1) (tr.). *Le Tcheou-Li ou Rites des Tcheou [Chou].* 3 vols. Imp. Nat., Paris, 1851. (Photographically reproduced Wêntienko, Peiping, 1930.)

BIOT, E. (14). 'Sur la Direction de l'Aiguille Aimantée en Chine, et sur les Aurores Boréales observées dans ce même Pays.' *CRAS*, 1844, **19**.

BLACKER, C. (1). 'Ōhashi Totsuan.' *TAS/J*, 1959 (3rd ser.), **7**, 147.

BLOCHMANN, H. F. (1) (tr.). *The 'Ā'īn-i Akbarī' (Administration of the Mogul Emperor Akbar) of Abū'l Faẓl 'Allāmī.* Rouse, Calcutta, 1873. (Bibliotheca Indica, *NS*, nos. 149, 158, 163, 194, 227, 247 and 287.)

BOCK, E. (1). *Die Brille und ihrer Geschichte.* Safar, Vienna, 1903. Eng. summary by C. Barck, 'The History of Spectacles', *OC*, 1907, 1.

BODDE, D. (1). *China's First Unifier, a study of the Ch'in Dynasty as seen in the life of Li Ssu* (−280 **to** −208). Brill, Leiden, 1938. (Sinica Leidensia, no. 3.)

BODDE, D. (12). *Annual Customs and Festivals in Peking, as recorded in the 'Yenching Sui Shih Chi'* [by Tun Li-Chhen]. Vetch, Peiping, 1936. (Revs. J. J. L. Duyvendak, *TP*, 1937, **33**, 102; A. Waley, *FL*, 1936, **47**, 402.)

BODDE, D. (15). *Statesman, General and Patriot in Ancient China.* Amer. Oriental Soc. New Haven, Conn. 1940.

BODDE, D. (17). 'The Chinese Cosmic Magic known as "Watching for the Ethers".' Art. in *Studia Serica Bernhard Karlgren Dedicata*, p. 14. Ed. E. Glahn. Copenhagen, 1959.

BODDE, D. (18). 'Evidence for "Laws of Nature" in Chinese Thought.' *HJAS*, 1957, **20**, 709.

BOERSCHMANN, E. (3a). *China; Architecture and Landscape—a Journey through Twelve Provinces.* Studio, London, n.d. (1928–29). English edition of Boerschmann (3).

BOLL, F. (1). *Sphaera.* Teubner, Leipzig, 1904.

BOLL, F., BEZOLD, C. & GUNDEL, W. (1). (*a*) *Sternglaube, Sternreligion und Sternorakel.* Teubner, Leipzig, 1923. (*b*) *Sternglaube und Sterndeutung; die Gesch. ü. d. Wesen d. Astrologie.* Teubner, Leipzig, 1926.

BOSMANS, H. (3). 'l'Œuvre Scientifique d'Antoine Thomas de Namur, S.J. (+1644 to +1709).' *ASSB*, 1924, **44** (2ᵉ partie, Mémoires), 169; 1926, **46**, 154.

BOXER, C. R. (1) (ed.). *South China in the Sixteenth Century; being the Narratives of Galeote Pereira, Fr. Gaspar da Cruz, O.P., and Fr. Martin de Rada, O.E.S.A. (1550–1575).* Hakluyt Society, London. 1953. (Hakluyt Society Pubs. 2nd series, no. 106).

BOYER, C. B. (4). 'Aristotle's Physics.' *SAM*, 1950, **182** (no. 5), 48.

BOYLE, ROBERT (3). *The Aerial Noctiluca; or some New Phaenomena, and a Process of a Factitious Self-Shining Substance, in a Letter to a Friend, living in the Country.* Snowden, London, 1680.

BRAGG, SIR WM. (1). On Chinese 'Magic Mirrors'. *ILN*, 1932, **181**, 706.

BRETSCHNEIDER, E. (1). *Botanicon Sinicum; Notes on Chinese Botany from Native and Western Sources.* 3 vols. Trübner, London, 1882 (printed in Japan). (Repr. from *JRAS/NCB*, 1881, **16**.)

BREUSING, A. (1). *Die nautischen Instrumente bis zur Erfindung des Spiegelsextanten.* Bremen, 1890.

BREWSTER, D. (1). 'Account of a curious Chinese Mirror which reflects from its polished Face the Figures Embossed upon its Back.' *PMG*, 1832, **1**, 438.

BRINKLEY, F. (1). *Japan, its History, Arts and Literature.* 12 vols. Black, London 1903–4; Harvard Univ. Press, New York, 1904. Biography, *TP*, 1912, **13**, 660.

BROMEHEAD, C. E. N. (3). 'A Geological Museum of the Early Seventeenth Century.' *QJGS*, 1947, **103**, 65.

B[ROMEHEAD], C. [E.] N. (4). 'Alexander Neckam on the Compass Needle.' *GJ*, 1944, **104**, 63; *TM*, 1945, **50**, 139.

BROMEHEAD, C. E. N. (5). 'Ships' Loadstones.' *MMA*, 1948, **28**, 429.

BROWN, G. B. (1). 'Jets Musically Inclined.' *SPR*, 1938, **33**, 29.

BROWN, LLOYD A. (1). *The Story of Maps.* Little Brown, Boston, 1949.

BRUCE, J. P. (1) (tr.). *The Philosophy of Human Nature, translated from the Chinese, with notes.* Probsthain, London, 1922. (Chs. 42–48, inclusive, of *Chu Tzu Chhüan Shu*.)

BRUHL, ODETTE & LÉVI, S. (1). *Indian Temples.* Bombay, 1937; Calcutta, 1939.

BUCKLEY, H. (1). *A Short History of Physics.* Methuen, London, 1927.

BUKOFZER, M. (1). 'Präzisionsmessungen an primitiven Musikinstrumenten.' *ZP*, 1936, **99**, 643.

BULLING, A. (8). *The Decoration of Mirrors of the Han Period; a Chronology.* Artibus Asiae, Ascona, 1960 (Artibus Asiae Supplement ser. no. 20).

BURD, A. C. & LEE, A. J. (1). 'The Sonic Scattering Layer in the Sea.' *N*, 1951, **167**, 624.

BURKILL, I. H. (1). *A Dictionary of the Economic Products of the Malay Peninsula* (with contributions by W. Birtwhistle, F. W. Foxworthy, J. B. Scrivenor and J. G. Watson). 2 vols. Crown Agents for the Colonies, London, 1935.

BURNET, J. (1). *Early Greek Philosophy.* Black, London, 1908.

BURTON, E. H. (1) (tr.). 'Euclid's Optics.' *JAOPS*, 1945, **35**, 357.

BUSHELL, S. W. (2). *Chinese Art.* 2 vols. For Victoria and Albert Museum, HMSO, London, 1909. 2nd ed. 1914.

BYCHAWSKI, T. (1). 'The Measurements of a Degree Executed by the Arabs in the +9th Century.' Communication to the IXth International Congress of the History of Science, Barcelona, 1959. Abstract in *Guiones de las Comunicaciones*, p. 15.

CAJORI, F. (5). *A History of Physics, in its elementary branches, including the evolution of Physical Laboratories.* Macmillan, New York, 1899.

CANTON, JOHN (1). 'An easy Method of making a Phosphorus that will imbibe and emit Light like the Bononian Stone: with experiments and observations.' *PTRS*, 1768, **58**, 337.

CARPENTER, W. B. (1). 'On the Zoetrope and its Antecedents, e.g. the Anorthoscope.' *SIO*, 1868, **1**, 427; 1869, **2**, 24, 110.

CARTER, T. F. (1). *The Invention of Printing in China and its Spread Westward.* Columbia Univ. Press, New York, 1925, revised ed. 1931. 2nd ed. revised by L. Carrington Goodrich. Ronald, New York, 1955.

CARY, G. (1). *The Medieval Alexander.* Ed. D. J. A. Ross. Cambridge, 1956. (A study of the origins and versions of the Alexander-Romance; important for medieval ideas on flying-machine and diving bell or bathyscaphe.)

DE CASTRO, JOÃO (1). *Roteiro de Lisboã a Goa.* Annotated by João de Andrade Corro. Lisbon, 1882.

DE CASTRO, JOÃO (2). *Primo Roteiro da Costa da India desde Goa até Dio; narrando a viagem que fez o Vice-Rei D. Garcia de Noronha en socorro deste ultima Cidade, 1538–1539.* Köpke, Porto, 1843.

DE CASTRO, JOÃO (3). *Roteiro em que se contem a viagem que fizeran os Portuguezes no anno de 1541 partindo da nobre Cidade de Goa atee Soez que he no fim e stremidade do Mar Roxo ....* Paris, 1833.

CAVE, C. J. P. (1). 'The Orientation of Churches.' *ANTJ*, 1950, **30**, 47.

CHALMERS, J. (2). 'China and the Magnetic Compass.' *CR*, 1891, **19**, 52.

CHAMBERLAIN, B. H. (1). *Things Japanese.* Murray, London. 2nd ed. 1891; 3rd ed. 1898.

CHAMFRAULT, A. & UNG KANG-SAM (1). *Traité de Médecine Chinoise; d'après les Textes Chinois Anciens et Modernes.* Coquemard, Angoulême, 1954 and 1957.

Vol. 1. Traité, Acupuncture, Moxas, Massages, Saignées.

Vol. 2 (tr.). Les Livres Sacrés de Médecine Chinoise (*Nei Ching, Su Wên* and *Nei Ching, Ling Shu*).

CHAO YUAN-JEN (2). '[Chinese] Music.' Art. in *Symposium on Chinese Culture*. Ed. Sophia H. Chen Zen, p. 82. IPR, Shanghai, 1931.

CHAO YUAN-JEN (3). 'A Note on Chinese Scales and Music.' *OR*, 1957, **10**, 140.

CHAPIN, H. B. (1). 'Kyongju, ancient Capital of Silla' and 'Korea in Pictures.' *AH*, 1948, **1** (no. 4), 36.

CHAPMAN, S. (1). 'Edmond Halley and Geomagnetism.' *N*, 1943, **152**, 231; *TM*, 1943, **48**, 131.

CHAPMAN, S. & BARTELS, J. (1). *Geomagnetism*. Oxford, 1940.

CHAPMAN, S. & HARRADON, H. D. (1). 'Archaeologica Geomagnetica; Some Early Contributions to the History of Geomagnetism: I, The Letter of Petrus Peregrinus de Maricourt to Sygerus de Foucaucourt, Soldier, concerning the Magnet (+1269).' *TM*, 1943, **48**, 1, 3.

CHAPMAN, S. & HARRADON, H. D. (2). 'Archaeologica Geomagnetica; Some Early Contributions to the History of Geomagnetism: II and III, The "Treatise on the Sphere and the Art of Navigation" by Francisco Falero (+1535) and, The "Brief Compendium on the Sphere and Art of Navigating" by Martin Cortes (+1551).' *TM*, 1943, **48**, 77, 79. (Early observations on magnetic declination.)

CHATLEY, H. (1). MS. translation of the astronomical chapter (ch. 3, Thien Wên) of *Huai Nan Tzu*. Unpublished. (Cf. note in *O*, 1952, **72**, 84.)

CHATLEY, H. (3). 'Science in Old China.' *JRAS/NCB*, 1923, **54**, 65.

CHATLEY, H. (7). 'Fêng-Shui.' In *ES*, p. 175.

CHATLEY, H. (26). 'Chinese Mystical Philosophy in Modern Terms.' *CJ*, 1923, **1**, 112 and 212.

CHAVANNES, E. (1). *Les Mémoires Historiques de Se-Ma Ts'ien* [Ssuma Chhien]. 5 vols. Leroux, Paris, 1895–1905. (Photographically reproduced, in China, without imprint and undated.)

 1895 vol. 1 tr. *Shih Chi*, chs. 1, 2, 3, 4.

 1897 vol. 2 tr. *Shi Chih*, chs. 5, 6, 7, 8, 9, 10, 11, 12.

 1898 vol. 3 (i) tr. *Shih Chi*, chs. 13, 14, 15, 16, 17, 18, 19, 20, 21, 22.

    vol. 3 (ii) tr. *Shih Chi*, chs. 23, 24, 25, 26, 27, 28, 29, 30.

 1901 vol. 4 tr. *Shih Chi*, chs. 31, 32, 33, 34, 35, 36, 37, 38, 39, 40, 41, 42.

 1905 vol. 5 tr. *Shih Chi*, chs. 43, 44, 45, 46, 47.

CHAVANNES, E. (4) (tr.). *Voyages des Pèlerins Bouddhistes; Les Religieux Éminents qui allèrent chercher la Loi dans les Pays d'Occident; mémoire composé à l'époque de la grande dynastie T'ang par I-Tsing*. Leroux, Paris, 1894.

CHAVANNES, E. (9). *Mission Archéologique dans la Chine Septentrionale*. 2 vols. and portfolios of plates. Leroux, Paris, 1909–15. (Publ. de l'École France, d'Extr. Orient, no. 13.)

CHAVANNES, E. (11). *La Sculpture sur Pierre en Chine aux Temps des deux dynasties Han*. Leroux, Paris, 1893.

CHÊNG TÊ-KHUN (2) (tr.). 'Travels of the Emperor Mu.' *JRAS/NCB*, 1933, **64**, 124; 1934, **65**, 128.

CHESHIRE, H. F. (1). *Goh or Wei Chi; a Handbook of the Game and full Instructions; with introduction and critical notes by T. Komatsubara*. 1911.

CHHIEN LIN-CHAO (1). 'The Optics of the *Mo Ching*.' *Actes du VIIIe Congrès International de l'Histoire des Sciences*, p. 293. Florence, 1956.

CHHIU KHAI-MING (2). 'The Introduction of Spectacles into China.' *HJAS*, 1936, **1**, 186.

CHHU TA-KAO (2) (tr.). '*Tao Tê Ching*', a new translation. Buddhist Lodge, London, 1937.

CHIANG SHAO-YUAN (1). *Le Voyage dans la Chine Ancienne, considéré principalement sous son Aspect Magique et Religieux*. Commission Mixte des Œuvres Franco-Chinoises (Office de Publications), Shanghai, 1937. Transl. from Chinese by Fan Jen.

CHING PHEI-YUAN (1). 'Étude Comparative des diverses éditions du *Chouo Fou* [*Shuo Fu*].' *SSA*, 1946, no. 1.

[CIBOT, P. M.] (1). 'Essai sur le Passage de l'Écriture Hiéroglyphique à l'Écriture Alphabétique; ou sur la manière dont la première a pu conduire à la seconde.' *MCHSAMUC*, 1782, 8, 112.

[CIBOT, P. M.] (7). 'Observations de Physique et d'Histoire Naturelle faites par l'empereur Khang-Hsi.' (A paraphrase abridged translation of the *Khang-Hsi Chi Chia Ko Wu Pien*.) *MHSAMUC*, 1779, **4**, 452.

CLAGETT, M. (1). 'Some General Aspects of Physics in the Middle Ages.' *ISIS*, 1948, **39**, 29.

CLAGETT, M. (2). *The Science of Mechanics in the Middle Ages*. Univ. of Wisconsin Press, Madison, Wis., 1959; Oxford Univ. Press, London, 1959.

CLARK, R. E. D. (1). 'Will-o'-the-Wisp.' *SCSR*, 1942 (no. 90), 138.

CLARKE, J. & GEIKIE, A. (1). *Physical Science in the Time of Nero, being a Translation of the 'Quaestiones Naturales' of Seneca, with notes by Sir Archibald Geikie*. Macmillan, London, 1910.

CLOSSON, ERNEST (1). *History of the Piano*, tr. D. Ames. London, 1947.

COEDÈS, G. (1) (tr.). *Textes d'auteurs grecs et latins relatifs à l'Extrême Orient depuis le 4ème siècle avant J.C. jusqu'au 14ème siècle après J.C.* Leroux, Paris, 1910.

COOK, R. M. & BELSHÉ, J. C. (1). 'Archaeomagnetism; a preliminary report on Britain.' *AQ*, 1958, **32**, 167.

COOMARASWAMY, A. K. (1). 'Symplegades.' In Sarton Presentation Volume *Studies and Essays in the History of Science and Learning*, p. 465. Schuman, New York, 1944.

COOPER, D. & COOPER, R. (1). 'On the Luminosity of the Human Subject after Death.' *PMG*, 1838 (3rd ser.), **12**, 420.

CORNFORD, F. M. (2). *The Laws of Motion in Ancient Thought.* Inaug. Lect. Cambridge, 1931.

CORNFORD, F. M. (3) (tr.). *The 'Republic' of Plato.* Oxford, 1944.

CORNFORD, F. M. (7). 'Anaxagoras' Theory of Matter.' *CQ*, 1930, **24**, 14, 83.

COURANT, M. (2). 'Essai Historique sur la Musique classique des Chinois; avec un appendice relatif à la musique Coréenne.' In *Encyclopédie de la Musique et Dictionnaire du Conservatoire.* Ed. Lavignac & la Laurencie, pt. 1, vol. 1. Paris, 1912.

COUVREUR, F. S. (1) (tr.). *'Tch'ouen Ts'iou' [Chhun Chhiu] et 'Tso Tchouan' [Tso Chuan]; Texte Chinois avec Traduction Française.* 3 vols. Mission Press, Hochienfu, 1914.

COUVREUR, F. S. (3) (tr.). *'Liki' [Li Chi], ou Mémoires sur les Bienséances et les Cérémonies.* 2 vols. Hochienfu, 1913.

COXETER, H. S. M. (1). *Regular Polytopes.* Methuen, London, 1948.

COXETER, H. S. M., LONGUET-HIGGINS, M. S. & MILLER, J. C. P. (1). 'Uniform Polyhedra.' *PTRSA*, 1954, **246**, 401.

CRONIN, V. (1). *The Wise Man from the West* (biography of Matteo Ricci). Hart-Davis, London, 1955.

CROSBY, H. L. (1) (ed. and tr.). *Thomas Bradwardine's 'Tractatus de Proportionibus'; its Significance for the Development of Mathematical Physics.* Madison, Wis., 1955.

CROSSLEY-HOLLAND, P. C. (1). 'Chinese Music.' In *Grove's Dictionary of Music and Musicians.* Ed. E. Blom, vol. 2, pp. 219–48. London, 1954.

CULIN, S. (1). 'Chess and Playing-Cards; Catalogue of Games and Implements for Divination exhibited by the U.S. National Museum in connection with the Dept. of Archaeology and Palaeontology of the University of Pennsylvania at the Cotton States and International Exposition, Atlanta, Georgia, 1895.' *ARUSNM*, 1896, 671 (1898).

CULIN, S. (2). 'Chinese Games with Dice and Dominoes.' *ARUSNM*, 1893, 491.

CULIN, S. (3). 'The Game of Ma-Jong.' *BRMQ*, 1924, Oct.

DALLABELLA, J. A. (1). 'Sobre a Força Magnetica.' *MARSL*, 1797, **1**, 85.

DAMPIER-WHETHAM, W. C. D. (1). *A History of Science, and its Relations with Philosophy and Religion.* Cambridge, 1929.

DARBISHIRE, R. D. (1). Letter on Magic Mirrors. *N*, 1877, **16**, 142.

DAUJAT, J. (1). *Origines et Formation de la Théorie des Phénomènes Électriques et Magnétiques.* Hermann, Paris, 1945. 3 vols. (*ASI*, nos. 989, 990, 991.)

DAVIS, J. F. (1). *The Chinese; a general description of China and its Inhabitants.* 3 vols. Knight, London, 1844.

DAVIS, TENNEY L. (1). 'Count Michael Maier's Use of the Symbolism of Alchemy.' *JCE*, 1938, **15**, 403.

DAVISON, C. ST C. (2). 'Origin of the Foot-Measure.' *ENG*, 1957, **184**, 418.

DECHEVRENS, A. (1). 'Étude sur le Système Musical Chinois.' *SBIMG*, 1901, **2**, 484.

DEMBER, H. (1). 'Ostasiatische Zauberspiegel.' *OAZ*, 1933, **9** (**19**), 203.

DEMIÉVILLE, P. (1). 'Le Miroir Spirituel.' *S*, 1947, **1**, 112.

DEMIÉVILLE, P. (2). Review of Chang Hung-Chao (1), *Lapidarium Sinicum. BEFEO*, 1924, **24**, 276.

DICKINSON, H. W. (4). *A Short History of the Steam-Engine.* Cambridge, 1939.

DIELS–FREEMAN; FREEMAN, K. (1). *Ancilla to the Pre-Socratic Philosophers; a complete translation of the Fragments in Diels' 'Fragmente der Vorsokratiker'.* Blackwell, Oxford, 1948.

DIJKSTERHUIS, E. J. (1). *Simon Stevin.* 's-Gravenhage, 1943.

DIRCKS, H. (1). *The Life, Times, and Scientific Labours of the Second Marquis of Worcester, to which is added a Reprint of his 'Century of Inventions' (+1663), with a Commentary thereon.* Quaritch, London, 1865.

DREWS, R. A. (1). Letter on the Physics of the Boiling of Liquids. *SAM*, 1954, **191** (no. 5), 5.

DUBS, H. H. (2) (tr., with the assistance of Phan Lo-Chi and Jen Thai). *History of the Former Han Dynasty, by Pan Ku; a Critical Translation with Annotations.* 3 vols. Waverly, Baltimore, 1938–.

DUBS, H. H. (5). 'The Beginnings of Alchemy.' *ISIS*, 1947, **38**, 62.

DUBS, H. H. (7). *Hsün Tzu; the Moulder of Ancient Confucianism.* Probsthain, London, 1927.

DUCROS, H. (1). 'Études sur les Balances Égyptiennes.' *ASAE*, 1908, **9**, 32.

DUGAS, R. (1). *Histoire de la Mécanique.* Griffon (La Baconnière), Neufchâtel, 1950. Crit. rev. P. Costabel, *A/AIHS*, 1951, **4**, 783.

DUGAS, R. (2). *La Mécanique au XVIIème Siècle; des Antécédents Scholastiques à la Pensée Classique.* Griffon, Neuchâtel, 1954. Crit. rev. C. Truesdell, *ISIS*, 1956, **47**, 449.

DUHAMEL, J. P. F. (1). 'Exemples de quelques Circonstances qui peuvent produire des Embrasemens Spontanés.' *MRASP*, 1757, p. 150. (See also the *Histoire* section for the same year, p. 2, and the *Histoire* section for 1725, p. 4.)

DUHEM, J. (1). *Histoire des Idées Aéronautiques avant Montgolfier.* Inaug. Diss. Sorlot, Paris, 1943.

DUHEM, P. (2). *Les Origines de la Statique.* Hermann, Paris, 1905.

DUNCAN, G. S. (1). *A Bibliography of Glass.* Dawson, London, 1954.

DUYVENDAK, J. J. L. (1). 'Sailing Directions of Chinese Voyages' (a Bodleian Library MS.). *TP*, 1938, **34**, 230.

DUYVENDAK, J. J. L. (8). *China's Discovery of Africa.* Probsthain, London, 1949. (Lectures given at London University, Jan. 1947; rev. P. Paris, *TP*, 1951, **40**, 366.)

DUYVENDAK, J. J. L. (13). 'Early Chinese Studies in Holland.' *TP*, 1936, **32**, 293.

DUYVENDAK, J. J. L. (14). 'Simon Stevin's "Sailing-Chariot"' (and its Chinese antecedents). *TP*, 1942, **36**, 401.

DUYVENDAK, J. J. L. (18) (tr.). *'Tao Tê Ching', the Book of the Way and its Virtue.* Murray, London, 1954 (Wisdom of the East series). Crit. revs. P. Demiéville, *TP*, 1954, **43**, 95; D. Bodde, *JAOS*, 1954, **74**, 211.

DUYVENDAK, J. J. L. (19). 'Desultory Notes on the *Hsi Yang Chi* [Lo Mou-Têng's novel of +1597 based on the Voyages of Chêng Ho]' (concerns spectacles and bombards). *TP*, 1953, **42**, 1.

EASTLAKE, F. W. (2). 'The *sho* [*shêng*] or Chinese Reed Organ.' *CR*, 1882, **11**, 33.

EBERHARD, W. (6). 'Beiträge zur kosmologischen Spekulation Chinas in der Han-Zeit.' *BA*, 1933, **16**, 1–100.

ECKARDT, H. (1). 'Chinesische Musik, II. Vom Ende der Han-zeit bis zum Ende der Sui-zeit (+220 bis +618); der Einbruch westlicher Musik.' In *Die Musik in Geschichte und Gegenwart.* Ed. F. Blume, vol. 2, cols. 1205–7. Kassel and Basel, 1952. Cf. Robinson (3).

ECKARDT, H. (2). 'Chinesische Musik, III. Die Thang-zeit (+618 bis +907); die Rolle der westländischen (Hu-) Musik; die Zehn Orchester; die Musik der Zwei Abteilungen; Akademien und Konservatorien.' In *Die Musik in Geschichte und Gegenwart.* Ed. F. Blume, vol. 2, cols. 1207–16. Kassel and Basel, 1952.

EDER, M. (1). 'Lanterns and Lantern-Festivals in Peking.' *FLS*, 1947, **6** (no. 1).

EDKINS, J. (11). 'Ancient Physics.' *CR*, 1888, **16**, 73 and 370.

EDKINS, J. (12). 'Chinese Names for Boats and Boat Gear; with Remarks on the Chinese Use of the Mariner's Compass.' *JRAS/NCB*, 1877, **11**, 123. (Rev. *CR*, 1877, **6**, 128.)

EDKINS, J. (13). Note on the Magnetic Compass in China. *CR*, 1889, **18**, 197. (Abstracted anonymously, with Edkins (12), from an article in *NCH*, of about this date, in 'Is the Mariner's Compass a Chinese Invention?' *N*, 1891, **44**, 308.)

EDMUNDS, C. K. (1). 'Science among the Chinese.' *NCH*, 1911.

VAN EECKE, P. (1) (tr.). *Archimedes 'De Aequiponderantibus'.* Desclée de Brouwer, Paris and Antwerp, 1938.

EICHHORN, W. (5). 'Wang Chia's *Shih I Chi*.' *ZDMG*, 1952, **102** (N.F. **27**), 130.

EISLER, ROBERT (1). *The Royal Art of Astrology.* Joseph, London, 1946. (Crit. H. Chatley, *O*, 1947.)

EITEL, E. J. (2). *Fêng-Shui: Principles of the Natural Science of the Chinese.* Trübner, Hongkong and London, 1873. French tr. by L. de Milloué, *AMG*, 1880, **1**, 203.

EITNER, R. (1). *Biographisch-Bibliographisches Quellen-Lexikon d. Musiker und Musikgelehrten d. christlichen Zeitrechnung bis zur Mitte des 19. Jahrhunderts.* 10 vols. Breitkopf & Haertel, Leipzig, 1903.

D'ELIA, PASQUALE (2) (ed.). *Fonti Ricciane; Storia dell'Introduzione del Cristianesimo in Cina.* 3 vols. Libreria dello Stato, Rome, 1942–9. Cf. Trigault (1); Ricci (1).

ELLIS, A. J. (1). 'On the History of Musical Pitch.' *JRSA*, 1880, 294.

ERKES, E. (1) (tr.). 'Das Weltbild d. *Huai Nan Tzu*' (transl. of ch. 4). *OAZ*, 1918, **5**, 27.

ESCARRA, J. & GERMAIN, R. (1) (tr.). *La Conception de la Loi et des Théories des Légistes à la Veille des Ts'in* [*Chhin*] (tr. of chs. 7, 13, 14, 15 and 16 of Liang Chhi-Chhao (5), preface by G. Padoux). China Booksellers, Peking, 1926.

ESTERER, M. (1). *Chinas natürliche Ordnung und die Maschine.* Cotta, Stuttgart and Berlin, 1929. (Wege d. Technik series.)

FABER, E. (1). 'The Chinese Theory of Music.' *CR*, 1873, **1**, 324.

FALCONET, M. (1). 'Dissertation historique et critique sur ce que les Anciens ont cru de l'Aimant. *MAI/LTR*, 1723, **4**, 613, (read 1717).

FARMER, H. G. (1). *A History of Arabian Music to the 13th century.* London, 1929.

FARMER, H. G. (2). 'Reciprocal Influences in Music 'twixt Far and Middle East.' *JRAS*, 1934, 327.

FARMER, H. G. (3) 'The Origin of the Arabian Lute and Rebec.' *JRAS*, 1930, 777.

FARRINGTON, G. H. (1). *Fundamentals of Automatic Control*. Chapman & Hall, London, 1951. (Crit. J. Greig, *N*, 1953, **172**, 91.)

FEIFEL, E. (1) (tr.). *Pao Phu Tzu (Nei Phien)*, chs. 1–3. *MS*, 1941, **6**, 113.

FEIFEL, E. (2) (tr.). *Pao Phu Tzu (Nei Phien)*, ch. 4. *MS*, 1944, **9**, 1.

FEIFEL, E. (3) (tr.). *Pao Phu Tzu (Nei Phien)*, ch. 11. *MS*, 1946, **11**, 1.

FELDHAUS, F. M. (1). *Die Technik der Vorzeit, der Geschichtlichen Zeit, und der Naturvölker* (encyclopaedia). Engelmann, Leipzig and Berlin, 1914.

FELDHAUS, F. M. (2). *Die Technik d. Antike u. d. Mittelalter*. Athenaion, Potsdam, 1931. (Crit. H. T. Horwitz, *ZHWK*, 1933, **13** (N·F. **4**), 170.)

FELDHAUS, F. M. (16). *Studien z. Geschichte d. Glocken*. Berlin, 1911.

FELDHAUS, F. M. (17). 'Über d. Kennzeichen an Glocken der ältesten Periode.' *GTIG*, 1916, **3**, 100.

FELDHAUS, F. M. (20). *Die Maschine im Leben der Völker*. Birkhäuser, Basel, 1954.

FÊNG YU-LAN (1). *A History of Chinese Philosophy*. Vol. 1, *The Period of the Philosophers (from the beginnings to c. −100)*, tr. D. Bodde; Vetch, Peiping, 1937; Allen & Unwin, London, 1937. Vol. 2, *The Period of Classical Learning (from the −2nd century to the +20th century)*, tr. D. Bodde; Princeton Univ. Press, Princeton, N.J., 1953. At the same time, vol. 1 was reissued in uniform style by this publisher. Translations of parts of vol. 2 had appeared earlier in *HJAS*; see under Bodde. See also Fêng Yu-Lan (*1*). Crit. Chhen Jung-Chieh (Chan Wing-Tsit), *PEW*, 1954, **4**, 73; J. Needham, *SS*, 1955, **19**, 268.

FERGUSON, JOHN (2). Bibliographical Notes on Histories of Inventions and Books of Secrets, 2 vols. Glasgow, 1898; repr. Holland Press, London, 1959. (Papers collected from *TGAS*.)

FERNALD, H. E. (1). 'Ancient Chinese Musical Instruments.' *MUJ* (Philadelphia), 1936. (Repr. in Hsiao Chhien (1), pp. 395–440.)

FERRAND, G. (1). *Relations de Voyages et Textes Géographiques Arabes, Persans et Turcs relatifs à l'Extrême Orient, du 8ᵉ au 18ᵉ siècles, traduits, revus et annotés etc.* 2 vols. Leroux, Paris, 1913.

FERRAND, G. (3). 'Le K'ouen-Louen [Khun-Lun] et les Anciennes Navigations Interocéaniques dans les Mers du Sud.' *JA*, 1919 (11ᵉ sér.), **13**, 239, 431; **14**, 5, 201.

FILLIOZAT, J. (1). *La Doctrine Classique de la Médecine Indienne*. Imp. Nat., CNRS and Geuthner, Paris, 1949.

FLEET, J. F. (1). 'The *Yōjana* and the *Li*.' *JRAS*, 1906, **38**, 1011; 1912, 229, 462.

FLEET, J. F. (2). 'Some Hindu Values of the Dimensions of the Earth.' *JRAS*, 1912, 463.

FOKKER, A. D. (1). *Rekenkundige Bespiegeling der Muziek*. Gorinchem, 1945.

FOKKER, A. D. (2). *Just Intonation and the Combination of Harmonic Diatonic Melodic Groups*. Nijhoff, The Hague, 1949.

FOKKER, A. D. (3). *Les Mathématiques et la Musique; Trois Conférences*. Nijhoff, The Hague, 1947. See also *AMT*, 1953 (3ᵉ sér.), **10**, 1, 147, 161, 172.

FOKKER, A. D. (4) (with J. van Dyk & B. J. A. Pels). *Recherches Musicales, théoriques et pratiques*. Nijhoff, The Hague, 1951. See also *AMT*, 1953 (3ᵉ sér.), **10**, 1, 133, 147, 161, 172, 173.

FORBES, R. J. (2). *Man the Maker; a History of Technology and Engineering*. Schuman, New York, 1950. (Crit. rev. H. W. Dickinson & B. Gille, *A/AIHS*, 1951, **4**, 551.)

FORBES, R. J. (14). *Studies in Ancient Technology*. Vol. 5, *Leather in Antiquity; Sugar and its Substitutes in Antiquity; Glass*. Brill, Leiden, 1957.

FORBES, R. J. (15). *Studies in Ancient Technology*. Vol. 6, *Heat and Heating; Refrigeration, the art of cooling and producing cold; Lights and Lamps*. Brill, Leiden, 1958.

FORKE, A. (3) (tr.). *Me Ti [Mo Ti] des Sozialethikers und seiner Schüler philosophische Werke*. Berlin, 1922. (*MSOS*, Beibände, **23–25**.)

FORKE, A. (4) (tr.). '*Lun-Hêng*', *Philosophical Essays of Wang Chhung*. Vol. 1, 1907. Kelly & Walsh, Shanghai; Luzac, London; Harrassowitz, Leipzig. Vol. 2, 1911 (with the addition of Reimer, Berlin). (*MSOS*, Beibände, **10** and **14**.) (Crit. P. Pelliot, *JA*, 1912 (10ᵉ sér.), **20**, 156.)

FORKE, A. (9). *Geschichte d. neueren chinesischen Philosophie* (i.e. from the beginning of the Sung to modern times). de Gruyter, Hamburg, 1938. (Hansische Univ. Abhdl. a. d. Geb. d. Auslandskunde. no. 46 (ser. B, no. 25).)

FORKE, A. (12). *Geschichte d. mittelälterlichen chinesischen Philosophie* (i.e. from the beginning of the Former Han to the end of the Wu Tai). de Gruyter, Hamburg, 1934. (Hamburg. Univ. Abhdl. a. d. Geb. d. Auslandskunde, no. 41 (ser. B, no. 21).)

FORKE, A. (13). *Geschichte d. alten chinesischen Philosophie* (i.e. from antiquity to the beginning of the Former Han). de Gruyter, Hamburg, 1927. (Hamburg. Univ. Abhdl. a. d. Geb. d. Auslandskunde, no. 25 (ser. B, no 14).)

FORKE, A. (15). 'On Some Implements mentioned by Wang Chhung' (1. Fans, 2. Chopsticks, 3. Burning Glasses and Moon Mirrors). Appendix III to Forke (4).

FORKE, A. (17). 'Der Festungskrieg im alten China.' *OAZ*, 1919, **8**, 103. (Repr. from Forke (3), pp. 99 ff.)

FORSTER, L. (1). 'Translation: an Introduction.' In *Aspects of Translation*, ed. A. H. Smith, p. 1. Secker & Warburg, London, 1958. (University College, London, Communication Research Centre; Studies in Communication, no. 2.)

FOURNIER, G. (1). *Hydrographie*. Paris, 1643; repub. 1667.

FOX, H, M. (1). 'Lunar Periodicity in Reproduction.' *PRSB*, 1924, **95**, 523.

FRANKE, W. (2). 'Die Han-zeitlichen Felsengräber bei Chiating (West Szechuan).' *SSE*, 1948, **7**, 19.

FRANKEL, H. H. (1). *Catalogue of Translations from the Chinese Dynastic Histories for the Period +220 to +960*. Univ. Calif. Press, Berkeley and Los Angeles, 1957. (Inst. Internat. Studies, Univ. of California, East Asia Studies, Chinese Dynastic Histories Translations, Suppl. no. 1.)

FRASER, E. D. H. & LOCKHART, J. H. S. (1). *Index to the 'Tso Chuan'*. Oxford, 1930.

FRAZER, SIR J. G. (1). *The Golden Bough*, 3-vol. ed. Macmillan, London, 1900; superseded by 12-vol. ed. (here used), Macmillan, London, 1913–20. Abridged 1-vol. ed. Macmillan, London, 1923.

FREEMAN, K. (1). See Diels–Freeman.

FREEMAN, K. (2). *The Pre-Socratic Philosophers, a Companion to Diels, 'Fragmente der Vorsokratiker'*. Blackwell, Oxford, 1946.

FRÉMONT, C. (1). *Études Expérimentales de Technologie Industrielle, No. 10: Évolution des Méthodes et des Appareils employés pour l'Essai des Matériaux de Construction, d'après les Documents du Temps* (Renaissance onwards). (Internat. Congr. Strength of Materials, Paris, 1900.) Dunod, Paris, 1900.

FULKE, W. (1). *Uranomachia, seu Astrologorum Ludus*. Jones, London, 1571.

GALAMBOS, R. & GRIFFIN, D. R. (1). 'Avoidance of Obstacles by Bats.' *JEZ*, 1941, **86**, 481; 1942, **89**, 475.

GALE, E. M. (1) (tr.). *Discourses on Salt and Iron ('Yen Thieh Lun'), a Debate on State Control of Commerce and Industry in Ancient China, chapters 1–19*. Brill, Leiden, 1931. (Sinica Leidensia, no. 2.) (Crit. P. Pelliot, *TP*, 1932, 127.)

GALE, E. M., BOODBERG, P. A. & LIN, T. C. (1) (tr.). 'Discourses on Salt and Iron (*Yen Thieh Lun*), Chapters 20–28.' *JRAS/NCB*, 1934, **65**, 73.

LE GALL, S. (1). *Le Philosophe Tchou Hi, Sa Doctrine, son Influence*. T'ou-se-wei, Shanghai, 1894 (*VS*, no. 6). (Incl. tr. of part of ch. 49 of *Chu Tzu Chhüan Shu*.)

GALLAGHER, L. J. (1) (tr.). *China in the 16th Century; the Journals of Matthew Ricci, 1583–1610*. Random House, New York, 1953. (A complete translation, preceded by inadequate bibliographical details, of Nicholas Trigault's *De Christiana Expeditione apud Sinas* (1615). Based on an earlier publication: *The China that Was; China as discovered by the Jesuits at the close of the 16th Century: from the Latin of Nicholas Trigault*. Milwaukee, 1942.) Identifications of Chinese names in Yang Lien-Shêng (4). Crit. J. R. Ware, *ISIS*, 1954, **45**, 395.

GALPIN, F. W. (1). *The Music of the Sumerians*. Cambridge, 1937.

GALPIN, F. W. (2). *A Textbook of European Musical Instruments; their Origin, History and Character*. London, 1946.

GANDZ, S. (5). 'The Division of the Hour in Hebrew Literature.' *OSIS*, 1952, **10**, 10.

GARNER, H. M. (1). 'The Earliest Evidence of Chess in Western Literature: the Einsiedeln Verses' (*c. +1070*). *SP*, 1954, **29**, 734.

GARRISON, F. H. (2). 'History of Heating, Ventilation and Lighting.' *BNYAM*, 1927, **3**, 57.

GASSENDI, P. (1). *The Mirrour of True Nobility and Gentry, being the Life of N. C. Fabricius, Lord of Peiresk*. Tr. W. Rand. London, 1657.

GAUBIL, A. (1). Numerous contributions to *Observations Mathématiques, Astronomiques, Géographiques, Chronologiques et Physiques tirées des Anciens Livres Chinois ou faites nouvellement aux Indes et à la Chine par les Pères de la Compagnie de Jésus*, ed. E. Souciet. Rollin, Paris, 1729, vol. 1.

(*a*) Remarques sur l'Astronomie des Anciens Chinois en général, p. 1.

(*b*) Eclipses ⊙ Sexdecim in Historia aliisque veteribus Sinarum libris notatae et a Patre Ant. Gaubil e Soc. Jesu computate, p. 18. (The first is the *Shu Ching* eclipse attributed to −2155; then follows the *Shih Ching* eclipse attributed to −776; then five *Tso Chuan* eclipses (−720 to −495), then one of −382 and finally three Han ones.)

(*c*) Observations des Taches du Soleil, p. 33.

(*d*) Observation de l'Eclipse de ☾ du 22 Déc. 1722 à Canton, p. 44.

(*e*) Observatio Eclipsis Lunae totalis Pekini 22 Oct. 1725, p. 47.

(*f*) Occultations ou Eclipses des Etoiles Fixes par la lune, observées à Péking en 1725 & 1726, p. 59.

(*g*) Observations de Saturne, p. 69.

(*h*) Observations de Jupiter, p. 71.

(*i*) Observations de ♃ et de ses Satellites; Conjonctions ou Approximations de ♃ à des Étoiles Fixes, tirées des anciens livres d'Astronomie Chinoise (+73 to +1367), p. 72.

(*j*) Observations des Satellites de ♃, faites à Péking en 1724, p. 80.

(*k*) Observations de Mars, p. 95.

(*l*) Observations de Vénus, p. 98.

(*m*) Observations de Mercure, p. 101.

(*n*) Observations de la Comète de 1723 faites à Péking d'abord par des Chinois et ensuite par les PP. Gaubil & Jacques, p. 105.

(*o*) Observations géographiques (à) l'Ile de Poulo-Condor, p. 107.

(*p*) Plan de Canton, sa longitude et sa latitude, p. 123.

(*q*) Extrait du Journal du Voyage du P. Gaubil et du P. Jacques de Canton à Péking, p. 127.

(*r*) Plan (& Description) de Péking, p. 136.

(*s*) Situation de Poutala, demeure du grand Lama, des Sources du Gange et des pays circonvoisins, le tout tiré des Cartes Chinoises et Tartares, p. 138.

(*t*) Mémoire Géographique sur les Sources de l'Irtis et de l'Oby, sur le pays des Eleuthes et sur les Contrées qui sont au Nord et à l'Est de la Mer Caspienne, p. 141.

(*u*) Relation Chinoise contenant un itineraire de Péking à Tobol, et de Tobol au Pays des Tourgouts, p. 148.

(*v*) Remarques sur le Commencement de l'Année Chinoise, p. 182.

(*w*) Abrégé Chronologique de l'Histoire des Cinq Premiers Empereurs Mogols, p. 185.

(*x*) Observations Physiques (Lézard Volant à Poulo-Condor, Melon de Hami), p. 204.

(*y*) Observations sur la Variation de l'Aiman, p. 210.

(*z*) Observations Diverses, p. 223.

GAUBIL, A. (2). *Histoire Abrégée de l'Astronomie Chinoise.* (With Appendices 1, Des Cycles des Chinois; 2, Dissertation sur l'Éclipse Solaire rapportée dans le *Chou-King* [*Shu Ching*]; 3, Dissertation sur l'Éclipse du Soleil rapportée dans le *Chi-King* [*Shih Ching*]; 4, Dissertation sur la première Éclipse du Soleil rapportée dans le *Tchun-Tsieou* [*Chhun Chhiu*]; 5, Dissertation sur l'Éclipse du Soleil, observée en Chine l'an trente-et-unième de Jésus-Christ; 6, Pour l'Intelligence de la Table du *Yue-Ling* [*Yüeh Ling*]; 7, Sur les Koua; 8, Sur le Lo-Chou (recognition of Lo Shu as magic square).) In *Observations Mathématiques, Astronomiques, Géographiques, Chronologiques et Physiques, tirées des anciens Livres Chinois ou faites nouvellement aux Indes, à la Chine, et ailleurs, par les Pères de la Compagnie de Jésus*, ed. E. Souciet. Rollin, Paris, 1732, vol. 2.

GAUBIL, A. (11). *Description de la Ville de Pékin.* Ed. de l'Isle & Pingré, Paris, 1763, 1765. Russ. tr. by Stritter; Germ. tr. by Pallas; Eng. tr. (abridged), *PTRS*, 1758, **50**, 704.

GEIRINGER, K. (1). *Musical Instruments from the Stone Age to the Present Day.* Tr. B. Miall; ed. W. F. H. Blandford. London, 1945.

GERLAND, E. (1). *Geschichte d. Physik (erste Abt.); Von den ältesten Zeiten bis zum Ausgange des 18ten Jahrhunderts.* Oldenbourg, München and Berlin, 1913. (Geschichte d. Wissenschaften in Deutschland, vol. 24.)

GERLAND, E. (2). 'Zur Gesch. d. Kompasses.' *VDPG*, 1908, **10** (no. 10), 377 (*BDPG*, 1908, **6**).

GERLAND, E. & TRAUMÜLLER, F. (1). *Geschichte d. physikalischen Experimentierkunst.* Engelmann, Leipzig, 1899.

GHETALDI, MARINI (1). *Promotus Archimedis, seu, De Variis Corporum Generibus Gravitate et Magnitudine Comparatis.* Rome, 1603.

GIBSON, H. E. (1). 'Music and Musical Instruments of the Shang Dynasty.' *JRAS/NCB*, 1937, **68**, 8.

GILBERT, WILLIAM (1). *Tractatus sive Physiologia Nova de Magnete* .... Short, London, 1600, and several later editions. Eng. tr. Sylvanus P. Thompson *et al.* Chiswick Press, London, 1900; facsimile ed. Ed. Derek J. de S. Price. Basic Books, New York, 1958.

GILES, H. A. (1). *A Chinese Biographical Dictionary.* 2 vols. Kelly & Walsh, Shanghai, 1898; Quaritch, London, 1898. Supplementary Index by J. V. Gillis & Yü Ping-Yüeh, Peiping, 1936. Account must be taken of the numerous emendations published by von Zach (4) and Pelliot (34), but many mistakes remain. Cf. Pelliot (35).

GILES, H. A. (5). *Adversaria Sinica:*

1st series, no. 1, pp. 1–25. Kelly & Walsh, Shanghai, 1905.

no. 2, pp. 27–54. Kelly & Walsh, Shanghai, 1906.

no. 3, pp. 55–86. Kelly & Walsh, Shanghai, 1906.

no. 4, pp. 87–118. Kelly & Walsh, Shanghai, 1906.

no. 5, pp. 119–44. Kelly & Walsh, Shanghai, 1906.

no. 6, pp. 145–88. Kelly & Walsh, Shanghai, 1908.

no. 7, pp. 189–228. Kelly & Walsh, Shanghai, 1909.

no. 8, pp. 229–76. Kelly & Walsh, Shanghai, 1910.

no. 9, pp. 277–324. Kelly & Walsh, Shanghai, 1911.

no. 10, pp. 326–96. Kelly & Walsh, Shanghai, 1913.

no. 11, pp. 397–438 (with index). Kelly & Walsh, Shanghai, 1914.

2nd series, no. 1, pp. 1–60. Kelly & Walsh, Shanghai, 1915.

GILES, H. A. (6). 'Wei-Ch'i, or the Chinese Game of War.' In *Historic China and Other Sketches*, p. 330. London, 1882.

GILES, H. A. (11). *A History of Chinese Literature*. Heinemann, London, 1901.

GILES, H. A. (13). (tr.) *Strange Stories from a Chinese Studio* (transl. of Phu Sung-Ling's *Liao Chai Chih I*, +1679). 2nd ed. Kelly & Walsh, Shanghai, 1908.

GILES, L. (4) (tr.). *Taoist Teachings from the Book of Lieh Tzu*. Murray, London, 1912; 2nd ed. 1947.

GILES, L. (11) (tr.). *Sun Tzu on the Art of War ['Sun Tzu Ping Fa']; the oldest military Treatise in the World*. Luzac, London, 1910 (with original Chinese text). Repr. without notes, Nanfang, Chungking, 1945; also repr. in *Roots of Strategy*, ed. Phillips, T.R. (*q.v.*).

GIOVIO, PAOLO (1). *Historiarum Sui Temporis*. Florence, 1550-2; Strasbourg, 1556. Abridgement by V. Cartari, Venice, 1562.

GLANVILLE, S. R. K. (1). *Weights and Balances in Ancient Egypt*. Royal Institution, London, Nov. 1935.

VON GLASENAPP, H. (1). *La Philosophie Indienne, Initiation à son Histoire et à ses Doctrines*. Payot, Paris, 1951 (no index).

GODE, P. K. (2). 'Notes on the History of Glass Vessels and Glass Bangles in India, South Arabia and Central Asia.' *JOSP*, 1949, **1** (no. 1), 9.

GOODRICH, L. CARRINGTON (1). *Short History of the Chinese People*. Harper, New York, 1943.

GOODRICH, L. CARRINGTON (12). 'The Chinese *shêng* and Western Musical Instruments.' *CMAG*, 1941, **17**, 10, 11, 14.

GOODRICH, L. CARRINGTON & CHHÜ THUNG-TSU (1). 'Foreign Music at the Court of Sui Wên Ti.' *JAOS*, 1949, **69**, 148.

GOVI, M. (1). 'Les Miroirs Magiques des Chinois.' *ACP*, 1880 (5e sér.), **20**, 99. 'Nouvelles Expériences sur les Miroirs Chinois.' *ACP*, 1880 (5e sér.), **20**, 106.

GOWER, L. C. B. (1). *Looking at Chinese Justice*. Mimeographed report privately circulated after the visit of a delegation of jurists to China, April 1956.

GRAHAM, D. C. & DYE, D. S. (1). 'Ancient Chinese Glass; Beads from Koh Tombs in Western Szechuan.' *JWCBRS*, 1944 (ser. A), **15**, 34.

GRANET, M. (1). *Danses et Légendes de la Chine Ancienne*. 2 vols. Alcan, Paris, 1926.

GRANET, M. (2). *Fêtes et Chansons Anciennes de la Chine*. Alcan, Paris, 1926; 2nd ed. Leroux, Paris, 1929.

GRANET, M. (5). *La Pensée Chinoise*. Albin Michel, Paris, 1934. (Évol. de l'Hum. series, no. 25 *bis*.)

GRANTHAM, A. E. (1). *The Ming Tombs* (Shih San Ling). Wu Lai-Hsi, Peiping, 1926.

GREEFF, R. (1). *Die historische Entwicklung d. Brille*. Bergmann, Wiesbaden, 1913.

GREEFF, R. (2). *Die Erfindung der Augengläser*. Berlin, 1921.

GREGORY, J. C. (1). *A Short History of Atomism*. Black, London, 1931.

GRIFFIN, D. R. (1). *Listening in the Dark; the Acoustic Orientation of Bats and Men*. Yale Univ. Press, New Haven, 1958.

GRIFFIN, D. R. (2). 'The Navigation of Bats.' *SAM*, 1950, **183** (no. 2), 52. 'More about Bat "Radar".' *SAM*, 1958, **199** (no. 1), 40.

GRIFFIN, D. R. (3). 'Bird Sonar.' *SAM*, 1954, **190** (no. 3), 78.

DE GROOT, J. J. M. (2). *The Religious System of China*. Brill, Leiden, 1892.
  Vol. 1, Funeral rites and ideas of resurrection.
  Vols. 2, 3, Graves, tombs, and *fêng-shui*.
  Vol. 4, The soul, and nature-spirits.
  Vol. 5, Demonology and sorcery.
  Vol. 6, The animistic priesthood (*wu*).

GROVE, G. (1). *Dictionary of Music and Musicians*. 5 vols. Macmillan, London, 1950.

GRUBER, K. (1). *Das chinesische Schachspiel; Einführung mit Aufgaben und Parteien*. Siebenberg, Peking, 1937.

VAN GULIK, R. H. (1). *The Lore of the Chinese Lute*. Sophia University, Tokyo, 1940. (Monumenta Nipponica Monographs, no. 3.)

VAN GULIK, R. H. (6) (ed. and tr.). *'Thang Yin Pi Shih', Parallel Cases from under the Pear-Tree; a 13th-century Manual of Jurisprudence and Detection*. Brill, Leiden, 1956. (Sinica Leidensia, no. 10.)

GUNDEL, W. (2). *Dekane und Dekansternbilder*. Augustin, Glückstadt and Hamburg, 1936. (Stud. d. Bibl. Warburg, no. 19.)

GUNTHER, R. T. (1). *Early Science in Oxford*. 14 vols. Oxford, 1923-45. (The first pub. Oxford Historical Soc.; the rest privately printed for subscribers.)
  Vol. 1 1923 Chemistry, Mathematics, Physics and Surveying.
  Vol. 2 1923 Astronomy.
  Vol. 3 1925 Biological Sciences and Biological Collections.
  Vol. 4 1925 The [Oxford] Philosophical Society.

Vol. 5 1929 Chaucer and Messahalla on the Astrolabe.
Vol. 6 1930 Life and Work of Robert Hooke.
Vol. 7 1930 Life and Work of Robert Hooke (contd.).
Vol. 8 1931 Cutler Lectures of Robert Hooke (facsimile).
Vol. 9 1932 The *De Corde* of Richard Lower (facsimile), with introd. and tr. by K. J. Franklin.
Vol. 10 1935 Life and Work of Robert Hooke (contd.).
Vol. 11 1937 Oxford Colleges and their men of science.
Vol. 12 1939 Dr Plot and the Correspondence of the [Oxford] Philosophical Society.
Vol. 13 1938 Robert Hooke's *Micrographia* (facsimile).
Vol. 14 1945 Life and Letters of Edward Lhwyd.
GÜNTHER, S. (2). *Handbuch der Geophysik.* Enke, Stuttgart, 1897.
GUTKIND, E. A. (1). *Revolution of Environment.* Kegan Paul, London, 1946.

DE HAAN, BIERENS, D. (1). *'Van de Spiegeling der Singkonst' et 'Van de Molens'; deux Traités inédits* (of Simon Stevin). Amsterdam, 1884.
HACKIN, J. & HACKIN, J. R. (1). *Recherches archéologiques à Begram, 1937.* Mémoires de la Délégation archéologique française en Afghanistan, vol. 9. Paris, 1939.
HACKIN, J., HACKIN, J. R., CARL, J. & HAMELIN, P. (with the collaboration of J. Auboyer, V. Elisséeff, O. Kurz & P. Stern) (1). *Nouvelles Recherches archéologiques à Begram (ancienne Kāpiśi), 1939–40.* Mémoires de la Délégation archéologique française en Afghanistan, vol. 11. Paris, 1954. Crit. rev. P. S. Rawson, *JRAS,* 1957, 139.
HADDAD, SAMI I. & KHAIRALLAH, AMIN A. (1). 'A Forgotten Chapter in the History of the Circulation of the Blood.' *ASURG,* 1936, **104**, 1.
HAKLUYT, RICHARD (1). *The Principall Navigations, Voyages and Discoveries of the English Nation....* London, 1589; 2nd ed. much enlarged, 1598–1600; many times afterwards reprinted.
DU HALDE, J. B. (1). *Description Géographique, Historique, Chronologique, Politique et Physique de l'Empire de la Chine et de la Tartarie Chinoise.* 4 vols. Paris, 1735; The Hague, 1736. Eng. tr. R. Brookes, London, 1736, 1741.
HALL, A. R. (1). *Ballistics in the Seventeenth Century; a study in the Relations of Science and War, with reference principally to England.* Cambridge, 1951. Crit. T. S. Kuhn, *ISIS,* 1953, **44**, 284.
HALL, A. RIPLEY (1). 'The Early Significance of Chinese Mirrors.' *JAOS,* 1935, **55**, 182.
HALOUN, G. (2). Translations of *Kuan Tzu* and other ancient texts made with the present writer, unpub.
HALOUN, G. (6). 'Die Rekonstruktion der chinesischen Urgeschichte durch die Chinesen.' *JDZWT,* 1925, **3**, 243.
HALOUN, G. (7). 'Seit wann kannten die Chinesen die Tocharer oder Indogermanen überhaupt?' *AM,* 1924, **1**, 156.
HAMADA, KOSAKU & UMEHARA, SUEJI (1). *A Royal Tomb, 'Kinkan-Tsuka' or 'Gold-Crown' Tomb, at Keishu (Korea) and its Treasures.* 2 vols. text, 1 vol. plates. Sp. Rep. Serv. Antiq. Govt. Gen. Chosen, 1924, no. 3.
HARADA, YOSHITO & TAZAWA, KINGO (1). *Lo-Lang; a Report on the Excavation of Wang Hsü's Tomb in the Lo-Lang Province, an ancient Chinese Colony in Korea.* Tokyo University, Tokyo, 1930.
HARDEN, D. B. (1). 'Ancient Glass.' *AQ,* 1933, **7**, 419.
HARDING, R. E. M. (1). *A History of the Pianoforte.* Cambridge, 1940.
HARICH-SCHNEIDER, E. (1). 'The Present Condition of Japanese Court Music.' *MUQ,* 1953, **39**, 49.
HARLAND, W. A. (1). 'The Manufacture of Magnetic Needles and Vermilion [in China].' *JRAS/ NCB,* 1850, **1** (no. 2), 163. (Extracts from the *Thung Thien Hsiao.*)
DE HARLEZ, C. (5) (tr.). *Kuo Yü* (partial). *JA,* 1893 (9ᵉ sér.), **2**, 37, 373; 1894 (9ᵉ sér.), **3**, 5. Later parts published separately, Louvain, 1895.
HARRADON, H. D. (1). 'Some Early Contributions to the History of Geomagnetism; IV, The Letter of Georg Hartmann to Duke Albrecht of Prussia (+1544).' *TM,* 1943, **48**, 127. (Discovery of inclination.) See Chapman & Harradon (1, 2).
HARRADON, H. D. (2). 'Some Early Contributions to the History of Geomagnetism; VI, The Letter of Gerhard Mercator of Rupelmonde to Antonius Perrenotus, most venerable Bishop of Arras (+1546).' *TM,* 1943, **48**, 200. (First statement of the magnetic pole.)
HARRADON, H. D. & FERRAZ, J. DE SAMPAIO (1). 'Some Early Contributions to the History of Geomagnetism; V, The Shadow Instrument of Pedro Nunes (+1537).' *TM,* 1943, **48**, 197. (A magnetic compass combined with a vertical gnomon for ascertaining astronomical north for comparison with magnetic north.)
HARRADON, H. D. & FERRAZ, J. DE SAMPAIO (2). 'Some Early Contributions to the History of Geomagnetism; VII, Extracts on Magnetic Observations from the Log-Books of João de Castro (+1538, +1539 and +1541).' *TM,* 1944, **49**, 185. (Earliest observations of local variation.) See de Castro (1, 2, 3) and for a biography, Sanceau (2).

HARTNER, W. (7). 'Some Notes on Chinese Musical Art.' *ISIS*, 1938, **29**, 72.

HARTRIDGE, H. (1). 'Acoustic Control in the Flight of Bats.' *JOP*, 1920, **54**, 54; *N*, 1945, **156**, 490.

HARVEY, E. NEWTON (1). *A History of Luminescence from the Earliest Times until 1900.* American Philosophical Society, Philadelphia, 1957. (Amer. Philos. Soc. Memoirs, no. 44.)

HARVEY, E. NEWTON (2). *Bioluminescence.* New York, 1952.

HASHIMOTO, M. (1). 'On the Origin of the [Magnetic] Compass.' *MRDTB*, 1926, **1**, 69; originally partly in *TYG*, see Hashimoto (3). Crit. P. Pelliot, *TP*, 1929, **26**, 263; Anon. *ISIS*, 1930, **14**, 525; H. Maspero, *JA*, 1928, **212**, 159.

HAWKES, D. (1) (tr.). '*Chhu Tzhu*'; *the Songs of the South—an Ancient Chinese Anthology.* Oxford, 1959. (rev. J. Needham, *NSN*, 1959.)

HEATH, SIR THOMAS (6). *A History of Greek Mathematics.* 2 vols. Oxford, 1921.

VON HEINE-GELDERN, R. (1). 'Prehistoric Research in the Netherlands East Indies' (cultural connections between Indonesia and S.E. Europe). In *Science and Scientists in the Netherlands Indies.* Ed. P. Honig & F. Verdoorn. Board for the Netherlands Indies, Surinam and Curaçao; New York, 1945. (*Natuurwetenschappelijk Tijdschrift voor Nederlandsch Indie*, suppl. to **102**.)

HELLER, A. (1). *Geschichte d. Physik.* 2 vols. Enke, Stuttgart, 1882.

HELLER, J. F. (1). *Leuchten gefaulter Hölzer.* Deutsche Naturforscher-Versammlung, Berlin, 1843. 'Über das Leuchten in Pflanzen- und Tier-reiche.' *APPCM*, 1853, **6**, 44, 81, 121, 161, 201, 241.

HELLMANN, G. (5). 'Die Anfänge der magnetischen Beobachtungen.' *ZGEB*, 1897, **32**, 112.

VON HELMHOLTZ, HERMANN L. F. (1). *On the Sensations of Tone, as a Physiological Basis for the Theory of Music.* Tr. A. J. Ellis, orig. Germ. ed. 1877. Longmans Green, London, 1912.

HENNIG, R. (3). 'Ein Zusammenhang zwischen d. Magnetbergfabel und des Kenntnis d. Kompasses.' *AKG*, 1930, **20**, 350.

HENNIG, R. (4). *Terrae Incognitae; eine Zusammenstellung und kritische Bewertung der wichtigsten vor-columbischen Entdeckungsreisen an Hand der darüber vorliegenden Originalberichte.* 2nd ed. 4 vols. Brill, Leiden, 1944.

HENNIG, R. (5). 'Die Frühkenntnis der magnetischen Nordweisung'. *BGTI*, 1932, **21**, 25.

HENNIG, R. (6). 'Die Magnetbergsage und ihr naturwissenschaftlicher Hintergrund.' *GW*, 1935, **3**, 583.

HENNIG, R. (7). *Rätselhafte Lände.* Berlin, 1950.

HERMANN, H. (1). 'Chinesische Physik.' *UMN*, 1935, **41**, 21.

HERRMANN, A. (1). *Historical and Commercial Atlas of China.* Harvard-Yenching Institute, Cambridge, Mass., 1935.

HESSE, MARY B. (1). 'Action at a Distance in Classical Physics.' *ISIS*, 1955, **46**, 337.

HESSE, MARY B. (2). 'Models in Physics.' *BJPS*, 1953, **4**, 198.

HETHERINGTON, A. L. (1). *Chinese Ceramic Glazes.* Cambridge, 1937. Revised ed. Perkins, South Pasadena, Calif., 1948.

HETT, G. V. (1). 'Some [Confucian] Ceremonies at Seoul.' *GGM*, 1936, **3**, 179.

HIGHLEY, S. (1). Letter on Magic Mirrors. *N*, 1877, **16**, 132.

HIMLY, K. (2). 'Der Schachspiel d. Chinesen.' *ZDMG*, 1870, **24**, 172.

HIMLY, K. (3). 'Streifzüge in das Gebiet d. Gesch. d. Schachspiels.' *ZDMG*, 1872, **26**, 121; 1873, **27**, 121.

HIMLY, K. (4). 'Anmerkungen in Beziehung auf das Schach und andere Brettspiele.' *ZDMG*, 1887, **41**, 461.

HIMLY, K. (5). 'Morgenländisch oder Abendländisch?; Forschungen nach gewissen Spielausdrücken.' *ZDMG*, 1889, **43**, 415; 1890, **44**.

HIMLY, K. (6). 'Das japanische Schachspiel.' *ZDMG*, 1879, **33**, 672.

HIMLY, K. (7). 'The Chinese Game of Chess, as compared with that practised among Western Nations' (*hsiang chhi*). *JRAS/NCB*, 1870, **6**, 105.

HIRSCHBERG, J. (1). 'Geschichte d. Augenheilkunde.' In *Handbuch d. ges. Augenheilkunde*, vols. 12–15, ed. A. Graefe & T. Saemisch,

    Pt. I   vol. 12   Ancient Egypt, Greece, India, China, 1898.

    Pt. II   vol. 13   The Arabs, Mediaeval Europe, Sixteenth and Seventeenth Centuries, 1905–1908.

    Pt. III   vol. 14   Eighteenth and early Nineteenth Centuries, 4 vols., 1911–1915.

    Pt. IV   vol. 15   Modern, 3 vols., 1916–1918.

HIRTH, F. (1). *China and the Roman Orient.* Kelly & Walsh, Shanghai; G. Hirth, Leipzig and Munich, 1885. (Photographically reproduced in China with no imprint, 1939.)

HIRTH, F. (3). *Ancient History of China; to the end of the Chou Dynasty.* New York, 1908; 2nd ed. 1923.

HIRTH, F. (5). 'Chinese Metallic Mirrors.' In Boas Memorial Volume *Anthropological Papers written in honour of Franz Boas*, p. 208. Stechert, New York, 1906.

HIRTH, F. (6). 'Zur Geschichte d. Glases in China.' In *Chinesische Studien*, p. 62, Hirth (7).

HIRTH, F. (7). *Chinesische Studien.* Hirth, München and Leipzig, 1890.

HIRTH, F. (8). 'Origin of the Mariner's Compass in China.' *MO*, 1906, **16**, 321. (Approximately equivalent to the study of the same subject in Hirth, 3.)

HIRTH, F. & ROCKHILL, W. W. (1) (tr.). *Chau Ju-Kua; His work on the Chinese and Arab Trade in the 12th and 13th centuries, entitled 'Chu-Fan-Chi'*. Imp. Acad. Sci., St Petersburg, 1911. (Crit. G. Vacca, *RSO*, 1913, **6**, 209; P. Pelliot, *TP*, 1912, **13**, 446; E. Schaer, *AGNT*, 1913, **6**, 329; O. Franke, *OAZ*, 1913, **2**, 98; A. Vissière, *JA* 1914 (11ᵉ sér.), **3**, 196.)

HITCHINS, H. L. & MAY, W. E. (1). *From Lodestone to Gyro-Compass*. Philos. Lib., New York, 1953. (rev. D. H. D. Roller, *ISIS*, 1953, **44**, 303.)

HITTI, P. K. (1). *History of the Arabs*. 4th ed. Macmillan, London, 1949. 6th ed., 1956.

HO PING-YÜ (1). *The Astronomical Chapters of the 'Chin Shu'*. Inaug. Diss., Singapore, 1957.

HO PING-YÜ & NEEDHAM, JOSEPH (2). 'Theories of Categories in Early Mediaeval Chinese Alchemy' (with transl. of the *Tshan Thung Chhi Wu Hsiang Lei Pi Yao*, c. +7th cent.). *JWCI*, 1959, **22**, 173.

HO PING-YÜ & NEEDHAM, JOSEPH (3). 'The Laboratory Equipment of the Early Mediaeval Chinese Alchemists.' *AX*, 1959, **7**, 57.

HODGSON, W. C. (1). 'Echo-Sounding and the Pelagic Fisheries.' Min. of Agric. & Fisheries. *Fishery Investigation Series* (II), **17**, no. 4. HMSO, London, 1951.

HODGSON, W. C. & RICHARDSON, I. D. (1). 'The Cornish Pilchard Experiment [on the use of echo-sounding in fishing].' Min. of Agric. & Fisheries. *Fishery Investigation Series*, no. 2. HMSO, London, 1950.

HOGBEN, L. (1). *Mathematics for the Million*. Allen & Unwin, London, 1936; 2nd ed. 1937.

HOLLINGWORTH, H. G. (1). 'A short sketch of the Chinese game of chess called k'he [*chhi*] also seang k'he [*hsiang chhi*] to distinguish it from wei k'he [*wei chhi*], another game played by the Chinese.' *JRAS/NCB*, 1866, **3**, 107.

HOLT, H. F. W. (1). 'Notes on the Chinese Game of Chess.' *JRAS*, 1885, **17**, 352.

HOLTZ, V. (1). 'Japanisches Schachspiel.' *MDGNVO*, **1**, no. 5.

HOLZMAN, D. (1). 'Shen Kua and his *Mêng Chhi Pi Than*.' *TP*, 1958, **46**, 260.

HOMMEL, F. (1). 'Über d. Ursprung und d. Alter d. arabischen Sternnamen und insbesondere d. Mondstationen.' *ZDMG*, 1891, **45**, 616.

HOMMEL, R. P. (1). *China at Work; an illustrated Record of the Primitive Industries of China's Masses, whose Life is Toil, and thus an Account of Chinese Civilisation*. Bucks County Historical Society, Doylestown, Pa. 1937; John Day, New York, 1937.

HONEY, W. B. (1). *Glass; a Handbook and Guide to the Museum Collection*. Victoria and Albert Museum, London, 1946. 'Early Chinese Glass.' *BUM*, 1937, **71**, 211; *TOCS*, 1939, **17**, 35.

HOPPE, E. (1). 'Geschichte d. Physik.' In *Handbuch d. Physik*, I. Ed. H. Geiger & K. Scheel. Springer, Berlin and Leipzig, 1926.

HOPPE, E. (2). 'Magnetismus u. Elektrizität im klassischen Altertum.' *AGNT*, 1917, **8**, 92.

D'HORMON, A. (1) (ed.). *Lectures Chinoises*. École Franco-Chinoise, Peiping, 1945.

VON HORNBOSTEL, E. M. (1). On the cycle of blown fifths (Notes by P. G. Schmidt). *AN*, 1919, **14**, 569. 'Eine Tafel zur logarithmischen Darstellung von Zahlenverhältnissen.' *ZP*, 1921, **6**, 29, with corrigendum on p. 164.

VON HORNBOSTEL, E. M. (2). '*Chhao Thien Tzu* [Visiting the Son of Heaven]: eine chinesische Notation und ihre Ausführungen.' *AMW*, 1919, **1**, 477.

HORWITZ, H. T. (7). 'Beiträge z. Geschichte d. äussereuropäischen Technik.' *BGTI*, 1926, **16**, 290.

HOUGH, W. (1). 'Fire as an Agent in Human Culture.' *BUSNM*, 1926, no. 139.

HOUGH, W. (2). 'Collection of Heating and Lighting Utensils in the United States National Museum.' *BUSNM*, 1928, no. 141.

HSIA NAI (1). 'New Archaeological Discoveries.' *CREC*, 1952, **1** (no. 4), 13.

HSIANG TA & HUGHES, E. R. (1). 'Chinese Books in the Bodleian Library.' *BQR*, 1936, **8**, 227.

HSIAO CHHIEN (1). *A Harp with a Thousand Strings*. London, 1944.

HUANG MAN (WONG MAN) (1). 'The *Nei Ching*, the Chinese Canon of Medicine.' *CMJ*, 1950, **68**, 1 (originally Inaug. Diss. Cambridge.)

HUARD, P. & DURAND, M. (1). *Connaissance du Viêt-Nam*. École Française d'Extr. Orient, Hanoi, 1954; Imprimerie Nationale, Paris, 1954.

HUARD, P. & HUANG KUANG-MING (M. WONG) (1). 'La Notion de Cercle et la Science Chinoise.' *A/AIHS*, 1956, **9**, 111. (Mainly physiological and medical.)

HÜBOTTER, F. (1). *Die chinesische Medizin zu Beginn des XX. Jahrhunderts, und ihr historischer Entwicklungsgang*. Schindler, Leipzig, 1929. (China-Bibliothek d. Asia Major, no. 1.)

HUGHES, E. R. (1). *Chinese Philosophy in Classical Times*. Dent, London, 1942. (Everyman Library, no. 973.)

HUGHES, E. R. (2) (tr.). *The Great Learning and the Mean-in-Action*. Dent, London, 1942.

VON HUMBOLDT, A. (1). *Cosmos; a Sketch of a Physical Description of the Universe*. 5 vols., tr. E. Cotté, B. H. Paul & W. S. Dallas. Bohn, London, 1849–58.

VON HUMBOLDT, A. (2). *Asie Centrale, Recherches sur les Chaînes de Montagnes et la Climatologie comparée.* 3 vols. Gide, Paris, 1843.

VON HUMBOLDT, A. (3). *Examen critique de l'Histoire de la Géographie du Nouveau Continent et des Progrès de l'Astronomie Nautique au 15e et 16e Siècles.* 5 vols. Gide, Paris, 1836–39.

HUMMEL, A. W. (2) (ed.). *Eminent Chinese of the Ch'ing Period.* 2 vols. Library of Congress, Washington, 1944.

HUMMEL, A. W. (6). 'Astronomy and Geography in the Seventeenth Century [in China].' (On Hsiung Ming-Yü's work.) *ARLC/DO*, 1938, 226.

HUMMEL, A. W. (15). 'Dominoes in the Ming Period.' *ARLC/DO*, 1939, 265.

IBEL, T. (1). *Die Wage im Altertum und Mittelalter.* Inaug. Diss. Erlangen, 1908.

IRWIN, E. (1). 'An Account of the Game of Chess, as played by the Chinese.' *TRIA*, 1793, **5** (Antiq. Sect.), 53.

JACOB, G. & JENSEN, H. (1). *Das chinesische Schattentheater.* Stuttgart, 1933. (Das Orientalische Schattentheater, no. 3.)

JAKOB, M. (1). *Heat Transfer.* Wiley, New York, 1949.

JAL, A. (1). *Archéologie Navale.* 2 vols. Arthus Bertrand, Paris, 1840. (Crit. R. C. Anderson, *MMI*, 1920, **6**, 18; 1945, **31**, 160; A. B. Wood, 1919, **5**, 81.)

ABD AL-JALIL, J. M. (1). *Brève Histoire de la Littérature Arabe.* Maisonneuve, Paris, 1943; 2nd ed. 1947.

JEANS, J. H. (SIR JAMES) (2). *Science and Music.* Cambridge, 1937.

JENKINS, C. (1). 'Saint Augustine and Magic.' In *Science, Medicine and History.* 2 vols. Charles Singer Presentation Volume. Ed. E. A. Underwood, vol. 1, p. 132. Oxford, 1954.

JOHNSTON, R. F. (1). *Confucianism and Modern China.* Gollancz, London, 1934.

JONES, A. M. (1). *African Music in Northern Rhodesia and some other Places.* Rhodes-Livingstone Museum, Rhodesia, 1949. (Occasional Papers of the Museum, no. 4.)

JOVIUS, PAULUS. See Giovio, Paolo.

JULIEN, STANISLAS (3). 'Notice sur les Miroirs Magiques des Chinois et leur Fabrication; suivie de Documents Neufs sur l'invention de l'art d'imprimer à l'aide de planches en bois, de planches en pierre, et de types mobiles, huit, cinq, et quatre siècles, avant que l'Europe en fit usage.' *CRAS*, 1847, **24**, 999.

KALTENMARK, M. (2) (tr.). *Le 'Lie Sien Tchouan' [Lieh Hsien Chuan]; Biographies Légendaires des Immortels Taoistes de l'Antiquité.* Centre d'Études Sinologiques Franco-Chinois (Univ. Paris), Peking, 1953. Crit. P. Demiéville, *TP*, 1954, **43**, 104.

KAMMERER, ALBERT (1). 'La Découverte de la Chine par les Portugais au XVIème siècle et la Cartographie des Portulans.' *TP*, 1944, **39** (Suppl.), 122.

KARLGREN, B. (1). *Grammata Serica; Script and Phonetics in Chinese and Sino-Japanese.* BMFEA, 1940, **12**, 1. (Photographically reproduced as separate volume, Peiping, 1941.) Revised edition, *Grammata Serica Recensa*, Stockholm, 1957.

KARLGREN, B. (12) (tr.). 'The Book of Documents' (*Shu Ching*). *BMFEA*, 1950, **22**, 1.

KARLGREN, B. (14) (tr.). *The Book of Odes; Chinese Text, Transcription and Translation.* Museum of Far Eastern Antiquities, Stockholm, 1950. (A reprint of the text and translation only from his papers in *BMFEA*, **16** and **17**; the glosses will be found in **14**, **16** and **18**.)

KARRER, P. (1). *Organic Chemistry.* Elsevier, Amsterdam and New York, 1938. Tr. from the German by A. J. Mee.

KELLNER, L. (1). 'Alexander von Humboldt and the Organisation of International Collaboration in Geophysical Research.' *CPH*, 1959, **1**, 35. Also *SCI*, 1960, **95**, 252.

KIRCHNER, G. (1). 'Amber Inclusions.' *END*, 1950, **9**, 70.

KISA, A. (1). *Das Glas im Altertümer.* 3 vols. Hiersemann, Leipzig, 1908.

KLAPROTH, J. (1). '*Lettre à M. le Baron A. de Humboldt, sur l'Invention de la Boussole.* Dondey-Dupré, Paris, 1834. Germ. tr. A. Wittstein, Leipzig, 1884; résumés P. de Larenaudière, *BSG*, 1834, Oct.; Anon., *AJ*, 1834 (2nd ser.), **15**, 105.

KLEBS, L. (3). 'Die Reliefs und Malereien des neuen Reiches (18.–20. Dynastie, c. 1580–1100 v. Chr.): Material zur ägyptischen Kulturgeschichte.' Pt. I. 'Szenen aus dem Leben des Volkes.' *AHAW/PH*, 1934, no. 9.

KNIPPING, E. (1). 'Lokal-Attraction beobachtet auf dem Gipfel des Futarasan (Nantaisan) [in Japan].' *MDGNVO*, 1879, **2**, 35.

KOIZUMI, AKIO & HAMADA, KOSAKU (1). *The Tomb of the Painted Basket, and Two Other Tombs of Lo-Lang.* Archaeol. Res. Rep. no. 1. Soc. Stud. Korean Antiq. Seoul, 1934.

KOOP, A. J. (1). *Early Chinese Bronzes.* Benn, London, 1924.

KOYRÉ, A. (3). 'Galileo and the Scientific Revolution of the Seventeenth Century.' *PHR*, 1943, **52**, 333.

KRAMER, J. B. (1). 'The Early History of Magnetism.' *TNS*, 1934, **14**, 183.

KRAMERS, R. P. (1) (tr.). '*Khung Tzu Chia Yu*': *the School Sayings of Confucius* (chs. 1–10). Brill, Leiden, 1950. (Sinica Leidensia, no. 7.)

KU PAO-KU (1) (tr.). *Deux Sophistes Chinois; Houei Che [Hui Shih] et Kong-souen Long [Kungsun Lung]*. Presses Univ. de France (Imp. Nat.), Paris, 1953. (Biblioth. de l'Instit. des Hautes Études Chinoises, no. 8.) (Crit. P. Demiéville, *TP*, 1954, **43**, 108.) 'Notes Complémentaires sur "Deux Sophistes Chinois"', in *Mélanges pub. par l'Inst. des Htes. Études Chinoises*, 1957, vol. 1 (Biblioth. de l'Instit. des Htes. Études Chinoises, no. 11).

KUNST, J. (1). *Musicologica*. Indisch Instituut, Amsterdam, 1950.

KUNST, J. (2). *Around von Hornbostel's Theory of the Cycle of Blown Fifths*. Indisch Instituut, Amsterdam, 1948. (Koninklijke Vereeniging Indisch Instituut, Mededeelingen no. 76; Afd. Volkenkunde, no. 27.)

KUNST, J. (3). 'A Hypothesis about the Origin of the Gong.' *ETH*, 1947, **12** (no. 1/2), 79; amplified in 1947, **12** (no. 4), 147; 1949, **14** (no. 2/4), 160.

KUTTNER, F. A. (1). 'The Musical Significance of Archaic Chinese Jades of the *pi* disc type.' *AA*, 1953, **16**, 25.

KUTTNER, F. A. (2). 'Acoustical Skills and Techniques in Early Chinese History.' Unpub. MS.

KUTTNER, F. A. (3). 'A "Pythagorean" Tone System in China, antedating the early Greek Achievements.' Unpub. MS.

KUWABARA, JITSUZO (1). "On Phu Shou-Kêng, a man of the Western Regions, who was the Superintendent of the Trading Ships' Office in Chhüan-Chou towards the end of the Sung Dynasty, together with a general sketch of the Trade of the Arabs in China during the Thang and Sung eras.' *MRDTB*, 1928, **2**, 1; 1935, **7**, 1 (rev. P. Pelliot, *TP*, 1929, **26**, 364; S.E[lisséev], *HJAS*, 1936, **1**, 265). Chinese translation by Chhen Yü-Ching, Chung-hua, Peking, 1954.

KUWAKI, AYAO (1). 'The Physical Sciences in Japan, from the time of the first contact with the Occident until the time of the Meiji Restoration.' In *Scientific Japan, Past and Present*. Ed. Shinjo Shinzo, p. 243. IIIrd Pan-Pacific Science Congress, Tokyo, 1926.

DE LACOUPERIE, TERRIEN (3). 'On the Ancient History of Glass and Coal, and the Legend of Nu Kua's Coloured Stones in China.' *TP*, 1891, **2**, 234.

DE LACOUPERIE, T. (4). *The Calendar Plant of China, the Cosmic Tree, and the Date Palm of Babylonia*. Nutt & Luzac, London, 1890.

LALOY, L. (1). *Aristoxène de Tarente*. Paris, 1904.

LALOY, L. (2). *La Musique Chinoise*. Paris, 1910.

LANGE, H. (1). 'Die Kenntnis d. Missweisung oder magnetische Deklination bei dem Londoner Geoffrey Chaucer [+1380]." *FF*, 1935, **11**, 156.

LANGHORNE, J. & W. (1) (tr.). *Plutarch's 'Lives'*. London, 1770, 1823.

LANSER, O. (1). 'Zur Geschichte d. hydrometrischen Messwesens.' *BTG*, 1953, **15**, 25.

VON D. LASA, T. (1). *Zur Geschichte u. Literatur des Schachspiels*. Leipzig, 1897.

LASSWITZ, K. (1). *Geschichte d. Atomistik in Mittelalter bis Newton*. Leipzig, 1890; 2nd ed. 1926.

LAUFER, B. (1). *Sino-Iranica; Chinese Contributions to the History of Civilisation in Ancient Iran*. *FMNHP/AS*, 1919, **15**, no. 3 (Pub. no. 201) (rev. and crit. Chang Hung-Chao, *MGSC*, 1925 (ser. B), no. 5).

LAUFER, B. (3). *Chinese Pottery of the Han Dynasty*. (Pub. of the East Asiatic Cttee. of the Amer. Mus. Nat. Hist.). Brill, Leiden, 1909. (Reprinted Tientsin, 1940.)

LAUFER, B. (8). *Jade; a Study in Chinese Archaeology and Religion*. *FMNHP/AS*, 1912. Repub. in book form, Perkins, Westwood & Hawley, South Pasadena, 1946 (rev. P. Pelliot, *TP*, 1912, **13**, 434.)

LAUFER, B. (10). 'The Beginnings of Porcelain in China.' *FMNHP/AS*, 1917, **15**, no. 2 (Pub. no. 192) (includes description of +2nd-century cast-iron funerary cooking-stove).

LAUFER, B. (12). 'The Diamond; a study in Chinese and Hellenistic Folk-Lore.' *FMNHP/AS*, 1915, **15**, no. 1 (Pub. no. 184).

LAUFER, B. (14). 'Optical Lenses' (in China and India). *TP*, 1915, **16**, 169 and 562.

LAUFER, B. (16). 'Zur Geschichte d. Brille.' *MGGM*, 1907, **6**, 379.

LAUFER, B. (17). 'Historical Jottings on Amber in Asia.' *MAAA*, 1906, **1**, 211.

LAUFER, B. (18). 'The Prehistory of Television' (legends about mirrors showing events at long distance and future time). *SM*, 1928, **27**, 455.

LAUFER, B. (26). 'Chinese Pigeon Whistles.' *SAM*, 1908, 394.

LAYARD, A. H. (1). *Discoveries among the Ruins of Nineveh and Babylon*. London, 1845.

LECOMTE, LOUIS (1). *Nouveaux Mémoires sur l'État présent de la Chine*. Anisson, Paris, 1696. (Eng. tr. *Memoirs and Observations Topographical, Physical, Mathematical, Mechanical, Natural, Civil and Ecclesiastical, made in a late journey through the Empire of China, and published in several letters, particularly upon the Chinese Pottery and Varnishing, the Silk and other Manufactures, the Pearl Fishing, the History of Plants and Animals, etc. translated from the Paris edition, etc.* 2nd ed. London, 1698. Germ. tr. Frankfurt, 1699–1700.)

LEGGE, J. (1) (tr.). *The Texts of Confucianism, translated*: Pt. I. The '*Shu king*', *the religious portions of the* '*Shih Ching*', *the* '*Hsiao Ching*'. Oxford, 1879. (*SBE*, no. 3; reprinted in various eds. Com. Press, Shanghai.) For the full version of the *Shu Ching* see Legge (10).

LEGGE, J. (2) (tr.). *The Chinese Classics, etc.*: Vol. 1. *Confucian Analects, The Great Learning, and the Doctrine of the Mean.* Legge, Hongkong, 1861; Trübner, London, 1861.

LEGGE, J. (3) (tr.). *The Chinese Classics, etc.*: Vol. 2. *The Works of Mencius.* Legge, Hongkong, 1861; Trübner, London, 1861.

LEGGE, J. (5) (tr.). *The Texts of Taoism.* (Contains (*a*) *Tao Tê Ching*, (*b*) *Chuang Tzu*, (*c*) *Thai Shang Kan Ying Phien*, (*d*) *Chhing Ching Ching*, (*e*) *Yin Fu Ching*, (*f*) *Jih Yung Ching*.) 2 vols. Oxford, 1891; photolitho reprint, 1927. (*SBE*, nos. 39 and 40.)

LEGGE, J. (7) (tr.). *The Texts of Confucianism*: Pt. III. The '*Li Chi*'. 2 vols. Oxford, 1885; reprint, 1926. (*SBE*, nos. 27 and 28.)

LEGGE, J. (8) (tr.). *The Chinese Classics, etc.*: Vol. 4, Pts. 1 and 2. '*Shih Ching*'; *The Book of Poetry.* 1. The First Part of the *Shih Ching*; or, the Lessons from the States; and the Prolegomena. 2. The Second, Third and Fourth Parts of the *Shih Ching*; or the Minor Odes of the Kingdom, the Greater Odes of the Kingdom, the Sacrificial Odes and Praise-Songs; and the Indexes. Lane Crawford, Hongkong, 1871; Trübner, London, 1871. Repr., without notes, Com. Press, Shanghai, n.d.

LEGGE, J. (9) (tr.). *The Texts of Confucianism*: Pt. II. The '*Yi King*' [*I Ching*]. Oxford, 1882, 1899. (*SBE*, no. 16.)

LEGGE, J. (10) (tr.). *The Chinese Classics, etc.*: Vol. 3, Pts. 1 and 2. The '*Shoo King*' (*Shu Ching*). Legge, Hongkong, 1865; Trübner, London, 1865.

LEGGE, J. (11). *The Chinese Classics, etc.*: Vol. 5, Pts. 1 and 2. The '*Ch'un Ts'ew*' with the '*Tso Chuen*' (*Chhun Chhiu* and *Tso Chuan*). Lane Crawford, Hongkong, 1872; Trübner, London, 1872.

LEJEUNE, A. (1). 'Les Lois de la Réflexion dans l'Optique de Ptolémée.' *AC*, 1947, **15**, 241.

LEJEUNE, A. (2). 'Les Tables de Réfraction de Ptolémée.' *ASSB*, 1946, **60**, 93.

LEVIS, J. H. (1). *Foundations of Chinese Musical Art.* Vetch, Peiping, 1936.

LI HUI (1). 'A comparative study of the "Jew's Harp" among the Aborigines of Formosa and East Asia.' *AS/BIE*, 1956, **1**, 137.

LI SHU-HUA (2). 'Origine de la Boussole, II; Aimant et Boussole.' *ISIS*, 1954, **45**, 175. Engl. tr. with the addition of Chinese characters, *CHJ/T*, 1956, **1** (no. 1), 81.

LI SHU-HUA (3). 'Première Mention de l'Application de la Boussole à la Navigation.' *ORE*, 1954, **1**, 6.

LIAO WÊN-KUEI (1) (tr.). *The Complete Works of Han Fei Tzu; a Classic of Chinese Legalism.* 2 vols. Probsthain, London, 1939.

LIBES, A. (1). *Histoire philosophique des Progrès de la Physique.* 4 vols. Courcier, Paris, 1810–13.

LIBRI-CARRUCCI, G. B. I. T. (1). *Histoire des Sciences Mathématiques en Italie depuis la Renaissance des Lettres jusqu'à la Fin du 17ème Siècle.* 4 vols. Renouard, Paris, 1838–40.

LIESEGANG, F. P. (1). 'Der Missionar und China-geograph Martin Martini als erster Lichtbildredner.' *PRO*, 1937, **2**, 112.

LILLEY, S. (2). 'Attitudes to the Nature of Heat about the Beginning of the Nineteenth Century.' *A/AIHS*, 1948, **1**, 630.

LIN YÜ-THANG (1) (tr.). *The Wisdom of Lao Tzu* [*and Chuang Tzu*], *translated, edited, and with an introduction and notes.* Random House, New York, 1948.

VAN DER LINDE, A. (1). *Geschichte u. Litteratur d. Schachspiels.* Springer, Berlin, 1874.

VAN DER LINDE, A. (2). *Quellenstudien z. Gesch. d. Schachspiels.* Springer, Berlin, 1881.

VON LIPPMANN, E. O. (1). *Entstehung und Ausbreitung der Alchemie ... Ein Beitrag zur Kulturgeschichte.* Springer, Berlin, 1919.

VON LIPPMANN, E. O. (2). 'Geschichte d. Magnet-Nadel bis zur Erfindung des Kompasses (gegen +1300).' *QSGNM*, 1933, **3**, 1. Also separately published, Springer, Berlin, 1932.

VON LIPPMANN, E. O. (3). *Abhandlungen und Vorträge zur Geschichte d. Naturwissenschaft.* Veit, Leipzig, 1913. Has (*a*) 'Chemisches bei Marco Polo', p. 258; (*b*) 'Die spezifische Gewichtsbestimmung bei Archimedes', p. 168; (*c*) 'Zur Gesch. d. Saccharometers u. d. Senkspindel', pp. 171, 177, 183, etc.

LISSMANN, H. W. (1). 'Continuous Electrical Signals from the tail of a Fish *Gymnarchus niloticus*.' *N*, 1951, **167**, 201.

LISSMANN, H. W. (2). 'On the Function and Evolution of Electric Organs in Fish.' *JEB*, 1953, **35**, 156.

LISSMANN, H. W. & MACHIN, K. E. (1). 'The Mechanism of Object Location in *Gymnarchus niloticus* and similar Fish.' *JEB*, 1958, **35**, 451.

van der Lith, P. A. & Devic, L. M. (1) (tr.). *Le Livre des Merveilles de l'Inde* (the *'Aj'āib al-Hind* by Buzurj ibn Shahriyār al-Rāmhurmuzī, +953). Brill, Leiden, 1883.

Lu Gwei-Djen, Salaman, R. A. & Needham, Joseph (1). 'The Wheelwright's Art in Ancient China; I, The Invention of "Dishing".' *PHY*, 1959, **1**, 103.

McAdams, W. H. (1). *Heat Transmission.* McGraw-Hill, New York, 1942.

McGowan, D. J. (5). Note on 'the art of making luminous paint in the Celestial Empire' (i.e. on artificial phosphors in medieval China). *SC*, 1883, **2**, 698. (Abstract of a communication to *NCH*.)

McPhee, C. (1). 'The Five-Tone Gamelan Music of Bali.' *MUQ*, 1949, **35**, 250.

McPhee, C. (2). *A House in Bali.* New York, 1944.

Ma Hêng (1). *The Fifteen Different Classes of Measures as given in the 'Lü Li Chih' of the 'Sui Shu'*, tr. J. C. Ferguson. Privately printed, Peiping, 1932. (Ref. W. Eberhard, *OAZ*, 1933, **9** (**19**), 189.)

Machabey, A. (1). *Mémoire sur l'Histoire de la Balance et de la Balancerie.* Impr. Nat. Paris, 1949.

Mahdihassan, S. (10). 'The Chinese Origin of the Indian Terms for Climate and of the Arabic word for Magnet.' *PAKJS*, 1956, **8**, 127.

Mahillon, V. C. (1). *Catalogue descriptif et analytique du Musée instrumental du Conservatoire royal de Musique de Bruxelles* (Chinese Section). 4 vols. Hoste, Gand, 1886, 1893, 1912.

Maier, A. (1). *Die Impetustheorie der Scholastik.* Vienna, 1940.

Maier, A. (2). 'Der Funktionsbegriff in der Physik des 14. Jahrhunderts.' *DI*, 1946, **24**, 147. Repr. in (3).

Maier, A. (3). *Die Vorläufer Galileis im 14. Jahrhundert; Studien zur Naturphilosophie d. Spät-scholastik.* Ed. di Storia e Lett., Rome, 1949. Crit. rev. A. Koyré, *A/AIHS*, 1951, **4**, 769.

Maier, A. (4). *An der Grenze der Scholastik und Naturwissenschaft.* Essen, 1943 (rev. E. J. Dijksterhuis, *ISIS*, 1949, **40**).

Maier, A. (5). 'La Doctrine de Nicolas d'Oresme sur les *Configurationes Intensionum*.' *RSPT*, 1948, **32**, 52.

Maier, A. (6). 'Die Anfänge des physikalischen Denkens im 14 Jahrhundert.' *PN*, 1950, **1**.

Maier, A. (7). *Zwischen Philosophie und Mechanik.* Ed. di Storia e Lett., Rome, 1958. (Studien zur Naturphilosophie der Spätscholastik, no. 5.)

Maillard, M. (1). 'Note sur la Fabrication des Miroirs Magiques Chinois.' *CRAS*, 1853, **37**, 178.

Marcus, G. J. (1). 'The Navigation of the Norsemen.' *MMI*, 1953, **39**, 112.

Martin, T. Henri (1). *La Foudre, l'Électricité et le Magnétisme chez les Anciens.* Didier, Paris, 1866.

Martin, W. A. P. (3). *Hanlin Papers.* 2 vols. Vol. 1, Trübner, London, 1880, Harper, New York, 1880; vol. 2, Kelly & Walsh, Shanghai, 1894.

Martin, W. A. P. (5). 'Isis and Osiris; or, Oriental Dualism.' *CRR*, 1867. Repr. in Martin (3), vol. 1, p. 203.

Martin, W. A. P. (6). 'The Cartesian Philosophy before Descartes' (centrifugal cosmogony, lumini-ferous aether, etc., in Neo-Confucianism). *JPOS*, 1888, **2**, 121. Repr. in Martin (3), vol. 2, p. 207.

Mason, G. H. (1). 'The Costume of China.' Miller, London, 1800.

Mason, O. T. (1). 'Primitive Travel and Transportation.' *ARUSNM*, 1894, 237.

Maspero, H. (17). 'La Vie Privée en Chine à l'Époque des Han.' *RAA/AMG*, 1931, 185.

Masters, D. (1). *The Wonders of Salvage.* Lane, London, 1924.

May, W. E. (1). 'Historical Notes on the Deviation of the Compass.' *TM*, 1947, 217.

May, W. E. (2). 'The History of the Magnetic Compass.' *MMI*, 1952, **38**, 210.

May, W. E. (3). 'Hugues de Berze and the Mariner's Compass.' *MMI*, 1953, **39**, 103.

May, W. E. (4). 'The Birth of the Compass.' *JIN*, 1949, **2**, 259.

May, W. E. (5). 'Alexander Neckham (*c*. +1187) and the Pivoted Compass Needle.' *JIN*, 1955, **8**, 283.

Mayers, W. F. (1). *Chinese Reader's Manual.* Presbyterian Press, Shanghai, 1874; reprinted, 1924.

Mazaheri, A. (3). 'L'Origine Chinoise de la Balance "Romaine".' *AHES/AESC*, 1960, **15** (no. 5), 833.

Mazzarini, S. (1). *Aspetti Sociali del Quarto Secolo.* Rome, 1952.

Medhurst, W. H. (1) (tr.). *The 'Shoo King' [Shu Ching], or Historical Classic* (Ch. and Eng.). Mission Press, Shanghai, 1846.

Meibom, Marcus (ed.) (1). *Antiquae Musicae Auctores Septem.* Elzevir, Amsterdam, 1652. (including Aristides Quintilianus' *De Musica*, +1st cent., Nicomachus of Gerasa, *Encheiridion Harmonices*, +2nd cent., etc.)

de Mély, F. (1). *Les Lapidaires Chinois.* Vol. 1 of *Les Lapidaires de l'Antiquité et du Moyen Âge.* Leroux, Paris, 1896. (Contains facsimile reproduction of the mineralogical section of *Wakan Sanzai Zue* chs. 59 and 60.) Crit. rev. M. Berthelot, *JS*, 1896, 573.

DE MENDOZA, JUAN GONZALES (1). *Historia de las Cosas mas notables, Ritos y Costumbres del Gran Reyno de la China, sabidas assi por los libros de los mesmos Chinas, como por relacion de religiosos y oltras personas que an estado en el dicho Reyno.* Rome, 1585 (in Spanish). Eng. tr. Robert Parke, 1588 (1589), *The Historie of the Great & Mightie Kingdome of China and the Situation thereof; Togither with the Great Riches, Huge Citties, Politike Gouvernement and Rare Inventions in the same* [undertaken 'at the earnest request and encouragement of my worshipfull friend Master Richard Hakluyt, late of Oxforde ']. Reprinted in Spanish, Medina del Campo, 1595; Antwerp, 1596 and 1655; Ital. tr. Venice (3 editions), 1586; Fr. tr. Paris, 1588 and 1589; Germ. and Latin tr. Frankfurt, 1589. Ed. G. T. Staunton, Hakluyt Soc. Pub. 1853.

MERSENNE, MARIN (1). *Harmonie Universelle.* Paris, 1636.

MEYER, E. (1). 'Zur Geschichte d. Anwendungen der Festigkeitslehre im Maschinenbau....' *BGTI*, 1909, **1**, 108.

MEYERHOF, M. (1). 'Ibn al-Nafīs und seine Theorie d. Lungenkreislaufs.' *QSGNM*, 1935, **4**, 37.

MEYERHOF, M. (2). 'Ibn al-Nafīs (+ 13th century) and his Theory of the Lesser Circulation.' *ISIS*, 1935, **23**, 100.

DE MEYNARD, C. BARBIER & DE COURTEILLE, P. (1) (tr.). *Les Prairies d'Or* (the *Murūj al-Dhabab* of al-Mas'ūdī, +947). 9 vols. Paris, 1861–77.

MICHAELIS, G. A. (1). *Über das Leuchten der Ostsee nach eigener Beobachtungen.* Hamburg, 1830.

MICHEL, H. (6). 'Notes sur l'Histoire de la Boussole.' *CAM*, 1950, **5**, 1.

MIELI, ALDO (1). *La Science Arabe, et son Rôle dans l'Évolution Scientifique Mondiale.* Brill, Leiden, 1938.

MIKAMI, Y. (13). 'On Mayeno [Ryōtaku's] Description of the Parallelogram of Forces [in the MS. *Hon-yaku Undō-hō* (c. +1780)]. *NAW*, 1913, **11**, 76.

MILLER, K. (3). *Die Erdmessung im Altertum und ihr Schicksal.* Strecker & Schröder, Stuttgart, 1919.

MILLS, J. V. (1). 'Malaya in the *Wu Pei Chih* Charts.' *JRAS/M*, 1937, **15** (no. 3), 1.

MILLS, J. V. (4). MS. Translation of ch. 9 of the *Tung Hsi Yang Khao* (Studies on the Oceans East and West.) Unpub.

MILLS, J. V. (5). MS. Translation of *Shun Fêng Hsiang Sung* (Fair Winds for Escort). Bodleian Library, Laud Orient. MS. no. 145. Unpub.

MINAKATA, K. (2). 'Chinese Theories of the Origin of Amber.' *N*, 1895, **51**, 294.

MITCHELL, A. CRICHTON (1). 'Chapters in the History of Terrestrial Magnetism [I. The Discovery of Directivity].' *TM*, 1932, **37**, 105.

MITCHELL, A. CRICHTON (2). 'Chapters in the History of Terrestrial Magnetism [II. The Discovery of Declination].' *TM*, 1937, **42**, 241.

MITCHELL, A. CRICHTON (3). 'Chapters in the History of Terrestrial Magnetism [III. The Discovery of Dip].' *TM*, 1939, **44**, 77.

MITCHELL, A. CRICHTON (4). 'Chapters in the History of Terrestrial Magnetism [IV. The Development of Magnetic Science in Classical Antiquity].' *TM*, 1946, **51**, 323.

DE MOIDREY, J. & LOU [LU], F. (1). *Saecular Variations of Magnetic Elements in the Far East.* Étude no. 39 de l'Observatoire de Magnétisme Terrestre à Zi-ka-wei et Lu-kia-pong, Shanghai, 1932. (See also *Proc. Vth Pacific Science Congress*, Canada, 1933, vol. 3, p. 1853.)

VON MÖLLENDORFF, O. F. (1). 'Schachspiel d. Chinesen.' *MDGNVO*, 1876, **2**, 11.

MONTANDON, G. (1). *L'Ologénèse Culturelle; Traité d'Ethnologie Cyclo-Culturelle et d'Ergologie Systématique.* Payot, Paris, 1934.

MONTELL, G. (1). 'Thou-Hu: the Ancient Chinese Pitch-Pot Game.' *ETH*, 1940, **5** (no. 1–2), 70.

MOODY, E. A. (1). 'Galileo and Avempace [Ibn Bājjah]; the Dynamics of the Leaning Tower Experiment.' *JHI*, 1951, **12**, 163, 375.

MOODY, E. A. & CLAGETT, MARSHALL (1) (ed. and tr.). *The Mediaeval Science of Weights* ('*Scientia de Ponderibus*'); *Treatises ascribed to Euclid, Archimedes, Thābit ibn Qurra, Jordanus de Nemore, and Blasius of Parma.* Univ. of Wisconsin Press, Madison, Wis., 1952. (Revs. E. J. Dijksterhuis, *A/AIHS*, 1953, **6**, 504; O. Neugebauer, *SP*, 1953, **28**, 596.)

MORGAN, E. (1) (tr.). *Tao the Great Luminant; Essays from 'Huai Nan Tzu', with introductory articles, notes and analyses.* Kelly & Walsh, Shanghai, n.d. (1933?).

MOTZO, B. R. (1). '*Il Compasso da Navigare*'; opera Italiana della metà del Secolo XIII [+1253]. Univ. Cagliari, 1947. (Annali d. Fac. di Lett. e Filosofia, Univ. di Cagliari, no. 8.)

MOULE, A. C. (5). 'The Wonder of the Capital' (the Sung books *Tu Chhêng Chi Shêng* and *Mêng Liang Lu* about Hangchow). *NCR*, 1921, **3**, 12, 356.

MOULE, A. C. (7). 'The Chinese South-Pointing Carriage.' *TP*, 1924, **23**, 83. Chinese tr. by Chang Yin-Lin (5).

MOULE, A. C. (10). 'A List of the Musical and other Sound-producing Instruments of the Chinese.' *JRAS/NCB*, 1908, **39**, 1–162.

MOULE, A. C. (15). *Quinsai, with other Notes on Marco Polo.* Cambridge, 1957.

MOULE, A. C. & GALPIN, F. W. (1). 'A Western Organ in Mediaeval China.' *JRAS*, 1926, 193; 1928, 899.

MOULE, A. C. & YETTS, W. P. (1). *The Rulers of China, −221 to +1949; Chronological Tables compiled by A. C. Moule, with an Introductory Section on the Earlier Rulers, ca. −2100 to −249 by W. P. Yetts.* Routledge & Kegan Paul, London, 1957.

MOULE, G. E. (2). 'Notes on the Ting-Chi, or Half-Yearly Sacrifice to Confucius.' *JRAS/NCB*, 1900, **33**, 37.

MUKERJI, RADHAKAMUD (1). *Indian Shipping; a History of the Sea-Borne Trade and Maritime Activity of the Indians from the Earliest Times.* Longmans Green, Bombay and Calcutta, 1912.

MURAOKA, HANICHI (1). 'Erklärung d. "magischen" Eigenschaften des japanischer Bronzespiegels und seiner Herstellung.' *MDGNVO*, 1884, no. 31.

MURAOKA, HANICHI (2). 'Herstellung der japanischen "magischen" Spiegel und Erklärung der "magischen" Erscheinungen derselben.' *ANP* (Wiedemann's), 1884, **22**, 246.

MURAOKA, HANICHI (3). 'Über den japanischen "magischen" Spiegel.' *ANP* (Wiedemann's), 1885, **25**, 138.

MURAOKA, HANICHI (4). 'Über die Deformation der Metallplatten durch Schleifen.' *ANP* (Wiedemann's), 1886, **29**, 471.

MURRAY, H. J. R. (1). *A History of Chess.* Oxford, 1913.

MURRAY, H. J. R. (2). *A History of Board-Games other than Chess.* Oxford, 1952.

NEEDHAM, JOHN TURBERVILLE (2). 'Part of a letter from Mr Turberville Needham to James Parsons, M.D., F.R.S. of a new Mirror, which burns at 66 ft. distance, invented by Mr de Buffon, F.R.S. and Member of the Royal Academy of Sciences at Paris.' *PTRS*, 1747, **44**, 493.

NEEDHAM, JOSEPH (2). *A History of Embryology.* Cambridge, 1934. Revised ed. Cambridge, 1959; Abelard-Schuman, New York, 1959.

NEEDHAM, JOSEPH (31). 'Remarks on the History of Iron and Steel Technology in China' (with French translation: 'Remarques relatives à l'Histoire de la Sidérurgie Chinoise'). In *Actes du Colloque International 'Le Fer à travers les Âges'*, pp. 93, 103. Nancy, Oct. 1955. (*AEST*, 1956, Mémoire no. 16.)

NEEDHAM, JOSEPH (32). *The Development of Iron and Steel Technology in China.* Newcomen Soc. London, 1958. (Second Biennial Dickinson Memorial Lecture, Newcomen Society.)

NEEDHAM, JOSEPH (38). 'The Missing Link in Horological History; a Chinese Contribution.' *PRSA*, 1959, **250**, 147. (Wilkins Lecture, Royal Society.)

NEEDHAM, JOSEPH & ROBINSON, K. (1). 'Ondes et Particules dans la Pensée Scientifique Chinoise.' *SCIS*, 1960, **1** (no. 4), 65.

NEEDHAM, JOSEPH, WANG LING & PRICE, DEREK J. DE S. (1). *Heavenly Clockwork; the Great Astronomical Clocks of Medieval China.* Cambridge, 1960. (Antiquarian Horological Society Monographs, no. 1.) Prelim. pub. *AHOR*, 1956, **1**, 153.

NEEDHAM, JOSEPH, WANG LING & PRICE, DEREK J. DE S. (2). 'Chinese Astronomical Clockwork.' *N*, 1956, **177**, 600. Chinese tr. by Hsi Tsê-Tsung, *KHTP*, 1956 (no. 6), 100.

NEEDHAM, JOSEPH, WANG LING & PRICE, DEREK J. DE S. (3). 'Chinese Astronomical Clockwork.' *Actes du VIIIᵉ Congrès International d'Histoire des Sciences*, p. 325. Florence, 1956.

NEUBERG, F. (1). *Glass in Antiquity.* Art Trade Press, London, 1949.

NEUBERGER, A. (1). *The Technical Arts and Sciences of the Ancients.* Methuen, London, 1930. Tr. H. L. Brose from *Die Technik d. Altertums.* Voigtländer, Leipzig, 1919. (The English version inexcusably omits all the references to the literature.)

NEUGEBAUER, O. (4). 'The Study of Wretched Subjects' (a defence of the study of ancient and medieval pseudo-sciences for the unravelling of the threads of the growth of true science, and for the understanding of the mental climate of the early discoverers). *ISIS*, 1951, **42**, 111. Cf. Pagel (5).

NEUMANN, B. & KOTYGA, G. (1) (with the assistance of M. Rupprecht & H. Hoffman). 'Antike Gläser.' *ZAC*, 1925, **38**, 776, 857; 1927, **40**, 963; 1928, **41**, 203; 1929, **42**, 835.

NIEMANN, W. (1). 'J. F. Kammerer, der Erfinder der Phosphor-Zundhölzer.' *AGNT*, 1918, **8**, 206.

NORLIND, T. (1). 'History of Chinese Musical Instruments.' *STMF*, 1933, 48.

O'DEA, W. T. (1). *Darkness into Daylight.* London, 1948.

O'DEA, W. T. (2). *The Social History of Lighting.* Routledge & Kegan Paul, London, 1958.

OGDEN, C. K. & WOOD, JAMES (1). 'Sound and Colour', and 'Colour-Harmony.' *CM*, 1921, **11** (no. 1), 9, 20 (decennial number).

OLIVER, G. H. (1). 'The History of the Invention and Discovery of Spectacles.' *BMJ*, 1913, 1049.

OLIVER, G. H. (2). *History of the Invention and Discovery of Spectacles.* London, 1913.

O'MALLEY, C. D. (1). 'A Latin Translation (+1547) of Ibn al-Nafīs, related to the Problem of the circulation of the Blood.' *JHMAS*, 1957, **12**, 248. Abstract in *Actes du VIIIᵉ Congrès International d'Histoire des Sciences*, p. 716. Florence, 1956.

ORE, OYSTEIN (1). *Cardano, the Gambling Scholar.* Princeton Univ. Press, Princeton, N.J., 1953.
OSANN, G. W. (1). 'Über einige neue Lichtsauger von vorzüglicher Stärke.' *AGNL*, 1825, **5**, 88.

PAGEL, W. (4). 'William Harvey; Some Neglected Aspects of Medical History.' *JWCI*, 1944, **7**, 144.
PAGEL, W. (5). 'The Vindication of "Rubbish".' *MHJ*, 1945. Cf. Neugebauer (4).
PAGET, SIR RICHARD (2). *Human Speech.* Kegan Paul, London, 1930.
PAPANASTASIOU, C. E. (1). *Les Théories sur la Nature de la Lumière de Descartes à nos Jours.* Paris, 1935.
DE PAREJA, BARTHOLOMÉ RAMOS (1). *De Musica Tractatus, explicit Musica Practica.* Bologna, 1482.
PARKER, E. H. (1). 'Glass in China.' *CR*, 1886, **15**, 372; 1887, **16**, 48 & 129; 1888, **17**, 114; 1889, **18**, 196, 197.
PARKER, E. H. (2). 'The Early Laos and China.' *CR*, 1890, **19**, 67.
PARKER, E. H. (3). 'The Old Thai or Shan Empire of Western Yunnan.' *CR*, 1891, **20**, 337.
PARNELL, J. (1). Letter on Magic Mirrors. *N*, 1877, **16**, 227.
PARTINGTON, J. R. (2). 'The Origins of the Atomic Theory.' *ANS*, 1939, **4**, 245.
PAUSCHMANN, G. (1). 'Zur Geschichte d. linsenlosen Abbildung.' *AGNT*, 1919, **9**, 86.
PECK, A. L. (3). 'Anaxagoras and the Parts.' *CQ*, 1926, **20**, 57.
PECK, A. L. (4). 'Anaxagoras; Predication as a Problem in Physics.' *CQ*, 1931, **25**, 27, 112.
DE PEIRESC, C. N. FABRI. See Gassendi.
PELLIOT, P. (2a). 'Les Grands Voyages Maritimes Chinois au Début du 15ᵉ Siecle' (review of Duyven-dak, 10). *TP*, 1933, **30**, 237.
PELLIOT, P. (2b). 'Notes additionelles sur Tcheng Houo [Chêng Ho] et sur ses Voyages.' *TP*, 1934, **31**, 274.
PELLIOT, P. (2c). 'Encore à Propos des Voyages de Tcheng Houo [Chêng Ho].' *TP*, 1936, **32**, 210.
PELLIOT, P. (9). 'Mémoire sur les Coutumes de Cambodge' (a translation of Chou Ta-Kuan's *Chen-La Fêng Thu Chi*). *BEFEO*, 1902, **2**, 123. Revised version; Paris, 1951, see Pelliot (33).
PELLIOT, P. (17). 'Deux Itinéraires de Chine à l'Inde à la Fin du 8ᵉ Siècle.' *BEFEO*, 1904, **4**, 131.
PELLIOT, P. (23). Review of Lo Chen-Yü (2) and Tomioka (1), on bronze mirrors. *TP*, 1921, **20**, 142.
PELLIOT, P. (30). Note on Han relations with South-East Asian countries, with tr. of a passage from *CHS*, ch. 28 B, in review of Hirth & Rockhill. *TP*, 1912, **13**, 446 (457).
PELLIOT, P. (33) (tr.). *Mémoire sur les Coutumes de Cambodge de Tcheou Ta-Kouan [Chou Ta-Kuan]; Version Nouvelle, suivie d'un Commentaire inachevé.* Maisonneuve, Paris, 1951. (Œuvres Posthumes, no. 3.)
PELLIOT, P. (43). Criticism of Laufer (8) on Jade, with note on glass technology in China. *TP*, 1912, **13**, 434.
PELLIOT, P. (44). 'Un Fragment du *Suvarṇaprabhāsa Sūtra* en Iranien Oriental.' *MSLP*, 1913, **18**, 89.
PELLIOT, P. (45). 'Les Franciscains en Chine au 16ᵉ et au 17ᵉ Siècles.' *TP*, 1938, **34**, 191.
PERSON, M. (1). 'Observations faites sur des Miroirs Chinois dits Magiques.' *CRAS*, 1847, **24**, 1111.
PESCHEL, O. (1). *Abhandlungen zur Erd- und Völker-kunde.* 2 vols., ed. J. Löwenberg. Leipzig, 1877, 1878.
PEYRARD, F. (1) (tr.). *Traité de l'Équilibre des Plans ou de leurs Centres de Gravité* (Part of the *Œuvres d'Archimède*). Paris, 1807.
PFISTER, L. (1). *Notices Biographiques et Bibliographiques sur les Jésuites de l'Ancienne Mission de Chine (+1552 to +1773).* 2 vols. Mission Press, Shanghai, 1932 (*VS* no. 59).
PFIZMAIER, A. (58) (tr.). 'Ungewöhnliche Erscheinungen und Zufälle in China um die Zeiten der Südlichen Sung.' *SWAW/PH*, 1875, **79**, 362. (Tr. chs. 30–4 (*Wu Hsing Chih*) of (*Liu*) *Sung Shu*.)
PFIZMAIER, A. (98) (tr.). 'Die Anwendung und d. Zufälligkeiten des Feuers in d. alten China.' *SWAW/PH*, 1870, **65**, 767, 777, 786, 799. (Tr. chs. 868, 869 (fire and fire-wells), 870 (lamps, candles and torches), 871 (coal), of *Thai-Phing Yü Lan*.)
PHELPS, D. L. (1). 'The Place of Music in the Platonic and Confucian Systems of Moral Education.' *JRAS/NCB*, 1928, **59**, 128.
PICKEN, L. E. R. (1) 'The Music of Far Eastern Asia, I. China.' In *New Oxford History of Music*, vol. 1, pp. 83, 190. Oxford, 1957.
PICKEN, L. E. R. (2). 'The Music of Far Eastern Asia, II. Countries other than China' (Mongolia, Sinkiang, Tibet, Korea, the Miao peoples, the Lo-lo and Min-chia peoples, the Nagas, Annam, Cambodia, Siam, Burma, Java, Sumatra and Nias, Bali and other islands of the Indonesian archipelago). In *New Oxford History of Music*, vol. 1, pp. 135, 190. Oxford, 1957.
PICKEN, L. E. R. (3). 'Chinese Music.' Lecture delivered before the Britain–China Friendship Association (Cambridge Branch), 3rd June 1954.
PICKEN, L. E. R. (4). 'Twelve Ritual Melodies of the Thang Dynasty.' In *Studia Memoriae Bela Bartók Sacra*, p. 147. National Academy, Budapest, 1956.
PICKEN, L. E. R. (5). 'Chiang Khuei's *Nine Songs for Yüeh* [+1202].' *MUQ*, 1957, **43**, 201.

PICKEN, L. E. R. (6). 'The Origin of the Short Lute.' *GSJ*, 1955, **8**, 1.

PINES, S. (1). *Beiträge z. islamischen Atomlehre*. Berlin, 1936.

PINES, S. (2). 'Les Précurseurs Mussulmans de la Théorie de l'Impétus.' *A*, 1938.

PINOT, V. (1). *La Chine et la Formation de l'Esprit Philosophique en France (+1640 to +1740)*. Geuthner, Paris, 1932.

PLEDGE, H. T. (1). *Science since +1500*. HMSO, London, 1939.

POGGENDORFF, J. C. (1). *Geschichte d. Physik*. Barth, Leipzig, 1879.

DELLA PORTA, J. B. (GIAMBATTISTA) (1). *Magia Naturalis*. Naples, 1558, 1589; Antwerp, 1561. Eng. tr. by R. Gaywood, Young & Speed, London, 1658; Wright, London, 1669. Fascimile edition of the 1658 edition, ed. D. J. de S. Price, Basic Books, New York, 1957. Bibliography in John Ferguson (2).

PRENER, J. S. & SULLENGER, D. B. 'Phosphors.' *SAM*, 1954, **191** (no. 4), 62.

PRIESTLEY, JOSEPH (1). *History and Present State of Discoveries relating to Vision, Light and Colours*. Johnson, London, 1772.

PRINSEP, J. (1). On a Chinese 'Magic Mirror'. *JRAS/B*, 1832, **1**, 242.

PRZYŁUSKI, J. (3). 'La Divination par l'Aiguille Flottante et par l'Araignée dans la Chine Méridionale.' *TP*, 1914, **15**, 214.

DA RADA, MARTÍN (1). 'Narrative of his Mission to Fukien (June–Oct. 1575).' 'Relation of the things of China, which is properly called Taybin [Ta Ming].' Tr. and ed. Boxer (1).

RAKUSEN, C. P. (1). 'History of Chinese Spectacles.' *CMJ*, 1938, **53**, 379. 'Optics in China' (only on opticians' practice in the treaty ports). *NYSOAYB*, 1930, 361.

RAMES, BARTOLO. See de Pareja.

RAMOS, BARTOLO. See de Pareja.

RASMUSSEN, O. D. (1). *Old Chinese Spectacles*. North China Press, Tientsin, 1915. 2nd ed., enlarged. *Chinese Eyesight and Spectacles*. Pr. pr. Tonbridge, 1949.

RASMUSSEN, S. E. (1). *Towns and Buildings*. Univ. Press, Liverpool, 1951 (translated from the Danish of 1949 by Eve Wendt). Original edition different in many ways.

READ, BERNARD E. (1) (with LIU JU-CHHIANG). *Chinese Medicinal Plants from the 'Pên Ts'ao Kang Mu' A.D. 1596... a Botanical, Chemical and Pharmacological Reference List*. (Publication of the Peking Nat. Hist. Bull.). French Bookstore, Peiping, 1936 (chs. 12–37 of *Pên Tshao Kang Mu*) (rev. W. T. Swingle, *ARLC/DO*, 1937, 191).

READ, BERNARD E. (2) (with LI YÜ-THIEN). *Chinese Materia Medica; Animal Drugs*.

		Serial nos.	Corresp. with chaps. of *Pên Tshao Kang Mu*
Pt. I	Domestic Animals	322–349	50
II	Wild Animals	350–387	51 *A* and *B*
III	Rodentia	388–399	51 *B*
IV	Monkeys and Supernatural Beings	400–407	51 *B*
V	Man as a Medicine	408–444	52

PNHB, 1931, **5** (no. 4), 37–80; **6** (no. 1), 1–102. (Sep. issued, French Bookstore, Peiping, 1931.)

READ, BERNARD, E. (3) (with LI YÜ-THIEN). *Chinese Materia Medica; Avian Drugs*.

Pt. VI	Birds	245–321	47, 48, 49

PNHB, 1932, **6** (no. 4), 1–101. (Sep. issued, French Bookstore, Peiping, 1932.)

READ, BERNARD E. (4) (with LI YÜ-THIEN). *Chinese Materia Medica; Dragon and Snake Drugs*.

Pt. VII	Reptiles	102–127	43

PNHB, 1934, **8** (no. 4), 297–357. (Sep. issued, French Bookstore, Peiping, 1934.)

READ, BERNARD E. (5) (with YU CHING-MEI). *Chinese Materia Medica; Turtle and Shellfish Drugs*.

Pt. VIII	Reptiles and Invertebrates	199–244	45, 46

PNHB (Suppl.), 1939, 1–136. (Sep. issued, French Bookstore, Peiping, 1937.)

READ, BERNARD E. (6) (with YU CHING-MEI). *Chinese Materia Medica; Fish Drugs*.

Pt. IX	Fishes (incl. some amphibia, octopoda and crustacea)	128–198	44

	Serial nos.	Corresp. with chaps. of *Pên Tshao Kang Mu*

PNHB (Suppl.), 1939. (Sep. issued, French Bookstore, Peiping, n.d. prob. 1939.)

READ, BERNARD E. (7) (with YU CHING-MEI). *Chinese Materia Medica; Insect Drugs.*
Pt. X Insects (incl. arachnidae, etc.) 1–101 39, 40, 41, 42
PNHB (Suppl.), 1941. (Sep. issued, Lynn, Peiping, 1941.)

READ, BERNARD E. (8). *Famine Foods listed in the 'Chiu Huang Pên Tshao'.* Lester Institute, Shanghai, 1946.

READ, BERNARD E. & PAK, C. (PAK KYEBYŎNG) (1). *A Compendium of Minerals and Stones used in Chinese Medicine, from the 'Pên Ts'ao Kang Mu'.* PNHB, 1928, **3** (no. 2), i–vii, 1–120. (Revised and enlarged, issued separately, French Bookstore, Peiping, 1936 (2nd ed.).) Serial nos. 1–135, corresp. with chs. of *Pên Tshao Kang Mu*, 8, 9, 10, 11.

REIN, J. J. (1). *Industries of Japan; together with an Account of its Agriculture, Forestry, Arts and Commerce.* Hodder & Stoughton, London, 1889.

REISCHAUER, E. O. (2) (tr.). *Ennin's Diary; the Record of a Pilgrimage to China in Search of the Law* (the *Nittō Guhō Junrei Gyōki*). Ronald Press, New York, 1955.

REISCHAUER, E. O. (3). *Ennin's Travels in Thang China.* Ronald Press, New York, 1955.

RENOU, L. & FILLIOZAT, J. (1). *L'Inde Classique; Manuel des Études Indiennes.*
Vol. 1, with the collaboration of P. Meile, A. M. Esnoul & L. Silburn. Payot, Paris, 1947.
Vol. 2, with the collaboration of P. Demiéville, O. Lacombe, & P. Meile. École Française d'Extrême Orient, Hanoi, 1953; Impr. Nationale, Paris, 1953.

DE RHODES, ALEXANDRE (1). *Dictionarum Annnamiticum Lusitanum et Latinum ope sacrae congregationis de propaganda fide in lucem editum....* Typ. Sacr. Congreg. Rome, 1651, 1667.

RICCIOLI, J. B. (1). *Geographia et Hydrographia Reformata.* Bologna, 1661.

RICHARDSON, J. C. (1). 'On the ignition of petroleum by the heat of quicklime in contact with water; one of the proposed explanations of Greek fire—an experimental demonstration. *N*, 1927, **120**, 165.

RIDGEWAY, SIR WM. (1). 'The Game of Polis and Plato's Republic.' *JHS*, 1896, **16**, 288.

RIDLEY, MARKE (1). *A Short Treatise of Magneticall Bodies and Motions.* Okes, London, 1613.

RITCHIE, P. D. (1). 'Spectrographic Studies on Ancient Glass; Chinese Glass from the pre-Han to Thang times.' *TSFFA*, 1937, **5**, 209; **6**, 155.

ROBINS, F. W. (3). *The Story of the Lamp.* Oxford, 1939.

ROBINSON, K. (1). *A Critical Study of Ju Dzai-Yü's [Chu Tsai-Yü's] Account of the System of the Lü-Lü or Twelve Musical Tubes in Ancient China.* Inaug. Diss. Oxford, 1948.

ROBINSON, K. (2). 'A Possible Use of Music for Divination.' Unpub. paper.

ROBINSON, K. (3). 'Chinesische Musik, I. Geschichtliche Entwicklung von der Frühzeit (Shang-dynastie) bis zum Ende der Han-Zeit (−1523 bis +206).' Germ. tr. by H. Eckardt. In *Die Musik in Geschichte und Gegenwart*. Ed. F. Blume, vol. 2, cols. 1195–1205. Kassel and Basel, 1952. Cf. Eckardt (1).

ROBINSON, K. (4). 'Ichthy-Acoustics.' *ACSS*, 1953, 67.

ROBINSON, K. (5). 'New Thoughts on Ancient Chinese Music.' *ACSS*, 1954, 30.

ROCKHILL, W. W. (1). 'Notes on the Relations and Trade of China with the Eastern Archipelago and the Coast of the Indian Ocean during the 15th Century.' *TP*, 1914, **15**, 419; 1915, **16**, 61.

ROCKHILL, W. W. (2). 'Notes on the Ethnology of Tibet.' *ARUSNM*, 1893, 669.

VON ROHR, M. (2). 'Contributions to the History of the Spectacle Trade from the earliest times to Thomas Young.' *TOPS*, 1923, **25**, 41.

RONCHI, V. (1). *Storia della Luce.* Zanichelli, Bologna, 1939. French tr. *Histoire de la Lumière*, by J. Taton. Sevpen, Paris, 1956. (Pub. Bibl. Gén. de l'École Prat. des Htes. Études, VIe Section.)

RONCHI, V. (2). 'Sul Contributo di Ibn al-Haitham alle Teorie della Visione e della Luce.' *Proceedings of the VIIth International Congress of the History of Science*, p. 516. Jerusalem, 1953.

RONCHI, V. (3). '"Ciò che si vede" coincide con "Ciò che c'e"?' *AFGR*, 1957, **12**, 350. (*PINO*, ser. 2, no. 772.)

ROSEN, E. (2). 'The Invention of Eyeglasses [Spectacles].' *JHMAS*, 1956, **11**, 13, 183.

ROSEN, E. (3). 'Carlo Dati on the Invention of Eyeglasses [Spectacles].' *ISIS*, 1953, **44**, 4.

DES ROTOURS, R. (1) (tr.). *Traité des Fonctionnaires et Traité de l'Armée, traduits de la Nouvelle Histoire des T'ang* (chs. 46–50). 2 vols. Brill, Leiden, 1948. (Bibl. de l'Inst. des Hautes Études Chinoises, no. 6.) (rev. P. Demiéville, *JA*, 1950, **238**, 395).

DES ROTOURS, R. (2) (tr.). *Traité des Examens* (translation of chs. 44 and 45 of the *Hsin Thang Shu*). Leroux, Paris, 1932. (Bibl. de l'Inst. des Hautes Études Chinoises, no. 2.)

ROUSSIER, P. J. (1). *Mémoire sur la Musique des Anciens, où l'on expose le Principe des Proportions authentiques, dites de Pythagore, et de divers Systèmes de Musique chez les Grecs, les Chinois et les Égyptiens; avec un Parallèle entre le Système des Égyptiens et celui des Modernes.* Paris, 1770.

RUDOLPH, R. C. (3). 'The Antiquity of Thou-Hu.' *AQ*, 1950, **24**, 175.

RUDOLPH, R. C. & WÊN YU (1). *Han Tomb Art of West China; a Collection of First and Second Century Reliefs.* Univ. of Calif. Press, Berkeley and Los Angeles, 1957 (rev. W. P. Yetts, *JRAS*, 1953, 72).

RUFUS, W. C. (2). 'Astronomy in Korea.' *JRAS/KB*, 1936, **26**, 1. Sep. pub as *Korean Astronomy*. Literary Department, Chosen Christian College, Seoul (Eng. Pub. no. 3), 1936.

RUNCORN, S. K. (1). 'The Permanent Magnetisation of Rocks.' *END*, 1955, **14**, 152.

RUPP, H. (1). *Die Leuchtmassen und ihre Verwendung.* Berlin, 1937.

SACHS, C. (1). *The Rise of Music in the Ancient World; East and West.* New York, 1943.

SACHS, C. (2). *A History of Musical Instruments.* New York, 1940.

SAMBURSKY, S. (1). *The Physical World of the Greeks.* Tr. from the Hebrew edition by M. Dagut. Routledge & Kegan Paul, London, 1956.

SAMBURSKY, S. (2). *The Physics of the Stoics.* Routledge & Kegan Paul, London, 1959.

SANCTORIUS, S. (1). *Medicina Statica: being the Aphorisms of Sanctorius, translated into English with large Explanations, wherein is given a Mechanical Account of the Animal Oeconomy, and of the Efficacy of the Non-Naturals, either in bringing about or removing its Disorders; also with an Introduction concerning Mechanical Knowledge, and the Grounds of Certainty in Physick, by John Quincy.* Newton, London, 1712.

DE SANDE, ÉDOUART, S.J. (1). Letters from China, in *Sommaire des Lettres du Japon et de la Chine de l'an MDLXXXIX et MDXC.* Paris, 1592.

VAN DER SANDE, G. A. J. (1). *Nova Guinea.* 3 vols. Leiden, 1907.

SANDERS, L. (1). 'Evolution of the Pivot, with special reference to Weighing Instruments.' *TNS*, 1944, **24**, 81.

SARTON, GEORGE (1). *Introduction to the History of Science.* Vol. 1, 1927; Vol. 2, 1931 (2 parts); Vol. 3, 1947 (2 parts). Williams & Wilkins, Baltimore. (Carnegie Institution Pub. no. 376.)

SARTON, GEORGE (2). 'Simon Stevin of Bruges; the first explanation of Decimal Fractions and Measures (+1585); together with a history of the decimal idea, and a facsimile of Stevin's *Disme*.' *ISIS*, 1934, **21**, 241; 1935, **23**, 153.

SARTON, GEORGE (7). 'Chinese Glass at the Beginning of the Confucian Age.' *ISIS*, 1936, **25**, 73.

SARTON, G. & WARE, J. R. (1). 'Were the Ancient Chinese Weights and Measures related to Musical Instruments?' *ISIS*, 1947, **37**, 73.

DE SAUSSURE, L. (11). 'Les Origines de l'Astronomie Chinoise; La Règle des *cho-ti* [*shê-thi*].' *TP*, 1911, **12**, 347. Repr. as [F] in (1).

DE SAUSSURE, L. (16 *a, b, c, d*). 'Le Système Astronomique des Chinois.' *ASPN*, 1919 (5ᵉ sér. **1**), **124**, 186, 561; 1920 (5ᵉ sér. **2**), **125**, 214, 325. (*a*) Introduction; (i) Description du Système, (ii) Preuves de l'Antiquité du Système; (*b*) (iii) Rôle Fondamental de l'Étoile Polaire, (iv) La Théorie des Cinq Eléments, (v) Changements Dynastiques et Réformes de la Doctrine; (*c*) (vi) Le Symbolisme Zoaire, (vii) Les Anciens Mois Turcs; (*d*) (viii) Le Calendrier, (ix) Le Cycle Sexagésimal et la Chronologie, (x) Les Erreurs de la Critique. Conclusion.

DE SAUSSURE, L. (35). 'L'Origine de la Rose des Vents et l'Invention de la Boussole.' *ASPN*, 1923 (5ᵉ sér.), **5** (nos. 3 and 4). Sep. pub. Luzac, London, 1923 and reprinted in Ferrand (6), vol. 3, pp. 31 ff. Emendations by P. Pelliot, *TP*, 1924, 52.

DE SAVIGNAC, J. (1). 'La Rosée Solaire de l'Ancienne Égypte.' *LNC*, 1954, **6**, 345. (Mélanges Roger Goossens.)

SCHAEFFNER, A. (1). *Origine des Instruments de Musique.* Payot, Paris, 1936.

SCHAFER, E. H. (5). 'Notes on Mica in Medieval China.' *TP*, 1955, **43**, 265.

SCHINDLER, B. (4). 'Preliminary Account of the Work of Henri Maspero concerning the Chinese Documents on Wood and Paper discovered by Sir Aurel Stein on his third expedition in Central Asia'. *AM*, 1950, **1**, 216.

SCHLEGEL, G. (4). *Chinesische Bräuche u. Spiele in Europa.* Breslau, 1869.

SCHLEGEL, G. (5). *Uranographie Chinoise, etc.* 2 vols. with star-maps in separate folder. Brill, Leiden, 1875. (Crit. J. Bertrand, *JS*, 1875, 557; S. Günther, *VAG*, 1877, **12**, 28. Reply by G. Schlegel, *BNI*, 1880 (4ᵉ volg.), **4**, 350.)

SCHLEGEL, G. (7). *Problèmes Géographiques; les Peuples Étrangers chez les Historiens Chinois.*
  (*a*) Fu-Sang Kuo (ident. Sakhalin and the Ainu). *TP*, 1892, **3**, 101.
  (*b*) Wên-Shen Kuo (ident. Kuriles). *Ibid.* p. 490.
  (*c*) Nü Kuo (ident. Kuriles). *Ibid.* p. 495.

(d) Hsiao-Jen Kuo (ident. Kuriles and the Ainu). *TP*, 1893, **4**, 323.
(e) Ta-Han Kuo (ident. Kamchatka and the Chukchi) and Liu-Kuei Kuo. *Ibid.* p. 334.
(f) Ta-Jen Kuo (ident. islands between Korea and Japan) and Chhang-Jen Kuo. *Ibid.* p. 343.
(g) Chün-Tzu Kuo (ident. Korea, Silla). *Ibid.* p. 348.
(h) Pai-Min Kuo (ident. Korean Ainu). *Ibid.* p. 355.
(i) Chhing-Chhiu Kuo (ident. Korea). *Ibid.* p. 402.
(j) Hei-Chih Kuo (ident. Amur Tungus). *Ibid.* p. 405.
(k) Hsüan-Ku Kuo (ident. Siberian Giliak). *Ibid.* p. 410.
(l) Lo-Min Kuo and Chiao-Min Kuo (ident. Okhotsk coast peoples). *Ibid.* p. 413.
(m) Ni-Li Kuo (ident. Kamchatka and the Chukchi). *TP*, 1894, **5**, 179.
(n) Pei-Ming Kuo (ident. Behring Straits islands). *Ibid.* p. 201.
(o) Yu-I Kuo (ident. Kamchatka tribes). *Ibid.* p. 213.
(p) Han-Ming Kuo (ident. Kuriles). *Ibid.* p. 218.
(q) Wu-Ming Kuo (ident. Okhotsk coast peoples). *Ibid.* p. 224.
(r) San Hsien Shan (the magical islands in the Eastern Sea, perhaps partly Japan). *TP*, 1895, **6**, 1.
(s) Liu-Chu Kuo (the Liu-Chhiu islands, partly confused with Thaiwan, Formosa). *Ibid.* p. 165.
(t) Nü-Jen Kuo (legendary, also in Japanese fable). *Ibid.* p. 247.
A volume of these reprints, collected, but lacking the original pagination, is in the Library of the Royal Geographical Society. Chinese transl. under name Hsi Lo-Ko. (rev. F. de Mély, *JS*, 1904.)

SCHLEGEL, G. (8). Note on the Ancient History of Glass and Coal. *TP*, 1891, **2**, 178.
SCHLESINGER, K. (1). *The Greek 'Aulos'*. London, 1939.
SCHLESINGER, K. (2). Articles on 'Accordion', '*Chêng*' (i.e. *Shêng*), 'Free Reed Vibrator', 'Harmonium', 'Organ', and 'Reed Instruments'. *EB*, 1910 ed.
SCHOLES, P. A. (1) (ed.). *The Oxford Companion to Music*. 7th ed. Oxford, 1947; 8th ed. 1950.
SCHÜCK, [K. W.] A. (1). *Der Kompass*. 2 vols. Pr. pr. Hamburg, 1911–1915. (The second volume, *Sagen von der Erfindung des Kompasses; Magnet, Calamita, Bussole, Kompass; Die Vorgänger des Kompasses*, contains a good deal on the Chinese material, in so far as it could be evaluated at the time, mainly a long account of 18th- and 19th-century European views about it. This had seen preliminary publication in *DNAT*, 1891, **40** (nos. 51 and 52).)
SCHÜCK, K. W. A. (2). 'Gedanken über die Zeit d. ersten Benutzung des Kompasses in nördl. Europa.' *AGNT*, 1910, **3**, 127.
SCHÜCK, K. W. A. (3). 'Zur Einführung des Kompasses in die nordwest-europäischer Nautik.' *AGNT*, 1913, **4**, 40.
SCHÜCK, K. W. A. (4). 'Zur Entwicklung der Einteilungen der Chinesischen Schiffs- und der Gaukler-Bussole.' *MGMNW*, 1917, **16**, 7.
SCHÜCK, [K.W.] A. (5). 'Erwähnung eines Vorgängers des Kompasses in Deutschland um die Mitte des 13. Jahrhunderts.' *MGMNW*, 1914, **13**, 333.
SCHÜCK, [K. W.] A. (6). 'Die Vorgänger des Kompasses.' *CZOM*, 1911, **32**, 1.
SCOTT, J. (1). 'On the Burning-Mirrors of Archimedes, with some Propositions relating to the Concentration of Light produced by Reflectors of different Forms.' *TRSE*, 1868, **25**, 123.
SEEGER, R. J. (1). 'The Beginnings of Physics.' *JWAS*, 1934, **24**, 501; 1935, **25**, 341.
SÉGUIER, M. (1). Note on St Julien (3). *CRAS*, 1847, **24**, 1001.
SELIGMAN, C. G. (5). 'Early Chinese Glass.' *TOCS*, 1942, **18**, 19.
SELIGMAN, C. G. & BECK, H. C. (1). 'Far Eastern Glass; some Western Origins.' *BMFEA*, 1938, **10**, 1.
SELIGMAN, C. G., RITCHIE, P. D. & BECK, H. C. (1). 'Chinese Glass.' *N*, 1936, **138**, 721.
SHIH SHÊNG-HAN (1). *A Preliminary Survey of the book 'Chhi Min Yao Shu'; an Agricultural Encyclopaedia of the +6th Century*. Science Press, Peking, 1958.
SHRYOCK, J. K. (1). *Origin and Development of the State Cult of Confucius*. Appleton-Century, New York, 1932.
SINGER, C. (9). 'Steps leading to the Invention of the First Optical Apparatus.' In *Studies in the History and Method of Science*, vol. 2, p. 385. Oxford, 1921. Reissued, 1955.
SINGER, C., HOLMYARD, E. J., HALL, A. R. & WILLIAMS, T. I. (1) (ed.). *A History of Technology*. 5 vols. Oxford, 1954–58. (revs. M. I. Finley, *EHR*, 1959, **12**, 120; J. Needham, *CAMR*, 1957, 299; 1959, 227.)
SINGER, D. W. (1). *Giordano Bruno; His Life and Thought, with an annotated Translation of his Work 'On the Infinite Universe and Worlds'*. Schuman, New York, 1950.
SKINNER, F. G. (1). 'Measures and Weights [in Ancient Civilisations].' In *A History of Technology*. Ed. C. Singer, E. J. Holmyard & A. R. Hall, vol. 1, p. 774. Oxford, 1954.
SLOBODCHIKOV, L. A. (1). 'Le Jeu d'Échecs des Chinois.' *BUA*, 1945 (3e sér.), **6**, 1.
SMITH, A. H. (1) (ed.). *A Guide to the Exhibition illustrating Greek and Roman Life*. British Museum Trustees, London, 1920.

SMITH, C. S. (3). 'A Sixteenth-Century Decimal System of Weights.' *ISIS*, 1955, **46**, 354.

SMITH, D. E. (1). *History of Mathematics*. Vol. 1. *General Survey of the History of Elementary Mathematics*, 1923. Vol. 2. *Special Topics of Elementary Mathematics*, 1925. Ginn, New York.

SÖKELAND, H. (1). 'Ancient Desemers or Steelyards.' *ARSI*, 1900, 551. (From *VBGE*, 1900.)

SØLVER, C. V. (1). 'Leidarsteinn: the Compass of the Vikings.' *OL*, 1946, **10**, 293.

SOOTHILL, W. E. (5) (posthumous). *The Hall of Light; a Study of Early Chinese Kingship*. Lutterworth, London, 1951. (On the Ming Thang, and contains discussion of the *Pu Thien Ko*.)

SOULIÉ DE MORANT, G. (1). *La Musique en Chine*. Leroux, Paris, 1911.

SPINDEN, H. J. (1). *Ancient Civilisations of Mexico and Central America*. Amer. Mus. Nat. Hist., New York, 1946.

STAUNTON, SIR GEORGE LEONARD (1). *An Authentic Account of an Embassy from the King of Great Britain to the Emperor of China ... taken chiefly from the Papers of H.E. the Earl of Macartney, K.B. etc. ... '* 2 vols. Bulmer & Nicol, London, 1797; repr. 1798. Abridged ed. 1 vol. Stockdale, London, 1797.

STAUNTON, SIR GEORGE THOMAS (2). *Notes on Proceedings and Occurrences during the British Embassy to Peking in 1816 [Lord Amherst's]*. London, 1824.

STEELE, J. (1) (tr.). *The 'I Li', or Book of Etiquette and Ceremonial*. 2 vols. London, 1917.

STEVIN, SIMON (1). *Beghinseln der Weegkonst*. Leiden, 1585.

STEVIN, SIMON (2). *Hypomnemata Mathematica*. Leiden, 1608.

STONE, J. F. S. & THOMAS, L. C. (1). 'The Use and Distribution of Faience in the Ancient East and Prehistoric Europe.' *PPHS*, 1956, **22**, 37.

STONER, E. C. (1). *Magnetism and Matter*. Methuen, London, 1934.

STRANGE, H. O. H. (1) (tr.). 'Die Monographie über Wang Mang' (ch. 99 of the *Chhien Han Shu*). *AKML*, 1938, **22** (no. 3). (Crit. J. J. L. Duyvendak, *TP*, 1939, **35**, 407.)

STRAUB, H. (1). *Die Geschichte d. Bauingenieurkunst; ein Überblick von der Antike bis in die Neuzeit*. Birkhäuser, Basel, 1949. Eng. tr. by E. Rockwell. *A History of Civil Engineering*. Leonard Hill, London, 1952.

SUN ZEN E-TU & DE FRANCIS, J. (1). *Chinese Social History; Translations of Selected Studies*. Amer. Council of Learned Societies, Washington, D.C., 1956. (ACLS Studies in Chinese and Related Civilisations, no. 7.)

SWALLOW, R. W. *Ancient Chinese Bronze Mirrors*. Vetch, Peiping, 1937.

TASCH, P. (1). 'Quantitative Measurements and the Greek Atomists.' *ISIS*, 1948, **38**, 185.

TAYLOR, E. G. R. (6). 'The South-Pointing Needle.' *IM*, 1951, **8**, 1.

TAYLOR, E. G. R. (8). *The Haven-Finding Art; a History of Navigation from Odysseus to Captain Cook*. Hollis & Carter, London, 1956.

TAYLOR, E. G. R. (9). 'The Oldest Mediterranean Pilot.' *JIN*, 1951, **4**, 81.

TAYLOR, F. SHERWOOD (1). 'The Origin of the Thermometer.' *ANS*, 1942, **5**, 129.

TAYLOR, F. SHERWOOD (4). *A History of Industrial Chemistry*. Heinemann, London, 1957.

TAYLOR, L. W. (1). *Physics, the Pioneer Science*. Houghton Mifflin, Boston, 1941.

TEMKIN, O. (2). 'Was Servetus influenced by Ibn al-Nafîs?' *BIHM*, 1940, **8**, 731.

TESTUT, C. (1). *Mémento du Pesage; Les Instruments de Pesage (et) leur histoire à travers les Âges*. Hermann, Paris, 1946.

THELLIER, E. (1). 'Sur l'aimantation des Terres Cuites et ses Applications géophysiques.' Inaug. Diss. Paris, 1938 and *AIPG*, 1938, **16**, 157.

THELLIER, E. (2). 'Sur l'Intensité du Champ magnétique terrestre en France à l'Époque Gallo-Romaine.' *CRAS*, 1946, **222**, 905; *JPH*, 1951, **12**, 205.

THELLIER, E. & THELLIER, D. (1). 'Sur la Direction du Champ magnétique terrestre retrouvée sur des Parois de Fours des Époques Punique et Romaine à Carthage.' *CRAS*, 1951, **233**, 1476.

THOMPSON, SYLVANUS P. (1). 'The Rose of the Winds; the Origin and Development of the Compass-Card.' *PBA*, 1913, **6**, 1.

THOMPSON, SYLVANUS P. (2) (tr.). *William Gilbert of Colchester, Physician of London, 'On the Magnet, Magnetick Bodies also, and on the great Magnet the Earth; a new Physiology, demonstrated by many arguments and experiments'*. Chiswick Press, London, 1900. Facsimile reproduction, ed. Derek J. de S. Price. Basic Books, New York, 1958.

THOMPSON, SYLVANUS P. (3). Letter on Magic Mirrors. *N*, 1877, **16**, 163.

THOMPSON, SYLVANUS P. (4). *Peregrinus and his 'Epistola'*. London, 1907.

THORNDIKE, L. (1). *A History of Magic and Experimental Science*. 8 vols. Columbia Univ. Press, New York: vols. 1 and 2, 1923; 3 and 4, 1934; 5 and 6, 1941; 7 and 8, 1958 (rev. W. Pagel, *BIHM*, 1959, **33**, 84).

THUROT, C. (1). 'Recherches Historiques sur le Principe d'Archimède.' *RA*, 1868 (2e sér.), **18**, 389; 1869 (2e sér.), **19**, 42, 111, 284, 345; **20**, 14.

TIMOSHENKO, S. P. (1). *History of Strength of Materials; with a brief account of the History of Theory of Elasticity and Theory of Structures.* McGraw Hill, New York and London, 1953.

TOLANSKY, S. (1). 'Multiple-Beam Interferometry.' *END*, 1950, **9**, 196. *Multiple-Beam Interferometry of Surfaces and Films.* Oxford, 1948.

TREFZGER, H. (1). 'Das Musikleben der Thang-zeit.' *SA*, 1938, **13**.

TRIEWALD, MÅRTEN (1). *A Short Description of the Fire and Air Machine at the Dannemora Mines.* Tr. by Are Waerland from the Swedish edition, Schneider, Stockholm, 1734; Heffer, Cambridge, 1928. (Extra Publications of the Newcomen Society, no. 1.)

TRIGAULT, NICHOLAS (1). *De Christiana Expeditione apud Sinas.* Vienna, 1615; Augsburg, 1615. Fr. tr.: *Histoire de l'Expédition Chrétienne au Royaume de la Chine, entreprise par les PP. de la Compagnie de Jésus, comprise en cinq livres … tirée des Commentaires du P. Matthieu Riccius, etc.* Lyon, 1616; Lille, 1617; Paris, 1618. Eng. tr. (partial): *A Discourse of the Kingdome of China, taken out of Ricius and Trigautius.* In *Purchas his Pilgrimes.* London, 1625, vol. 3, p. 380. Eng. tr. (full); see Gallagher (1). Trigault's book was based on Ricci's *I Commentarj della Cina* which it follows very closely, even verbally, by chapter and paragraph, introducing some changes and amplifications, however. Ricci's book remained unprinted until 1911, when it was edited by Venturi (1) with Ricci's letters; it has since been more elaborately and sumptuously edited alone by d'Elia (2).

TROMBE, F. (1). 'The Utilisation of Solar Energy.' *R*, 1948, **1**, 393.

TSENG CHU-SÊN (TJAN TJOE-SOM) (1). '*Po Hu T'ung*'; *The Comprehensive Discussions in the White Tiger Hall; a Contribution to the History of Classical Studies in the Han Period.* 2 vols. Brill, Leiden, 1949, 1952. (Sinica Leidensia, vol. 6.)

TSIEN LING-CHAO. See Chhien Lin-Chao.

TU CHING-MING (1). 'Hsiang Chhi: Chinese Chess.' *CREC*, 1954, **3** (no. 5), 43.

TURNER, W. E. S. (1). 'Studies in ancient Glasses and Glass-making Processes; III. The Chronology of Glass-making Constituents.' *TSGT*, 1956, **40**, 39.

TURNER, W. E. S. (2). 'Studies in ancient Glasses and Glass-making Processes; IV, The Chemical Composition of Ancient Glasses.' *TSGT*, 1956, **40**, 162.

TURNER, W. E. S. (3). 'Studies in ancient Glasses and Glass-making Processes; V, Raw Materials and Melting Processes.' *TSGT*, 1956, **40**, 277.

TURNER, W. E. S. (4). (*a*) 'Studies in ancient Glasses and Glass-making Processes; II, The Composition, Weathering Characteristics and Historical Significance of some Assyrian Glasses of the −8th to −6th centuries from Nimrud.' *TSGT*, 1954, **38**, 445. (*b*) 'Glass Fragments from Nimrud of the −8th to the −6th Century.' *IRAQ*, 1955, **17**, 57.

TURNER, W. E. S. (5). 'Ancient Sealing-wax-red Glasses.' *JEA*, 1957, **43**, 110.

TURNER, W. E. S. (6). 'Die Leistungen der alten Glasmacher und ihre Grenzen.' *GLB*, 1957, **30**, 257.

TURNER, W. E. S. (7). 'Glas bei unseren Vorfahren.' *GLB*, 1955, **28**, 255.

TWITCHETT, D. C. & CHRISTIE, A. H. (1). 'A Mediaeval Burmese Orchestra' (presented to the Thang Court in +802). *AM*, 1959, **7**, 176.

UCCELLI, A. (1) (ed.) (with the collaboration of G. SOMIGLI, G. STROBINO, E. CLAUSETTI, G. ALBENGA, I. GISMONDI, G. CANESTRINI, E. GIANNI, & R. GIACOMELLI). *Storia della Tecnica dal Medio Evo ai nostri Giorni.* Hoeppli, Milan, 1945.

UMEHARA, SUEJI. (1). 'A study of the bronze *chhun* [upward-facing bell with suspended clapper].' *MS*, 1956, **15**, 142.

USHER, A. P. (1). *A History of Mechanical Inventions.* McGraw-Hill, New York, 1929; 2nd ed. revised Harvard Univ. Press, Cambridge, Mass., 1954 (rev. Lynn White, *ISIS*, 1955, **46**, 290).

DE VAUX, CARRA (4) (tr.). *Le Livre de l'Avertissement et de la Revision* (a translation of al-Mas'ūdī's *Kitāb al-Tanbīh wa'l-Ishrāf*). Paris, 1896. (Collection d'Ouvrages publiée par la Soc. Asiat.)

VÁVRA, J. R. (1). *Five Thousand Years of Glass-making; the History of Glass.* Orig. ed. Artia, Prague, 1954 (in Czech). Eng. tr. by I. R. Gottheiner. Heffer, Cambridge, 1954.

VEITH, I. (1) (tr.). '*Huang Ti Nei Ching Su Wên*'; *the Yellow Emperor's Classic of Internal Medicine*, *chs. 1–34 translated from the Chinese, with an Introductory Study.* Williams & Wilkins, Baltimore, 1949 (revs. J. R. H[ightower], *HJAS*, 1951, **14**, 306; J. R. Ware, *BIHM*, 1950, **24**, 487; author's reply, *BIHM*, 1951, **25**, 86; see also W. Hartner, *ISIS*, 1951, **42**, 265.) Circumspection must be exercised in the use of this translation. Cf. Chamfrault & Ung Kang-Sam (1).

VERNET, J. (1). 'Influencias Musulmanas en el Origen de la Cartografía Náutica.' *PRSG*, 1953, ser. B, no. 289.

VIRDUNG, SEBASTIAN (1). *Musica getutscht.* Basel, 1511. (Repr. in Publ. d. ält. prakt. u. theoret. Musikwerke (Berlin), no. 13.)

DE VISSER, M. W. (1). 'Fire and Ignes Fatui in China and Japan.' *MSOS*, 1914, **17**, 97.

VOLPICELLI, Z. (1). 'Chinese Chess' (*Wei Chhi*). *JRAS/NCB*, 1894, **26**, 80.

VOLPICELLI, Z. (2). 'Chinese Chess' (*Hsiang Chhi*). *JRAS/NCB*, 1888, **23**, 248.

VOST, W. (1). 'The Lineal Measures of Fa-Hsien and Hsüan-Chuang.' *JRAS*, 1903, 65.

WAGENER, G. (1). 'Bemerkungen ü. d. Theorie d. chinesischen Musik und ihren Zusammenhang mit d. Philosophie.' *MDGNVO*, 1879, **2**, 42.

WAGENER, G. & NINAGAWA, N. (1). 'Geschichtliches ü. Maas- und Gewichts-Systeme in China und Japan.' *MDGNVO*, 1879, **2**, 41, 61.

WALEY, A. (1) (tr.). *The Book of Songs*. Allen & Unwin, London, 1937.

WALEY, A. (4) (tr.). *The Way and its Power; a study of the 'Tao Tê Ching' and its Place in Chinese Thought.* Allen & Unwin, London, 1934. (Crit. Wu Ching-Hsiung, *TH*, 1935, **1**, 225.)

WALEY, A. (5) (tr.). *The Analects of Confucius*. Allen & Unwin, London, 1938.

WALEY, A. (6). *Three Ways of Thought in Ancient China*. Allen & Unwin, London, 1939.

WALEY, A. (8). 'The Book of Changes.' *BMFEA*, 1934, **5**, 121.

WALEY, A. (10). *The Travels of an Alchemist* (Chhiu Chhang-Chhun's journey to the court of Chingiz Khan). Routledge, London, 1931. (Broadway Travellers series.)

WALEY, A. (11). *The Temple, and other Poems*. Allen & Unwin, London, 1923.

WALEY, A. (14). 'Notes on Chinese Alchemy.' *BLSOAS*, 1930, **6**, 1.

WANG CHI-MIN & WU LIEN-TÊ (1). *History of Chinese Medicine*. Nat. Quarantine Service, Shanghai, 1936 (1st ed. 1932).

WANG KUANG-CHI (1). 'Musikalische Beziehungen zwischen China und dem Westen im Laufe d. Jahrtausende.' In Paul Kahle Festschrift, *Studien z. Gesch. u. Kultur des nahen und fernen Ostens*. Ed. W. Heffening & W. Kirfel, p. 217. Brill, Leiden, 1935.

WANG YÜ-CHHÜAN (2). 'Relics of the State of Chhu.' *CPICT*, 1953 (August) (ref. Yuan Thung-Li, communication to the 23rd International Congress of Orientalists, Cambridge, 1954).

WARREN, SIR CHARLES (1). *The Ancient Cubit*. London, 1903.

WATANABE, N. (1). 'Saecular Variation in the Direction of Geomagnetism as the Standard Scale for Geomagneto-chronology in Japan.' *N*, 1958, **182**, 383.

WATANABE, N. (2). 'The Direction of Remanent Magnetism of Baked Earth and its Application to Chronology for Anthropology and Archaeology in Japan; an Introduction to Geomagneto-chronology.' *JFSUT*, 1959 (Sect. V), **2**, 1–188.

WATERS, D. W. (7). 'The Lubber's Point.' *MMI*, 1952, **38**, 224.

WATSON, E. C. (1). 'A Sixteenth-Century Spectacle Shop.' *PASP*, 1946, **58**, 297.

WEIL, GOTTHOLD (1). *Der Königslöse* (choosing kings by lot). Berlin, 1929.

WELLER, F. (1). '*Yōjana* und *Li* bei Fa-Hsien.' *ZDMG*, 1920, **74**, 225.

WELTFISH, G. (1). 'Coiled Gambling Baskets of the Pawnee and other Plains Tribes.' *IN*, 1930, **7**, 277.

WERNER, E. T. C. (2). *Descriptive Sociology* [Herbert Spencer's] ; *Chinese*. Williams & Norgate, London, 1910.

WERNER, E. T. C. (3). *Chinese Weapons*. Royal Asiatic Society (North China Branch), Shanghai, 1932.

WERNER, O. (1). *Zur Physik Leonardo da Vincis*. Inaug. Diss. Berlin, 1910. (Abstr. *GTIG*, 1916, **3**, 239.)

WESTWATER, J. W. (1). 'The Boiling of Liquids.' *SAM*, 1954, **190** (no. 6), 64.

WHEWELL, WILLIAM (1). *History of the Inductive Sciences*. Parker, London, 1847. 3 vols. (Crit. G. Sarton, *A/AIHS*, 1950, **3**, 11.)

WHITE, LYNN (1). 'Technology and Invention in the Middle Ages.' *SP*, 1940, **15**, 141.

WHITEHEAD, A. N. (2). *Adventures of Ideas*. Cambridge, 1933. Repr. 1938.

WHITTAKER, SIR EDMUND T. (1). *A History of the Theories of Aether and Electricity*. 1st ed. 1910. New ed. 2 vols. Nelson, London, 1951.

WIEDEMANN, E. (3). 'Zu ibn al-Haitham's Optik. *AGNT*, 1911, **3**, 1.

WIEDEMANN, E. (4). 'Über die Camera Obscura bei ibn al-Haitham.' *SPMSE*, 1914, **46**, 155; *VDPG*, 1910, **12**, 177; *JFP*, 1910, **24**, 12.

WIEDEMANN, E. (5). 'Zur Kenntnis d. Phosphoreszenz bei den Muslimen.' *AGNT*, 1909, **1**, 156.

WIEDEMANN, E. (8). 'Zur Gesch. des Kompasses bei den Arabern.' *VDPG*, 1907, **9** (no. 24), 764 [*BDPG*, 1907, **5**]; 1909, **11** (nos. 10–11), 262 [*BDPG*, 1909, **7**].

WIEDEMANN, E. (11). 'Beiträge z. Gesch. d. Naturwiss., XV; Über die Bestimmung der Zusammensetzung von Legierungen.' *SPMSE*, 1908, **40**, 105.

WIEDEMANN, E. (12). 'Beiträge z. Gesch. d. Naturwiss., XVI; Über die Lehre von Schwimmen, die Hebelgesetze und die Konstruktion des Qaraṣṭūn.' *SPMSE*, 1908, **40**, 133.

WIEDEMANN, E. (16). 'Über Fluorescenz und Phosphorescenz.' *ANP*, 1888, **34**, 446.

WIEDEMANN, E. (17). 'Beiträge z. Gesch. d. Naturwiss., XLII; Zwei naturwissenschaftliche Stellen aus dem Werk von Ibn Ḥazm über die Liebe, über das Sehen und den Magneten.' *SPMSE*, 1915, **47**, 93.

WIEDEMANN, E. (18). 'Beiträge z. Gesch. d. Naturwiss., XLIV; Kleine Mitteilungen.' *SPMSE*, 1915, **47**, 121.

WIEDEMANN, E. & HAUSER, F. (3). 'Byzantinische und arabische akustische Instrumente.' *AGNT*, 1918, **8**, 140.

WIEGER, L. (1). *Textes Historiques*. 2 vols. (Ch. and Fr.). Mission Press, Hsienhsien, 1929.

WIEGER, L. (2). *Textes Philosophiques*. (Ch. and Fr.). Mission Press, Hsienhsien, 1930.

WIEGER, L. (3). *La Chine à travers les Ages; Précis, Index Biographique et Index Bibliographique*. Mission Press, Hsienhsien, 1924. (Eng. tr. E. T. C. Werner.)

WIEGER, L. (6). *Taoisme*. Vol. 1. *Bibliographie Générale*: (1) Le Canon (Patrologie); (2) Les Index Officiels et Privés. Mission Press, Hsienhsien, 1911. (Crit. by P. Pelliot, *JA*, 1912 (10ᵉ sér.) **20**, 141.)

WIEGER, L. (7). *Taoisme*. Vol. 2. *Les Pères du Système Taoiste* (tr. selections of Lao Tzu, Chuang Tzu, Lieh Tzu). Mission Press, Hsienhsien, 1913.

WILHELM, RICHARD (2) (tr.). '*I Ging*' [*I Ching*]; *Das Buch der Wandlungen*. 2 vols. (3 books, pagination of 1 and 2 continuous in first volume). Diederichs, Jena, 1924. (Eng. tr. C. F. Baynes (2 vols.). Bollingen-Pantheon, New York, 1950.) See Vol. 2, p. 308.

WILHELM, RICHARD (3) (tr.). *Frühling u. Herbst d. Lü Bu-We* (the *Lü Shih Chhun Chhiu*). Diederichs, Jena, 1928.

WILHELM, RICHARD (4) (tr.). '*Liä Dsi*' [*Lieh Tzu*]; *Das Wahre Buch vom Quellenden Urgrund*, '*Tschung Hü Dschen Ging*': *Die Lehren der Philosophen Liä Yü-Kou und Yang Dschu*. Diederichs, Jena, 1921.

WILHELM, RICHARD (6) (tr.). '*Li Gi*', *das Buch der Sitte des* '*älteren und jüngeren Dai*' (i.e. both *Li Chi* and *Ta Tai Li Chi*). Diederichs, Jena, 1930.

WILKINSON, J. G. (1). *A Popular Account of the Ancient Egyptians*. 2 vols. Murray, London, 1854.

WILKINSON, W. H. (1). *A Manual of Chinese Chess*. NCH, Shanghai, 1893. (Repr. from *NCH*.)

WILKINSON, W. H. (2). 'The Chinese Origin of Playing-Cards.' *AAN*, 1895, **8**, 61.

WILLIAMS, S. WELLS (1). *The Middle Kingdom; A survey of the Geography, Government, Education, Social Life, Arts, Religion, etc. of the Chinese Empire and its Inhabitants*. 2 vols. Wiley, New York, 1848; later eds. 1861, 1900; London, 1883.

WINTER, H. (1). 'Der Stand der Kompassforschung mit Bezug auf Europa.' *FF*, 1936, **12**, 287.

WINTER, H. (2). 'Die Nautik der Wikinger und ihre Bedeutung f. d. Entwicklung d. europäischen Seefahrt.' *HGB*, 1937, **62**, 173.

WINTER, H. (3). 'Who Invented the [Mariner's] Compass?' *MMI*, 1937, **23**, 95.

WINTER, H. (4). 'Petrus Peregrinus von Maricourt und die magnetische Missweisung' (Declination; not known to him). *FF*, 1935, **11**, 304; *AHMM*, 1935, **63**, 352.

WINTER, H. (5). 'Die Erkenntnis der magnetischen Missweisung und ihr Einfluss über die Karto-graphie'. *Comptes Rendus du Congrès International de Géographie*, Amsterdam, 1938.

WINTER, H. J. J. (3). 'The Arabic Achievement in Physics.' *END*, 1950, **9**, 76.

WINTER, H. J. J. (4). 'The Optical Researches of Ibn al-Haitham.' *CEN*, 1954, **3**, 190.

WINTER, H. J. J. & ARAFAT, W. (2). 'Ibn al-Haitham on the Paraboloid Focussing Mirror.' *JRAS/B*, 1949, **15**, 25.

WINTER, H. J. J. & ARAFAT, W. (3). 'A Discourse on the Concave Spherical Mirror by Ibn al-Haitham.' *JRAS/B*, 1950, **16**, 1.

WITTFOGEL, K. A., FÊNG CHIA-SHÊNG et al. (1). *History of Chinese Society (Liao), +907 to +1125*. *TAPS*, 1948, **36**, 1–650. (revs. P. Demiéville, *TP*, 1950, **39**, 347; E. Balazs, *PA*, 1950, **23**, 318.)

WOEPCKE, F. (4). Abū'l-Wafā's tessellations. *JA*, 1855 (5ᵉ sér.), 5, 309. (Part of Woepcke, 2.)

WOLF, A. (1) (with the co-operation of F. Dannemann & A. Armitage). *A History of Science, Technology and Philosophy in the 16th and 17th Centuries*. Allen & Unwin, London, 1935. 2nd ed., revised by D. McKie, London, 1950.

WOLF, A. (2). *A History of Science, Technology and Philosophy in the 18th century*. Allen & Unwin, London, 1938. 2nd ed., revised by D. McKie, London, 1952.

WOLF, R. (3). *Handbuch d. Mathematik, Physik, Geodäsie und Astronomie*. 2 vols. Schulthess, Zürich, 1869 to 1872.

WOOLLEY, L. (3). *The Royal Cemetery [at Ur]; a Report on the Predynastic and Sargonid Graves....* London, 1934. (Ur Excavations, no. 2. Publ. of the Joint Exped. of the Brit. Mus. and of the Mus. of the Univ. of Pennsylvania to Mesopotamia.)

WOOTTON, A. C. (1). *Chronicles of Pharmacy*. 2 vols. Macmillan, London, 1910.

WORCESTER, MARQUIS OF (EDWARD SOMERSET). See DIRCKS (1). Bibliography in John Ferguson (2).

WRIGHT, T. (2) (ed.). *Alexander Neckam* '*De Naturis Rerum*'. Rolls Series. HMSO, London, 1863.

WU LU-CHHIANG & DAVIS, T. L. (1) (tr.). 'An Ancient Chinese Treatise on Alchemy entitled *Tshan Thung Chhi*.' *ISIS*, 1932, **18**, 210.

WU SHIH-CHHANG (1). *A Short History of Chinese Prose Literature* (in the press).

WÜRSCHMIDT, J. (1). 'Zur Theorie d. Camera Obscura bei Ibn al-Haitham.' *SPMSE*, 1914, **46**, 151. 'Zur Gesch. d. Theorie u. Praxis d. Camera Obscura.' *ZMNWU*, 1915, **46**, 466.

WYLIE, A. (1). *Notes on Chinese Literature*. 1st ed. Shanghai, 1867. Ed. here used Vetch, Peiping, 1939 (photographed from the Shanghai 1922 ed.).

WYLIE, A. (5). *Chinese Researches*. Shanghai, 1897. (Photographically reproduced, Wêntienko, Peiping, 1936.)

WYLIE, A. (11). 'The Magnetic Compass in China.' *NCH*, 1859, 15 March. Repr. in Wylie (5), Sci. Sect.

YABUUCHI, KIYOSHI (1). 'Indian and Arabian Astronomy in China.' In Silver Jubilee Volume of the Zinbun Kagaku Kenkyusyo, p. 585. Kyoto University, Kyoto, 1954.

YAMAZAKI, Y. (1). 'The Origin of the Chinese Abacus.' *MRDTB*, 1959, **18**, 91.

YANG LIEN-SHÊNG (2). 'An Additional Note on the Ancient Game *Liu-Po*.' *HJAS*, 1952, **15**, 124.

YANG LIEN-SHÊNG (5). 'Notes on the Economic History of the Chin Dynasty.' *HJAS*, 1945, **9**, 107 (with tr. of *Chin Shu*, ch. 26).

YANG YIN-LIU (1). 'Recovering Ancient Chinese Music.' *PC*, 1956 (no. 1), 26.

YETTS, W. P. (5). *The Cull Chinese Bronzes*. Courtauld Institute, London, 1939.

YOUNG, S. & GARNER, SIR HARRY M. (1). 'An Analysis of Chinese Blue-and-White [Porcelain]' with 'The Use of Imported and Native Cobalt in Chinese Blue-and-White [Porcelain].' *ORA*, 1956 (n.s.), **2** (no. 2).

YULE, H. & BURNELL, A. C. (1). *Hobson-Jobson: being a Glossary of Anglo-Indian Colloquial Words and Phrases ....*' Murray, London, 1886.

VON ZACH, E. (6). *Die Chinesische Anthologie; Übersetzungen aus dem 'Wên Hsüan'*. 2 vols. Ed. I. M. Fang. Harvard Univ. Press, Cambridge, Mass., 1958. (Harvard-Yenching Studies, no. 18.)

ZELLER, E. (1). *Stoics, Epicureans and Sceptics*. Longmans Green, London, 1870.

ADDENDA TO BIBLIOGRAPHY C

AITKEN, M. J. (2). *Physics and Archaeology*. Interscience, London, 1961.

VON ARNIM, J. (1). *Stoicorum Veterum Fragmenta*. Leipzig, 1903.

BARROW, J. (1). *Travels in China*. London, 1804. German tr. 1804; French tr. 1805; Dutch tr. 1809.

BAUER, L. A. (1). 'On the Saecular Variation of a Free Magnetic Needle.' *PHYR*, 1899, **3**, 34.

VAN BERGEIJK, W. A., PIERCE, J. R. & DAVID, E. E. (1). *Waves and the Ear*. Heinemann, London, 1961.

BITTER, F. (1). *Magnets; the Education of a Physicist*. Heinemann, London, 1960.

BOAS, M. & HALL, A. R. (1). 'Newton's "Mechanical Principles".' *JHI*, 1959, **20**, 167.

BUTTERFIELD, H. (1). *The Origins of Modern Science, +1300 to 1800*. Bell, London, 1949.

CHARLESTON, R. J. (1). 'Lead in Glass.' *AMY*, 1960, **3**, 1.

CHHEN SHIH-HSIANG (1). 'In Search of the Beginnings of Chinese Literary Criticism.' *SOS*, 1951, **11**, 45.

DANIÉLOU, A. (1). *Traité de Musicologie Comparée*. Hermann, Paris, 1959. (Actualités Scientifiques et Industrielles, no. 1265.)

DAUMAS, M. (2) (ed.). *Histoire de la Science; des Origines au XXe Siècle*. Gallimard, Paris, 1957. (Encyclopédie de la Pléiade series.)

DAWSON, H. CHRISTOPHER (1). *Progress and Religion; an Historical Enquiry*. Sheed & Ward, London, 1929.

DRACHMANN, A. G. (7). 'Ancient Oil Mills and Presses.' *KDVS/AKM*, 1932, **1**, no. 1.

FORBES, R. J. (19). 'Power [in the Mediterranean Civilisations and the Middle Ages].' Art. in *A History of Technology*, ed. C. Singer *et al*. Oxford, 1956, vol. 2, p. 589.

FRACASTORIUS, HIERONYMUS (1). *De Sympathia et Antipathia Rerum*, with *De Contagione et contagiosis Morbis et Curatione*. Venice, 1546, and a number of later editions, e.g. Lyons, 1554.

FRANKE, H. (2). *Beiträge z. Kulturgeschichte Chinas unter der Mongolenherrschaft*. (Complete translation and annotation of the *Shan Chü Hsin Hua* by Yang Yü, +1360.) *AKML*, 1956, **32**, 1–160.

FRANKE, H. (14). 'Some Aspects of Chinese Private Historiography in the +13th and +14th Centuries.' Art. in 'Historians of China and Japan', ed. W. G. Beazley & E. G. Pulleyblank. London, 1961, p. 115.

GERINI, G. E. (1). 'Researches on Ptolemy's Geography of Eastern Asia (Further India and Indo-Malay Peninsula).' Royal Asiatic Society and Royal Geographic Society, 1909. (Asiatic Society Monographs, no. 1.)

GIBBON, EDW. (1). *The History of the Decline and Fall of the Roman Empire*. Strahan, London, 1790. 12 vols.

GRIFFIN, D. R. (4). *Echoes of Bats and Men.* Heinemann, London, 1960.

DE GUIGNES, [C. L. J.] (1). 'Idée générale du Commerce et des Liaisons que les Chinois ont eus avec les Nations Occidentales.' *MAI/LTR*, 1784 (1793), **46**, 534.

HALL, M. BOAS & HALL, A. R. (1). 'Newton's Electric Spirit; Four Oddities.' *ISIS*, 1959, **50**, 473.

HARTNER, W. (11). 'Remarques sur l'Historiographie et l'Histoire de la Science du Moyen Âge, en particulier au 14ᵉ et au 15ᵉ Siècle.' Lecture at the IXth International Congress of the History of Science, Barcelona, 1959. *Textos de las Ponencias*, p. 43; *Actes*, p. 69.

HASKINS, C. H. (1). *Studies in the History of Mediaeval Science.* Harvard Univ. Press, Cambridge, Mass., 1927.

HAUSER, F. (1). *Über d.* Kitāb fī al-Ḥiyal (*das Werk ü. d. sinnreichen Anordnungen*) *d. Banū Mūsa* [+803 to +873]. Mencke, Erlangen, 1922. Abhdl. z. Gesch. d. Naturwiss. u. Med. no. 1.)

HESSE, MARY B. (3). *Forces and Fields; the Concept of Action at a Distance in the History of Physics.* Nelson, London, 1961.

KLEMM, F. (1). *Technik; eine Geschichte ihrer Probleme.* Alber, Freiburg and München, 1954. Eng. tr. by Dorothea W. Singer, *A History of Western Technology.* Allen & Unwin, London, 1959.

MALLIK, D. N. (1). *Optical Theories.* Cambridge, 1917; 2nd ed. 1921.

MARCUS, G. J. (4). 'The Early Norse Traffic to Iceland.' *MMI*, 1960, **46**, 179.

NEEDHAM, JOSEPH (8). 'Geographical Distribution of English Ceremonial Folk-Dances.' *JEFDS*, 1936, **3**, 1.

NEEDHAM, JOSEPH (34). 'The Translation of Old Chinese Scientific and Technical Texts.' Art. in *Aspects of Translation*, ed. A. H. Smith. Secker & Warburg, London, 1958, p. 65. (Studies in Communication, no. 2.) (And *BABEL*, 1958, **4** (no. 1), 8.)

NEEDHAM, JOSEPH (39). 'The Chinese Contributions to the Development of the Mariner's Compass.' In *Resumo das Comunicações do Congresso Internacional de História dos Descobrimentos.* Lisbon, 1960, p. 273. And *SCI*, 1961.

PFIZMAIER, A. (94) (tr.). 'Beiträge z. Geschichte d. Perlen.' *SWAW/PH*, 1867, **57**, 617, 629. (Tr. chs. 802 (in part), 803, *Thai Phing Yü Lan.*)

PI, H. T. (1). 'The History of Spectacles in China.' *CMJ*, 1928, **42**, 742.

POHL, H. A. (1). 'Non-uniform Electric Fields.' *SAM*, 1960, **203** (no. 6), 106.

PURCELL, V. (3). *The Chinese in Malaya.* Oxford, 1948.

SANCEAU, E. (2). *Knight of the Renaissance; Dom João de Castro, soldier, sailor, scientist, and Viceroy of India,* +1500 to +1548. Hutchinson, London, n.d. (1949).

STEIN, R. A. (1). 'Le Lin-Yi; sa localisation, sa contribution à la formation du Champa, et ses Liens avec la Chine.' *HH*, 1947, **2** (nos. 1-3), 1-335.

THOMSON, T. (1). *History of Chemistry.* Colburn & Bentley, London, 1830.

WRIGHT, A. F. (5). On Teleological Assumptions in the History of Science. *AHR*, 1957, **62**, 918.

GENERAL INDEX

by MURIEL MOYLE

NOTES

(1) Articles (such as 'the', 'al-', etc.) occurring at the beginning of an entry, and prefixes (such as 'de', 'van', etc.) are ignored in the alphabetical sequence. Saints appear among all letters of the alphabet according to their proper names. Styles such as Mr, Dr, if occurring in book titles or phrases, are ignored; if with proper names, printed following them.

(2) The various parts of hyphenated words are treated as separate words in the alphabetical sequence. It should be remembered that, in accordance with the conventions adopted, some Chinese proper names are written as separate syllables while others are written as one word.

(3) In the arrangement of Chinese words, Chh- and Hs- follow normal alphabetical sequence, and *ŭ* is treated as equivalent to *u*.

(4) References to footnotes are not given except for certain special subjects with which the text does not deal. They are indicated by brackets containing the superscript letter of the footnote.

(5) Explanatory words in brackets indicating fields of work are added for Chinese scientific and technological persons (and occasionally for some of other cultures), but not for political or military figures (except kings and princes).

Homer, 237
Hommel, R. P., 80, 122
'Homoeomereity', 92 (e)
Honan, 46, 50
Honey, W. B., 104
Honorius of Autun (+11th century), 4
Hooke, Robert, 12, 28, 114
Hoppe, E. (1), 245
von Hornbostel, E. M., 182
Horwitz, H. T., 38, 85
Hot-air balloons, 124 (c)
Hou chhi ('observing the *chhi*'), 137, 186 ff., 189
Hou Chou Shu. See Chou Shu
Hou Han Shu (History of the Later Han Dynasty), 188, 322, 328
Hou Khuei (legendary musician). *See* Khuei
Hou Shan Than Tshung (Collected Discussions at Hou Shan), 94
Hou Thien system, 262, 296–7
Hour-glass, 290 (a)
Hsi (game of chess, or divination by chess), 315
Hsi Ching Tsa Chi (Miscellaneous Records of the Western Capital), 91, 123
Hsi Khang (poet and Taoist, +223 to +262), 132
Hsi-Yang Chhao Kung Tien Lu (Record of the Tributary Countries of the Western Oceans), 285
Hsi-Yang Chi. See San Pao Thai-Chien Hsia Hsi-Yang Chi Thung Su Yen I
Hsia inch, 221 (c)
Hsia Jih Chi (Records of Leisure Hours), 121
Hsiahou Tun (general), 327
Hsiang (chess term), 319 ff., 325
Hsiang (drum), 150
Hsiang ('facing'), 157
Hsiang chhi. See Chess *and* Image-chess
Hsiang Ching (Image-chess Manual), 320
Hsiang Ching Fa Thi I (The Substance and Main Idea of the Image-chess Manual), 320
Hsiang Shan Yeh Lu (Rustic Notes from Hsiang-shan), 76
Hsiang Yen Lu (The Smoke of Hunan Hearths), 209
Hsiang Yin Shih Yo Phu, 131, 152
Hsiao (pan-pipes), 146 ff., 153 (c), 182 ff.
Hsiao Yen. *See* Wu Ti (Liang emperor)
Hsieh Ho-Chhing (Sung geomantic writer), 304, 333
Hsieh Tsai-Hang (writer, +1573 to +1619), 236
Hsien chih (local topographies), 54 (c)
Hsien Thien system, 296–7, 301
Hsien Tsung (Thang emperor, *r.* +805 to +820), 259
Hsienmên Shih Fa (The Shaman Method of Using the Diviner's Board), 269
Hsin Ching (Canon of the Core of the Earth, geomantic book), 346
Hsin Ching (popular name for the *Prajñā-pāra-mitā Sūtra*), 121
Hsin Thang Shu (New History of the Thang Dynasty), 115–16
Hsing Chha Shêng Lan (Triumphant Visions of the Starry Raft), 313
Hsing Lung Shêng (reed organ), 211

Hsing Yün-Lu (calendar expert, +1573 to +1620), 190
Hsintu Fang (+6th-century mathematician and surveyor), 35, 189
Hsiu (equatorial lunar mansions), 258, 262, 265, 312, 322 ff.
transference to azimuth, 265
Hsiung Ming-Yü (astronomer and geographer, *fl.* +1601 to +1648), 288
Hsiung San-Pa. *See* de Ursis, Sabbathin
Hsü Chhuan (+986), 257
Hsü Chien (Thang encyclopaedist, *c.* +700), 127
Hsü Chih-Mo (Ming geomancer, *c.* +1570), 242, 309
Hsü Chih-O (Sung high official), 76
Hsü Ching (*fl.* +1124), 280
Hsü Han Shu (Addenda to the Han History), 116
Hsü Shen (Han lexicographer, +121), 256
Hsü Yo (Later Han mathematician), 4, 259
Hsüan (globular flute), 145 ff.
Hsüan chen ('mysterious' or 'suspended' needle), 259 (b), 273, 274, 275
Hsüan Nü, 303
Hsüan shih ('mysterious stone'), 251
Hsüan-Ying (Thang Buddhist monk and scholar), 105, 106
Hsüeh Chhêng (mechanician, +538), 35
Hsün Chho (historian, *fl.* +312), 185
Hsün Hsü (acoustics expert, *d.* +289), 185
Hsün Tzu (Book of Master Hsün), 34, 59
Hu chhin (violin), 217 (b)
Hu Lung (army commander, +15th century), 119
Hu Shou (Fox-Head Books), 307–8
Hu Wei (Chhing scholar and geographer), 297
Hua Shu (Book of the Transformations (in Nature)), 92, 116, 206
Huai Nan Tzu (The Book of (the Prince of) Huai-Nan), 15, 23, 62, 73, 79, 88, 90, 106, 111, 159, 161, 169, 181, 218, 232, 257, 265, 281
Huai Nan Wan Pi Shu (The Ten Thousand Infallible Arts of the Prince of Huai-Nan), 69, 91, 93, 237, 254 (a), 278, 281, 316
Huai-Ping (monk and engineer, *c.* +1064), 40–1
Huai River, 162
Huan Wên (+347 to +373), 209
Huang (reed, of musical instrument), 211
Huang-chung (pipe, bell, or note), 45, 173, 178–9, 187–8, 200–1
dimensions, 212–14
Huang Kan (Sung philosopher), 73
Huang Shêng-Tsêng (+1520), 285
Huang Shih Kung (the Old Gentleman of the Yellow Stone), 326
Huang Ti (legendary Yellow Emperor), 64, 178–9, 181, 259, 270
Huang Ti Nei Ching, Su Wên (The Yellow Emperor's Manual of Internal Medicine; the Plain Questions), 5–6
Huang-Yu Hsin Yo Thu Chi (New illustrated Record of Musical Matters of the Huang-Yu reign-period), 25
Hui Shih (logician), 92
Hui-Yuan (Thang monk), 40

夏	HSIA kingdom (legendary?)		*c.* −2000 to *c.* −1520
商	SHANG (YIN) kingdom		*c.* −1520 to *c.* −1030
周	CHOU dynasty (Feudal Age)	⎧ Early Chou period	*c.* −1030 to −722
		⎨ Chhun Chhiu period 春秋	−722 to −480
		⎩ Warring States (Chan Kuo) period 戰國	−480 to −221
First Unification	秦 CHHIN dynasty		−221 to −207
漢 HAN dynasty	⎧ Chhien Han (Earlier or Western)		−202 to +9
	⎨ Hsin interregnum		+9 to +23
	⎩ Hou Han (Later or Eastern)		+25 to +220
三國 SAN KUO (Three Kingdoms period)			+221 to +265
First Partition	蜀 SHU (HAN)	+221 to +264	
	魏 WEI	+220 to +264	
	吳 WU	+222 to +280	
Second Unification	晉 CHIN dynasty: Western		+265 to +317
	Eastern		+317 to +420
	劉宋 (Liu) SUNG dynasty		+420 to +479
Second Partition	Northern and Southern Dynasties (Nan Pei chhao)		
	齊 CHHI dynasty		+479 to +502
	梁 LIANG dynasty		+502 to +557
	陳 CHHEN dynasty		+557 to +589
	魏 ⎧ Northern (Thopa) WEI dynasty		+386 to +535
	⎨ Western (Thopa) WEI dynasty		+535 to +556
	⎩ Eastern (Thopa) WEI dynasty		+534 to +550
	北齊 Northern CHHI dynasty		+550 to +577
	北周 Northern CHOU (Hsienpi) dynasty		+557 to +581
Third Unification	隋 SUI dynasty		+581 to +618
	唐 THANG dynasty		+618 to +906
Third Partition	五代 WU TAI (Five Dynasty period) (Later Liang, Later Thang (Turkic), Later Chin (Turkic), Later Han (Turkic) and Later Chou)		+907 to +960
	遼 LIAO (Chhitan Tartar) dynasty		+907 to +1124
	West LIAO dynasty (Qarā-Khiṭāi)		+1124 to +1211
	西夏 Hsi Hsia (Tangut Tibetan) state		+986 to +1227
Fourth Unification	宋 Northern SUNG dynasty		+960 to +1126
	宋 Southern SUNG dynasty		+1127 to +1279
	金 CHIN (Jurchen Tartar) dynasty		+1115 to +1234
	元 YUAN (Mongol) dynasty		+1260 to +1368
	明 MING dynasty		+1368 to +1644
	清 CHHING (Manchu) dynasty		+1644 to +1911
	民國 Republic		+1912

N.B. When no modifying term in brackets is given, the dynasty was purely Chinese. Where the overlapping of dynasties and independent states becomes particularly confused, the tables of Wieger (1) will be found useful. For such periods, especially the Second and Third Partitions, the best guide is Eberhard (9). During the Eastern Chin period there were no less than eighteen independent States (Hunnish, Tibetan, Hsienpi, Turkic, etc.) in the north. The term 'Liu chhao' (Six Dynasties) is often used by historians of literature. It refers to the south and covers the period from the beginning of the +3rd to the end of the +6th centuries, including (San Kuo) Wu, Chin, (Liu) Sung, Chhi, Liang and Chhen.

SUMMARY OF THE CONTENTS OF VOLUME 4

PHYSICS AND PHYSICAL TECHNOLOGY

Part 1, Physics

With the collaboration of Wang Ling and the special co-operation of Kenneth Robinson

Part 2, Mechanical Engineering

With the collaboration of Wang Ling

Part 3, Engineering and Nautics

With the collaboration of Wang Ling